THE PRINCIPLES OF ELECTROCHEMISTRY

By

DUNCAN A. MacINNES

THE ROCKEFELLER INSTITUTE

DOVER PUBLICATIONS, INC.

NEW YORK

Published in Canada by General Publishing Company, Ltd., 30 Lesmill Road, Don Mills, Toronto, Ontario.
Published in the United Kingdom by Constable and Company, Ltd., 10 Orange Street, London WC 2.

This Dover edition, first published in 1961, is an unabridged and corrected republication of the second (1947) printing of the work originally published in 1939. This Dover edition is published by special arrangement with the Reinhold Publishing Corporation, publisher of the original edition.

International Standard Book Number: 0-486-60052-1

Manufactured in the United States of America.
Dover Publications, Inc.
180 Varick Street
New York, N. Y. 10014

Preface to Dover Edition

The author has welcomed the suggestion that this book be republished in an inexpensive edition. He has been assured that although first published in 1939, and out of print for a number of years, there is still a demand for it. Possibly this is due to the fact that it deals with the more solid accomplishments of electrochemistry. Treatments of the basic principles naturally tend to require less change and revision than do expositions of the newer and more speculative parts of a science.

The preparation of the present edition has provided an opportunity to make a limited number of corrections and additions. Misprints have been corrected, although some may still remain. Where space permitted at the end of chapters, information on work since 1939, as reported in books and articles, is given. The note added to page 95 discusses the important work by L. G. Longsworth which has contradicted the interpretation of the transference measurements in terms of hydration of ions as presented on pages 91–95.

A more thoroughly revised edition of this work has been under way for some time. Work has gone slowly, partly because of the enormous increase in the amount of related published material which must be considered, but especially because the author has been actively engaged in time-consuming research. This projected edition will include a more adequate treatment of electrophoresis which was just coming over the scientific horizon in 1939. After some soul searching, the author decided to omit a discussion of polarography from the original edition because the theoretical background of the subject was then not sufficiently developed. A chapter on that subject will be included in the proposed revision. There will also be a more extensive treatment of over-voltage and passivity.

DUNCAN A. MACINNES.

New York City
January, 1961.

Preface to First Edition

This book has been written with the idea of furnishing an account of theoretical electrochemistry as it is today, and to satisfy an inner urge of the author to see the subject he is interested in as a logical, connected whole. The subject of electrochemistry has, however, become so extended that it is possible for one person to accomplish such a task only by severely limiting the range of topics considered, and the manner of their presentation. As to range, nearly all chemistry, and much of physics, is strictly speaking, electrochemistry, so that the selection of subjects must of necessity be arbitrary. Furthermore, the desire to confine the book to what the French term the *actualités* means that most of the historical background of electrochemistry must be slighted. This mode of treatment results in a certain amount of injustice in assigning adequate credit to the early workers in the field. By using the most accurate data to illustrate a principle, reference must usually be made to the work of the more recent investigators rather than to that of the earlier men whose researches furnished the basis on which the principle was originally stated. In any case, the assignment of "priority" to individuals is a task for which the author feels he has little competence and knows that he has little interest.

Another policy in writing the book has been the attempt to base the deduction of all equations on first principles. What actually constitutes such principles is, to an extent, a matter of individual preference. Any attempt at definition would immediately lead one into the field of the professional philosopher. Such an intrusion the author is, above everything, anxious to avoid. He feels, however, that the attempt to build from the ground up has been accomplished in most of the subjects considered. Exceptions are, however, the extension of the Debye-Hückel theory, and the application of the interionic attraction theory to electrolytic conductance. In the latter case the fundamentals lie in the field of statistical mechanics, which cannot be adequately treated short of a book the size of this one, and which, in any case, would not be written by the author.

An additional restriction of subject matter is to the more solid accomplishments of theoretical electrochemistry. While we must look for much of the advance in the future to material that is, at the moment, in the speculative stage, the appropriate place for publication of such material is in the current journals and monographs. What constitutes accomplishment and what is speculation must, however, depend upon personal judgment, for which the writer cannot avoid responsibility. One omission, somewhat reluctanctly made, is that of a discussion of the generalized theory of acids and bases. Though the ideas involved are undoubtedly useful in many connections the electrochemical aspects of the subject do not appear to be sufficiently advanced to fit in with the rest of the book.

An effort has been made to choose the best of the available experimental data and to obtain therefrom the most accurate values of derived physical constants, such as, for instance, activity coefficients and standard potentials. Much time and effort have been expended on this portion of the work. Such values of physical constants cannot be expected to be final as new experiments are constantly being made and standards of accuracy are steadily improving. Also some allowance must be made for the fallibility of the results of all human endeavor. It is of interest, however, that by far the greater portion of the experimental values quoted in the book have been obtained during the past ten or fifteen years, and are, usually, of a much higher order of accuracy than those that preceded them. It is to be hoped therefore that readers will find that, for the field covered, the book will be an addition and supplement to other critical compilations of data and physical constants.

It is also to be hoped that adequate citations of the original literature have been included in the text. Although very extensive bibliographies were prepared for each subject dealt with, it has been thought best, in order to avoid confusion, to include only strictly relevant references to journals and books.

The author's point of view has, of course, been influenced by all that he has heard and all that he has read. One influence, however, of which he is particularly conscious, and for which sufficient acknowledgment has not been made in the chapters which follow, is that of the publications of Dr. E. A. Guggenheim, whose discussions of thermodynamics, and of fundamental topics of electrochemistry, have been found by the author to be both stimulating and useful.

The writing of the book was started in collaboration with Dr. Edgar R. Smith, now of the Bureau of Standards. Changes of occupation and of geographical location of both of us made a continuation of the collaboration difficult, but the author wishes to acknowledge, gratefully, this able assistance, and the pleasant associations that went with it. The author is greatly indebted to his associates, Dr. Theodore Shedlovsky, Dr. Lewis G. Longsworth, and Dr. Donald Belcher, who have given invaluable assistance and criticism in the preparation of most of the chapters of the book. As the reader will see, a fair portion of this volume deals with the results of investigations made in collaboration with these associates. Other parts of the volume have taken form and content as a result of long and profitable discussions. Without this aid and encouragement the book would never have appeared. In addition the author wishes to acknowledge the assistance of Dr. Mary L. Miller who has served in the various capacities of bibliographer, computer, draughtsman, typist and proofreader. Her suggestions have at all times been constructive and helpful, and have had much effect on the final form of the pages which follow.

DUNCAN A. MACINNES.

New York City
January, 1939.

Contents

List of Symbols

A	Area, Debye-Hückel constant
A	Area of Conductor
a	Activity, a_i, "distance of closest approach"
B	Constant in Hückel equation (Equation 31, Chapter 8)
C	Concentration in mols per liter, capacity
C	Concentration in equivalents per liter, normality
D	Dielectric constant
d	Differential
E	Electromotive force
E_0	Constant in pH scale
e	Base of natural logarithms
e	Electron
F	Free energy (Helmholtz), force
F	Fugacity
F	Faraday
f	Fraction of a faraday, "rational" activity coefficient
f	Activity coefficient on concentration, *C*, basis
g	Acceleration due to gravity
H	Heat content
H	Joule heat
h	Height
I	Current in amperes
J	Cell constant
K	Constant
K	Conductance
K	Mass action constant
k	Boltzmann's constant
L	Length
L	Specific conductance
l	Length
M	Molecular weight
m	Molality, *i.e.*, mols per 1000 grams of solvent, electric moment
N	Number of equivalents, normality
N	Mol fraction
N	Avogadro's number
n	Number of mols, number of ions, number in general

P Pressure, molar polarization
p Partial pressure, increment of titrating reagent
Q Heat absorbed
q Number of coulombs
R Gas constant
R' Gas constant times 2.3026
r Radius, distance from ion
S Entropy, solubility
s Solubility product
s Time in seconds
T Absolute temperature
t Temperature in degrees Centigrade, transference number
U Total energy, ion constituent mobility
U Ion mobility, electrophoretic mobility
U ΣCU^+ (Equation 22, Chapter 13)
u Undissociated
V Volume
V ΣCU^- (Equation 22, Chapter 13)
v Velocity
W Work done by system
x Variable
Z Gibbs free energy
z Valence
α Degree of dissociation, polarizability
β Constant in Debye-Hückel equation (Equation 25, Chapter 7)
Γ Specific resistance
γ Activity coefficient on molar, m, basis
Δ Increment
δ Percentage error, dielectric increment, thickness of Helmholtz layer
∂ Partial differential
ϵ Magnitude of charge on electron or proton
ζ Zeta potential
η Viscosity
θ Constant in Onsager's equation (Equation 17, Chapter 18), degree of association
κ Constant in Debye-Hückel theory (Equation 8, Chapter 7)
Λ Equivalent conductance
Λ_m Molar conductance
λ Conductance of ion constituent, wave length
μ Chemical potential, dipole moment
ν Number of ions from dissociation of one mol, volume of titrating reagent, frequency

ξ	Quantity in Planck's equation (Equation 27, Chapter 13)
ρ	Density, charge density
Σ	Summation
σ	Surface charge density, constant in Onsager's equation (Equation 17, Chapter 18)
τ	True transference number, time of relaxation
ψ	Interionic potential
ω	Ionic strength
ln	Logarithm to the base e
log	Logarithm to the base 10
[ ]	Concentration or molality when inclosing chemical formula
( )	Activity when inclosing chemical formula

Values of Universal Constants and of Factors

Faraday	**F**	96,500 int. coul., gram equiv.$^{-1}$
Electronic charge	ϵ	4.770×10^{-10} abs. e. s. u.
Electronic charge	ϵ	$1.5910_8 \times 10^{-20}$ abs. e. m. u.
Avogadro's number	**N**	$6.06_4 \times 10^{23}$
Zero degrees Centigrade on absolute scale		$273.1_3°$
Volume of perfect gas, at 0° and 1 atm.		22.4141×10^3 cm.^{3} mol^{-1}
Gas constant	R	8.3151 joules,mols^{-1}, deg.$^{-1}$
Boltzmann constant	k	1.3712×10^{-16} erg. deg.$^{-1}$
Electrical equivalent of heat	J	4.1835 int. joules cal.$^{-1}$
$298.1_3 \times 2.3026 \times R/\mathbf{F})$ log ( )		0.05915 log ( )

Chapter 1
Introduction

According to the theories which are held at the present time, matter is largely, if not entirely, electrical in nature. This conception makes all chemistry a branch of the science of electricity, or, in other words, all chemistry is electrochemistry. It would, however, be of little use to define electrochemistry to include such a wide field as this. According to the usual understanding of the term, electrochemistry includes the study of chemical reactions brought about by electrical energy, and of chemical reactions giving rise to electrical energy, *i.e.*, the chemical changes which occur in systems when electrical energy either appears in or disappears from their surroundings. This is accompanied by a study of the effect of chemical composition on certain electrical properties of substances, such as the dielectric constant and electrical conductance. It is well to recall, however, that there is no definite line of demarcation of electrochemistry from inorganic, organic and physical chemistry, since electrochemistry is part of the structure, and is, as has just been mentioned, in the larger sense of the term, the foundation of all of the branches of the science of chemistry. If it is difficult to draw sharp lines of division of electrochemistry from other branches of chemistry, it is equally difficult to separate electrochemistry from the science of physics. However, the following considerations will give a rough idea of such a division.

Since electrochemistry is mostly concerned with the appearance or disappearance of electrical energy in the surroundings when chemical reactions occur in a system, it is necessary that the system itself should consist, for the greater part at least, of electrical conductors. These can be roughly divided into three types, namely: (a) metallic conductors, (b) electrolytic conductors, and (c) gaseous conductors. The first of these has been studied as a portion of physics, and the investigation of the second is a branch of physical chemistry or electrochemistry. The very complicated phenomena of conduction through gases have been explored to the greater extent by physicists.

The most important electrochemical phenomena take place at the boundaries joining two of these types of conductors, mentioned above, or two conductors of the same type but of different chemical composition. The possible forms of such conducting boundaries are metal-metal, metal-electrolyte, electrolyte-electrolyte, metal-gas, and electrolyte-gas. Of these the study of phenomena at metal-electrolyte and electrolyte-electrolyte boundaries belong definitely to the domain of electrochemistry. Concerning the others, the phenomena at the surfaces connecting electrolytes and gases have received very little attention from investigators. The remaining types of conduction at boundaries, *i.e.*, metal-metal and metal-gas, have been considered to belong to the field of physics and will receive only casual mention in the following pages.

Any actual conducting system usually involves more than one type of conductor and several forms of boundary, and there can be two or more examples of each type. Much clearness and simplification may, however, be gained by considering the phenomena occurring in each of the conductors and boundaries separately, even though these phenomena may have decided influence on each other.

Metallic and Electrolytic Conduction of Electricity. The solids and liquids through which electrical energy will pass readily can be divided into two types: metallic conductors and electrolytic conductors, sometimes referred to respectively as conductors of the first and second class. The movement of electrical energy through a metallic conductor is carried on without transfer of matter in the material of the conductor, because, as we shall see, the carriers of the energy are smaller than atoms. All metals and metallic alloys possess this type of conductance, and a few chemical compounds, such as lead peroxide and manganese dioxide, show it also. The passage of current in metallic conductors is made evident by the evolution of heat and by magnetic effects. In general, no changes in composition are observed as a result of the passage of electrical energy through a metallic conductor.

The theory universally held at the present time is that conduction in metals is due to the movement of electrons. These are the units of negative electricity, and have a mass of about 1/1800 of that of the hydrogen ion, and a charge of 4.770×10^{-10} electrostatic units of electricity. A current of electricity in a metallic conductor is therefore due to a stream of electrons moving in the contrary direction to what is usually known as the "direction of the current." More precisely, a galvanic current is due to superposing of a definite shift of the electrons in one direction on the random, or unordered, motion of the electrons in the metal.

Electrolytic Conductance, Ionic Theory. In electrolytic conductance the carriers of electricity are of atomic or molecular size and possess both positive and negative charges. For negative carriers these charges have been found to be either of the same magnitude as those of the electron, or a simple multiple of this charge. The carriers with positive charges are of opposite sign but have the same magnitudes, that is to say, the positive charges on these carriers could, theoretically at least, be exactly neutralized by an integral number of electrons.

The carriers of electricity in electrolytic conductors are called "ions," a Greek term meaning "wanderers." An ion of sodium, for example, is an atom of sodium which has acquired a positive charge by losing an electron. The profound differences in properties between the charged ions and the atoms of the metal are largely, but not solely, due to this charge. The solvent also plays a role in this change of properties, by what is known as "solvation," that is, by forming aggregates around the charged ion. A negatively charged chloride ion, on the other hand, is an atom of chlorine which has gained an additional electron, and has probably also attracted a portion of the solvent.

Svante Arrhenius [1] in 1887 was the first to advance and to give quantitative experimental support to the theory that a considerable portion of the material in an electrolytic conductor is in the form of free ions. According to this conception an aqueous solution of sodium chloride, for instance, contains the salt largely in the form of the charged particles; sodium ion, Na^+, and chloride ion, Cl^-, which are present whether the solution is carrying a current or not, and move under the influence of electric forces. This view of the conductance of electrolytes is very generally accepted at the present time. The details of the theory and its experimental foundations will be dealt with in later chapters.

Nearly every conductor can be placed definitely in the metallic or electrolytic class. There are, however, a very few conductors, such as solutions of alkali metals in liquid ammonia, that belong in neither category, and must be considered as intermediate between the two.

In general, metals and alloys become better conductors when the temperature is lowered. With several of the softer metals the property of "superconduction" appears suddenly at very low temperatures (around − 268° C. or 5° absolute).[2] The electrical resistance, which is already very small at these low temperatures, practically disappears when the temperature is lowered below a critical value. Most electrolytic conductors, on the other hand, increase in conductivity as the temperature is raised, that is to say, electrolytic conductors have positive temperature coefficients of conductivity. This difference of temperature coefficient is occasionally of use in deciding whether a conductor is metallic or electrolytic in nature. Metallic conductors are, with few exceptions, far better conductors of electricity than are electrolytic conductors.

An Outline of the History of Electrochemistry. Although it is the aim of this book to describe electrochemistry as nearly as possible as it

[1] S. Arrhenius, *Z. physik Chem.*, 1, 631 (1887).
[2] H. K. Onnes, *Koninklije Akad. Wetenschappen, Amsterdam, Proc.*, 13, 1274 (1911), 14, 113,818 (1912).

exists in the minds of the investigators now at work, a short summary of the history of the subject may be of interest. The method adopted in writing this treatise necessarily does less than justice to workers who made important discoveries on which the later development of the science has been based. Another way in which earlier workers tend to get slighted has been the desire of the author to give the later, and usually more accurate, data in illustration of a principle or theory, rather than the figures on which the principle or theory may have been based. The following few pages may offset these unintentional injustices by at least the mention of some of the names of investigators who made the basic discoveries on which the science now rests.

Early History of Electricity, 1600 to 1799. In 1600, William Gilbert (court physician to Queen Elizabeth) published the first scientific treatise on electricity.[3] He called substances which attracted particles when rubbed, "electrics," (from the Greek word for amber) and those that do not show that property, "non-electrics." Robert Boyle also studied frictional electricity, and observed the phenomenon of electrical sparks. The first revolving frictional machine was constructed by Guerke, using a ball of sulphur. Guerke's machine was improved by Newton (1675) who substituted glass for the sulphur and made other contributions to the knowledge of electricity. Grey studied electrical conduction, and divided substances into conductors and non-conductors. He was able to convey electricity from rubbed glass a distance of nearly nine hundred feet along a string supported on silk threads. Du Faye (1732) discovered that the conduction of fibres is due to their moisture content, and that they are better conductors when wet. This worker also distinguished between what he called "resinous" and "vitreous" electricity, which were later called "positive" and "negative" by Franklin. In 1744 Ludolf was able to ignite ether with an electric spark.

A very important advance was the invention of the Leyden Jar, the precursor of the modern electric condenser, by von Kleist, in 1745. The Leyden Jar became very popular because of the shocks that could be obtained from it. As pointed out by Singer,[4] "In the same year it was shewn by itinerant exhibitors in almost every part of Europe."

One of the great pioneers in the development of electrical science was Benjamin Franklin (1706-1790). One of his important researches dealt with the theory of the Leyden Jar. By ingenious experiments he showed that the electrical charges rested, not on the inner and outer metallic coatings, but rather in the glass separating them. Another epoch-

[3] "De Magnite."

[4] G. J. Singer, "Elements of Electricity and Electrochemistry," Longman, Hurst, Rees, Orme and Brown, London, 1814.

making discovery was that of identifying lightning with frictional electricity by means of his famous "kite experiment."

In Franklin's time there were four known sources of electricity: frictional, atmospheric (lightning), pyroelectricity (due to heating or cooling of certain substances) and animal electricity, as from the torpedo fish. All these sources yield what is somewhat inaccurately described as "static" electricity. Since as we now know, the phenomena observed in connection with "static" electricity are due to relatively small electric charges at high intensities it is not surprising that only scattered observations were made that indicated that electricity could be associated with chemical action.

The Beginnings of Electrochemistry, (1790-1800). In 1786, or perhaps considerably earlier, Galvani, a lecturer at the University of Bologna, noticed that freshly prepared frogs' legs twitched as if alive, when a nearby frictional electric machine was operating. The observation that the suspension of certain of these animals on an iron railing by copper hooks produced similar results led him to the invention of his "metallic arc." This consisted of two metallic conductors, one of which was placed in contact with a frog's nerve and the other with a muscle. On making contact of the two metals the muscles would twitch. Galvani believed the system to be analogous to a Leyden Jar, the muscles and nerves corresponding to the two coatings of the jar and the metals serving simply as a discharging rod. Later, however, Alessandro Volta observed that there were very marked effects if the metals were different and but weak ones if they were the same. He therefore rejected Galvani's theory and reached the conclusion that the source of the electricity was either at the contacts of the metals with the nerve and muscle or at the point of contact of the two metals, and finally decided in favor of the latter. He divided conductors into two classes, corresponding to what are now known as metallic and electrolytic conductors.

Volta's greatest contribution was, however, the discovery, in 1796, of the voltaic pile, which consisted of a series of units, each made from sheets of dissimilar metals such as zinc and silver separated by wet cloth. Volta showed that metals could be arranged in an "electromotive series" so that each became positive when placed in contact with the one next below it in the series. Although, as has already been mentioned, Volta considered that the source of the electric energy was at the surface of contact of the metals, this theory was thrown in doubt when it was discovered that chemical action accompanied the operation of the pile. It is of interest that the question of the seat of the potential of the galvanic cell is not, even today, finally settled. Many improvements of the voltaic pile were made. It is, of course, the precursor of the modern galvanic cell.

Shortly after the discovery of the voltaic pile Nicholson and Carlisle demonstrated, in 1800, the electrolytic decomposition of water. These investigators found that hydrogen and oxygen were evolved at the surface of gold and platinum wires if they were connected with the terminals of a pile and dipped in water.

A worker who was responsible for a number of the earlier developments of electrochemistry was Sir Humphry Davy. His most famous electrochemical experiment, made in 1807, was the preparation of metallic potassium from solid potassium hydroxide by electrolysis. Davy's greatest service to electrochemistry was, possibly, that he prepared the way for Michael Faraday to whom electrochemistry owes more than to any other single person. Faraday stated, in 1835, what is now known as Faraday's Law, which is fully discussed in Chapter 2. He was apparently the first to have clear ideas concerning the quantity and intensity of electricity, *i.e.*, the quantities now measured in terms of amperes and volts. We owe to Faraday many of the terms, such as *ion, cation, anion, electrode, electrolyte, etc.*, in common use today.

Early Ideas Concerning Electrolytic Conductance. It was early observed that during what we now term an electrolysis there are chemical reactions at the two electrodes whereas the liquid between the electrodes remains unaltered. An explanation which satisfied the scientific world for many decades was advanced by C. J. D. Grotthuss in 1805. He believed that chains of substances, which may be represented by AB, existed between the electrodes. When a current passed the atoms simply shifted partners by alternate decompositions and recombinations, with the liberation of A at one electrode and B at the other. According to this theory ionization, as we now understand it, exists only during the time that the current is passing. The theory was criticized by Clausius in 1857. According to his views "the current does not decompose the molecules but only guides those that are momentarily free." This idea was supported by the experimental work of Wilhelm Hittorf (1824-1914) who showed that one deduction from Grotthuss' theory, *i.e.*, that all components taking part in electrolytic conduction must have the same mobilities, is contrary to fact. Hittorf's method for determining the relative mobilities of ion constituents is described in Chapter 3. Clausius was the immediate precursor of Arrhenius (1859-1927)[5] whose theory, *i.e.*, that salts in aqueous solutions are largely dissociated into free ions, was largely instrumental in founding the science of physical chemistry and of modern electrochemistry. Arrhenius' ideas were some of the most fruitful that have been advanced in the history of science. Although later work has shown that his theory is, as is shown later in this book, incorrect

[5] In a very readable address, *J. Am. Chem. Soc.*, **34**, 353 (1912), Arrhenius has given an intimate and interesting account of the early days of the ionic theory.

in details, his assumption as to the presence of free ions is still held to be valid, in fact the more recent theories indicate a larger proportion of such ions than he deduced from his ideas. Early and important workers in the ionic theory were Jacobus Henricus van't Hoff (1852-1911) who laid the foundation for the thermodynamic theory of solutions, Wilhelm Ostwald (1853-1932) who was influential as investigator, writer and editor, and Walter Nernst (1864-1941), who with his students made both experimental and theoretical investigations into nearly every branch of electrochemistry. With these investigators, however, we enter the modern period of electrochemistry which it is the purpose of this book to describe.

Chapter 2

Faraday's Law and Coulometers

As has been seen, nearly all conductors are either metallic or electrolytic in nature, that is to say, the transport of electricity in such conductors is entirely due to electrons or entirely due to ions. From the point of view of the subject of electrochemistry a large proportion of the most interesting and important phenomena occur when electric currents pass from one type of conductor to the other, since at that point the type of carrier must change suddenly, either from ions to electrons or the reverse. Thus it is evident that at such a boundary there must be, after the passage of current, an accumulation of atoms which have either lost or gained electrons, depending on the direction of the current. To put the matter in more familiar chemical terms, the passage of current across a metallic-electrolytic boundary results in chemical reactions of an "oxidizing" or "reducing" nature, these expressions having their more general sense of the increasing or decreasing of the positive and negative valencies, respectively. A few examples will make these ideas clear. When current is passed in a positive direction across a boundary consisting of metallic silver immersed in a solution of a silver salt, such as silver nitrate, electrons move away from the boundary in a direction opposite to the "direction of the current" in the metal, and silver ions with positive charges go into solution; that is, the reaction:

$$Ag = Ag^{+} + e^{-} \tag{1}$$

in which e^{-} represents the negative electron, occurs when the current flows. The material of an electrode, however, does not necessarily enter into the reaction. If an inert electrode (such as platinum or carbon) is placed in a solution containing ferrous ions, and current is passed in a positive direction, a stream of electrons flows away from the boundary. In this case the source of the electrons is the ferrous ions, each one of which, by the loss of an electron, is changed into a ferric ion, a reaction which can be represented by the equation,

$$Fe^{++} = Fe^{+++} + e^{-} \tag{2}$$

Reactions such as (1) and (2) can also be made to occur in the reverse direction as follows:

$$Ag^{+} + e^{-} = Ag, \quad \text{and} \quad Fe^{+++} + e^{-} = Fe^{++}$$

in which cases the electrons flow through the metallically conducting electrode to the boundary and are absorbed by the ions.

An important generalization can be made at this point, as follows: *The passage of an electric current from a metallic to an electrolytic conductor, or vice versa, is always accompanied by a chemical reaction.* This follows directly from our theories as to the nature of metallic and electrolytic conductors. The only possible way that atoms can furnish electrons to metallic conductors, or *vice versa,* is by a change of valence, *i.e.,* by an oxidation or a reduction. Reactions such as (1) and (2) which include electrons, are called *electrochemical reactions.* In order for an electric current to flow through an electrolytically conducting system, the current must, in general, enter and leave by metallic conductors or by substances having metallic conductance. These conductors are known as *electrodes.* The electrode by which the current enters the electrolyte is known as an *anode* and the one by which it leaves, as a *cathode.* A system containing an anode and cathode is represented diagrammatically in Fig. 1. Strictly speaking, these metallic electrodes can only lead the

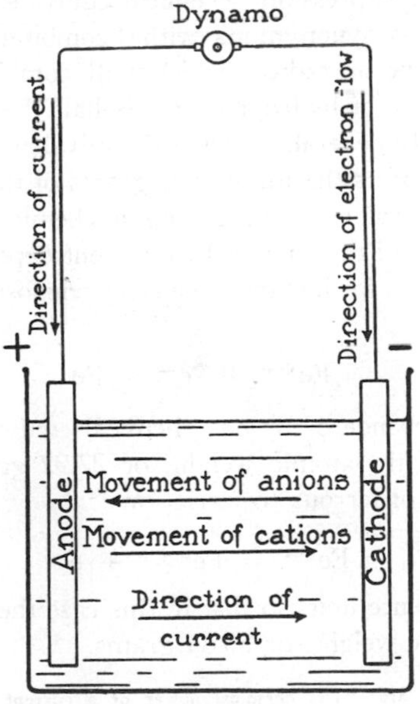

Fig. 1.

electrical energy to and from the metal-electrolyte boundaries. Electrical energy can be led into and out of an electrolytically conducting system by means of gaseous or electrolytic conductors, but it is in accord with general practice to limit the terms *electrode, anode,* and *cathode* to metallic conductors. As has already been mentioned, the transport of electric current through an electrolytic conductor is due to ions, carrying either positive or negative charges. As indicated in Fig. 1, positively charged ions are called "cations" and move toward the cathode, whereas "anions" carry negative charges and drift, when current is flowing, toward the anode.

Faraday's Law. It is now desirable to discuss the relation of the quantity of electricity flowing across metal-electrolyte boundaries to the amount of chemical change there produced. This question was first investigated by Michael Faraday in 1833. He reached the following conclusion:

The magnitude of the chemical effect, in chemical equivalents, is the same at each of the metallic-electrolytic boundaries in an electric circuit and is determined solely by the amount of electricity passed.[1]

For an understanding of this important law, known as *Faraday's law,* it is necessary to consider the terms "chemical equivalent" and "amount of electricity." The expression "chemical equivalent" may be regarded in this connection as synonymous with "combining weight," *i.e.,* the weight of an element or radical which will combine with one atomic weight, 1.0078 grams, of hydrogen, or one-half the atomic weight, 8.00 grams, of oxygen. In general, a chemical equivalent or combining weight of a substance represents the number of grams of the substance involved in a change of one valence unit during a chemical or electrochemical reaction. The value of the chemical equivalent depends upon the nature of the reaction. Thus, for instance, the electrodeposition of iron from the ferrous state

$$Fe^{++} + 2e^- = Fe$$

involves a valence change of 2; one chemical equivalent of iron, in this reaction, is one-half the atomic weight, or 27.92 grams. On the other hand, the oxidation of ferrous to ferric ions

$$Fe^{++} = Fe^{+++} + e^-$$

requires but one valence unit, so that in this case the chemical equivalent of iron is one atomic weight, or 55.84 grams.

[1] Faraday's own words are: "The chemical power of a current of electricity is in direct proportion to the absolute quantity of electricity which passes." (Faraday's Experimental Researches in Electricity, Everyman's Library.) At that time (1833) the distinction between quantity of electricity and electrical energy was not clearly drawn. As a result Faraday's law was severely criticized by some of his contemporaries, especially by Berzelius, who could not believe that the same quantity of electricity would separate the constituents of different compounds having different amounts of energy associated with their formation. However, as will be seen later in this book, the energy changes associated with different electrochemical reactions can have very different magnitudes, even though the same *quantity* of electricity is involved, since the electrical energy depends on the potential difference as well as on the quantity of electricity.

The amount, or quantity, of electricity passing through a circuit is measured in *coulombs,* which are, in turn, the product of the *amperes* and time in seconds.[2] At an international electrical conference in London (1908) the *ampere* was adopted as one of the fundamental units. The ampere is defined as the steady current which, when passed through a solution of silver nitrate in water, under definite conditions to be described later, deposits silver at the rate of 0.00111800 gram per second. This value of the ampere is one-tenth of the c.g.s. (electromagnetic) unit within a few parts in one hundred thousand. The instrument used for measuring current in terms of deposited silver is called a silver coulometer or a silver voltameter.[3]

A little consideration will show that Faraday's law follows directly and necessarily, if our conceptions concerning conduction in metallic and electrolytic conductors are correct. If ϵ is the magnitude of the charge on an electron in coulombs, and **N** is the number of electrons released or absorbed when one chemical equivalent of a substance reacts at an electrode, then the product

$$\mathbf{F} = \mathbf{N}\epsilon \tag{3}$$

is the total amount of electricity, in coulombs, per chemical equivalent of substance reacting at the electrode. As each ion of a univalent substance carries one unit of charge, **N** must obviously be the number of atoms in a gram-atom, or of molecules in a gram-molecule. This number is a universal constant known as "Avogadro's number," and has been found by a variety of closely agreeing methods to have the value 6.064×10^{23}. The quantity **F**, known as the *faraday,* since it is a product of two universal constants, is also a constant. Its value is however known more accurately than either Avogadro's number or the electronic charge, of which it is the product. Some of the researches which have been carried out for the purpose of establishing the value of the faraday will be described later in this chapter.

If, therefore, one faraday of electricity passes through a system, one chemical equivalent of substance reacts at every metal-electrolyte boundary in the system. For a different amount of electricity the formula is, obviously,

$$q = \mathbf{F}N \tag{4}$$

in which q is the amount of electricity passed, in coulombs, and N is the number of chemical equivalents.

[2] This is true only if the current in amperes is constant; otherwise the coulombs, q, are the integral $q = \int I\, ds$, in which I represents amperes and s time in seconds.

[3] The term voltameter has been longer in use, but the expression "coulometer," suggested by T. W. Richards, is more descriptive of the use of the instrument and will be employed in this book.

To apply to specific examples the principles just discussed let us consider what happens at each of the electrodes when a current consisting of four electrons passes in turn through (a) a cell consisting of silver electrodes in a silver nitrate solution, (b) copper electrodes in a copper sulphate solution, (c) platinum electrodes in a solution containing a mixture of ferrous and ferric chlorides, and (d) platinum electrodes in an acidified water solution. The arrangement is shown diagrammatically in Fig. 2. When the electrons moving in the reverse direction to "the

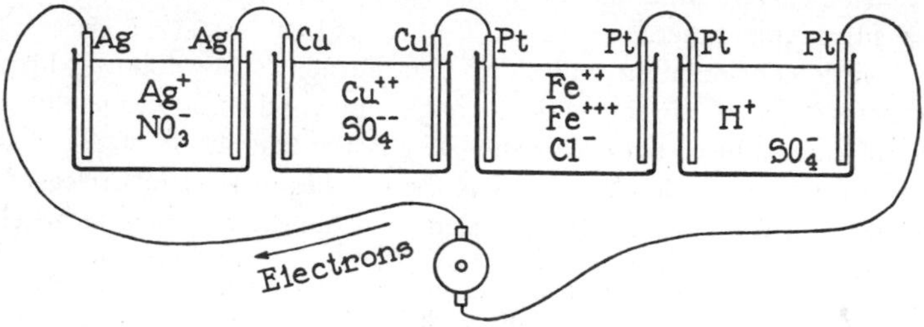

Fig. 2.

direction of the current" reach the metal-electrolyte boundary in the first cell, the reaction

$$4e^- + 4Ag^+ = 4Ag \tag{5}$$

takes place, *i.e.*, each electron neutralizes the positive charge on a silver ion giving an atom of metallic silver. After the current has passed through the electrolyte (the mechanism of this conduction does not concern us at the present time, and will be treated in a later chapter) the same number of electrons is released by the reverse of reaction (5)

$$4Ag = 4e^- + 4Ag^+$$

At the electrodes of the cell containing the copper sulphate solution the current passes the boundaries by means of the reactions

$$4e^- + 2Cu^{++} = 2Cu \quad \text{and} \quad 2Cu = 2Cu^{++} + 4e^-$$

In the third cell there occurs another type of electrochemical reaction which does not involve the material of the electrodes, which serve simply as carriers of electrons. The reactions taking place in the mixture of ferrous and ferric chlorides and involving four electrons are:

$$4Fe^{+++} + 4e^- = 4Fe^{++} \quad \text{and} \quad 4Fe^{++} = 4Fe^{+++} + 4e^-$$

resulting in a reduction at the cathode which is quantitatively reversed at the anode. Still another type of electrochemical reaction, involving gaseous reaction products, occurs in the fourth cell. At the cathode we have

$$4e^- + 4H^+ = 2H_2$$

while at the anode there is a more complicated reaction:

$$4OH^- = O_2 + 2H_2O + 4e^-$$

This last example shows, incidentally, that the reactions at the two electrodes of a cell need not be the reverse of each other. However, one is always reducing and the other always oxidizing in nature.

General Discussion of Faraday's Law. One important point to be observed in connection with Faraday's law is that the amount of electricity passing a boundary determines the *total* number of chemical equivalents entering into reaction at that point, and not simply the number of equivalents of a single electrochemical reaction. For instance the deposition of zinc from an aqueous solution of one of its salts:

$$Zn^{++} + 2e^- = Zn$$

is usually accompanied by the evolution of gaseous hydrogen:

$$2H^+ + 2e^- = H_2$$

The *sum* of the electrochemical equivalents liberated is determined by Faraday's law, which, however, gives no information as to the relative amounts, which are influenced by temperature, composition of electrolyte, etc. Equation (4) can therefore be written

$$q = \mathbf{F}(N_1 + N_2 + \cdots) = \mathbf{F}\Sigma N$$

in which N_1, N_2, etc. represent the numbers of equivalents of different substances. It is, in fact, comparatively rarely that a single electrochemical reaction takes place at an electrode, as will be seen later in this chapter.

Also, Faraday's law can obviously have nothing to say concerning which ions in an electrolyte will take part in an electrochemical reaction. In fact, the current can be carried through an electrolyte by ions which do not enter into the electrochemical reactions by which the current enters and leaves the electrolyte. For instance, in the electrolysis of a sodium chloride solution with inert, *i.e.*, non-reacting, electrodes, the current is carried, except for an almost negligible amount, by the sodium ions, Na^+, and the chloride ions, Cl^-. However, for low current densities, the electrochemical reactions at the anode and cathode are

$$2OH^- = \tfrac{1}{2}O_2 + H_2O + 2e^- \quad \text{and} \quad 2H^+ + 2e^- = H_2$$

In pure water, and in solutions of neutral substances, the hydrogen and hydroxyl ions, H^+ and OH^-, are present in such slight amounts that their presence has, for moderate concentrations of the solute, almost no effect on the conductivity. These small concentrations of the ions of water are, however, in this case the ones that enter into the electrode reactions to the exclusion of ions present in much higher amounts.

The Determination of the Value of the Faraday. In order to determine the value of the faraday, **F**, in equations (3) and (4), it is, theoretically, only necessary to find the total number of chemical equivalents reacting at any electrode when a given number of coulombs pass. Practically, however, to obtain an accurate value of this important constant it is desirable that a single electrochemical reaction take place at the electrode under observation. Also, the reaction product must be of a nature which permits an accurate quantitative determination.

Up to the present time only two electrochemical reactions, the deposition of silver from a silver nitrate solution and the liberation of free iodine from a potassium iodide solution, have been found to fulfill these conditions to a precision of 0.01 per cent. The value of the faraday in general use is based on careful studies of these reactions. In the former case the amount of silver deposited is determined by weighing, and in the latter case the amount of iodine liberated is determined by titration with a standard solution of arsenious acid. It is also necessary to measure the number of coulombs used in the electrochemical reaction. It will be recalled that, as a practical unit, the coulomb is defined as the amount of electricity necessary to deposit 0.00111800 gram of silver from a silver nitrate solution, under certain definite conditions. The silver so deposited is found to be 99.996 per cent pure, the other 0.004 per cent of the weight being due to the inclusion of a slight quantity of the silver nitrate solution in the deposited crystals of the silver. Since the atomic weight of silver is 107.880, the value of the faraday would evidently be $107.880/0.00111800 = 96494$ coulombs per equivalent if the silver deposit were 100 per cent pure, while if the 0.004 per cent of impurity is taken into consideration the value becomes 96497. However, the atomic weight of silver is probably not known to this precision.

By the electrolysis of a solution of iodine in potassium iodide solution, between inert electrodes, iodine is set free at the anode and free iodine is converted to iodide ion at the cathode. The anode reaction is:

$$2I^- = I_2 + 2e^-$$

and the cathode reaction is the reverse of this. Under certain well-studied conditions this single electrochemical reaction has been found to take place. The amount of free iodine produced at the anode or consumed at the cathode can be determined very accurately by titration. The value of the faraday can therefore be determined by measuring the number of coulombs required to liberate, at the anode, or consume, at the cathode, one gram atomic weight of iodine (126.92 grams). Bates and Vinal[4] have carried out this determination with great care, using an apparatus for the electrolysis which will be described in the next section, and measuring the coulombs used by means of the silver coulometer, which will also be described in more detail. On the basis of the values obtained by the silver and the iodine determinations, it has been usual to accept, for general use, the round number, **F** = **96500** coulombs per equivalent, as the best value of the faraday. The ratio of silver to iodine in the experiments just described is 0.85016, whereas the atomic weight determinations by Baxter[5] and others give 0.849984. There is therefore a discrepancy of 0.02 per cent, which remains unexplained.

Coulometers. As already mentioned, any instrument designed to measure a quantity of electricity by a determination of its electrochemical effect is called a coulometer. There have been many types of coulometers proposed, but only those which possess exceptional accuracy, or convenience in practical use, will be discussed.

The Silver Coulometer. The silver coulometer consists, essentially, of a platinum dish or crucible as cathode, and a silver anode with a silver nitrate solution as electrolyte. Surrounding the anode is, in the latest types of the instrument, a porous cup of ceramic material, for reasons to be explained below. A convenient form of silver coulometer is shown diagrammatically in Fig. 3. A dish, *Pt*, holds a solution of silver nitrate

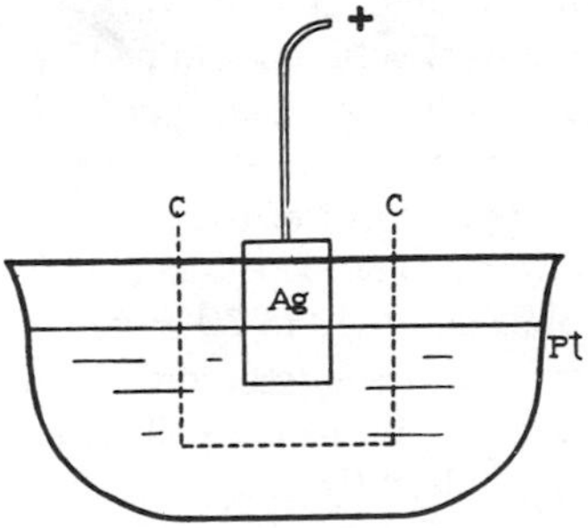

Fig. 3. Diagram of a Silver Coulometer.

and serves as the cathode on which the silver is deposited. The current enters the solution by means of the silver anode, *Ag*, and surrounding

[4] S. J. Bates and G. W. Vinal, *J. Am. Chem. Soc.*, **36**, 916 (1914).
[5] G. P. Baxter, *J. Am. Chem. Soc.*, **32**, 1591 (1910).

the anode is a cup of porous ceramic material represented by *C—C*. A number of painstaking investigations on the silver coulometer have been carried out by the national laboratories of several countries, as well as by independent investigators.[6] This extensive study is due to the fact that the international, or practical, ampere and coulomb are defined, as we have seen, in terms of measurements made with the silver coulometer. It has been found that the silver coulometer is subject to disturbing effects when used without definite precautions. During electrolysis the anode disintegrates to some extent so that particles become detached and drop off. At the same time a dense "anode slime" is formed, the composition of which is still in doubt. The porous cup protects the cathode by catching the particles which fall from the anode, and it also obstructs the diffusion of the anode slime into the cathode chamber. In an older form of this coulometer the cathode was protected from these disturbing anode products by enclosing the anode in a bag made of filter paper, but it has since been found that the presence of any kind of organic material in the silver nitrate solution causes, by its reducing action, the formation of colloidal silver which, being positively charged, deposits on the cathode. These particles of colloidal silver deposit more than one atom of silver for each unit charge, and so, for a given quantity of electricity they cause the deposit of silver to be heavier than corresponds to the simple reaction

$$Ag^+ + e^- = Ag$$

The silver nitrate used should be very pure. The solution cannot even be filtered through filter paper without the formation of some colloidal material. If impure silver nitrate is used the deposit of silver crystals appears irregular and striated, and may be too heavy by more than 0.1 per cent. The weight of solution retained by inclusion in the crystals does not average more than 0.004 per cent. If the following specifications are observed, the silver coulometer will probably give results accurate to 0.05 per cent if the weight of the silver deposit is over one gram. For higher precision the original papers should be consulted.

The solution should contain 10 to 20 grams of silver nitrate (purified by recrystallization from acid solution, and by fusion) in 100 cc. of distilled water. The anode should be pure silver and contained in a porous cup, which should be sufficiently fine-grained to hold back the anode slime without introducing too high a resistance. The cathode should be a crucible or bowl of platinum (although gold may be used) and its surface should not be roughened. The cathode current density should not be greater than 0.01 ampere per cm.^{2}, and the anode current density not greater than 0.05 ampere per cm.^{2} When the electrolysis is finished

[6] An excellent summary of this work is given by E. B. Rosa and G. W. Vinal, *Bur. Standards, Bull.*, 13, 479 (1916-17).

the cathode dish should be washed out with distilled water by means of a siphon or a pipette, until the washings give no test for silver. The cathode dish with the silver deposit should then be dried at about 150° C. Heating over a free flame in an effort to expel inclusions has been found to result in the formation of a layer of platinum black which, during the next electrolysis, adsorbs a serious amount of solution. In weighing the cathode a similar platinum dish should be used as a tare.

The Iodine Coulometer. The iodine coulometer, as developed by Washburn and Bates,[7] has a precision equal to that of the silver coulom-

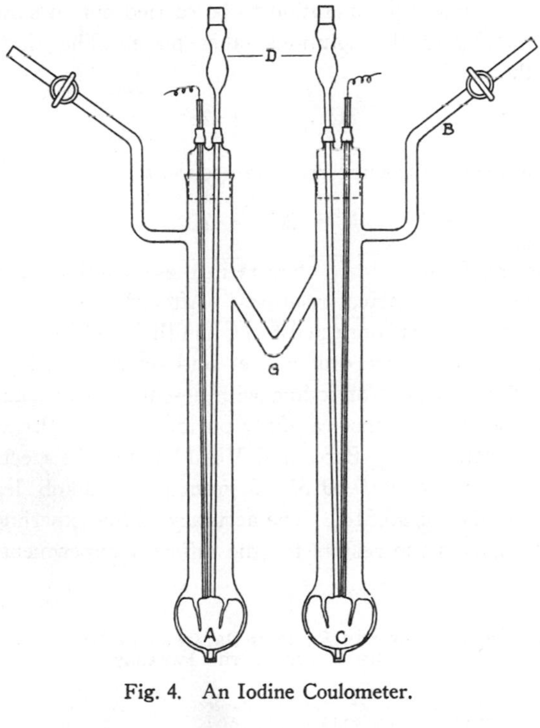

Fig. 4. An Iodine Coulometer.

eter, and in addition, a single electrochemical reaction occurs in opposite directions at the anode and cathode, so that the results of a determination can be checked in one apparatus by means of the two electrode reactions. However, this coulometer is more elaborate and difficult to manipulate than the silver coulometer and requires carefully standardized solutions. Its chief importance consists of its use in the determination of

[7] E. W. Washburn and S. J. Bates, *J. Am. Chem. Soc.*, **34**, 1341 (1912).

the value of the faraday, as has been previously explained. As shown in Fig. 4 the iodine coulometer consists of two vertical limbs containing the electrodes. These tubes are connected by a V-shaped tube, *G*. The tubes *B* and *D* serve for filling and emptying the apparatus. The electrodes, *A* and *C*, are of platinum-iridium foil, which is not attacked by iodine. The limbs are first filled to a little above the V-shaped tube with a 10 per cent potassium iodide solution. Sufficient concentrated potassium iodide solution to cover the electrode is then poured through *D* into the anode side, *A*, and in the same way the cathode at *C* is covered with a standardized solution of iodine in potassium iodide solution. A leveling bulb attached to *B* permits this operation to be carried out so slowly that no appreciable mixing of the solutions takes place. The electrochemical reaction at the anode is

$$2I^- = I_2 + 2e^-$$

and at the cathode the reverse reaction occurs, *viz.,*

$$I_2 + 2e^- = 2I^-$$

The equilibrium, $I_2 + I^- = I_3^-$, exists in these solutions, but this does not affect the stoichiometrical relations. After the electrolysis is completed, a delivery tube is connected to *D*, and the anode and cathode portions of the electrolytes are drawn over into separate flasks. The two portions are then titrated for iodine with arsenious acid solution which has been standardized against carefully purified iodine. By comparison with the silver coulometer, Bates and Vinal [8] found the electrochemical equivalent of iodine to be 0.00131505 gram per coulomb, leading to a value of the faraday of 96,514. The accuracy of the experimental work may be judged from the results for the different experiments given in Table I.

TABLE I. SUMMARY OF THE RESULTS OBTAINED BY BATES AND VINAL FOR THE VALUE OF THE FARADAY

Grams of silver	Grams of iodine	Number of coulombs	Ratio of Ag/I	Milligrams iodine per coulomb	Value of the faraday
4.09903	4.82224	3666.39	0.85002_6	1.31526	96498
4.39711	5.17273	3933.01	0.85005_6	1.31521	96502
4.10523	4.82851	3671.94	0.85020_5	1.31498	96518
4.12310	4.84942	3687.92	0.85022_6	1.31495	96521
4.10475	4.82860	3671.51	0.85009_1	1.31515	96506
4.18424	4.92130	3742.61	0.85023_0	1.31494	96521
4.10027	4.82247	3667 50	0.85024_2	1.31492	96523
4.10516	4.82844	3671.88	0.85020_4	1.31498	96519
		Mean	0.85016	1.31505	96514

[8] S. J. Bates and G. W. Vinal, *loc. cit.*

The Copper Coulometer. In this coulometer, Fig. 5, which is easily constructed and accurate enough for many types of work, copper is deposited on a cathode of sheet copper suspended between two anodes made of plates of the same material. The electrolyte is an acidified copper sulphate solution. A solution which yields good results consists of 125 grams of crystallized copper sulphate ($CuSO_4 \cdot 5H_2O$), 50 grams of concentrated sulphuric acid, and 50 grams of ethyl alcohol, made up

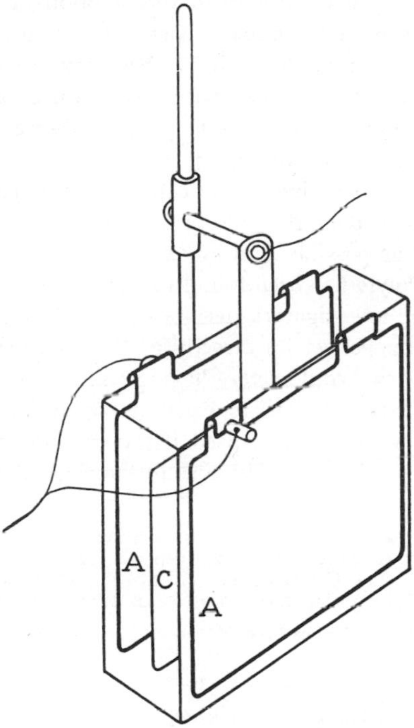

Fig. 5. A Copper Coulometer.

with distilled water to one liter. The copper coulometer is subject to certain sources of error, as has been shown by the work of Richards, Collins, and Heimrod.[9] These investigators found that an acid solution of cupric sulphate slowly dissolves metallic copper, according to the reaction

$$CuSO_4 + Cu = Cu_2SO_4 \quad \text{or} \quad Cu + Cu^{++} = 2Cu^+ \tag{6}$$

[9] T. W. Richards, E. Collins and G. W. Heimrod, *Proc. Am. Acad.*, **35**, 123 (1899-1900).

thus causing a decrease in weight of the electrode. On the other hand, a copper plate immersed in neutral copper sulphate gains in weight, because the cuprous sulphate formed according to reaction (6) hydrolyzes in neutral solution and cuprous oxide is precipitated on the plate, as follows:

$$Cu_2SO_4 + H_2O = Cu_2O + H_2SO_4 \quad \text{or} \quad 2Cu^+ + H_2O = Cu_2O + 2H^+$$

The latter source of error is readily overcome by the addition of acid to the electrolyte. The error due to the solution of the copper to form cuprous ions can be reduced and a correction for it can be applied, but it cannot be entirely eliminated. It has been found that the rate at which reaction (6) proceeds increases with temperature, as does also the concentration of cuprous ions necessary to bring the reaction to equilibrium. Furthermore the oxygen of the air re-oxidizes cuprous to cupric ions, thus increasing the solution of the copper by the same reaction. From these facts it is evident that the error can be reduced by keeping the solution as cold as possible, and by excluding air. The addition of alcohol to the electrolyte has been found to decrease its solvent action on the electrodes. The investigators just mentioned also noted that the rate of solution of the copper varies directly as the area of the plate. By connecting in series two coulometers having cathodes of different areas they were able, by extrapolation, to calculate what increase in weight the electric current would have caused if the cathode were a plate of zero area, so that none of it could dissolve in the electrolyte. In Table II are

TABLE II. COMPARISON OF THE COPPER WITH THE SILVER COULOMETER. TEMPERATURE OF SILVER COULOMETER, 15° TO 25°. TEMPERATURE OF COPPER COULOMETER, −2° TO 0°. COPPER COULOMETER IN AN ATMOSPHERE OF HYDROGEN

Grams of copper: deposited on cathode, 50 cm.² in area	Grams of copper: deposited on cathode, 25 cm.² in area	Grams of copper: corrected to cathode of zero area	Grams of silver, corrected for impurity	Computed atomic weight of copper Ag = 107.88
0.83036	0.83064	0.83092	2.8197	63.58
0.63407	0.63449	0.63491	2.1556	63.55
0.69956	0.70029	0.70102	2.3768	63.64
0.84341	0.84375	0.84409	2.8638	63.59
0.87458	0.87455	0.87462	2.9687	63.57
0.69379	0.69392	0.69405	2.3549	63.59
			Average	63.59

given results obtained by Richards, Collins, and Heimrod in their careful comparison of the copper and silver coulometers. They connected a silver coulometer in series with two copper coulometers, one of which had a cathode 50 cm.² in area and the other a cathode 25 cm.² in area. The copper coulometers were kept at about 0° C. in an atmosphere of hydrogen, and at the end of an experiment the weight of silver deposited was compared with the weight of copper as calculated for a plate of zero area. In the fourth column of the table the weights of the silver deposits are corrected by an experimentally determined factor, due to the fact that the silver nitrate solution used was contaminated with filter paper. Since copper is bivalent, its atomic weight is equivalent to two atomic weights of silver. At the time that this work was performed (1899) the value 107.93 was taken as the atomic weight of silver, and the figures given in the fifth column have been re-calculated to correspond to 107.88, which is the value accepted at the present time. The value of the atomic weight of copper now accepted is 63.57. Thus Faraday's law has been shown to hold for copper deposition within 0.03 per cent.

The Water Coulometer. A number of aqueous solutions, when electrolyzed between inert electrodes, evolve oxygen at the anode and hydrogen at the cathode. It is therefore possible to use, as a measure of the coulombs passed, the loss in weight of the solution, the volume of either gas, or the combined volumes of both gases. In the most accurate

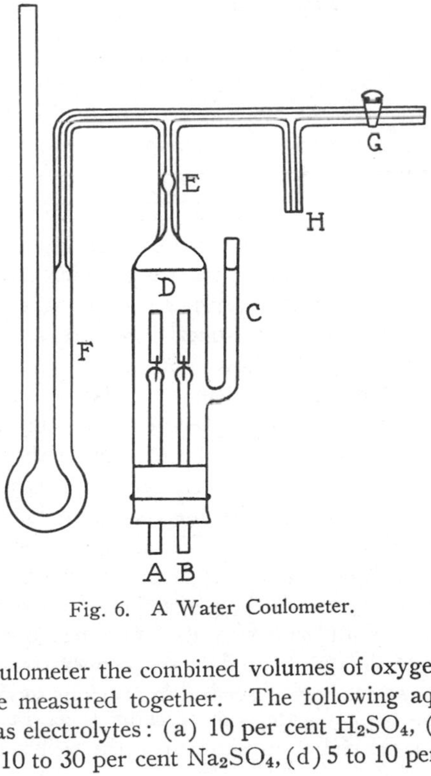

Fig. 6. A Water Coulometer.

type of water coulometer the combined volumes of oxygen and hydrogen gas liberated are measured together. The following aqueous solutions have been used as electrolytes: (a) 10 per cent H_2SO_4, (b) 10 to 25 per cent NaOH, (c) 10 to 30 per cent Na_2SO_4, (d) 5 to 10 per cent $K_2Cr_2O_7$.

An apparatus recommended by Lehfeldt[10] is shown in Fig. 6. The electrodes are of platinum sealed into the electrode tubes *A* and *B*. The side tube, *C*, is used for filling the electrolysis chamber, *D*. The small bulb, *E*, in the capillary outlet is a trap to catch spray. *F* is a manometer by means of which the pressure of the gas can be brought to that of the atmosphere with which the stopcock, *G*, allows communication when desired. The gas evolved during a determination is led through *H* into a gas burette, or other measuring device. One faraday of electricity will liberate 8 grams of oxygen and 1.0078 grams of hydrogen, which have densities, under standard conditions, of 0.0014290 and 0.0008986, respectively. The volume of gas evolved per faraday should therefore be

$$8/0.0014290 + 1.008/0.0008986 = 16810 \text{ cc.}$$

The water decomposed has a volume of about 10 cc., so that the increase in volume, per faraday, should be 16800 cc. at 0° C. and one atmosphere. Hence the volume increase per coulomb will be

$$16800/96500 = 0.1741 \text{ cc.}$$

Lehfeldt found, as an average of 16 determinations, a volume increase of 0.17394 cc. per coulomb. Instead of having the tube graduated in cubic centimeters, it may be more convenient and accurate to allow the gas to displace mercury into a weighing bottle and to determine the volume from the weight of the mercury.

Tests of Faraday's Law Under Varying Conditions. We have already seen that, if disturbing effects are taken into account, Faraday's law applies to all electrochemical reactions which have been carefully studied. The tests so far mentioned, however, have all been made at ordinary temperature, under atmospheric pressure, and in aqueous solutions. A number of researches have been carried out to find out whether variations in the nature of the solvent, or variations in the physical conditions, such as temperature and pressure, have any influence on the constant in Faraday's law. No real variation in the constant has yet been observed. There are, to be sure, many apparent deviations from the law, such as that observed with the copper coulometer, which gives a deposit at the cathode which is lighter than the computed value. In this case, as has been seen, the cause of the discrepancy has been found to be the occurrence of a disturbing reaction. In every similar case a simple explanation of the apparent deviation has been readily found. The comparison of the iodine, and of the copper coulometer, with the silver coulometer, as has been described in previous sections, affords precise evidence, for these reactions at least, that Faraday's law is inde-

[10] R. A. Lehfeldt, *Phil. Mag.*, (6) **15**, 614 (1908).

pendent of the nature of the dissolved substance, the time, and the current strength. That the law is also independent of the nature of the solvent is demonstrated by some experiments made by Kahlenberg,[11] who electrolyzed solutions of silver nitrate between a silver anode and a platinum cathode in a variety of solvents, including pyridine, quinoline, aniline, and benzonitrile, and obtained, in every case, a silver deposit which was in accord with that calculated by Faraday's law. Wilcox [12] electrolyzed a solution of silver nitrate in pyridine at −55° C. and obtained a silver deposit having the same weight as that in an ordinary silver coulometer connected in series. Richards and Stull [13] have found in some careful experiments that a cell containing fused silver nitrate as electrolyte at 250° C. gives the same weight of silver deposit, within 0.005 per cent, as does a silver coulometer containing an aqueous solution of silver nitrate in series. The experimental results are given in Table III.

TABLE III. COMPARISON OF SILVER COULOMETERS CONTAINING AQUEOUS SOLUTIONS OF, AND FUSED, SILVER NITRATE

From fused silver nitrate — Grams of silver deposited at 250° C.			From aqueous solution. Grams of silver deposited at 20° C.	Difference	
Observed	Weight of impurity milligrams	Corrected		In milligrams	In per cent
1.14958	0.39	1.14919	1.14916	0.03	0.003
1.12264	0.69	1.12195	1.12185	0.10	0.009
1.10242	0.42	1.10200	1.10198	0.02	0.002
				Average	0.005

The fact that change of pressure has no influence on Faraday's law is shown by the work of Cohen [14] who connected two silver coulometers in series, one as a reference coulometer at atmospheric pressure and the other, in successive experiments, at 500, 1000, and 1500 atmospheres. The results which are given in Table IV show that the same amount of silver was deposited in each coulometer.

TABLE IV. COMPARISON OF SILVER COULOMETERS AT ONE ATMOSPHERE AND UNDER HIGH PRESSURES

Pressure	1 atm.	500 atm.	1000 atm.	1500 atm.
Grams of Ag deposited in reference coulometer	1.5543	1.4002	0.6338	1.6408
Grams of Ag deposited in coulometer under pressure	1.5544	1.4002	0.6338	1.6408

An interesting verification of Faraday's law for the electrolysis of solid salts of heavy metals has been obtained by Tubandt and Eggert.[15]

[11] L. Kahlenberg, *J. Phys. Chem.*, **4**, 349 (1900).
[12] W. G. Wilcox, *J. Phys. Chem.*, **13**, 383 (1909).
[13] T. W. Richards and W. N. Stull, *Proc. Amer. Acad.*, **38**, 409 (1902).
[14] E. Cohen, *Z. Elektrochem.*, **19**, 132 (1913).
[15] C. Tubandt and S. Eggert, *Z. anorg. allgem. Chem.*, **110**, 196 (1920).

These investigators electrolyzed (without fusion) solid silver iodide, bromide, chloride, and nitrate, between a silver anode and a platinum cathode. The weight of silver deposited in each case was found to agree with that in a reference silver coulometer.

Since, as we have seen, the value of the faraday, **F**, is the product of Avogadro's number and the charge on the electron, a demonstrated example of a deviation from the law would involve the variation of one or the other of these fundamental constants, and is, therefore, not to be expected.

Apparent Deviations from Faraday's Law. Although Faraday's law is valid for all electrolytes at metal-electrolyte boundaries, apparent deviations can occur for several reasons, some of which have already been mentioned and others will be brought up from time to time later in this book. Some of the more important apparent deviations from the law are as follows:

Simultaneous electrode reactions. "Current efficiency." If, as is frequently the case, several electrochemical reactions occur simultaneously at an electrode, Faraday's law will be found to hold only if the total number of equivalents which have entered into reaction are used in the computation. Failure to include all the reactions at the electrode will thus result in an apparent deviation from the law. The ratio of the number of equivalents of a single electrode product to the total possible number computed by Faraday's law is called the *current efficiency* with respect to the electrochemical reaction in question. For instance, in the electrodeposition of zinc from an aqueous solution of one of its salts, hydrogen is always evolved. The ratio of the number of equivalents of zinc deposited to the total number of chemical equivalents (zinc and hydrogen) is the current efficiency of the deposition of zinc. Thus, as we have seen, a current efficiency of less than 100 per cent does not indicate a failure in the application of the law, but only that all the electrochemical reactions have not been included in the computation.

Interaction of anode and cathode products. Since the products of the reactions at the anode of any electrolytic cell have been reduced and those at the cathode have been oxidized, it is evident that mixing of the electrode products may result in chemical reactions which will be in the direction of lowering the efficiency of both electrode processes, and the observed amounts of the products of the electrode reactions will consequently be less than the computed values. For instance, in the electrolysis of water containing a salt the reaction at the anode is usually,

$$H_2O = \tfrac{1}{2}O_2 + 2H^+ + 2e^-$$

and at the cathode it is

$$2e^- + 2H_2O = H_2 + 2OH^-$$

That is to say, at the anode there is an equivalent of an acid, and at the cathode an equal amount of a base, formed for every faraday passed through the solution. Evidently if the solutions around the two electrodes are allowed to mix, the current efficiency of the production of these substances will be lowered.

Electrolytic reversal of electrode processes. If the product of one electrode in a cell is carried by diffusion or other means to the other electrode the product may be wholly or in part restored to its original condition. An oxidized anode product is likely to be reduced if it is allowed to get to the cathode, and *vice versa.* The passage of current through a solution containing a mixture of ferrous and ferric chlorides is an instructive example of this effect. As has been seen, the electrode reactions are

$$Fe^{++} = Fe^{+++} + e^-$$

at the anode, and

$$e^- + Fe^{+++} = Fe^{++}$$

at the cathode. If the electrolyte is stirred the anode oxidation products are quantitatively reduced at the cathode, and no permanent chemical change results from the passage of current. All intermediate steps between this zero current efficiency and one hundred per cent current efficiency can be obtained by allowing mixing of anode and cathode products or by completely separating them.

The examples given above are the simplest possible. Many more complicated phenomena may occur. For instance, the mixing of the anode and cathode products may not result in the re-formation of the substances originally present. New compounds may form, and these may be further reacted upon at the electrodes. It is such complications as these that in many cases limit the practical application of electrochemical methods and in other cases extend their interest and possibilities.

Chapter 3

Electrolytic Conductance and the "Classical" Theories of Dissociation

As has been seen, the phenomenon of the passage of electricity in electrolytic conductors is characterized by the movement of matter, that is to say, by particles larger than electrons, in contrast to metallic conductance in which the movement of electrons alone is involved. Electrolytic conductors are, further, unlike metals in that the carriers of electricity have both positive and negative charges. In a solution of sodium chloride in water, for instance, a portion of the current is maintained by the movement of sodium ions in a positive direction, and another portion is due to chloride ions traveling in a negative direction. The proportions of the total current which are carried by the different varieties of ions will be considered in the next chapter. In this chapter the total effect of the presence of ions on the conductance will be studied. The theoretical discussion will be largely deferred until both types of data, *i.e.*, total conductance and the proportions of the conductance due to each variety of ions, have been considered.

With the exception of metals dissolved in metals, and a few others, conducting solutions are electrolytic conductors. Many solid salts possess electrolytic conductivity, particularly at elevated temperatures. It is a characteristic of fused salts and of fused salt mixtures.

It is first of all necessary to define the units in which conductance and related quantities are expressed.

Specific Conductance, etc. *Definitions.* The *conductance,* K, of a conductor of electricity is the reciprocal of its electrical resistance, R, or

$$\text{K} = 1/\text{R} \qquad (1)$$

The resistance of a conductor is a characteristic property by virtue of which the energy of an electric current passing through it is converted into heat. The heat, H, in joules, developed in the conductor may be computed by the formula

$$\text{H} = I^2\text{R}s \qquad (2)$$

in which I is the current in amperes and s is the time in seconds. The resistance of a homogeneous substance of uniform cross-sectional area, A, and length, l, can be found from the equation

$$\mathrm{R} = \Gamma \frac{l}{A} \tag{3}$$

in which the term, Γ, is a constant called the *specific resistance.* This constant, which is characteristic of a substance under given physical conditions, is numerically equal to the resistance between two opposite sides of a unit cube (usually one cubic centimeter) of the substance. Resistances are stated in terms of the *ohm,* which is a primary electrical unit. The *international ohm* is the resistance offered to an unvarying electric current by a column of mercury, at the temperature of melting ice, 14.4521 grams in mass, of a constant cross-sectional area, and of a length of 106.300 centimeters.

Since conductance is the reciprocal of resistance its values may be roughly regarded as measures of the relative ease with which electricity can pass through conductors. From (1) and (3) we obtain

$$\mathrm{K} = \frac{1}{\mathrm{R}} = \frac{A}{\Gamma l} = \mathrm{L}\frac{A}{l} \tag{4}$$

in which L is the *specific conductance.* According to equation (4) then, the conductance of a conductor of uniform cross-section is directly proportional to the area of the cross-section and inversely proportional to the length of the conductor, and also directly proportional to a constant, L, which, being the reciprocal of the specific resistance, is a property of the conductor. The values of conductances are recorded in reciprocal ohms (sometimes called *mhos*). The specific conductance of a substance is evidently the conductance between opposite sides of a cube of the material, one centimeter in each dimension.

Ohm's Law. For any homogeneous metallic or electrolytic conductor the relation between the resistance, R, and the current in amperes, I, is, for a direct current, as follows:

$$I = E/\mathrm{R} \tag{5}$$

in which E is the *electromotive force* (which may be conveniently abbreviated to *emf*), the unit of which is the *volt.* The relation indicated by equation (5) is known as *Ohm's law.*

The Measurement of the Conductance of Solutions. Kohlrausch's Method. The resistances, and naturally also the conductances, of a conducting system can be readily obtained by means of the Wheatstone bridge, a diagram of which is given in Fig. 1. Current from the source of electrical energy S passes to A where the circuit divides, one part going through the unknown resistance R_x and the known adjustable

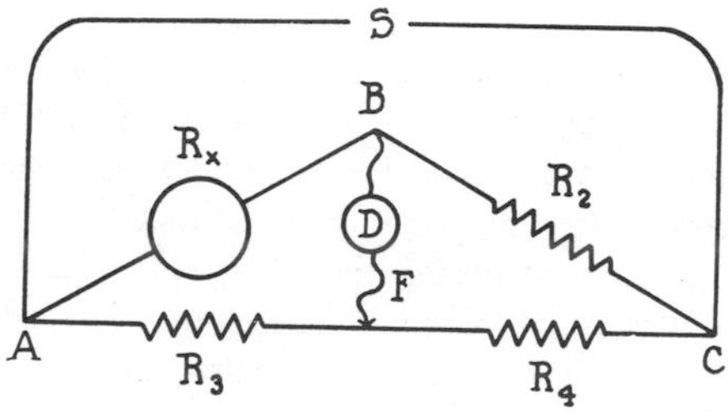

Fig. 1. The Wheatstone Bridge.

resistance R_2. The other portion of the current passes through two resistances, R_3 and R_4. Some form of current indicator, D, such as a galvanometer (or telephone if alternating current is used) is connected between the point F and the junction B of the resistances R_x and R_2. It can be readily shown that when the detector D indicates that no current is passing between B and F the following relation holds:

$$R_x : R_3 = R_2 : R_4 \tag{6}$$

To measure the resistance of a metallic conductor by this principle requires, in general, no special precautions if a direct current is used. With electrolytic conductors the procedure is not so simple. The electrolytic conductor whose resistance is to be measured must be contained in a vessel into which are inserted electrodes having metallic conduc-

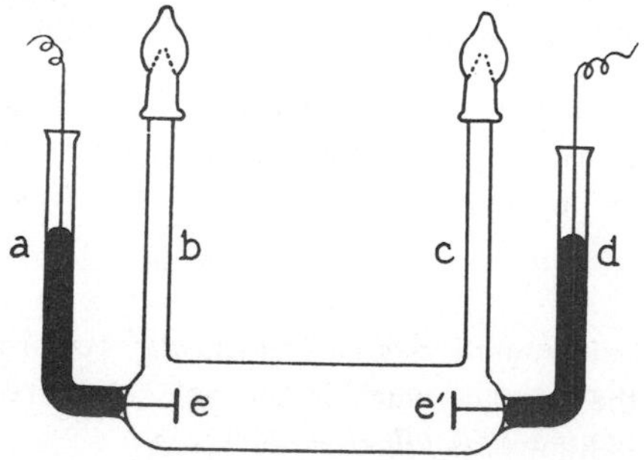

Fig. 2. A Conductance Cell.

tance. A typical modern cell used for the measurement of the conductance of solutions is shown in Fig. 2. The cell is filled with solution through the tubes *b* and *c*. Electrical contact is made with the electrodes *e* and *e′* through wires sealed through the glass wall which connect with mercury in the tubes *a* and *d*.

A type of cell designed by Shedlovsky [1] for dealing with very dilute solutions is shown in Fig. 3. The flask *F* is used for preparing a solution. A portion of the solution can then be forced by gas pressure through the tube *S* into the conductance cell proper, which is the space

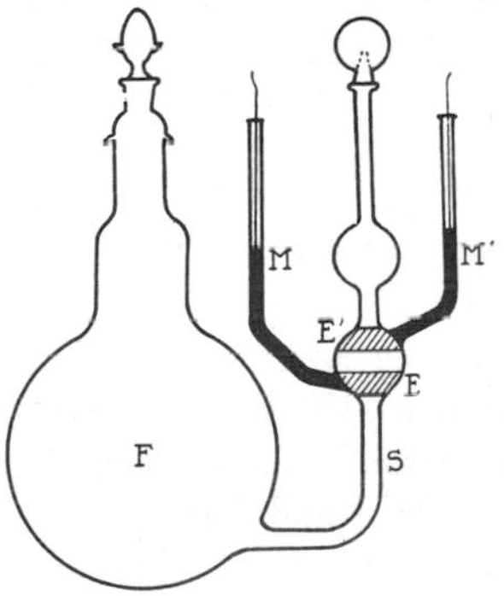

Fig. 3. A Conductance Cell for Dilute Solutions.

between the conical electrodes *E* and *E′*. The electrical leads to these electrodes are made through mercury in the tubes *M* and *M′*.

As we have seen in Chapter 1, any current passing from an electrode to an electrolyte causes a chemical reaction. Therefore, unless special precautions are taken, the current used in making the measurements will produce changes in the chemical composition of the solution. For instance, if the resistance of an aqueous salt solution is being found, the passage of a continuous current will cause the formation of oxygen and an acid at one electrode and of hydrogen and a base at the other. These changes not only produce real variations in the conductance of the solution but, in addition, potentials are set up at the electrodes which invalidate the conductance measurements. The term *polarization* is used, rather indiscriminately, to denote not only the concentration and chemical changes due to the electrolysis, but also the potentials at the electrodes which result from these changes. To overcome the difficulties due to polarization it is usual to make use of alternating current in the determinations. By this device the electrolysis produced by the current flow for an instant in a positive direction tends to be imme-

[1] T. Shedlovsky, *J. Am. Chem. Soc.*, **54**, 1411 (1932).

diately reversed by the passage of the current in a negative direction, and *vice versa.* This reversal of polarization effects by each change of direction of the current does not, however, by any means occur with all types of electrodes. It is almost universal practice to use platinum electrodes that are "platinized," *i. e.,* that are covered with an adherent coat of finely divided platinum. This has the effect of greatly increasing the effective surface of the electrode, and in addition the material has a catalytic effect on some of the electrochemical reactions involved. Obviously, then, the current indicator D must be capable of detecting small alternating currents. For this purpose a telephone is suitable, silence or a minimum of sound indicating that the adjustment of resistances represented by equation (6) has been reached (unless there are sources of error to be discussed below). This adaptation of the Wheatstone bridge to the measurement of the conductance of electrolytes is known as "Kohlrausch's method."

The use of alternating current for the measurements, while largely overcoming the polarization effects just alluded to, introduces several complications into the measurements of conductivity. In order that no current will flow between B and F of Fig. 1, it is not only necessary, using alternating or oscillating currents, that the relation of equation (6) for the resistances hold, but that the *reactances* of the different portions of the circuit be adjusted also. These reactances are due to inductances and capacities in the various portions of the circuit. These may produce surges through the telephone or other current indicator even when the resistances have been adjusted to meet the requirements of equation (6). Wheatstone bridges for measuring the conductances of electrolytes, designed to avoid errors from reactances, have been described by Washburn and his associates,[2] by Morgan and Lammert,[3] Jones and Josephs,[4] Shedlovsky,[5] and others. An excellent account of the alternating current theory involved is given in a book by B. Hague.[6]

The cells shown in Figs. 2 and 3 have been designed to avoid a rather insidious error that can readily creep into conductance work. If the electrodes of the cell shown in Fig. 3 were placed in the body of the solution, as is shown in Fig. 4 (A), there would be a capacity effect between the leads, shown diagrammatically at T, the vertical lines being a symbol for capacity. This effect is constant, and can be completely compensated for by placing an adjustable capacity across the

[2] E. W. Washburn and J. E. Bell, *J. Am. Chem. Soc.,* **35,** 177 (1913); Washburn, *ibid.,* **38,** 2431 (1916); E. W. Washburn and K. Parker, *ibid.,* **39,** 235 (1917).

[3] J. L. R. Morgan and O. M. Lammert, *ibid.,* **48,** 1220 (1926).

[4] G. Jones and R. C. Josephs, *ibid.,* **50,** 1049 (1930).

[5] T. Shedlovsky, *ibid.,* **52,** 1793 (1928).

[6] B. Hague, "Alternating Current Bridge Methods," Sir Isaac Pitman and Sons, Ltd., London (1930). 2nd ed.

opposite arm of the Wheatstone bridge. There is, however, another capacity effect, indicated at K in the figure, between the leads and the solution. This capacity is in series with the resistance r of the solution. The arrangement is shown schematically in Fig. 4 (B), the serrated lines representing resistances. Although this more complicated condition can also be balanced by adjusting the condenser in the opposite

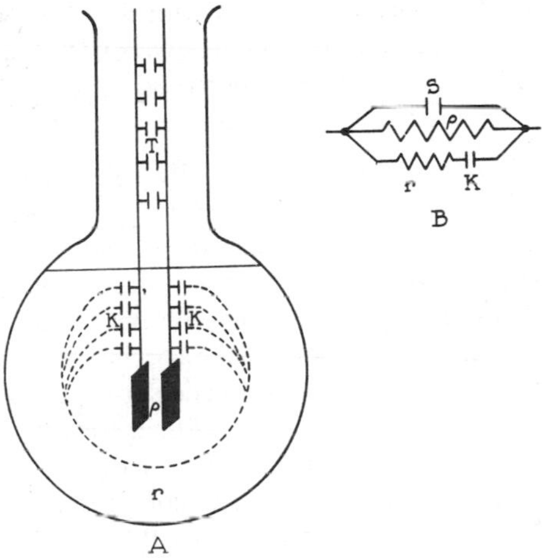

Fig. 4. Capacitance Effects in a Conductance Cell.

arm of the bridge, the apparent resistance readings will be a function of the frequency of the current used in the measurements. With the cell shown in Fig. 3, capacities are reduced by a wider spacing of the leads, and the capacity-resistance paths have been reduced to nearly zero, since they pass through the thermostat liquid, which should be oil. Resistance measurements on solutions in this cell and in the cell shown in Fig. 2 have been found to be independent of the frequency, within the limits of audibility of tones heard in a telephone.

This incomplete discussion of the effect of capacities in and around a conductance cell is included to indicate the nature of the errors which may be encountered with improperly designed cells. Similar errors may arise from wrongly arranged Wheatstone bridges. Careful conductance measurements should not be attempted without an understanding of the phenomena connected with alternating currents. The author has somewhat reluctantly left out a more complete discussion of the subject.

It would, however, take an undue amount of space, and is adequately treated in the references just given.

The question has been repeatedly raised as to whether Kohlrausch's method for determining conductances, involving as it does electrodes where potential differences are possible, and high frequency currents, gives the same values for conductances as would be obtained by a method not involving polarizable electrodes and using direct current. Careful experiments to test this matter have been made by Eastman,[7] who measured the conductances of solutions with direct and high frequency current. The two methods agreed within about 0.01 per cent; the slight differences, possibly due to experimental error, were in the direction of a decrease of conductance when high frequency was used.

The Cell Constant. Since conductivity cells, such as that shown in Fig. 2, are not constructed with uniform length and cross-section, equation (4) can evidently not be used to compute the specific conductance L of a solution after its conductance, K, in a given cell has been determined. This difficulty is overcome by obtaining the *cell constant* by calibrating the cell with a solution the specific conductance of which is already known. A few solutions, particularly potassium chloride at various concentrations and over a range of temperatures, have been carefully measured in cells of definite uniform length and area, by Kohlrausch, Holborn and Dieselhorst,[8] and more recently by H. C. and E. W. Parker.[9] From such measurements the specific conductance, L, of the solution measured may be computed by equation (4). With one of the solutions of known specific conductance, L, the cell constant, J, may be found by means of the formula

$$L = JK = J/R \tag{7}$$

in which K is the conductance and R the resistance of the solution as measured in the cell under consideration. From equation (4), it will be seen that the constant J has the dimensions of l/A, and could be computed from this relation if the two electrodes were parallel and completely filled the ends of a vessel of uniform cross-sectional area A.

Since, however, previous determinations of the specific conductances of the potassium chloride solutions used in calibrating cells were not in complete agreement, Jones and Bradshaw [10] have redetermined the values of those constants for some of the more important solutions, using a different method from the one just described. Instead of using

[7] E. D. Eastman, *J. Am. Chem. Soc.*, **42**, 1648 (1920).
[8] F. Kohlrausch, L. Holborn and H. Dieselhorst, *Wied. Ann.*, **64**, 425 (1898).
[9] H. C. and E. W. Parker, *J. Am. Chem. Soc.*, **46**, 312 (1924).
[10] G. Jones and B. C. Bradshaw, *J. Am. Chem. Soc.*, **55**, 1780 (1933).

a cell of known dimensions they measured the electrical resistance of cells when filled with mercury at 0°. With the density of mercury at that temperature its specific conductance (L = 10629.63) may be computed from the definition of the ohm given on p. 41. However, the difference between this specific conductance and that of a solution of potassium chloride is so great that it is inconvenient to measure both in the same cell. This difficulty is overcome by the use of two cells, in one of which mercury and strong sulphuric acid are measured, and in the other, the same solution of sulphuric acid and the standard potassium chloride. Jones and Bradshaw's results are summarized in Table I. A "demal" solution, in terms of which the concentration is

TABLE I. SPECIFIC CONDUCTANCES OF STANDARD POTASSIUM CHLORIDE SOLUTIONS

Concentration demal	Grams KCl per 1000 grams solution in vacuum	Specific conductances 0°	18°	25°
1	71.1352	0.06517_6	0.09783_8	0.11134_2
0.1	7.41913	0.007137_9	0.011166_7	0.012856_0
0.01	0.745263	0.0007736_4	0.0012205_2	0.0014087_7

expressed, is defined as a solution containing a gram mol of salt dissolved in a cubic decimeter of solution at *zero degrees*. Such a solution, at zero degrees only, is equal to a normal solution times 1.000027, the factor by which a liter differs from a cubic decimeter. The data on conductivity given in this book have been determined, or recomputed, on the basis of these new values for the conductivities of the 0.1 demal potassium chloride solution listed in Table I.

Molar and Equivalent Conductances. A quantity much used in computations and in tables of constants is the *molar conductance*, Λ_m. It can be computed from the specific conductance, L, and the concentration, C, of the solute, which is usually stated in mols per liter of solution, by means of the formula

$$\Lambda_m = 1000 \frac{L}{C} \tag{8}$$

The molar conductance is, physically, the conductance of the amount of solution that contains one mol of the solute when measured between parallel electrodes which are one centimeter apart and large enough in area to include the necessary volume of solution. The arrangement is shown diagrammatically in Fig. 8. The molar conductance is numerically equal to the number of amperes that would pass through such a cell if a potential difference of one volt were applied across the electrodes, polarization and other disturbing effects being excluded. The *equivalent conductance,* Λ, is similarly the conductance of a solution containing one gram equivalent of a solute under the conditions defined above and is obtained from the equation

$$\Lambda = \frac{1000\text{L}}{\text{c}} \tag{8a}$$

in which c is the concentration in equivalents. For solutions containing only univalent ions the molar and equivalent conductances have, of course, the same values.

The Change of Equivalent Conductance with Dilution. If a gram equivalent of a salt is dissolved in, for instance, enough solvent to yield a liter of solution, this resulting solution will have, under given physical conditions, a definite equivalent conductance. The value of this conductance will depend upon the number of electrically charged ions in the solution and the mobilities, U, of these ions, *i.e.,* their velocities in a potential gradient of one volt per centimeter. More strictly speaking the mobilities are the components of velocity, in the direction of the electric force, superimposed on the random motion of the ions. The number and mobilities of the ions vary from solute to solute, and from solvent to solvent, and are, further, influenced by such factors as pressure and temperature. A variation of the equivalent conductance due to a change of one of the variables mentioned may be caused by a change in the number of the ions present or in the mobilities of the ions, and the nature of the ions may change also. We have, at the present time, no method that can distinguish with certainty between changes in conductance due to these various causes. In many cases it is probable that all these types of change occur simultaneously. Much of what follows in this book will be concerned with this subject.

The most interesting and important changes of equivalent conductance are those which are observed when the solutions are diluted or concentrated, the other variables, such as temperature and pressure, being kept constant. A very large amount of experimental data has been accumulated in this field. Curves, each representative of the behavior of a large class of solutions, are given in Figs. 5 to 7. In these diagrams the equivalent conductance Λ is plotted against the logarithm of the dilution, V, which is the number of liters of solution containing one gram equivalent of solute. Logarithms are used only because they bring the dilutions, which vary through a wide range of values, into convenient plots.

Fig. 5, in which the equivalent conductance, Λ, of aqueous solutions of potassium chloride at 25° is plotted as ordinates against the logarithms of the dilution, V, represents, in general trend at least, the variation of the equivalent conductance with dilution of aqueous solutions of all salts, and of strong acids and bases. These equivalent con-

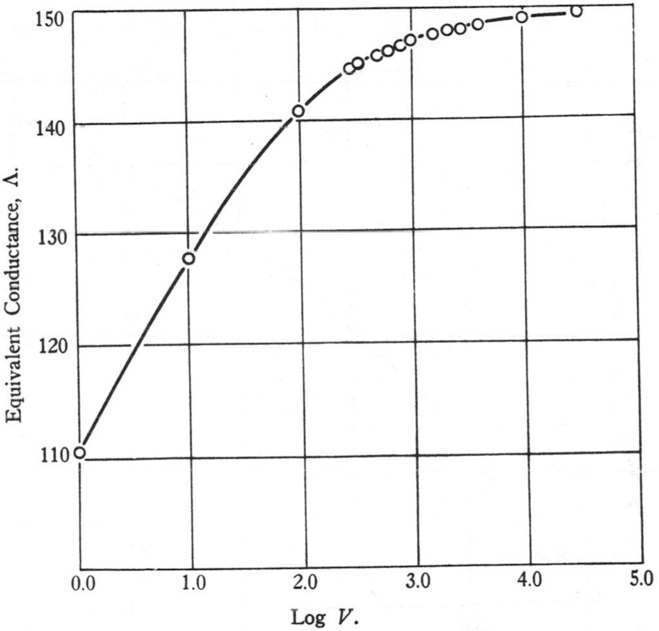

Fig. 5. Equivalent Conductance, Λ, of Potassium Chloride in Water at 25°.

T. Shedlovsky, *J. Am. Chem. Soc.*, **54**, 1411 (1932). G. Jones and B. C. Bradshaw, *ibid.*, **55**, 1780 (1933).

ductances at first increase rapidly with dilution and then more slowly approach what appears to be a maximum limiting value as the dilution is indefinitely increased. Careful work by Weiland,[11] Kraus,[12] and more recently by Shedlovsky,[13] indicates, however, that the equivalent conductances of these solutions continue to increase, though at a continuously decreasing rate, as the dilution is carried as far as is experimentally possible.

[11] H. J. Weiland, *J. Am. Chem. Soc.*, **40**, 131 (1918).
[12] C. A. Kraus and W. C. Bray, *ibid.*, **35**, 1315 (1913).
[13] T. Shedlovsky, *ibid.*, **54**, 1411 (1932).

Fig. 6, in which the equivalent conductance of sodium iodide in ethyl alcohol at 25° is plotted as ordinates against the logarithms of the dilution, represents the behavior of many salts in solvents other than water. It will be seen that although the equivalent conductance is apparently approaching a maximum, the latter is much farther removed from the experimentally determined points than is the case with water solutions at corresponding dilutions.

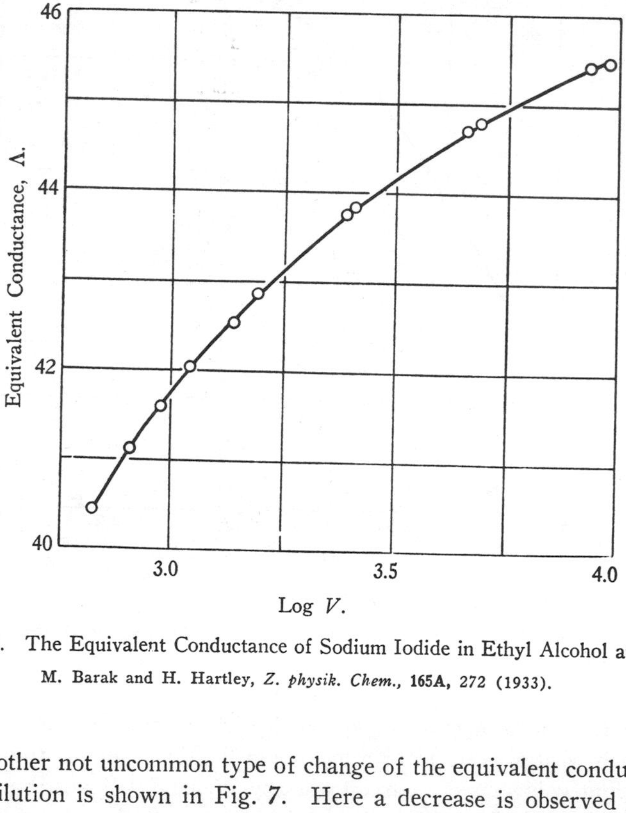

Fig. 6. The Equivalent Conductance of Sodium Iodide in Ethyl Alcohol at 25°.
M. Barak and H. Hartley, *Z. physik. Chem.*, **165A**, 272 (1933).

Another not uncommon type of change of the equivalent conductance with dilution is shown in Fig. 7. Here a decrease is observed in the equivalent conductance with the dilution, V, for small values of the latter, and followed after a minimum by the more usual increase, such as is shown in Figs. 5 and 6. Still more complex behavior has also been observed.

To anticipate the explanations of these phenomena, to be discussed more fully later in this volume, an important factor in determining the shapes of these curves is the dielectric constant of the solvent. If it is large, as is the case with water, the attraction of particles with unlike charge for each other will be relatively low, and they will be largely kept separate by thermal vibration. Dissolved electrolytes in such solutions will consist in great part of ions, and their equivalent conductances will be relatively large. On the other hand if the dielectric constant of the solvent is low, particles carrying opposite charges will have

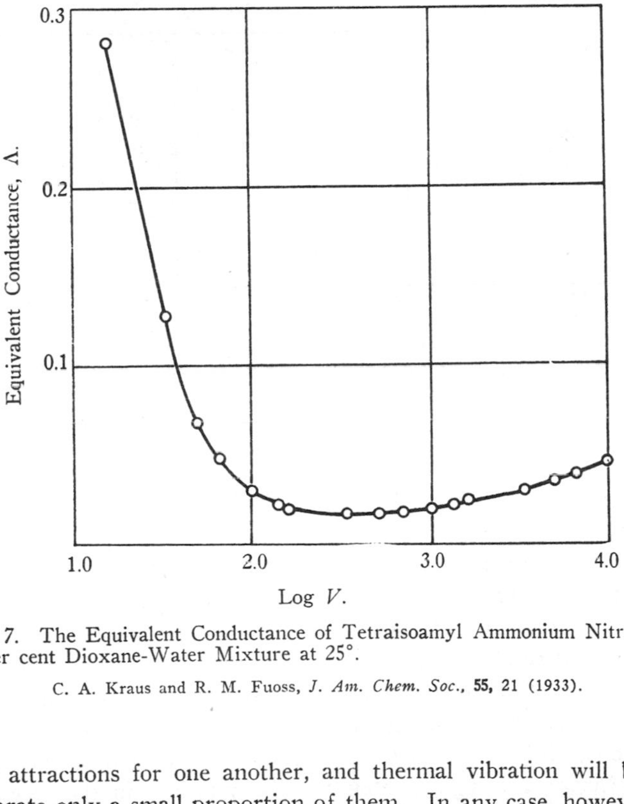

Fig. 7. The Equivalent Conductance of Tetraisoamyl Ammonium Nitrate in a 95.99 per cent Dioxane-Water Mixture at 25°.

C. A. Kraus and R. M. Fuoss, *J. Am. Chem. Soc.*, **55,** 21 (1933).

strong attractions for one another, and thermal vibration will be able to separate only a small proportion of them. In any case, however, the proportion remaining ionized will be greater at high dilutions, since the chance that an ion will encounter another of opposite charge is smaller than when the solvent is more concentrated. It is for this reason that equivalent conductances generally increase with dilution. In solutions with solvents of very low dielectric constants the phenomena are probably complicated by the presence of complexes involving the undissociated electrolyte and the ions. It will be shown in Chapter 19 that modern theories are sufficient to account, quantitatively, for a large proportion of the phenomena that have been observed in this field.

Some General Statements Concerning Electrolytic Conductance. Let us imagine a solution containing a gram molecular weight of an electrolyte (salt, base, or acid) placed in a cell, such as is shown in Fig. 8, having electrodes one centimeter apart and of sufficient area

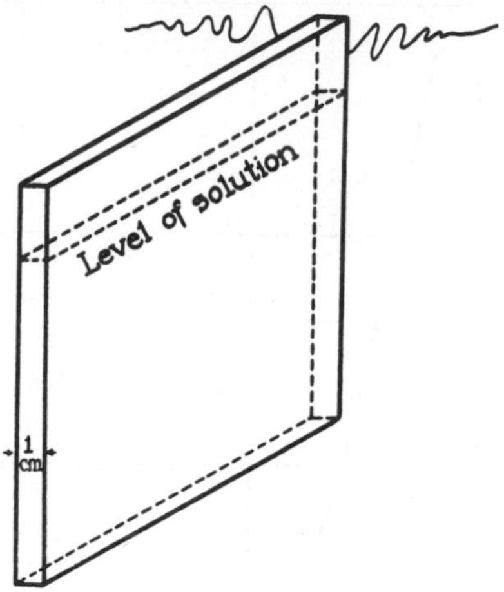

Fig. 8. Diagrammatic Conductance Cell.

to include a volume large enough to contain the solution. With a potential difference of one volt impressed across the electrodes and assuming zero or balanced electrode potentials, the current through the cell will be equal numerically to the molar conductance Λ_m. This conductance can depend upon three factors only, (a) the number, (b) the electrical charges and (c) the velocities of the ions present in the solution. The charge on each ion is equal to that of a proton, ϵ, or to a multiple, $z\epsilon$, of that charge, z being an integer. The velocities in the case under consideration will be equal to the mobilities, U_1, U_2, U_3, ..., of the ions, *i. e.*, the velocities under a potential gradient of one volt per centimeter. The number of each ion in the solution will be represented by n_1, n_2, ... The general equation for the molar conductance is therefore,

$$\Lambda_m = n_1\text{U}_1z_1\epsilon + n_2\text{U}_2z_2\epsilon + \cdots \quad (9)$$

Restricting the discussion to electrolytes which furnish only two species of ions, equation (9) reduces to

$$\Lambda_m = n^+U^+z^+\epsilon + n^-U^-z^-\epsilon \tag{10}$$

in which n^+, n^-, U^+, U^- are respectively the numbers and mobilities of the positive and negative ions. Now the maximum number of positive or negative ions, n^+ and n^-, that can arise from the dissociation of a gram molecular weight of a salt is **N**, *i.e.*, Avogadro's number, times the number, ν, of ions of the given type produced by the dissociation of one molecule of the salt. However, the electrolyte may yield smaller numbers than these maxima. Thus for calcium chloride the factor, ν^+, is unity for the Ca^{++} ion and two for the Cl^- ion. The maximum values of n^+ and n^- in equation (10) are therefore ν^+ **N** and ν^- **N**, and if the salt is completely dissociated

$$\Lambda_m = \mathbf{N}\epsilon(\nu^+z^+U^+ + \nu^-z^-U^-) \tag{11}$$

If, however, the salt is considered to be incompletely broken up into ions, the factor, α, called the "degree of dissociation" must be introduced, giving

$$\begin{aligned}\Lambda_m &= \mathbf{N}\epsilon\alpha(\nu^+z^+U^+ + \nu^-z^-U^-) \\ &= \mathbf{F}\alpha\,(\nu^+z^+U^+ + \nu^-z^-U^-)\end{aligned} \tag{12}$$

since the product $\mathbf{N}\epsilon$ is equal to the faraday, **F**. It is the presence of the factor α that has made the study of solutions of even the simplest electrolytes a difficult matter. Many methods have been proposed for determining this factor, some of which will be described later in this book.

Since the solution is electrically neutral the number of units of positive charge must be equal to the number of units of negative charge, so that in equation (12)

$$\mathbf{N}\alpha\nu^+z^+ = \mathbf{N}\alpha\nu^-z^-$$

and

$$\nu^+z^+ = \nu^-z^-$$

One gram mol of a salt contains νz equivalents, so that the relation between the molar conductance, Λ_m, and equivalent conductance, Λ, is

$$\Lambda_m = \nu z\Lambda \tag{13}$$

The equation for the equivalent conductance of a binary electrolyte thus takes the simple form

$$\Lambda = \mathbf{F}\alpha(u^+ + u^-) \tag{14}$$

A few additional remarks should be made with respect to the mobilities, u_1, u_2, ..., of the ions. If the ions were in free space the presence of a field of electric force would result in an accelerated motion, *i. e.,* the velocity would increase with time. In a solution, however, the opposition to the motion of the ions is large, with the result that a constant velocity proportional to the potential gradient is established very nearly instantly. If this were not true Ohm's law would not be found to hold for solutions of electrolytes, since the resistance of such solutions would not be found to be independent of the time during which the measuring potential is applied.

The "Classical" Ionic Theory of Arrhenius. Arrhenius,[14] in 1887, proposed a method for computing the degree of dissociation, α, which has had an enormous influence on the progress of physical chemistry and upon related sciences. According to his method the degree of dissociation is found by means of the simple formula,

$$\alpha = \Lambda/\Lambda_0 \tag{15}$$

in which Λ is the equivalent conductance at the concentration in question, and Λ_0 is the limiting or "infinite dilution" value of that conductance.

Because of its historic importance and the fact that all of the more recent theories of electrolytes are an outgrowth of Arrhenius' original statement of the theory of electrolytic dissociation in the form given in equation (15), the rest of this chapter will be devoted to a survey of some of the evidence for and against the theory.

In the use of equation (15) by the earlier workers Λ_0 was obtained by measuring the equivalent conductance at increasing dilutions until the effect of further dilution made but a slight effect on that quantity. A plot showing a series of such measurements for KCl is given in Fig. 5. Later investigators used the results of such measurements for the purpose of extrapolating to values at infinite dilution by means of empirical equations. Although, as later developments have shown, equation (15) gives a nearly correct result for the degree of dissociation with a limited type of electrolytes, its application to highly dissociated electrolytes is attended with uncertainty. For one thing it has not been sufficiently recognized until recently that the equation involves

[14] S. A. Arrhenius, *Z. physik. Chem.,* **1,** 631 (1887).

the assumption that the ion mobilities, u_1, u_2, ..., are constant from the concentration in question to infinite dilution. This can readily be seen for a simple binary electrolyte by dividing the value of Λ from equation (14) by the corresponding value, Λ_0, at infinite dilution, where the degree of dissociation, α, can safely be assumed to be unity, as follows:

$$\frac{\Lambda}{\Lambda_0} = \frac{\alpha \mathbf{F}(u^+ + u^-)}{\mathbf{F}(u_0^+ + u_0^-)} \tag{16}$$

Here u_0^+ and u_0^- are the values of the mobilities at infinite dilution. It is evident that for equation (15) to be true the ratio of the sum of the ion mobilities must not change with the concentration, since otherwise the terms containing mobilities will not cancel out. In that case equation (15) will yield an incorrect value of the degree of dissociation, α. It will be shown in Chapter 4, from evidence due to transference measurements, that only in exceptional cases can the mobilities be independent of the concentration.

The general acceptance of the Arrhenius method for computing the degree of dissociation was, however, due to the agreement, usually not very precise, of computations by that method and by methods depending upon the colligative properties of solutions, *e. g.*, freezing point and vapor pressure measurements, for the details of which a textbook on physical chemistry may be consulted. In addition, however, it was found that upon the basis of Arrhenius' theory the law of mass action apparently holds for electrolytes which are only slightly dissociated. The thermodynamic basis for the law of mass action will be discussed in Chapter 6. The following simple treatment contains assumptions which have been shown to be incorrect by more recent work, which is discussed later in this book.

The dissociation of a univalent electrolyte which ionizes to give the positive ion C^+ and the negative ion A^- according to the equation

$$CA = C^+ + A^-$$

would be expected to follow the law of mass action. When Arrhenius proposed his theory, and until comparatively recently, this implied that the expression

$$\frac{[C^+][A^-]}{[CA]}$$

should be a constant, independent of the total concentration, *C*, of the electrolyte. The terms $[C^+]$, $[A^-]$ and $[CA]$ etc. represent respectively the concentrations of the ions C^+, A^- and of the undissociated substance CA. Since $[C^+]$ and $[A^-]$ are both equal to αC, C being the total concentration, the undissociated portion [CA] will be equal to $(1-\alpha)\,C$. Upon substitution, the expression reduces to

$$\frac{\alpha^2 C}{(1 - \alpha)} = \mathbf{K} \tag{17}$$

K being a constant if the law of mass action is followed. Equation (17) is known as "Ostwald's dilution law" and the term **K** is known as the "ionization constant." If we further assume from equation (15) that $\alpha = \Lambda/\Lambda_0$ equation (17) reduces to

$$\frac{\Lambda^2 C}{\Lambda_0(\Lambda_0 - \Lambda)} = \mathbf{K} \tag{18}$$

How well this Ostwald dilution law holds for solutions of acetic acid is shown in Table II. In this table the first column contains

TABLE II. THE IONIZATION CONSTANT OF ACETIC ACID FROM OSTWALD'S DILUTION LAW, $\Lambda_0=390.71$

Concentration Equivalents per liter $C\times10^3$	Equivalent conductance Λ	Ionization constant $\mathbf{K}\times10^5$
0.028014	210.38	1.760
0.15321	112.05	1.767
1.02831	48.146	1.781
2.41400	32.217	1.789
5.91153	20.962	1.798
12.829	14.375	1.803
50.000	7.358	1.808
52.303	7.202	1.811

concentrations, the second the corresponding equivalent conductances, and the third the values of **K** computed from the Ostwald dilution law. The data are from the work of MacInnes and Shedlovsky.[15] In obtaining these data the attempt was made to include every precaution that would insure accuracy. It will be observed that although **K** is fairly constant it shows a small but unmistakable drift as the concentration changes. Values of **K** which are constant over wider ranges of concentration could be obtained, as has been done by Kendall,[16] by using a different value of Λ_0 from that given in the table. However, evidence of the correctness of the Λ_0 value used in this computation will be given in Chapter 18.

Acetic acid is a relatively weak acid, *i. e.*, its solutions conduct rather poorly. Let us examine the application of the Ostwald dilution law

[15] D. A. MacInnes and T. Shedlovsky, *J. Am. Chem. Soc.*, **54**, 1429 (1932).
[16] J. Kendall, *J. Chem. Soc.*, **101**, 1275 (1912).

to a considerably stronger acid, *e. g.*, chloroacetic. The data and the corresponding values of **K** are given in Table III and are from the

TABLE III. THE IONIZATION "CONSTANT" OF CHLOROACETIC ACID FROM OSTWALD'S DILUTION LAW, $\Lambda_0=389.5$

Concentration equivalents per liter $C\times10^3$	Equivalent conductance Λ	Ionization constant $K\times10^3$
0.11010	362.10	1.353
0.30271	328.92	1.388
0.58987	295.58	1.409
1.3231	246.15	1.436
2.8211	197.14	1.463
3.8124	177.98	1.466
7.4620	139.85	1.501
14.043	109.00	1.527
20.179	93.83	1.543

work of Shedlovsky, Brown, and MacInnes.[17] Here it will be seen that **K** is much less constant than is the case for acetic acid and increases steadily with the concentration in the range given in the table.

TABLE IV. THE APPARENT FAILURE OF THE LAW OF MASS ACTION FOR A STRONG ELECTROLYTE, HYDROCHLORIC ACID, $\Lambda_0=426.16$

Concentration equivalents per liter $C\times10^4$	Equivalent conductance Λ	The "constant" K of Ostwald's dilution law
0.28408	425.13	0.0116
0.81181	424.87	0.02666
1.7743	423.94	0.03355
3.1863	423.55	0.05139
5.9146	422.54	0.05995
7.5404	421.78	0.07169
15.768	420.00	0.1059
18.766	419.76	0.1212
25.614	418.44	0.1363
29.943	418.10	0.1523

Finally in Table IV are given the data obtained for the conductance measurements on the "strong" acid, hydrochloric,[18] together with the corresponding values of **K**, which in this case varies with the concentration in order of magnitude.

The shifts of **K** values with the concentration, as given in the three examples just discussed, are quite typical of the observations on weak, intermediate, and strong electrolytes. It is found that the variation of **K** with the concentration increases with the strength of the electrolyte. The conclusion arrived at from the study of these data is that the Ostwald dilution law is only true in the limiting case, *i. e.*, for an infinitely weak electrolyte. We shall find later that such constancy as is observed

[17] T. Shedlovsky, A. S. Brown, and D. A. MacInnes, *Trans. Electrochem. Soc.*, **66**, 165, (1934).
[18] T. Shedlovsky, *J. Am. Chem. Soc.*, **54**, 1411 (1932).

in the **K** values is due to the compensation of two assumptions, both tacitly made in the derivation of the Ostwald expression.

The complete failure of highly conducting solutions to follow the mass law if their degrees of dissociation are computed by means of the Arrhenius assumption (equation (15)) was found very difficult to explain by the early proponents of the ionic theory. It was known as "the anomaly of the strong electrolyte." Many ingenious attempts to resolve the difficulty were made. These have, in general, only historic interest.

As has just been made clear, the Arrhenius theory provided an inadequate picture of the phenomena occurring in solutions of electrolytes. The theory did, however, serve as a basis for further research. As has been mentioned, one of the tacit assumptions of the theory is that ions have mobilities that do not change with the concentration of the electrolyte from which they arise. A test of the validity of this assumption may be obtained with accurate data on transference numbers, a subject that will be discussed in the next chapter.

Transference numbers will also be found useful in obtaining precise values of the "activities" of ion constituents. It was another of Arrhenius' tacit assumptions that ion concentrations may be used without error in the law of mass action. To investigate the limits of validity of that assumption, and to lay a foundation for the modern interionic attraction theory of solutions, it is necessary to consider the thermodynamics of solutions, and of the galvanic cell, subjects which are discussed in Chapters 5 and 6.

Chapter 4

Electrical Transference

When an electric current is passed through a solution of an electrolyte the various ions present carry different proportions of the current. These proportions are called the *transference numbers* of the ions. (In British publications they are usually referred to as *transport numbers.*) The relation between the transference number, t^+, of the positive ion of a binary electrolyte and the ion mobilities, U^+ and U^-, of the electrolyte may be seen as follows: From equation (8a)

$$\Lambda = \frac{1000L}{C}$$

and equation (14) both of Chapter 3

$$\Lambda = \mathbf{F}\alpha(U^+ + U^-)$$

we may obtain the following equation for the specific conductance

$$L = 0.001C\alpha\mathbf{F}(U^+ + U^-) \tag{1}$$

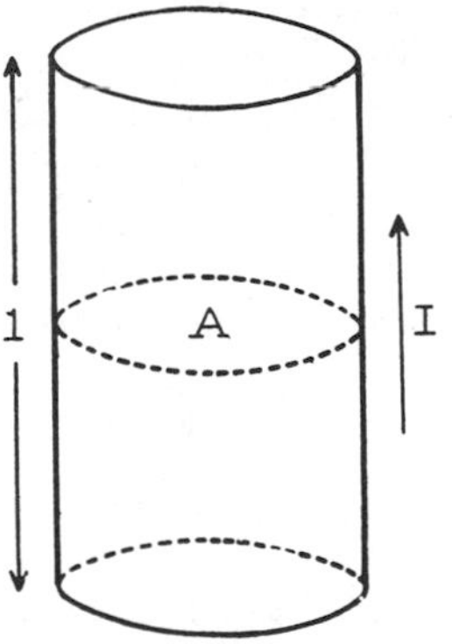

Fig. 1.

If an electromotive force E is impressed upon a tube, Fig. 1, of length l and area A, the electric current I through a solution with the specific conductance L in the tube will be, from equations (4) and (5) Chapter 3

$$I = E\,\mathrm{L}\,A/l = 0.001\,E\,A\,c\,\alpha\,\mathbf{F}\,(\mathrm{u}^+ + \mathrm{u}^-)/l \qquad (2)$$

Of this current $0.001EAc\alpha\mathbf{F}\mathrm{u}^+/l$ is carried by the positive ion. The proportion of the total current, *i.e.*, the transference number t^+, of that ion is therefore

$$t^+ = \frac{0.001EAc\alpha\mathbf{F}\mathrm{u}^+/l}{0.001EAc\alpha\mathbf{F}(\mathrm{u}^+ + \mathrm{u}^-)/l} = \frac{\mathrm{u}^+}{\mathrm{u}^+ + \mathrm{u}^-} \qquad (3)$$

That is to say, the transference number of the positive ion is the ratio of the mobility of the positive ion to the sum of the mobilities of the positive and negative ions. The transference number of the negative ion can be similarly obtained and the relation between the two numbers is evidently

$$t^+ + t^- = 1 \qquad (4)$$

Much use of transference numbers has been made in the development of electrochemistry. The chief methods for their determination are (*a*) the Hittorf method, (*b*) the moving boundary method and (*c*) the electromotive force method. Of these the first two will be considered in this chapter.

It is, however, useful in our discussion to distinguish between an *ion* and an *ion constituent*. The term ion constituent is used to denote the ion-forming portion of an electrolyte without reference to the extent to which it may actually exist in the dissociated state. The term ion is used with reference to the electrically charged substances which have resulted from the dissociation of the electrolyte. Thus in a solution of acetic acid all of the hydrogen capable of dissociating as hydrogen ion is considered to be hydrogen ion constituent. Of this, the actual concentration of hydrogen ion is, of course, only a small proportion. The relation between the mobility of an ion u^+ or u^-, and of an ion constituent, U^+ or U^-, is given by the expressions

$$U^+ = \alpha\mathrm{u}^+ \quad \text{and} \quad U^- = \alpha\mathrm{u}^- \qquad (5)$$

In terms of ion constituents equation (14) of Chapter 3 becomes

$$\Lambda = \mathbf{F}(U^+ + U^-) \qquad (5a)$$

and equation (3) takes the form

$$t^+ = \frac{U^+}{U^+ + U^-} \qquad (5b)$$

From these relations and equation (4)

$$U^+ = t^+\Lambda/\mathbf{F} \quad \text{and} \quad U^- = t^-\Lambda/\mathbf{F} \tag{6}$$

follow directly. The mobility of an ion constituent can thus be computed from the experimentally determined quantities t, Λ and **F**. Ion mobilities, on the other hand, involve a knowledge of the degree of dissociation, α, which cannot be obtained without a theory of the ionization of the electrolyte concerned. Though we shall be more concerned with ion constituents than with ions, the latter term will be sometimes used when ion constituent is meant, to avoid awkward repetition.

The *equivalent conductance of an ion constituent,* λ, is a quantity of which use will be made particularly in Chapter 18. For a binary electrolyte it may be defined by the relation

$$\Lambda = \lambda^+ + \lambda^-$$

in which λ^+ and λ^- are the equivalent conductances of the positive and negative ion constituents, respectively. With the aid of equations (5a) and (6) these quantities are also contained in relations

$$\lambda^+ = \mathbf{F}U^+ \qquad \lambda^- = \mathbf{F}U^- \tag{6a}$$

$$\lambda^+ = t^+\Lambda \qquad \lambda^- = t^-\Lambda \tag{6b}$$

The Hittorf Method for Determining Transference Numbers. If two silver electrodes, *A* and *C* of Fig. 2, are placed in a solution of silver nitrate contained in the tube *B-B* and an electric current is passed between them in the direction indicated, the electrode reactions are, as we have seen,

$$\text{Ag} = \text{Ag}^+ + \text{e}^- \tag{7a}$$

at the anode *A*, and

$$\text{Ag}^+ + \text{e}^- = \text{Ag} \tag{7b}$$

at the cathode *C, i.e.,* for every faraday of current passed, a full equivalent of silver ion constituent will appear around the anode and a like amount of the same constituent will disappear from the region of the cathode. Between these electrodes, however, the current is only partly carried by silver ions, the conductance being shared in this case by the negatively charged nitrate ions. Since the silver ions do not carry all the current they do not move away from the region of the anode as fast as they form. Silver ions must therefore accumulate around that electrode; and conversely, since silver ions are not brought by conductance into the region of the cathode as fast as they are discharged, their concentration must diminish in that region. Both these phenomena are observed experimentally. The movement of the nitrate ions in the reverse direction to that of the silver ions results in the carrying of the remainder of the current and in maintaining electrical neutrality.

The electrolysis just considered may be carried out and the resulting concentration changes at least roughly determined as is shown in Fig.

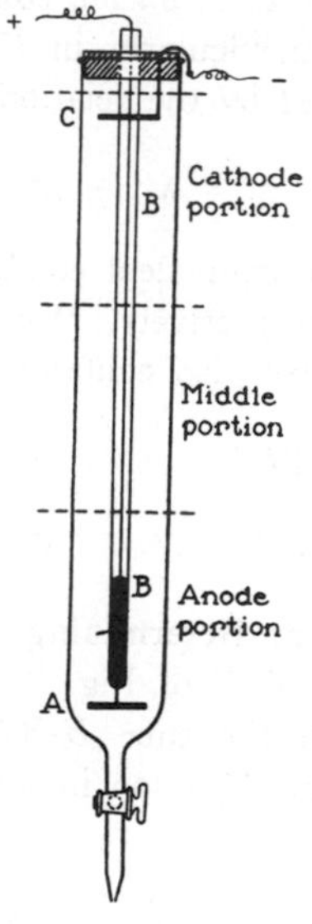

Fig. 2. Simple Hittorf Transference Apparatus.

2. The procedure to be described is a determination of a transference number by the Hittorf[1] method. The solution can be arbitrarily divided into three portions, as shown in the diagram, called respectively the anode, middle, and cathode portions. On passing a current the anode portion will become more concentrated, and the cathode portion more dilute. The middle portion will retain its original concentration. After the electrolysis the separate portions may be drawn off one after

[1] J. W. Hittorf, *Pogg. Ann.*, **89**, 177 (1853); **98**, 1 (1856); **103**, 1 (1858); **106**, 337 (1859).

another through the stopcock at the bottom of the apparatus. A more elaborate and exact procedure is described later in this chapter. First let us consider the changes that occur at the anode, *A*, and in the solution surrounding it when a faraday of electricity is caused to pass through the apparatus. These changes are (a) the formation of an equivalent of silver ion constituent according to equation (7a), (b) the passage out of the anode portion of silver ion carrying the proportion t_{Ag} of the faraday of electricity, and (c) the entrance into the anode portion of nitrate ion carrying the proportion t_{NO_3} of the current as negative ion, the passage of negative ion in a negative direction being equivalent to the movement of positive ions in a positive direction. Since the silver and nitrate ions carry all the current, the sum of the fractions of current carried by the two ions must equal unity, or

$$t_{Ag} + t_{NO_3} = 1$$

The changes in the solution surrounding the anode can be summarized as follows:

Gained	Lost	Net change
1 equiv. Ag^+ ion	$t_{Ag} = (1 - t_{NO_3})$ equiv. Ag^+ ion	t_{NO_3} equiv. Ag^+ ion gained
t_{NO_3} equiv. NO_3^- ion		t_{NO_3} equiv. NO_3^- ion gained

Thus it can be seen that the passage of a faraday of electricity through the solution does not, in this case, cause an increase of a whole equivalent of silver ions in the neighborhood of the anode, but only the proportion t_{NO_3} of that amount, t_{NO_3} being the transference number of the nitrate ion. This increase in silver ion is accompanied by an equivalent increase of the nitrate ion so that the solution remains electrically neutral.

A corresponding summary of the changes at the cathode follows:

Gained	Lost	Net change
$t_{Ag} = (1 - t_{NO_3})$ equiv. Ag^+ ion	1 equiv. Ag^+ ion (by deposition)	t_{NO_3} equiv. Ag^+ ion lost
	t_{NO_3} equiv. NO_3^- ion	t_{NO_3} equiv. NO_3^- ion lost

It is evident that the changes at the cathode are, in this case, the precise reverse of the changes at the anode. The net effect of the changes at both electrodes is the transfer of t_{NO_3} equivalent, per faraday, of silver nitrate from the cathode portion to the anode portion, since the anode portion gains and the cathode portion loses that amount of substance.

A summary of the changes in the middle portion is hardly necessary but will be given:

Gained	Lost	Net change
t_{Ag} equiv. Ag^+ ion	t_{Ag} equiv. Ag^+ ion	0
t_{NO_3} equiv. NO_3^- ion	t_{NO_3} equiv. NO_3^- ion	0

That is to say, the middle portion does not change in composition during the passage of the current. A homogeneous electrolytic conductor thus resembles a metallic conductor in that no physical or chemical change, other than heating, results from the passage of the current.

It is evident that if the concentration of electrolyte changes in the region of an electrode during the passage of electricity there will be accompanying volume changes. For that reason the "portions" must not refer to given volumes of electrolyte. To obtain the *Hittorf transference numbers* the changes in concentration, such as are considered above, are referred to the weight of solvent (usually, of course, water) which the electrode portion contains.[2] In actual experiments electrode portions are withdrawn which are large enough to include some of the solution which has not changed in concentration. This is in order to be sure that all the solution which has changed in composition is included in the electrode portion. Analysis of an electrode portion yields the amount of solvent as well as of solute. From the composition of the solution before the electrolysis the amount of solute originally associated with this amount of solvent can be computed, from which the changes due to the current flow can be found by simple subtraction. A typical computation from experimental data will serve to make this point clear.

A transference apparatus was filled with a 0.1 molal solution of silver nitrate (*i.e.,* 0.1 gram equivalent of that salt dissolved in 1000 grams of water) and a current was passed until a silver coulometer connected in series showed a weight increase of its cathode of 0.2158 gram, after which 25.00 grams of anode portion (including all that had changed in composition) were withdrawn. By analysis this was found to contain 0.5955 gram of silver nitrate. The middle portion was found to be unchanged. To obtain the transference number from these data we proceed as follows. Of the 25.00 grams of anode portion 24.41 grams were water, which originally contained $24.41 \times 0.001 \times 0.1 = 0.002441$ equivalent of silver nitrate, since the solution contained 0.1 equivalent per 1000 grams of water. After the electrolysis the same amount of water contained $0.5955/169.9 = 0.003505$ equivalent, an increase of 0.001064 equivalent. Since the total number of faradays

[2] The so-called "true" transference numbers are considered on page 91.

passed through the solution was 0.2158/107.88 = 0.00200, the increase of the number of equivalents per faraday is 0.001064/0.00200 = 0.532 which is, according to the discussion given above, the transference number of the nitrate ion constituent in silver nitrate at this concentration. Thus the nitrate ion constituent carries somewhat more than half the total current.

Another way of arriving at the result is as follows. The passage of 0.002 faraday will result in the formation of this fraction of an equivalent of silver ion in the anode portion. However, the observed increase is only 0.001064; therefore, 0.000936 equivalent must have migrated out of the portion carrying a proportion of the current equal to 0.000936/0.00200 = 0.468, which is the transference number of the silver ion at this concentration.

A Formula for the Computation of Transference Numbers. The computation of Hittorf transference numbers from the experimental data can be put into a simple formula. Let N_O and N_F represent the original and final number of equivalents of an ion associated with a given weight of solvent. Now if N_E is the number of equivalents of this ion added to the solvent by the electrode reaction, and tN_E the number lost by ionic migration (t being a transference number and N_E being also the number of faradays of current passed) then the change due to the passage of the current must be equal to either side of the equation

$$N_F - N_O = N_E - N_E t \qquad (8)$$

and therefore the transference number is equal to

$$t = \frac{N_O - N_F + N_E}{N_E} \qquad (9)$$

This compact formula for computing results of Hittorf transference experiments is due to Washburn.[3]

Transference Measurements by the Hittorf Method. The early transference measurements by the Hittorf method have been summarized by McBain,[4] and by Noyes and Falk.[5] In general this early work contained enough sources of error to make its use uncertain in studying the properties of salt solutions. However, recent researches by Jones

[3] E. W. Washburn, "Principles of Physical Chemistry," 276, McGraw-Hill Book Co., New York, 1921.

[4] J. W. McBain, *Proc. Washington Acad. Sci.*, 9, 1 (1907).

[5] A. A. Noyes and K. G. Falk, *J. Am. Chem. Soc.*, 33, 1436 (1911).

and Dole[6] on aqueous solutions of barium chloride, by MacInnes and Dole[7] on solutions of potassium chloride, and by Jones and Bradshaw[8] on lithium chloride have all had the benefit of modern developments in technique, and are of a good order of accuracy.

All these researches have been made with a type of apparatus devised by Washburn[9] which is shown, with some minor changes, in Fig. 3.

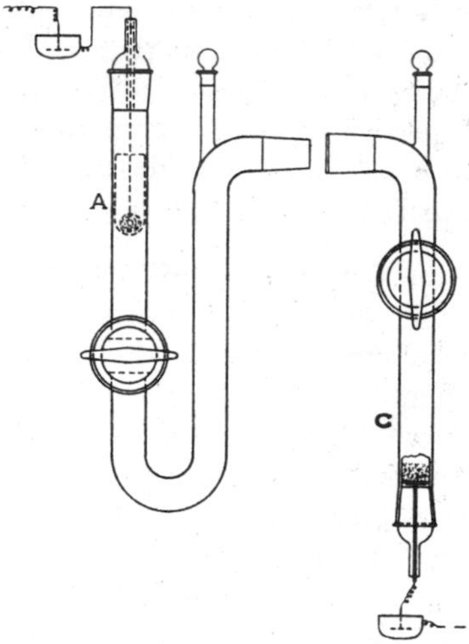

Fig. 3. Washburn's Hittorf Transference Apparatus.

The long bent tube of uniform bore is divided into three parts by two stopcocks, which must have the same bore as the rest of the tube. The electrodes are of silver wire for the anodes, and silver wire with a coating of silver chloride (produced electrolytically) for the cathodes. If the solution in the vessel is a chloride, the reactions occurring are, at the anode, *A*,

$$Ag + Cl^- = AgCl + e^-$$

and at the cathode, *C*, the reverse reaction

$$Ag + e^- = Ag + Cl^-$$

As in this case the solution around the anode becomes more dilute during the electrolysis, it is placed at the upper end of the apparatus,

[6] G. Jones and M. Dole, *ibid.*, **51**, 1073 (1929).
[7] D. A. MacInnes and M. Dole, *ibid.*, **53**, 1357 (1931).
[8] G. Jones and B. C. Bradshaw, *ibid.*, **54**, 138 (1932).
[9] E. W. Washburn, *ibid.*, **31**, 322 (1909).

and the opposite arrangement is made for the cathode end. In use the apparatus is first filled with solution, the electrodes are inserted, and current is passed until the desired concentration changes in the electrode portions are produced. One or more silver coulometers [10] are connected in series. Vibrations of the apparatus must be avoided since these can produce mixing of the solutions around the electrode with the middle portions.

After the electrolysis the stopcocks are turned, thus isolating the electrode portions. Then three middle portions are removed from the apparatus with pipettes, one portion from the solution next to the anode portion and one next to the cathode portion, and one from the region between the last two mentioned. These must, on analysis, be found to be unchanged in composition, otherwise mixing with the electrode portions has taken place, and the determination is a failure. The weight of each of the electrode portions is next obtained by weighing the section of the apparatus containing it, and again after pouring out the liquid and drying that part of the apparatus. A sample of the electrode portion is then taken and analyzed. Such an analysis must be of high order of accuracy if results of any value are to be obtained since they depend upon the relatively small difference between two analytical results.

The data obtained by MacInnes and Dole [11] in a series of measurements of the transference numbers of potassium chloride are given in Table I. The table is largely self-explanatory. Two figures for a

TABLE I. DETERMINATIONS BY THE HITTORF METHOD OF TRANSFERENCE NUMBERS OF POTASSIUM CHLORIDE AT 25°

Concentration, equivalents per liter, c	0.02	0.05	0.1	0.5	1.0	3.0
Wt. Ag in coulometers	0.16024 .16043	0.3217 .3215	0.6136 .6135	1.9769 1.9767	2.4837 2.4833	2.7760 2.7756
Wt. anode portion	117.79	116.18	117.51	119.48	121.41	131.10
Wt. cathode portion	120.99	120.34	120.17	122.93	125.66	135.30
Per cent KCl in anode portion	0.10336	0.27963	0.56662	3.1151	6.5099	19.207
Per cent KCl in anode middle portion	.14932	.37299	.74219	3.6531	7.1478	19.777
Per cent KCl in middle portion	.14948	.37297	....	3.6537	7.1474	......
Per cent KCl in cathode middle portion	.14939	.37302	.74217	3.6543	7.1485	19.775
Per cent KCl in cathode portion	.19398	.46294	.91369	4.1784	7.7668	20.327
KCl transferred anode	.05428	.1090	.20768	0.6680	0.8343	0.9331
KCl transferred cathode	.05408	.1086	.20763	.6698	.8380	.9309
Trans. No. t_K (anode)	.4902	.4904	.4898	.4890	.4861	.4864
Trans. No. t_K (cathode)	.4884	.4884	.4897	.4903	.4882	.4853
Mean value t_K	.4893	.4894	.4898	.4896	.4871	.4858

[10] Page 29.

[11] D. A. MacInnes and M. Dole, *loc. cit.*

weight or analysis refer to duplicate determinations. It should be noted that the concentration of potassium chloride in three middle portions is the same, within the experimental error, for each determination. They have retained the concentration of the original solution. Around the anode the concentration has decreased and around the cathode it has increased.

One example will illustrate the method of computation. Take, for instance, the figures for the anode portion at 1.0 normal. Here 121.41 grams of anode portion was found to be 6.5100 per cent potassium chloride, that is to say, 7.9039 grams KCl and 113.51 grams water. This water originally contained the same proportion of potassium chloride as the middle portions, or 7.1479 per cent. The amount of salt, x, originally associated with the water may therefore be obtained from the proportion

$$x : 113.51 = 7.1479 : 92.852$$

from which $x = 8.7382$ grams. Thus $8.7382 - 7.9039 = 0.8343$ gram, or 0.011189 equivalent, of potassium has left the anode portion. Since the silver coulometers show that $2.4835/107.88 = 0.02302$ faraday has passed through the solution, $0.011189/0.0230200 = 0.4861$ equivalent of potassium ion constituent leaves the anode portion per faraday. This is the transference number of potassium ion at 1.0 normal within the experimental error of the determination.

In spite of its simplicity accurate results are very difficult to get with the Hittorf method. The main difficulties are, first, the necessity for avoiding mixing of the electrode and middle portions during an electrolysis, which may take from sixteen to twenty-four hours, and secondly, the need for extremely accurate analyses of the solutions, since the method depends essentially on small differences between large quantities. For these reasons the more recent accurate data on transference numbers have been obtained, in greatest part, by means of the more complicated, but more speedily and accurately carried out, method of moving boundaries, which will be next described.

The Moving Boundary Method for Determining Transference Numbers. A means of obtaining transference numbers which has proved, in recent years, to be of greater precision than the Hittorf procedure is the method of moving boundaries. The phenomenon which makes the measurements possible is as follows. If a potassium chloride solution is placed in a tube above a cadmium chloride solution, as is shown in Fig. 4a, and electric current is passed in the direction indicated, the boundary between the two solutions will become sharp and will move up the tube. The motion of the boundary can be readily followed because the two solutions have different indices of refraction. By measuring the volume swept through by the boundary during the passage of a given quantity of electricity the transference number of the potassium ion can be obtained with the aid of a formula to be derived below. The method may be used to determine the transference number of any ion, provided a following or "indicator" ion with the necessary properties can be found.

The fundamental equation connecting the transference number with the quantities measured during a determination may be obtained as follows. The derivation is essentially that used by Miller.[12] Let Fig. 4b represent the section of the tube in which there are two solutions

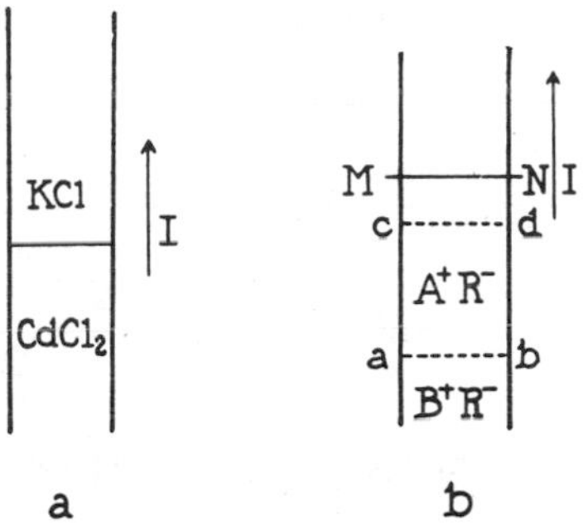

Fig. 4.

of electrolytes, A^+R^- and B^+R^-, with a common negative ion constituent R^-, between which there is the boundary *a—b*. If the constituents A^+ and B^+ have been properly chosen this boundary will move, when current is passed through the tube, to another position, say *c—d*. The effect of the passage of the current will be to sweep, out of the region between *a—b* and *c—d*, all of the ion constituent A^+ whether A^+ is or is not all present as free ions. In other words, the effect of the passage of current is to replace the solution of electrolyte A^+R^- in the region between the two positions of the boundary by another solution containing B^+R^-. Thus the flow of a certain number of coulombs causes a number of equivalents of the ion constituent A^+, equal to

[12] W. L. Miller, *Z. physik. Chem.*, 69, 436 (1909).

that originally contained in the volume between the two positions of the boundaries *a—b* and *c—d*, to pass a fixed plane, as for instance *M-N*, in the unchanged portion of the solution. This will also be the number of equivalents (faradays) of current carried across the plane *M-N* by the positive ion. Let V be the volume in liters swept through by the boundary when one faraday of current passes through the tube. The number of equivalents of the ion constituent A^+ passing the plane *M-N* will thus be $V\text{c}_{AR}$, in which c_{AR} is the concentration of electrolyte A^+R^- in equivalents per liter. Therefore

$$t_{A^+} = V\text{c}_{AR} \tag{10}$$

in which t_{A^+} is the transference number of the ion constituent A. If another, usually smaller, number of coulombs, **f**, is passed through the tube the boundary will pass through another volume, v, which is related to V by

$$\frac{v}{V} = \frac{\mathbf{f}}{\mathbf{F}} \tag{11}$$

Eliminating V from equations (10) and (11)

$$t_{A^+} = \frac{v\text{c}_{AR}\mathbf{F}}{\mathbf{f}} \tag{12}$$

If the current, I, is constant then $\mathbf{f} = sI$, in which s is the number of seconds. Dropping the subscripts

$$t = \frac{v\text{c}\mathbf{F}}{sI} \tag{13}$$

which is the fundamental equation for the moving boundary method.

In this derivation of equation (13) a number of assumptions have been made, the principal of which are as follows: (a) that there are no disturbing effects due to interdiffusion or mixing of the solutions meeting at the boundary, (b) that the motion of the boundary is uninfluenced by the nature or concentration of the following or "indicator" ion constituent (B^+ of Fig. 4b), and (c) that there are no volume changes in the apparatus that affect the motion of the boundary. It will be shown that if the determinations are properly carried out these assumptions are either fully justified or that the necessary corrections can be accurately made.

The early researches in this field were as follows. Lodge [13] was the first to investigate the possibility of observing, directly, the motion of an ion constituent in an electric field. The movement of a boundary between two solutions, one of them having a colored ion, was studied by Whetham,[14] and Nernst.[15] Masson [16] clearly indicated the conditions necessary for quantitative work with moving boundaries. References to later work will be made in the following paragraphs.

In order to observe the motion of a boundary between two ion species it is, of course, necessary to form the boundary. An invariable condition is that the lighter solution must be on top. The junction between the solution containing the leading ion constituent and the following or indicator ion constituent must be made with only a slight amount of mixing or diffusion, though, as we shall see, there is a

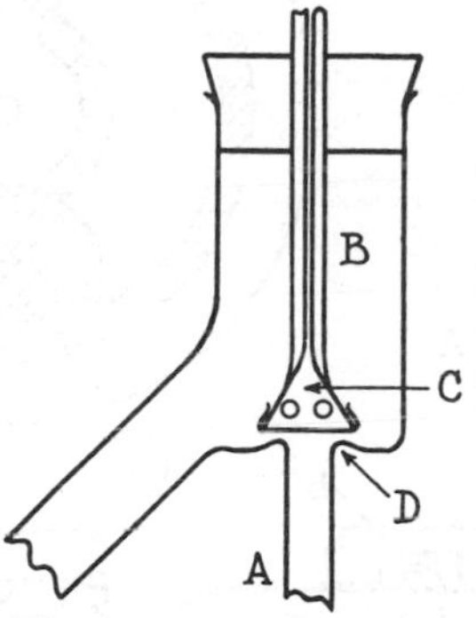

Fig. 5. Denison and Steele's Device for Forming a Moving Boundary.

"restoring" effect which overcomes, at least to a limited extent, the result of such disturbances. The first successful boundaries in which one or both of the solutions was not set in a jelly, and which were not of the "autogenic" type to be described below, were made by Denison and Steele [17] by means of the scheme shown in Fig. 5. The solution whose transference number was desired was placed in Tube *A*. This was separated from the indicator solution in vessel *B* by pressing a membrane of parchment paper stretched over the hollow cone *C*

[13] Oliver Lodge, "Report of the British Association for the Advancement of Science," 389, Birmingham, 1886.

[14] W. C. D. Whetham, *Phil. Trans.*, **A 184**, 337 (1893).

[15] W. Nernst, Z. *Elektrochem.*, **3**, 308 (1897).

[16] H. Masson, *Phil. Trans.*, **A 192**, 331 (1899).

[17] R. B. Denison and B. D. Steele, *Phil. Trans.*, **205**, 449 (1906); *Z. physik. Chem.*, **57**, 110 (1906).

onto the shoulder D. On impressing a potential in the appropriate direction, current passed around the edges of and through the paper, causing the boundary to move down the tube, after which the cone could be cautiously lifted to the position shown. This device was simplified by MacInnes and Smith,[18] who replaced the cone and parchment paper by a flattened glass rod and a disk of soft rubber. It was found that enough current leaked around the edge of the disk between the rubber and the glass shoulder when they were in contact to allow the boundary to move down the tube before the rod and disk were lifted

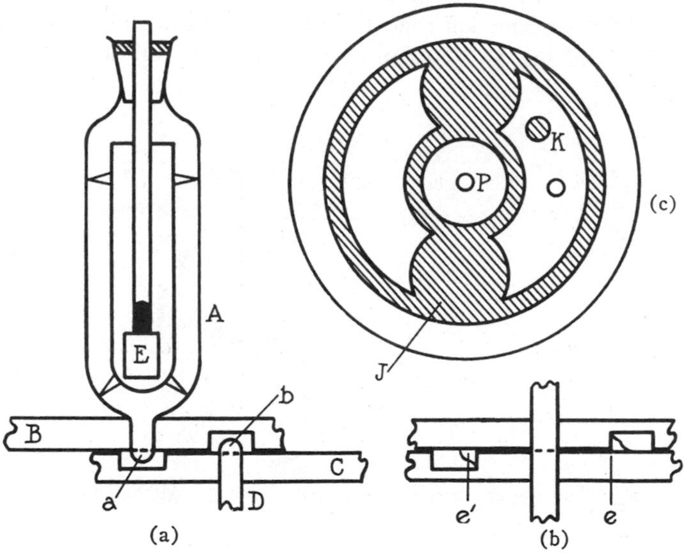

Fig. 6. Device for Forming a "Sheared Boundary."

from the shoulder D. Adequate separation of the leading and indicator ion constituents was, however, not easy to obtain with these arrangements.

An apparatus by means of which the two solutions could be kept quite separate until a boundary was formed, and with which a sharper initial boundary could be obtained, was developed by MacInnes and Brighton.[19] This device has been used in most of the recent work. The principle is shown in Fig. 6a. The electrode vessel A is fitted into

[18] D. A. MacInnes and E. R. Smith, *J. Am. Chem. Soc.*, **45**, 2246 (1923).
[19] D. A. MacInnes and T. B. Brighton, *J. Am. Chem. Soc.*, **47**, 994 (1925).

the disk *B* and the graduated tube *D* into a corresponding disk *C*, the two disks having plane surfaces where they meet. The electrode vessel *A* is filled with the "indicator" solution and is closed with the stopper which carries the electrode in such a way that a drop of the solution *a* hangs from the open end of the electrode tube. The graduated tube *D* is filled with a slight excess of the solution under examination so that a

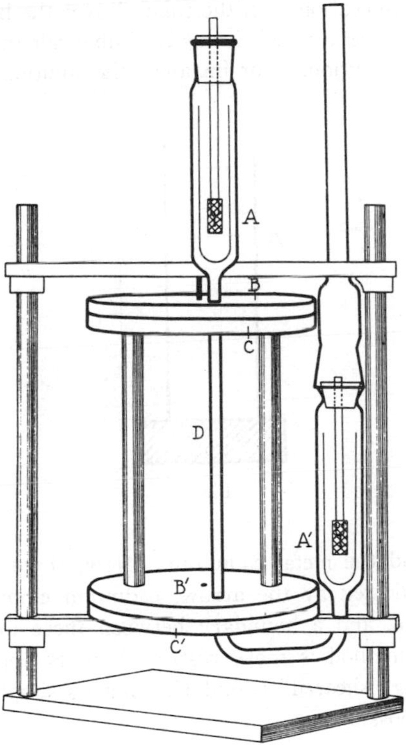

Fig. 7. Moving Boundary Apparatus.

drop *b* protrudes. Now if the disk *B* slides over the disk *C* the excess amounts *e, e'* of both solutions are sheared away and a boundary, very little disturbed by mixing or diffusion, will result when the tubes are in place over each other, as shown in Fig. 6b. The complete disks *B* and *C* are of the design shown in Fig. 6c and are arranged to hold two tubes which may or may not be electrically connected. With these disks the shearing motion just described takes place about the central pivot *P*. The end of the tube is shown by the circle *K*.

A complete apparatus for moving boundary measurements is shown in Fig. 7. It includes two electrode vessels *A* and *A'*, the graduated tube *D* and two sets of disks, *B*, *C* and *B'*, *C'*. All the disks are similar to the one shown in Fig. 6c. The upper pair of disks is used for descending boundaries and the lower pair for rising boundaries.

In addition there is the "autogenic" boundary first used by Franklin and Cady.[20] The principle of this method is shown in Fig. 8. The solution under observation is in the tube *A* over the bottom of which is placed a disk of a metal which forms a soluble salt in combination with the anion of this solution. For instance, the solution *A* may be potas-

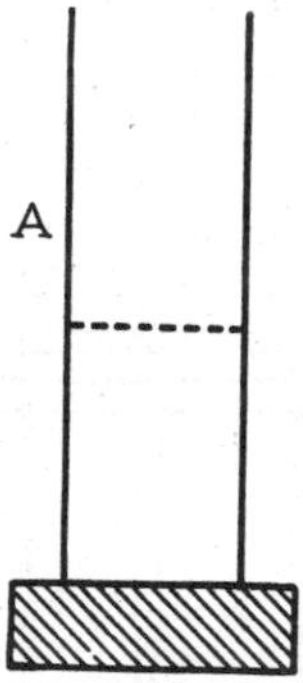

Fig. 8.

sium chloride and the metal cadmium. Now, if current is passed in the direction indicated by the arrow, cadmium chloride will form at the metal surface, and a boundary between these ions will move up the tube. This method is more restricted in its application than the method employing "sheared" boundaries, but its simplicity recommends it for use when possible.

As already stated, measurements are made with both rising and descending boundaries. In the first case the indicator solution must be heavier, and in the latter case lighter, than the solution under observation. In addition the mobility of the indicator ion constituent must always be lower than that of the leading ion. It must, obviously, be possible to follow the motion of the boundary. For this purpose the difference of refractive index of the two solutions in contact can usually be used, though occasionally the difference in color of the two ions is of service.

[20] E. G. Franklin and H. P. Cady, *J. Am. Chem. Soc.*, **26**, 499 (1904).

From the equation

$$t = \frac{v\text{c}\mathbf{F}}{sI} \tag{13}$$

it is evident that to obtain a transference number it is necessary to determine the volume swept through by the boundary, v, during the time in seconds, s, the concentration c, and the current I. It is theoretically possible to determine the products sI with a coulometer, but only in exceptional cases is the number of coulombs large enough to obtain in that manner. In the more recent work a device has been used which keeps the current, I, constant, to about 0.01 per cent, in spite of the steadily increasing resistance of the column of electrolyte in the apparatus.[21] Transference numbers can also be determined without measurement of current and time by employing two boundaries, one for the cation constituent and one for the anion constituent. For the first we have

$$t_+ = \frac{v_+\text{c}\mathbf{F}}{sI} \tag{14}$$

where t_+ is the cation transference number and v_+ the volume swept through by the cation boundary, and

$$t_- = \frac{v_-\text{c}\mathbf{F}}{sI} \tag{15}$$

which contains the corresponding quantities for the anion constituent. Since

$$t_+ + t_- = 1$$

and the product sI is the same in both cases, equations (14) and (15) yield

$$t_+ = \frac{v_+}{v_+ + v_-} \tag{16}$$

Most of the early measurements were made using this principle. It has, however, several disadvantages. In the first place it is, of course, essential that the two boundaries move at the correct rates. A necessary but not conclusive test that must be applied to these rates is whether the volumes swept out, per coulomb of current passed, are constant. If not constant it is probable that mixing of the observed and indicator solutions is taking place. The most convenient method for determining whether a boundary rate is constant is to keep the current constant, and if this current is measured all the data necessary

[21] L. G. Longsworth and D. A. MacInnes, *J. Optical Soc. Am.*, **19**, 50 (1929); *Chem. Rev.*, **11**, 171 (1932).

for using equation (13) are available. Also with the two-boundary method it is necessary to select two indicator solutions, which, as we have seen, must have certain definite properties, and are not always available. For these reasons all the recent work has involved the use of one boundary, though the relation $t_+ + t_- = 1$ has been used to test these single boundary measurements.

The boundary when formed by one of the methods already described cannot be perfectly sharp. Also, diffusion and mixing of the two solutions in contact must tend to take place, the latter being due to convection currents set up from the electrical heating of the solutions in the tube. It would therefore appear that the assumption made in deriving equation (13), that appreciable mixing does not occur, is unjustified. This would be true except that there is an "adjusting effect" which operates to overcome the result of such mixing. This

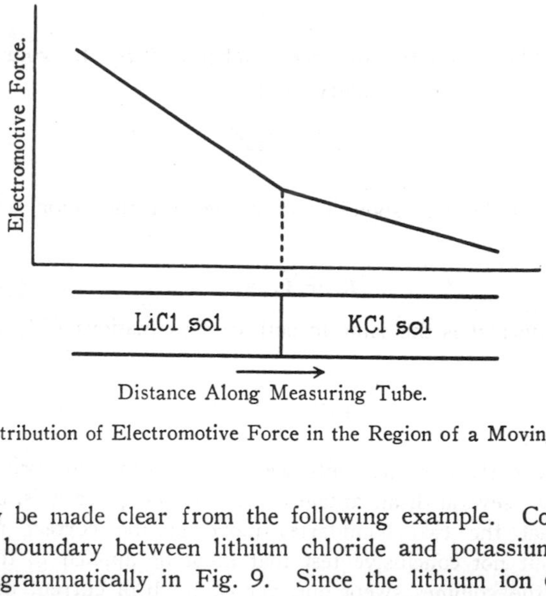

Fig. 9. Distribution of Electromotive Force in the Region of a Moving Boundary.

effect may be made clear from the following example. Consider the case of a boundary between lithium chloride and potassium chloride, shown diagrammatically in Fig. 9. Since the lithium ion constituent has a lower mobility than the potassium ion constituent, and in addition, as we shall see, the lithium chloride solution is the more dilute, the passage of current will cause a greater potential drop in the former than in the latter. This is also shown diagrammatically in the figure, where values of the electromotive force are plotted as ordinates and distances along the measuring tube as abscissae. Now if some of the relatively fast-moving potassium ions diffuse or are carried by convection into the lithium chloride region they will encounter a high potential gradient and will be rapidly sent forward to the boundary. On the other hand, if lithium ions diffuse into the potassium chloride region they will move more slowly than the potassium ions and will eventually be overtaken by the boundary. Thus there is an active mechanism at work tending to keep the boundary sharp. After appreciable mixing or diffusion there will not, of course, be a sharp break in the potential gradient, such as is shown in Fig. 9. However, for this adjusting effect to be operative it is necessary only for the gradient to be steeper behind a given solution layer than in front of it.

That such a mechanism really functions has been shown by MacInnes and Cowperthwaite [22] and by MacInnes and Longsworth.[23] In these investigations it was found on stopping the current during a moving boundary experiment, that the initially sharp boundary disappeared when two colorless solutions were in contact, or could be seen, when

TABLE II. TEST OF "ADJUSTING" EFFECT. RESULT OF INTERRUPTING CURRENT IN DETERMINATION OF THE TRANSFERENCE NUMBER OF THE CHLORIDE ION IN 0.01 NORMAL NaCl

Tube distance centimeters	Time seconds	Transference number t_-
0	0	
1	326	0.6083
2	651	.6091
3	976	.6094
Current interrupted 8 minutes		
4	1303	.6088
Current interrupted 16 minutes		
5	1628	.6092
6	1955	.6088
Current interrupted 32 minutes		
7	2282	.6088
8	2610	.6086
9	2938	.6085
10	3264	.6087

one of the solutions was colored, to be replaced by a zone in which the solutions had diffused one into the other. However, when the current was impressed once more a sharp boundary reappeared after a period, depending upon how long the current had been interrupted. The results of the second investigation mentioned are given in Table II.

[22] D. A. MacInnes and I. A. Cowperthwaite, *Proc. Nat. Acad. Sci.*, **15**, 18 (1929).

[23] D. A. MacInnes and L. G. Longsworth, *Chem. Rev.*, **11**, 171 (1932).

The boundary studied was between 0.01 normal sodium chloride and the sodium salt of tetraiodofluorescein, the bright red color of which made the effects of mixing and diffusion visible. As shown in the table, the current was interrupted during the experiment for periods of 8, 16 and 32 minutes. During each interruption the boundary, which had been extremely sharp, became hazy due to diffusion. In the case of the longest interruption the color of the indicator solution was visible about 3 mm. above the position that the boundary occupied when the current was stopped. After each interruption the boundary slowly regained its original sharpness. The most surprising observation from this experiment is, however, that the diffuse zone between the two solutions moved, on starting the current, with the same velocity as the fully formed boundary. This is shown by the constancy of the transference numbers, given in the third column of the table. These numbers are computed from equation (13), $t = vc\mathbf{F}/sI$, making allowance, of course, in the time s for the period in which the current was off. Part of the time during which the boundary was sweeping through the volume v it consisted of a diffuse mixture of the two solutions in contact. It is thus evident that the boundary does not have to be a mathematical plane in order to be useful for measuring transference numbers. As a matter of fact all moving boundaries consist of more or less diffuse regions in which mixing and diffusion have occurred. The "thickness" of the boundary and related topics are discussed in a paper by MacInnes and Longsworth.[24]

So far in our discussion nothing has been said about the concentration of the following or indicator solution. There is, as a matter of fact, a relation between the leading and indicator concentrations, automatically produced by means of the adjusting mechanism just described. This mechanism will cause a change in the initial indicator concentration to a new value determined by the properties of the two solutions in contact. To obtain a relation between the concentrations of the leading and indicator solutions we may proceed as follows. Consider, Fig. 10, a boundary that has moved from a position near *a-b* to another position, *c-d*. As has already been made clear, the electrolyte A^+R^- which originally filled the volume between the two positions of the boundary has been replaced by another electrolyte B^+R^-. Now it is found that the concentration c_{BR} of B^+R^- between *a-b* and *c-d* will in general be somewhat different from the initial concentration c'_{BR} which remains the same as that behind the region of the starting position of the boundary. If, under the new conditions, an additional

[24] D. A. MacInnes and L. G. Longsworth, *Chem. Rev.*, **11**, 171 (1932).

faraday of current is passed, the boundary at c-d will move to c'-d' sweeping through a volume V. From equation (10) we have

$$t_{A+} = c_{AR} V$$

Now the number of equivalents per faraday of ion constituent B^+ passing the plane c-d and filling the volume V between that plane and

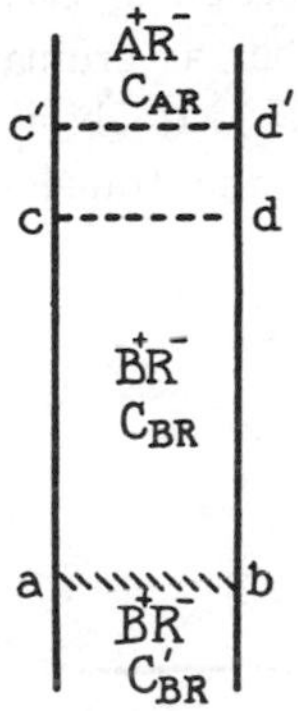

Fig. 10.

c'-d' is also the transference number, t_{B+}, of the positive ion constituent in the electrolyte B^+R^-. Therefore we have the additional relation

$$t_{B+} = c_{BR} V$$

Elimination of V from the last two equations gives the important relation

$$\frac{t_{A^+}}{c_{AR}} = \frac{t_{B^+}}{c_{BR}} \tag{17}$$

connecting the transference numbers and concentrations of the leading and indicator ion constituents of a moving boundary. It is important to notice that this equation is independent of any assumption concerning the ionization of the solutions. The reason why the "adjusted" concentration c_{BR} differs from the initial concentration c'_{BR} is that the indicator ion constituent must move at the same rate as the leading ion constituent. This, in general can occur at only one concentration of the indicator solution for each concentration of the solution containing the leading ion constituent, and usually requires a shift of concentration at the plane a-b.

Although this theory indicates no limits to the difference between the initial concentration of the indicator c'_{BR} and the adjusted concentration of c_{BR} it has been found experimentally that c'_{BR} must not be too far from that required for c_{BR} from equation (17). This matter has been studied by MacInnes and Smith.[25] These workers found that the observed values of the transference number when plotted against the concentration c'_{BR} of the indicator solution gave curves of the form shown in Fig. 11. The flat portion of each curve included the correct value of the transference number as ordinate, and the "adjusted" concentration c_{BR} of equation (17) as abscissa. The adjustment range in which corrected values of the transference number are obtained

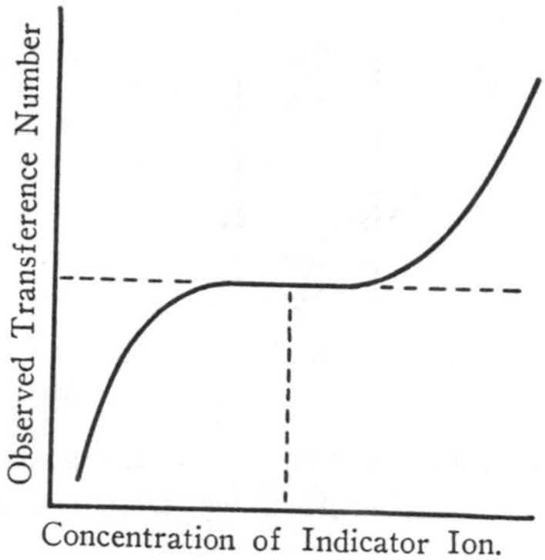

Fig. 11. Effect of the Indicator Concentration on the Observed Transference Number.

depends upon factors which are not well understood. The range is greater for dilute solutions than for concentrated, and for rising boundaries than for descending boundaries. The range of adjustment also appears to be greater for small bore tubes than for larger ones. One cause of failure of the indicator solution to adjust to the concentration demanded by equation (17) is as follows. If, with a rising boundary, such as is shown in Fig. 10, the initial concentration of the indicator solution c'_{BR} is lower than the adjusted concentration c_{BR} the latter will, in general, be denser than the former, and the two solutions will tend to mix. Mixing from the same cause will occur with a descending boundary if the initial concentration of the indicator solution is greater than that of the adjusted concentration.

[25] D. A. MacInnes and E. R. Smith, *J. Am. Chem. Soc.*, **45**, 2246 (1923); **46**, 1398 (1924).

Since, as has already been pointed out, in many cases the concentrations of the leading and following solutions must be adjusted to fit equation (17) it appears that the two transference numbers must be known ahead of the experimental determination of one of them. However, the adjustment need only be within 5 to 10 per cent, so that the necessary information is usually available either from previous measurements or can be roughly computed from other properties of the solutions.

In the development of the moving boundary method it was pointed out by early workers that the electrochemical reactions which occur at the electrodes on the passage of current are accompanied by volume changes which may affect the observed displacement of the boundary. Denison and Steele considered the effect to be negligible, though their conclusion was subsequently shown to be incorrect by Lewis [26] whose discussion of the matter is essentially that given below.

The Hittorf transference number may be defined as the number of equivalents of a given ion constituent which, on passage of one faraday of electricity, cross a plane fixed with respect to the solvent, usually, of course, water. In a determination by the moving boundary method the position of a boundary is fixed with respect to the graduations of the tube. Hence, in order to obtain a value of a transference number comparable with that found by the Hittorf method, the motion of the water with respect to the tube must be computed.

As an example we may take a rising boundary between potassium and barium chlorides with a silver anode, the end of the apparatus containing this electrode being closed and the other end open. The condition at the beginning of the experiment may be represented diagrammatically by Fig. 12a. In this figure x denotes the position of an "average" water particle at some point that the boundary does not pass.

On the passage of one faraday the boundary will move from $a - b$ to $c - d$, Fig. 12b, and the "average" water particle will move from x to x'. If we measure a boundary at the beginning of its motion with reference to the point x and at the end with reference to point x', we shall be obtaining its movement with reference to a given amount of water, and shall thus obtain the Hittorf transference number. In detail the volume changes will be as follows. During the passage of one faraday, t_K equivalents of potassium ion will move out of the region between x and the closed electrode. The corresponding loss in volume in this region is

[26] G. N. Lewis, *J. Am. Chem. Soc.*, 32, 862 (1910).

$t_K \bar{V}_K^{KCl}$, in which $\bar{V}_K^{KCl}$ is the partial molal volume[27] of an equivalent of potassium ion in potassium chloride at the concentration of the leading solution. At the electrode one equivalent of silver with a volume of V_{Ag} will disappear and one equivalent of silver chloride with a volume V_{AgCl} will be formed. This will involve the disappearance from the solution of one equivalent of chloride ion, with a volume $\bar{V}_{Cl}^{BaCl_2}$, *i.e.*, the partial molal volume of the chloride ion constituent in the solution

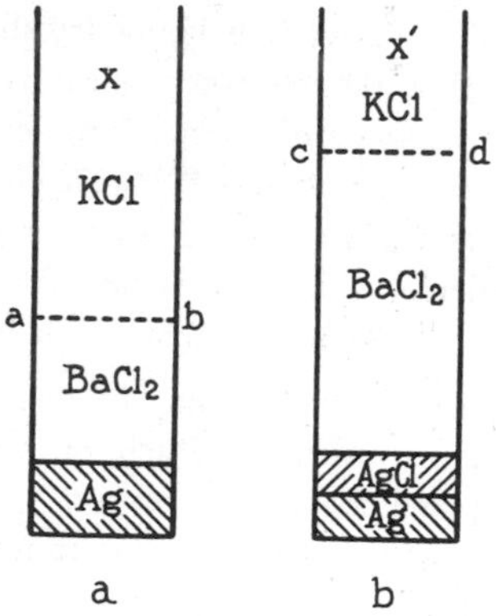

Fig. 12.

of barium chloride. As the boundary moves from $a - b$ to $c - d$ one equivalent of chloride ion will move downward across the boundary, resulting in the volume change $\bar{V}_{Cl}^{BaCl_2} - \bar{V}_{Cl}^{KCl}$. It is evident that an

[27] The partial molal volume, $\bar{V}_a$, of a component, a, of a solution, present at the molality, m_a, may be defined by the expression

$$\bar{V}_a = \left(\frac{\partial V}{\partial m_a}\right)_{T,b,c,\ldots} \tag{18}$$

in which V is the volume of the solution, the concentrations of the other components of the solution, b, c, etc., the temperature being kept constant. Physically it may be regarded as the increase (or decrease) of volume arising from the addition, at constant temperature, of a mol of the component to an infinite amount of the solution. To obtain a value of $\bar{V}$ for a salt in aqueous solution we may conveniently proceed as follows. Let ϕ, the *apparent* molal volume, be defined by

$$\phi = \frac{V - V_0}{m} \tag{18a}$$

in which V is the volume of an m molal solution containing 1000 grams of water and V_0 is the corresponding volume of pure water. Differentiating (18a) with respect to m yields

$$\frac{d\phi}{dm} = \frac{1}{m}\left(\frac{\partial V}{\partial m} - \frac{V}{m} + \frac{V_0}{m}\right) = \frac{1}{m}\left(\frac{\partial V}{\partial m} - \phi\right)$$

from which

$$\bar{V} = \frac{\partial V}{\partial m} = m\frac{d\phi}{dm} + \phi \tag{19}$$

entire equivalent of chloride ion will suffer this volume change since the transference number of this ion constituent *referred to the moving cation boundary* is unity. The net change in volume is

$$\Delta V = t_{Cl}\bar{V}_{Cl}^{KCl} - t_K\bar{V}_K^{KCl} - \bar{V}_{Cl}^{BaCl_2} + \bar{V}_{Cl}^{BaCl_2} - \bar{V}_{Cl}^{KCl} + V_{AgCl} - V_{Ag}$$

Since $t_K + t_{Cl} = 1$ and $\bar{V}_{Cl}^{KCl} + \bar{V}_K^{KCl} = \bar{V}_{KCl}$ this may be reduced to

$$\Delta V = V_{AgCl} - V_{Ag} - t_K\bar{V}_{KCl} \tag{20}$$

The observed volume through which a boundary sweeps can be seen from this typical example to be subject to a correction of ΔV. The Hittorf transference number is thus obtained from the equation

$$t = V\text{c} = (V' - \Delta V)\text{c} = t' - \text{c}\Delta V$$

in which V' and V are respectively the observed and corrected volumes and t' is the "observed" transference number, *i.e.*, the value computed from equation (13) without correction. Since ΔV is, in general, not large the correction is small for dilute solutions. In concentrated solutions it is, however, the chief factor which limits the accuracy of the results.

The validity of corrections made as described in the example given above has been demonstrated by two independent methods. The first method is that of Smith,[28] who made use of an electrolysis apparatus one electrode chamber of which could be disconnected and used as a pycnometer. The electrode chamber contained a silver electrode, covered

Values of ϕ may be computed from equation (18a) and density measurements, *i.e.*, $V = (1000 + \text{M}m)/\rho$ (in which M is the molecular weight of the solute and ρ is the density of the solution) and $V_0 = 1000/\rho_0$ in which ρ_0 is the density of the solvent. Now for solutions of a number of salts, values of ϕ are found to follow the empirical equation

$$\phi = \phi_0 + a\sqrt{m}$$

in which ϕ_0 and a are constants. By substitution in equation (19) this gives

$$\bar{V} = \frac{\partial V}{\partial m} = \phi_0 + {}^3/_2 a\sqrt{m}$$

Some of the resulting equations for the salts indicated by the subscripts, from the computations of Longsworth, *J. Am. Chem. Soc.*, **54**, 2741 (1932), are as follows:

$$\bar{V}_{LiCl} = 17.06 + 2.13\sqrt{m} \tag{19a}$$
$$\bar{V}_{KCl} = 26.65 + 3.31\sqrt{m} \tag{19b}$$
$$\bar{V}_{HCl} = 18.07 + 1.27\sqrt{m} \tag{19c}$$
$$\bar{V}_{CdCl_2} = 23.24 + 8.82\sqrt{m} \tag{19d}$$

[28] E. R. Smith, *Bur. Standards J. Research,* **8,** 457 (1932).

with silver chloride and potassium chloride solution. On passing current the chloride was reduced to silver and the potassium chloride increased in concentration. The increase in weight of the electrode portion was equal to that computed within the very small experimental error.

The other type of evidence that corrections of the type described lead to correct results has been reported by MacInnes and Longsworth.[29] Moving boundary measurements were made with relatively concentrated (0.512 and 1.0 normal) solutions of potassium chloride, using in each case two different electrode reactions. With a silver anode and a cation boundary the volume change is given by equation (20) whereas with a cadmium anode it is

$$\Delta V = \tfrac{1}{2}\bar{V}_{CdCl_2} - t_K\bar{V}_{KCl} - \tfrac{1}{2}V_{Cd}$$

TABLE III. THE INFLUENCE OF THE VOLUME CORRECTION IN THE DETERMINATION OF THE TRANSFERENCE NUMBER OF POTASSIUM CHLORIDE

Concentration of KCl equivalents per liter	Electrode	Transference Number observed	Correction $c\Delta V$	Transference Number corrected
0.512	Ag	0.4894	0.0007	0.4887
	Cd	.4854	−0.0035	.4889
1.00	Ag	.4893	0.0009	.4884
	Cd	.4817	−0.0063	.4880

The results are given in Table III. It will be seen that using a silver anode the uncorrected cation transference number is about two per cent higher, for the more concentrated solution, than that found with a cadmium anode, whereas the corrected values differ by less than 0.1 per cent. It thus seems entirely probable, from these two lines of evidence, that this manner of making corrections for solvent displacement leads to correct results.

The results of moving boundary determinations of transference numbers in which the modern developments of the method have been employed are given in Table IV, and are mainly due to the investigations of Longsworth. The figures in this table will be referred to a number of times in following chapters. The transference numbers are of use in interpreting the results of determinations of the potentials of concentration cells as activity coefficients which, in turn, may be used to test the validity of the thermodynamic aspects of the interionic attraction theory of electrolytes. In addition the transference numbers, alone, and with conductance measurements, are of utility in connection with tests of the interionic attraction theory of electrolytic conductance.

[29] D. A. MacInnes and L. G. Longsworth, *Chem. Rev.* 11, 171 (1932).

TABLE IV. CATION TRANSFERENCE NUMBERS AT 25° FOR AQUEOUS SOLUTIONS OF ELECTROLYTES DETERMINED BY THE MOVING BOUNDARY METHOD

Substance	Concentration, equivalents per liter, c 0.01	0.02	0.05	0.1	0.2	Reference
HCl	0.8251	0.8266	0.8292	0.8314	0.8337	*1*
$LiCl$	.3289	.3261	.3211	.3168	.3112	*1*
NH_4Cl	.4907	.4906	.4905	.4907	.4911	*2*
$NaCl$	.3918	.3902	.3876	.3854	.3821	*2*
KCl[5]	.4902	.4901	.4899	.4898	.4894	*1*
KBr	.4833	.4832	.4831	.4833	.4841	*1*
KI	.4884	.4883	.4882	.4883	.4887	*2*
KNO_3	.5084	.5087	.5093	.5103	.5120	*2*
$AgNO_3$	.4648	.4652	.4664	.4682	......	*3*
$NaC_2H_3O_2$	.5537	.5550	.5573	.5594	.5610	*2*
$CaCl_2$	.4264	.4220	.4140	.4060	.3953	*2*
Na_2SO_4	.3848	.3836	.3829	.3828	.3828	*2*
K_2SO_4	.4829	.4848	.4870	.4890	.4910	*6*
H_2SO_4	......	......	......	.8145	......	*4*
$LaCl_3$	.4625	.4576	.4482	.4375	.4233	*4*
$K_3Fe(CN)_6$	.4315	......	.4387	.4410	......	*6*
$Co(NH_3)_6Cl_3$	.5673	......	.5647	......	......	6

[1] L. G. Longsworth, *J. Am. Chem. Soc.*, **54**, 2741 (1932).
[2] L. G. Longsworth, *ibid.*, **57**, 1185 (1935).
[3] D. A. MacInnes and I. A. Cowperthwaite, *Chem. Rev.*, **11**, 210 (1932).
[4] L. G. Longsworth, private communication.
[5] Measured values of t_K for potassium chloride at lower concentrations are respectively, 0.490_5, 0.490_4 and 0.490_4 at concentrations, c, equal to 0.001, 0.002 and 0.005 normal. Values at 0.5 and 1.0 normal are given in Table V.
[6] G. S. Hartley and G. W. Donaldson, *Trans. Faraday Soc.*, **33**, 457 (1937).

Although we have assumed that the moving boundary and the Hittorf methods, when correctly used, yield the same values of transference numbers, there has been no adequate test of this assumption until recently. The only measurements of transference numbers by the Hittorf method available for this comparison, in which modern technique has

TABLE V. CATION TRANSFERENCE NUMBERS AT 25° OF AQUEOUS SOLUTIONS OF ELECTROLYTES DETERMINED BY THE HITTORF METHOD, COMPARED WITH THE RESULTS OF MOVING BOUNDARY DETERMINATIONS

Substance and Method		Concentration, equivalents per liter, c 0.01	0.02	0.05	0.10	0.20	0.50	1.0	Reference
KCl	*Hit.*	......	0.489_3	0.489_4	0.489_8	......	0.489_6	0.487_5	*1*
	m.b.	0.4902	.4901	.4899	.4898	0.4894	.4888	.4882	*4*
$LiCl$	*Hit.*	$.328_9$	$.326_9$	$.323_0$	$.318_7$	$.312_5$	$.300_6$	$.287_3$	*2*
	m.b.	.3289	.3261	.3211	.3168	.3112	......	......	*4*
$BaCl_2$	*Hit.*	$.440_5$	$.437_5$	$.431_7$	$.425_3$	$.416_2$	$.398_6$	$.379_2$	*3*

[1] D. A. MacInnes and M. Dole, *J. Am. Chem. Soc.*, **53**, 1357 (1931).
[2] G. Jones and B. C. Bradshaw, *ibid.*, **54**, 138 (1932).
[3] G. Jones and M. Dole, *ibid.*, **51**, 1073 (1929).
[4] Tables III and IV.

been used, are those of MacInnes and Dole [30] on potassium chloride solutions and of Jones and Bradshaw [31] on lithium chloride solutions. These data are given in Table V and for comparison the recent moving

[30] D. A. MacInnes and M. Dole, *J. Am. Chem. Soc.*, **53**, 1357 (1931).
[31] G. Jones and B. C. Bradshaw, *ibid.*, **54**, 138 (1932).

boundary data of Longsworth from Table IV. The agreement between values obtained by the two methods is probably within the limit of error and furnishes proof that the two methods measure the same property of a solution. The Hittorf transference numbers are probably the less accurate since that method, in spite of its apparent simplicity, is a difficult one and is subject to a variety of errors. The cation transference numbers of barium chloride, for which there are no comparable moving boundary values, are included in Table V for reference.

The Transference Numbers of Ion Constituents in Mixtures of Electrolytes. The moving boundary method can in certain cases be used to determine the transference numbers of the ion constituents in mixtures of electrolytes. The method used by Longsworth [32] for determining the transference numbers in mixtures of hydrochloric acid and potassium chloride is as follows.

The measuring tube was initially filled throughout with a mixture of uniform composition and was provided with a cadmium anode as shown in Fig. 13a. c_K denotes the concentration of the potassium

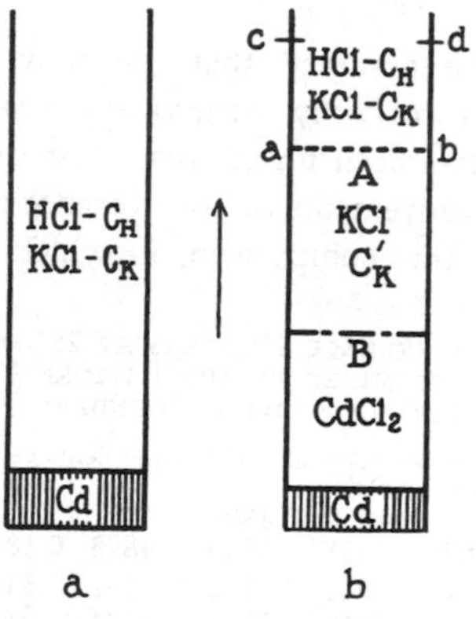

Fig. 13.

ion constituent and c_H that of the hydrogen ion constituent in the mixture. The concentration of the chloride ion, c_{Cl} (equal to $c_K + c_H$) was 0.1 normal throughout. After the current had passed for a time it was observed that the hydrogen ion constituent originally in the lower part of the tube had forged ahead of a portion of the potassium ion constituent. Two boundaries, A and B, as shown in Fig. 13b were

[32] L. G. Longsworth, *J. Am. Chem. Soc.*, **52**, 1897 (1930).

visible. The boundary *A* separates the mixture originally placed in the tube from a solution of pure potassium chloride at an "adjusted" concentration. Since no hydrogen enters through the plane *a-b*, but t_H equivalents pass out through the plane *c-d*, we have the relation

$$t_H = Vc_H \tag{22}$$

in which *V* is the volume between *a-b* and *c-d*, swept through when one faraday passes through the tube. t_K is obtained from the relation

$$t_H + t_K + t_{Cl} = 1 \tag{23}$$

after the chloride ion transference number, t_{Cl}, of the same mixture of electrolytes has been determined, in a separate experiment, from anion boundaries, using an appropriate mixture of potassium iodate and iodic acid as indicator. The data obtained using this method on three different hydrochloric acid-potassium chloride mixtures at a total concentration of 0.1 normal are given in Table VI.

TABLE VI. TRANSFERENCE NUMBERS OF IONS IN AQUEOUS HYDROCHLORIC ACID-POTASSIUM CHLORIDE MIXTURES AT 25°

Concentration equivalents per liter, KCl c_K	HCl c_H	Transference Numbers of ion— potassium t_K	hydrogen t_H
0.025	0.075	0.0503	0.7456
0.050	0.050	.1242	.6198
0.075	0.025	.2477	.4109

For an interpretation of the data obtained from the second boundary, *B*, the reader is referred to a discussion by MacInnes and Longsworth.[33]

The Motion of a Boundary Between Two Solutions of the Same Salt. It has already been mentioned that if a boundary is formed between two solutions of the same salt it will, in general, move when current is passed. The subject has been discussed theoretically by Kohlrausch,[34] Miller,[35] and von Laue,[36] and has been studied experimentally by Smith.[37]

A boundary between two concentrations of the same electrolyte, as for instance NaCl (c_1) : NaCl (c_2) will be a more or less broad diffusion band. If conditions are chosen so that it moves at a uniform speed with the passage of a constant current, the difference between

[33] D. A. MacInnes and L. G. Longsworth, *Chem. Rev.*, **11**, 171 (1932).
[34] F. Kohlrausch, *Ann. Physik*, **62**, 209 (1897).
[35] W. L. Miller, *Z. physik. Chem.*, **69**, 436 (1909).
[36] M. von Laue, *Z. anorg. u. allgem. Chem.*, **93**, 329 (1915).
[37] E. R. Smith, *Bur. Standards J. Research*, **8**, 457 (1932).

the transference numbers of one of the ion constituents of the solutions may be computed from the observed motion of the boundary. Let Fig. 14 represent the tube of uniform cross-section in which the boundary displacement occurs. If, on the passage of one faraday of electricity, **F**, in the direction indicated, the boundary moves from *a* - *b* to *c* - *d* through a volume V, the product $(c' - c'')V$, in which c′ and c″ are the concentrations of the electrolyte AR, must be equal to the difference

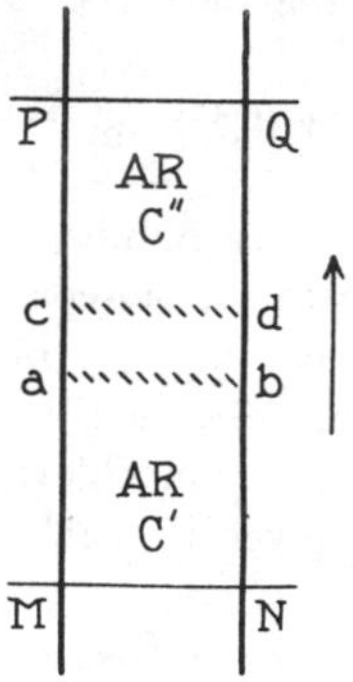

Fig. 14.

between the number of equivalents, t', of positive ion constituent passing across a plane M-N in the concentrated solution, and the number t'' passing across the plane P-Q. That is to say

$$(c' - c'')V = t' - t'' \tag{24}$$

For another quantity of electricity, **f**, the boundary displacement, v, will be $v = V\mathbf{f}/\mathbf{F}$ so that

$$v = \frac{(t' - t'')\mathbf{f}}{(c' - c'')\mathbf{F}} \tag{25}$$

The actual motion of the boundaries observed by Smith was in the predicted direction and of the right order of magnitude. This procedure could be used as a method for obtaining differences of transference numbers with concentration. It has not yet, however, reached the accuracy of the direct determinations.

The Influence of Complex and Intermediate Ions on Transference Numbers. In the foregoing discussion the tacit assumption has been made that the nature of the ions taking part in the conductance is known, at least that they can be represented by simple formulas such as Ag^+, NO_3^-, Na^+, Cl^-, etc. There is, however, the possibility that the ions are more complex and that the transference numbers of electrolytes as measured are merely resultant effects from the movements of these more complex ions. If, for instance, in solutions of sodium chloride some of the molecules were present in solution in pairs, $(NaCl)_2$, and these in turn should ionize as follows:

$$(NaCl)_2 = Na^+ + NaCl_2^-$$

the motion of these $NaCl_2^-$ ions in a reverse direction to the Na^+ ions would evidently decrease the amount of sodium ion constituent moving into the cathode portion as a result of the electrolysis. The presence of such complexes would make the study of salt solutions very difficult, and, fortunately, we have evidence, which will be outlined later, that in the great majority of cases the ions can, with an important reservation (*i.e.*, ionic hydration, see p. 91), be represented by the simple formulas.

With ions having multiple charges the possibilities of the formation of complex and intermediate ions are greater, and in many cases they undoubtedly exist. For instance, cadmium iodide might ionize in steps as follows:

$$CdI_2 = CdI^+ + I^- \quad \text{and} \quad CdI^+ = Cd^{++} + I^-$$

The ion of the type CdI^+ is called an *intermediate ion.* The effect of this kind of ionization would be to increase the directly measured or apparent transference number of the constituent Cd^{++}, since the complex ion we have postulated would carry iodine in the reverse direction to the normal motion of that ion and thus reduce the measured transference number of that ion. Another possibility, and the one which probably occurs in solutions of this salt, is the ionization of the polymerized salt in the form of complex ions, of which the following equations represent two of the many possibilities:

$$\text{(a)} \qquad (CdI_2)_2 = CdI_3^- + CdI^+$$

or

$$\text{(b)} \qquad (CdI_2)_2 = CdI_4^{--} + Cd^{++}$$

The problem of what ions are actually present has not been solved. It will be observed in Table VII that the measured transference number of the positive ion for this salt drops steadily from a value of 0.445 at 0.005 normal to below zero at 0.5 normal. This could be accounted for by either of the types of ionization (a) or (b). For instance, with type (b), if the ion CdI_4^{--} had a larger mobility than that of Cd^{++}, the expected decrease of cadmium ion constituent around the anode would be less than zero, as is observed for concentrations above 0.5 normal. As the concentration is diminished such complexes apparently break down, gradually giving simpler ions, and in very dilute solutions, the ionization probably takes the simpler form

$$CdI_2 = Cd^{++} + 2I^-$$

since the positive ion has a transference number not far removed from those of other cations at the same low concentrations.

TABLE VII. CATION TRANSFERENCE NUMBERS OF CADMIUM IODIDE SOLUTIONS AT 18° C.[1]

Concentration equivalents per liter	0.005	0.01	0.02	0.05	0.1	0.2	0.5	1.0
$t_{Cd^{++}}$	0.445	0.444	0.443	0.396	0.296	0.127	−0.003	−0.120

[1] Interpolated from the data of Redlich and Bukschnewski, *Z. physik. Chem.*, 37, 673 (1901).

The Compositions of Complex Ions from Transference Data. In the cases just discussed the compositions of the complex ions change rapidly from concentration to concentration. This makes the determination of their structure and the proportions in which they are present quite difficult, though methods for such determinations will be outlined later in this book, particularly for the polybasic acids. If the complexes are relatively stable, transference data may be very useful in determining their composition. The usefulness of such measurements is well illustrated by one of the classical experiments of Hittorf on a solution of potassium silver cyanide, $KAg(CN)_2$. The measurements were made in an apparatus similar to that shown in Fig. 2. The actual data for one experiment will be given for an example. Current was passed through the solution in the apparatus until 0.5462 gram of silver was deposited in a silver coulometer connected in series. Analysis of the cathode portion showed that it had decreased in silver content by 0.7645 gram. Of this amount, 0.5462 gram must have plated out on the cathode, in accord with Faraday's law, and the remainder, 0.2183 gram, or 0.002033 equivalent, must have migrated out of the cathode portion. Another analysis showed that the cathode portion had also decreased in content of the cyanide ion constituent, CN, by 0.1056 gram, or 0.004060 equivalent. Thus when one equivalent of silver ion migrated out of the region around the cathode, in the reverse direction to the normal movement of positively charged ions, it was accompanied by two equivalents of the cyanide radical. The complex anion therefore has the composition $Ag(CN)_2^-$.

In a similar way Hittorf[38] determined the compositions of other complex salts, such as $K_4Fe(CN)_6$ and Na_2PtCl_6. Reychler[39] has found that transference experiments on ammoniacal solutions of $AgNO_3$ and of $CuSO_4$ lead to the formulas $Ag(NH_3)_2^+$ and $Cu(NH_3)_4^{++}$ for these ammonio complex cations.

"True" Transference Numbers and Ionic Hydration. In addition to ionic complexes arising from association and partial dissociation of the solute in solutions of electrolytes there is also the possibility of *solvation,* that is to say, complex formation between the solute and the solvent. With aqueous solutions it is called *hydration.* There is much experimental evidence showing that electrolytes are hydrated in aqueous solutions.[40] One of the most important types of this evidence will be outlined below.

If, for instance, the positive ions carry more water into a solution surrounding a cathode than the negative ions carry out of it, the effect of the resulting movement of water will be a dilution of the solution around the cathode, and the measured Hittorf transference number of the cation will be smaller than would be the case if the ions were unhydrated, and moved at the same relative velocities. Similarly if the negative ions carry more water away from the cathode than the positive ions carry to it the effect on the Hittorf number will be in the opposite direction.

Transference numbers obtained by a method which is uninfluenced by the movement of water of hydration have been called "true"[41] transference numbers. The first attempt to obtain such numbers was made by Nernst and associates.[42] Successful measurements in this field have been carried out by Buchböck[43] and much more extensively by Washburn.[44] The procedure employed was essentially that of a Hittorf measurement. However, a second solute (usually a carbohydrate, such as sucrose or raffinose) is added to the aqueous solution, and, instead of referring the changes of salt concentration to the water, as in the computations for Hittorf transference numbers, the changes, both of salt and of water, are referred to the added solute. The apparatus used by Washburn has already been described. It is evident that if the added "reference" substance is uninfluenced by the passage of the

[38] W. Hittorf, *Pogg. Ann.*, **106,** 513 (1859).

[39] A. Reychler, *Bull. soc. chim. Belg.*, **28,** 215, 227 (1914).

[40] E. W. Washburn, *Tech. Quart.*, **21,** 360 (1908), has given an excellent summary of this evidence up to the year 1908.

[41] This term is somewhat unfortunate since it implies that the Hittorf transference numbers are false. The latter are, however, the values used in thermodynamic relations, and are measures of a perfectly definite, though somewhat complex, process. The term *true* in this connection is, however, of too general usage to make a change advisable.

[42] W. Nernst, Gerrard, and Oppermann, *Nachr. kgl. Ges. Wiss. Göttingen*, **56,** 86 (1900).

[43] G. Buchböck, *Z. physik. Chem.*, **55,** 563 (1906).

[44] E. W. Washburn, *J. Am. Chem. Soc.*, **31,** 322 (1909); E. W. Washburn and E. B. Millard, *ibid.*, **37,** 694 (1915).

current, the ratio of the reference substance to the water will be changed around the electrodes if water is dragged along by the moving ions. Now, if the concentration of reference substance can be accurately determined, the "true" transference numbers of the ion constituents and the increase, or decrease, of the number of mols of water, Δn, in a given electrode portion per faraday of electricity passed through the solution can be computed. In Washburn's experiments the changes of concentration of the reference substance were found by means of accurate measurements with a polariscope. The number Δn is evidently equal to the net effect of the water carrying by all the ions present. For a binary electrolyte the value of Δn for the cathode portion is equal to

$$\Delta n = \tau_c N_w^c - \tau_a N_w^a \tag{26}$$

in which τ_c and τ_a are respectively the "true" transference numbers of the cation and anion and N_w^c and N_w^a are the number of mols of water carried per equivalent of cation and anion respectively.

A summary of the experimental results taken from the paper by Washburn and Millard[45] is given in Table VIII. Several interesting

TABLE VIII. "TRUE" AND HITTORF CATION TRANSFERENCE NUMBERS AND TRANSFERENCE OF WATER FOR A SERIES OF CHLORIDES, AT 1.3 NORMAL

Electrolyte	Mols of water transferred from anode to cathode per faraday, Δn	"True" cation transference number, τ_c	Hittorf cation transference number, t_c
HCl	0.24 ± 0.04	0.844	0.82
CsCl	0.33 ± 0.1	0.491	0.485
KCl	0.60 ± 0.2	0.495	0.482
NaCl	0.76 ± 0.2	0.383	0.366
LiCl	1.50 ± 0.4	0.304	0.278

facts can be seen in these data. Since all these Δn values are positive the cations in these solutions carry more water than the negative (chloride) ion. Also, at the relatively high concentration of the experiments, 1.3 normal, the movement of the water has a decided effect on the transference number, as can be seen by comparing the corresponding "true" and Hittorf numbers computed from the same experimental data. The method is only suitable for obtaining differences in the extent of hydration, but these differences can be definitely stated as follows. Equation (26) can be put in the form

$$N_w^c = \frac{\Delta n}{\tau_c} + \frac{\tau_a}{\tau_c} N_a \tag{27}$$

by means of which the number of mols of water carried per mol of the cation is given as a function of the number carried per mol of anion.

[45] E. W. Washburn and E. B. Millard, *loc. cit.*

Putting the data of Table VIII in this form we have the equations given in Table IX, N_w^{H} for instance, being the number of mols of

TABLE IX

$$N_w^{\text{H}} = 0.28 + 0.185\ N_w^{\text{Cl}}$$

$$N_w^{\text{Gs}} = 0.67 + 1.03\ N_w^{\text{Cl}}$$

$$N_w^{\text{K}} = 1.3 + 1.02\ N_w^{\text{Cl}}$$

$$N_w^{\text{Na}} = 2.0 + 1.61\ N_w^{\text{Cl}}$$

$$N_w^{\text{Li}} = 4.7 + 2.29\ N_w^{\text{Cl}}$$

water carried per equivalent of hydrogen ion. Now if we arbitrarily assume (a) that the chloride ion is unhydrated, (b) that it carries four mols of water, and (c) that it carries eight mols, the corresponding

TABLE X. THE HYDRATION OF POSITIVE IONS CORRESPONDING TO DIFFERENT ASSUMED HYDRATIONS OF THE CHLORIDE ION

N_w^{Cl}	N_w^{H}	N_w^{Cs}	N_w^{K}	N_w^{Na}	N_w^{Li}
0	0.3	0.7	1.3	2.0	4.7
4	1.0	4.7	5.4	8.4	14.0
8	1.8	8.9	9.5	14.9	23.0

hydration values for the other ions are as given in Table X. Other methods for estimating the extent of hydration indicate that assumption (b) gives figures which are probably of the correct order of magnitude.

The use of the method just outlined for determining the amount of hydration involves the assumptions (a) that the reference substance does not move under an impressed emf, and (b) that it does not form compounds with the solute. The first assumption is capable of experimental test, and Buchböck made the validity of the second appear extremely probable since he found the same "true" transference number using various amounts of reference substance.

Taylor and Sawyer,[46] and Davies, Hassid and Taylor,[47] who have made transference experiments on sodium chloride solutions similar to those carried out by Washburn, using, however, urea as reference substance, obtain an average value of Δn at 0.5 normal of 1.08 which is greater than the value 0.76 found by Washburn at 1.3 normal. At 0°

[46] M. Taylor and E. W. Sawyer, *J. Chem. Soc.*, 2095 (1929).
[47] G. Davies, N. J. Hassid, and M. Taylor, *ibid.*, 2497 (1932).

there is evidence of more water being transported than at 25°, the values of Δn at 0.5 and 1.0 normal being 1.06 and 1.45 respectively. The result of this later work is, in general, a confirmation of Washburn's conclusions. It also gives some indication that the hydration increases with dilution and with decrease of temperature.

Recently Remy and Reisner [48] have succeeded in determining differences in ionic hydration by directly measuring the volume change due to the water of hydration when current is passed through a solution. The apparatus used by these workers is shown diagrammatically in Fig. 15. The solution was divided into halves by the parchment

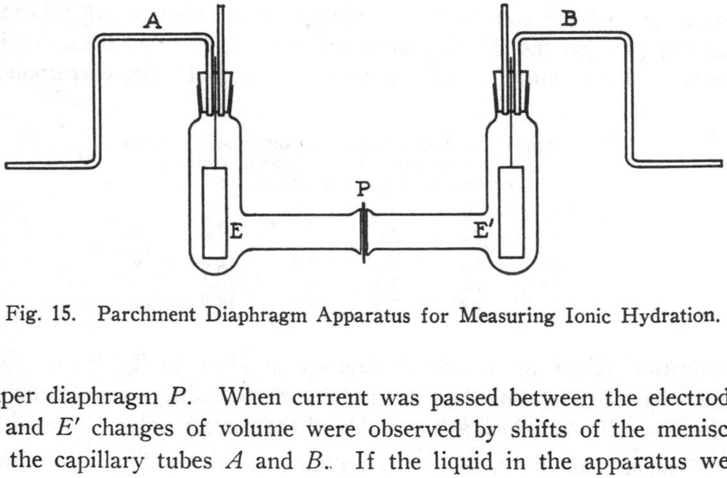

Fig. 15. Parchment Diaphragm Apparatus for Measuring Ionic Hydration.

paper diaphragm *P*. When current was passed between the electrodes *E* and *E'* changes of volume were observed by shifts of the meniscus in the capillary tubes *A* and *B*. If the liquid in the apparatus were pure water this change of volume would be due to electro-osmosis, a phenomenon discussed in Chapter 23. However, electro-osmosis decreases rapidly with increasing concentration of electrolyte, and is already very small at concentrations as low as 0.1 normal of most electrolytes. These authors consider that electro-osmosis is practically negligible at 1.0 normal at which concentration their measurements were made. If this is true the volume changes observed (when corrected for volume changes due to the electrode reaction) are due to the net amount of water carried by the ions. The results of these workers agree, at least in the order of magnitude of the effect, with those of Washburn and Buchböck. Table XI gives a comparison of values of Δn (the number

[48] H. Remy and H. Reisner, *Z. physik. Chem.*, **124**, 394 (1926). See also G. Baborovsky, *ibid.*, **129**, 129 (1927).

TABLE XI. MOLS WATER TRANSFERRED PER FARADAY, Δn, IN VARIOUS ELECTROLYTES AT 1.0 NORMAL

Method	Electrolyte HCl	KCl	NaCl	LiCl	Observers
Indifferent reference substance.	0.3	0.60	0.76	1.5	Buchböck, Washburn
Parchment paper.	0.4	0.47	1.3	1.54	Remy and Reisner

of mols transferred per faraday) as determined by the two methods. Considering the great difference of experimental technique they agree quite well.

[*Comment to 1961 Edition*] In an important paper, Longsworth* has shown that the interpretation Buchböck and Washburn (page 91 and following) give to their experimental work in terms of ionic hydration must be reconsidered. Longsworth studied the motion, under applied electrical potentials, of boundaries between solutions of salt in water and solutions of the same salt in mixtures of water and non-ionic substances such as raffinose, a typical boundary being

LiCl in water : LiCl in water + 2% raffinose

By utilizing a development of the "schlieren" technique mentioned on page 430, the motion of such a boundary can be accurately followed during the passage of the current, even though diffusion is taking place. Allowing for volume changes due to electrode reaction (as described on page 81 and following), values of Δn, the net number of water molecules transported per faraday of current, can be determined with greater accuracy than possible with the Hittorf transference method. The results show that, for a given salt, Δn is dependent on the nature of the added substance which, therefore, cannot, as Buchböck and Washburn assumed, be motionless during the passage of the current.

* L. G. Longsworth, *J. Am. Chem. Soc.*, **69**, 1288 (1947).

Chapter 5

The Principles of Thermodynamics and the Galvanic Cell

The development of electrochemistry has been greatly aided by an accompanying advance of the science of *thermodynamics,* which is a study of heat and its relation to other forms of energy. The application of thermodynamics to chemistry and to electrochemistry is very largely the work of Willard Gibbs.[1] The essentials of thermodynamics are contained in two laws, the first of which is usually known as the law of conservation of energy. The second law of thermodynamics cannot, unfortunately, be stated in terms of an equally familiar concept, and can be best expressed in connection with *entropy,* which will be presently discussed. A really adequate discussion of thermodynamics would require a book at least as extended as this one. An attempt is, however, made in this chapter and in parts of some of the following chapters to present the portions of thermodynamics that are of interest in the study of electrochemistry, particularly the parts of that subject connected with galvanic cells and the theory of solutions.

Two necessary conceptions are those of the *thermodynamic system* and its *surroundings.* A thermodynamic system may be any arbitrarily selected portion of matter or space. It may be, for instance, a volume of gas, a heat engine, or a galvanic cell. The surroundings are the immediate environment of the system, with which it may exchange energy. Important types of interaction between a thermodynamic system and its surroundings are the flow of heat from one to the other, or the action of work of one on the other, or both.

The First Law of Thermodynamics. When heat passes into a thermodynamic system it can perform two functions: (a) it can cause the system to do external work and (b) it can increase the total energy of the system. Stated mathematically

$$\Delta U = U_B - U_A = Q - W \tag{1}$$

in which Q is the heat absorbed by the system, ΔU is the increase in

[1] J. W. Gibbs, "The Collected Works of J. Willard Gibbs," Vol. 1, "Thermodynamics," Longmans, Green and Co., New York, 1928.

total energy, being the difference between the final U_B and initial U_A values of this energy, and W is the external work. For an infinitesimal process, equation (1) can be written in the form

$$dU = Q - W \tag{1a}$$

The manner in which this equation is written indicates that in such a process Q and W are not, in general, differentials of a definite function. The quantities U_A and U_B and therefore ΔU and dU, are properties of the system, and do not depend upon the way in which the system is brought from state A to state B, whereas Q and W do not have a similar independence. Equations (1) and (1a) contain the essentials of the first law of thermodynamics.

Some examples will make the physical meanings of the various thermodynamic quantities clear. The total energy U of a perfect gas is, by definition, independent of its volume and depends only upon its temperature. If a quantity of it is enclosed in a cylinder with a movable piston and is allowed to expand against an opposition, such as raising a weight or compressing a spring, and if at the same time the temperature of the gas is kept constant, the work done, W, must be equal to the heat absorbed through the walls of the container since

$$\Delta U = Q - W = 0 \quad \text{or} \quad Q - W$$

The isothermal expansion of a perfect gas is therefore a means for converting heat into work quantitatively.

Another example, more pertinent to the subject of electrochemistry, is the solution of a metal in an acid, as, for instance, the reaction of zinc with sulphuric acid

$$Zn + H_2SO_4 = ZnSO_4 + H_2 \tag{2}$$

involving a mol each of the reacting substances. If this change takes place in a calorimeter at atmospheric pressure there will be an evolution of heat $(-Q')$ and an amount of work $(+W')$ will be done due to the fact that the evolved hydrogen has to push back the air that is pressing upon it. Under these conditions

$$\Delta U = U_B - U_A = -Q' - W'$$

The change of total energy ΔU will, in this case, be negative. However, the work W' is far from being all the work that the reaction we are considering can do. A galvanic cell may be set up which uses this reaction as a source of electrical energy. If, as is shown diagrammatically in Fig. 1, a piece of zinc is placed in a solution containing sulphuric acid and zinc sulphate, it will serve as one electrode of the cell. The other electrode may be a chemically inert metallic conductor such as platinum. Now if the two electrodes are electrically connected the reaction (2) will take place as before, but with the hydrogen gas evolved from the platinum electrode instead of from the surface of the zinc. In addition, the electrical energy passing along the wire connecting

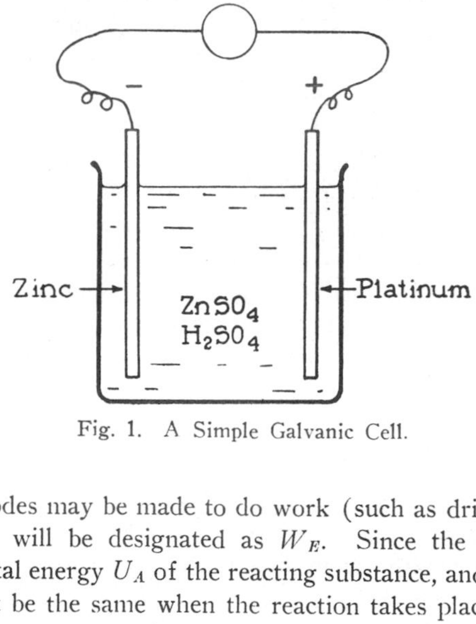

Fig. 1. A Simple Galvanic Cell.

the two electrodes may be made to do work (such as driving an electric motor) which will be designated as W_E. Since the difference ΔU, between the total energy U_A of the reacting substance, and that U_B of the products, must be the same when the reaction takes place as a galvanic cell as when the same reaction occurred in a calorimeter and delivered no electrical energy, we have

$$\Delta U = U_B - U_A = Q - (W' + W_E)$$

in which Q must have a different value from $-Q'$. The important point to notice is that the values of U_A and U_B, being properties of the system, are functions only of the state of the system at the beginning and end of the process, whereas the values of the work and the heat depend upon the way in which the process is carried out, *i.e.*, the path connecting states A and B.

The amount of electrical energy available for external work will depend to a certain extent upon the conditions under which the energy is withdrawn. If relatively large currents are taken from the galvanic cell there will be heat generated, due to electrical resistance, in all parts of the electrical circuit. However, if the electrical energy is withdrawn increasingly slowly these heat effects will diminish and will finally disappear, theoretically at least, if the process takes place infinitely slowly. Since the reactions at the two electrodes of the typical galvanic cell we are considering are

$$Zn = Zn^{++} + 2e^- \quad \text{and} \quad 2H^+ + 2c^- = H_2$$

two faradays of electricity pass through the external circuit for each mol of zinc reacting. Thus when the reaction shown in equation (2), which may be written in the ionic form

$$Zn + 2H^+ = Zn^{++} + H_2$$

takes place as a galvanic cell, the electrical energy available is $2\,E\,\mathbf{F}$, in which E is the electromotive force of the cell. In general,

$$\text{Electrical energy} = nE\mathbf{F} \tag{3}$$

in which n is the change in the number of valence units.

If a sufficiently high electromotive force of opposite polarity to that of the cell is impressed on the galvanic cell we have been considering, the cell reaction, equation (2), will take place in the reverse direction

$$H_2 + ZnSO_4 = H_2SO_4 + Zn \tag{2a}$$

For the case under consideration, the platinum electrode must be placed in an atmosphere of hydrogen. If the applied electromotive force is progressively lowered, reaction (2a) will take place more and more slowly and it will finally stop. It is found that the minimum potential difference for this reverse reaction is the same, within a very narrow limit, as the maximum electromotive force, E, of the direct operation of the cell. The cell will operate with maximum efficiency if it operates against an opposing electromotive force an infinitesimal amount less than the potential difference, E, of the cell, and the original state of the cell can be regained with the same efficiency on reversing the cell process by impressing on the cell a potential difference, dE, greater than E. The *reversible* electromotive force is therefore called E, and the operation of the cell is a reversible process under the condition just described. Reversible processes are of great importance in thermodynamics. If we consider a reversible process that is proceeding in one direction it is possible, by an infinitesimal change of some variable, to make it go in the reverse direction. All actual processes are irreversible since they must necessarily take place at a finite speed. A reversible process represents, however, an ideal or limiting condition to which the actual process may frequently be made to approach closely.

The Second Law of Thermodynamics. According to the first law of thermodynamics energy can neither be created nor destroyed. The sum of the energies in all forms remains the same after any process involving the change of energy from one form to another. The first law does not enable us to predict whether any given energy change will occur spontaneously, or whether a form of energy can, under a given set of conditions, be transformed completely into another form.

Qualitatively, we know that certain processes take place spontaneously only in one direction. Heat passes along a metal rod from the warmer to the colder end. Carbon unites with oxygen to form carbon dioxide spontaneously and the process cannot be readily reversed. An ordinary dry cell yields electrical energy, but once "run down" the original state of the cell is not obtained by any simple process. Evidently then it is desirable to consider the direction of energy changes as well as the conservation of energy which is the basis of the First Law.

Closely allied with the questions concerning the direction of energy changes is that of the availability of energy for given purposes. Mechanical work and electrical energy are, theoretically at least, capable of transformation one into the other with one hundred per cent efficiency. Heat cannot, however, be completely transformed into electrical energy or mechanical work except under conditions not ordinarily encountered.

Since heat from the combustion of fuel, or from the sun, is our most abundant source of energy it is of the greatest importance to find out the conditions under which heat can be made available as mechanical work or electrical energy, and the limitations underlying these transformations.

The second law of thermodynamics is concerned with the two questions just raised, *i.e.*, the direction of processes involving energy, and the relation of heat energy to available work. A complete statement of the second law will not be given at this point since it can be more clearly stated after further discussion. One possible statement of the law is, however, the following. It is impossible to carry out any process by which a quantity of heat is converted into an equivalent quantity of work without the simultaneous occurrence of some other change in the system or in the surroundings. On page 97 the quantitative transformation of heat into work was described. This was brought about by the expansion of a perfect gas in a cylinder, the pressure on the piston rod being exactly opposed by an external pressure. In this case, however, the gas occupied a greater volume at the end of the process than it did at the beginning, and, to bring the gas to its original condition as much or more work would have to be expended as was gained during the expansion. No net amount of useful work can be obtained from a process consisting of the expansion and compression, at constant temperature, of a gas which returns to its original state at the end of the process. The heat energy involved in the process must suffer a decrease of temperature if work is to be obtained from such a cycle. The following discussion will indicate why this is so.

For a further clarification of the relations between heat and work it is necessary to introduce the conception of *entropy* and be more precise than is usual about the conception of *temperature*. Of these the qualitative idea of temperature is, of course, familiar to everybody. The establishing of a scale of temperature is, however, a matter of some difficulty. In practice, temperatures are measured by the change of a property of a chosen substance, such as the increase or decrease of volume of mercury or alcohol, the change of pressure or volume of a gas, or the change of resistance of a metal. If these temperature scales assume the uniform change of such properties per degree of temperature change they will agree only at arbitrarily fixed points, such as the freezing and boiling points of water, and will disagree to various extents elsewhere. Fortunately, a thermodynamic temperature scale that is independent of the properties of any substance can be defined. With the aid of the Carnot cycle (named after its discoverer, Sadi Carnot [1796-1832]) it is possible to define such a scale in terms of measurable quantities of heat and work. In this cycle a "working substance," which may for instance be a gas, is caused to pass through four stages, all of them reversible, the last stage resulting in the return of the substance to its original condition. The gas is (a) first caused to absorb an amount of heat Q_1 from a heat reservoir at the temperature T_1, an expansion of the gas occurring so that the absorption takes place "isothermally," *i.e.*, at a constant temperature. The gas is then (b) heat insulated from its surroundings, and expanded "adiabatically" (*i.e.*, with no gain or loss of heat) with the result that its temperature changes to T_2. This stage is followed by (c) an isothermal compression in which the amount of heat Q_2 is given up. Finally (d) the gas is compressed adiabatically with the result that it returns to its state at the beginning of the cycle including the temperature T_1. In each of these stages there is an amount of work done by the gas, or on the gas, the net amount being

$$W = W_a + W_b - W_c - W_d \tag{4}$$

the subscripts indicating the stage as described above.

What has been accomplished by this cycle is the absorption of an amount of heat Q_1 at the higher temperature T_1 and the rejection of heat Q_2 at the lower temperature T_2. Since the working substance returns to its original state ($\Delta U = 0$) the net amount of work, W, accomplished during the cycle must, according to the first law of thermodynamics, be equal to the difference between the amounts of heat absorbed and rejected, *i.e.*,

$$W = Q_1 - Q_2 \tag{5}$$

If the temperatures are stated on the absolute thermodynamic scale the heats Q_1 *and* Q_2 *and the temperatures* T_1 *and* T_2 *are connected by the following relation:*

$$\frac{Q_1}{T_1} = \frac{Q_2}{T_2} \tag{6}$$

(It can be shown that this relation is also true if the temperature scale is based on the properties of a perfect gas.) From equations (5) and (6):

$$W = Q_1 \frac{T_1 - T_2}{T_1} \tag{7}$$

which indicates that the ratio $(T_1 - T_2)/T_1$ is a measure of the efficiency by which heat Q_1 can be transformed by a cyclical reversible process into the useful work W. It is evident that the complete transformation of heat into work can take place by such a process only if the lower temperature is absolute zero.

Now if the change of *entropy* ΔS of the working substance is defined by

$$\Delta S = \frac{Q}{T} \tag{8}$$

then the entropy changes for the four stages of the Carnot cycle are (a) $\Delta S_a = Q_1/T_1$, (b) $\Delta S_b = 0$, (c) $-\Delta S_c = -Q_2/T_2$, (d) $\Delta S_d = 0$. Thus, recalling equation (6), for the whole cycle we have

$$\Sigma \Delta S = 0 \tag{9}$$

That is to say, the entropy of the thermodynamic system represented by the working gas returns to its original value when it passes through a Carnot cycle. Equation (9) is valid for any reversible cyclical process.

The following general statements may be made about entropy. It resembles the total energy, U, in that it is a function of the state of a thermodynamic system. Only changes of entropy are of practical significance since absolute entropies are unknown. This arbitrariness may, however, be removed by choosing some standard condition as a point of reference. In general

$$dS = \frac{Q}{T} \tag{10}$$

for a reversible process and

$$dS > \frac{Q}{T} \tag{11}$$

for any naturally occurring (irreversible) process. If, in operating the Carnot cycle between the temperatures T_1 and T_2, the full value of maximum work W had not been developed, a larger value of Q_2 would have resulted, so that we would have

$$\Delta S_2 - \Delta S_1 = \frac{Q_2}{T_2} - \frac{Q_1}{T_1} > 0 \tag{12}$$

That is to say, if the heat energy Q_1 enters into a process involving a decrease of temperature, and the maximum work is not developed and stored in available form, there has been decrease of the power of the amount of energy originally represented by Q_1 to do work, which is another way of stating that the entropy has increased. If, however, all of the processes are reversible no such power of doing work is lost and the entropy remains constant. Equations (10) and (11) contain the essentials of the *Second Law* of *Thermodynamics*.

Now the mechanical work W against the surroundings is equal to

$$W = P\,dV \tag{13}$$

in which P is the pressure and V the volume, and, according to equation (10), for a reversible process

$$Q = T\,dS \tag{14}$$

Therefore equation (1a) may be written for such processes in the form

$$dU = T\,dS - P\,dV \tag{15}$$

This is an important equation which combines the first and second laws of thermodynamics for reversible processes.

There are other useful thermodynamic functions, in addition to those already mentioned. These are the Heat Content, H, the Helm-

TABLE I. THERMODYNAMIC NOTATION of VARIOUS AUTHORS

This book	U Total Energy	H Heat Content	F Helmholtz Free Energy	Z Gibbs Free Energy	μ Chemical Potential
J. W. Gibbs, "The collected works of J. Willard Gibbs," **1**, "Thermodynamics," Longmans Green and Co., New York, 1928.	ε Energy	χ Heat function for Constant Pressure	ψ	ζ	μ Potential
G. N. Lewis and M. Randall, "Thermodynamics," McGraw-Hill Book Co. New York, 1923.	E Internal Energy	H Heat Content	A Maximum Work	F Free Energy	$\bar{\mathrm{F}}$ Partial Molal Free Energy
W. Schottky, H. Ulich and C. Wagner, "Thermodynamik," VDI Buckhandlung, Berlin, 1929.	U Innere Energie	W Gibbssche Wärmefunktion	F Freie Energie	G Gibbssche thermodynamisches Potential	
J. R. Partington, "Chemical thermodynamics," Constable and Co., London, 1924.	U Intrinsic Energy	H Heat Function at Constant Pressure (Enthalpy)	F Free Energy	Z Thermodynamic Potential	μ Chemical Potential
E. A. Guggenheim, "Modern thermodynamics by the methods of Willard Gibbs," Methuen and Co., London, 1932.	E Total Energy	H Heat Content	F Helmholtz Free Energy	G Gibbs Free Energy	μ Chemical Potential
A. A. Noyes and M. S. Sherrill, "Chemical Principles," Macmillan Co., New York, 1938.	E Energy Content	H Heat Content	A Work Content	F Free Energy	$\tilde{F}$ Partial Molal Free Energy
M. Planck, "Vorlesunge über Thermodynamik," 9th ed., Walther de Gruyter and Co., Berlin and Leipzig, 1930.	U Gesamt-energie	W Gibbssche Wärmefunktion bei konstantem Druck	F Free Energie	$-T\Phi$	$-T\varphi$
M. Planck, "Treatise on thermodynamics," translated from the German by A. Ogg, Longmans Green and Co., New York, 1917.	U Internal Energy		F Free Energy	$-T\Psi$	
F. H. MacDougall, "Thermodynamics and Chemistry," 2nd ed., John Wiley and Sons, New York, 1926.	U Total Energy	H_p Heat Content at Constant Pressure	F Free Energy	Φ Thermodynamic Potential	μ Chemical Potential

holtz Free Energy, F, and the Gibbs Free Energy, Z. They are defined as follows:

$$H = U + PV \tag{16}$$

$$F = U - TS \tag{17}$$

$$Z = U - TS + PV = H - TS \tag{18}$$

In Table I the terms and symbols for thermodynamic functions as employed in this book are compared, for reference, with the corresponding terms and symbols employed by a number of other authors.

By differentiating equations (16), (17), and (18) and adding equation (15) to each of them the following differential equations are obtained,

$$dH = T\,dS + V\,dP \tag{19}$$

$$dF = -S\,dT - P\,dV \tag{20}$$

$$dZ = -S\,dT + V\,dP \tag{21}$$

the last of which will be particularly useful in later discussions. From it follow the two important partial differential equations:

$$\left(\frac{\partial Z}{\partial T}\right)_P = -S \tag{22}$$

and

$$\left(\frac{\partial Z}{\partial P}\right)_T = V \tag{23}$$

the subscripts indicating the variables held constant during the differentiation.

In order to illustrate the meaning of the quantities U, F, Z and H, let us consider a thermodynamic system which changes reversibly at a constant temperature from state A to state B. The total energy change will be

$$\Delta U = U_B - U_A \tag{24}$$

The corresponding change of Helmholtz's free energy, F, will be

$$\Delta F = F_B - F_A = \Delta U - T(S_B - S_A) = \Delta U - T\Delta S \tag{25}$$

since $T\Delta S$ is, for a reversible process, the heat absorbed, Q, the work done on the surroundings by the system is, by equation (1),

$$-\Delta F = W \tag{26}$$

that is to say, the decrease of the Helmholtz free energy of the system is equal to the work done on the surroundings in a reversible isothermal process. Work can, however, be divided into (a) mechanical work, in which a pressure P acts through a change of volume, ΔV, and (b) non-mechanical work, which includes electrical energy, surface energy, etc. Thus, at a constant pressure, P,

$$W = W_e + P(V_B - V_A) = W_e + P\Delta V \tag{27}$$

in which W_e represents non-mechanical work, and V_A and V_B the volumes of the system before and after the change. At constant temperature and pressure it follows that

$$-\Delta F = W_e + P\Delta V \tag{28}$$

Under these conditions, from equations (17) and (18),

$$Z_B - Z_A = \Delta Z = \Delta U - T\Delta S + P\Delta V = \Delta F + P\Delta V \tag{29}$$

which with equation (28) yields

$$-\Delta Z = W_e \tag{30}$$

The decrease of Gibbs free energy is therefore equal to the reversible non-mechanical work done by the system at constant temperature and pressure. In this book the only non-mechanical work considered is electrical energy. For a galvanic cell this has already been shown, page 99, to be equal to $En\mathbf{F}$, so that

$$-\Delta Z = En\mathbf{F} \tag{31}$$

in which E is the reversible potential of the cell, **F** the faraday and n the number of equivalents.

There remains the interpretation of the heat content, H. If the change from state A to state B is carried out at constant pressure and in such a way that only the mechanical portion of the work is done, then equation (1) takes the form

$$\Delta U = Q - P\Delta V \tag{32}$$

in which, as before, ΔU is the increase of total energy. Under like conditions the change of the function H from state A to state B is

$$H_B - H_A = \Delta H = \Delta U + P(V_B - V_A) = \Delta U + P\Delta V \tag{33}$$

Comparing equations (32) and (33) it can be seen that ΔH is the heat *absorbed* when the change from state A to state B takes place under the constant pressure P. It is the "heat of reaction" as measured in a constant pressure calorimeter, with a change of algebraic sign. Finally if we carry a system having the Gibbs free energy Z_A to a state at the same temperature where it has the value Z_B then from equation (18)

$$Z_B - Z_A = \Delta Z = \Delta H - T\Delta S \tag{34}$$

a useful equation to which reference will be made later on.

Further, and more concrete, examples of the use of the thermodynamic functions are contained in the discussion of galvanic cells immediately following.

The Galvanic Cell. Much of the remainder of this book will deal with galvanic cells. They are of interest in themselves as sources of electrical energy (primary cells), as arrangements for storing electrical energy (accumulators or storage batteries), and for furnishing definite reproducible values of electromotive force (standard cells). They are, however, of still more interest for theoretical electrochemistry. With the aid of the thermodynamic principles already outlined, and relations that will be derived from them, it is possible to arrive at many important conclusions concerning electrochemical reactions and solutions of electrolytes, from measurements of the potentials of galvanic cells. Thus, for instance, from such measurements at different temperatures, heats of reaction may be computed. They may also be used to obtain transference numbers, ionization constants, and other properties of solutions. Other galvanic cells are used to determine the acidity and alkalinity of solutions, to measure solubilities, and as aids in analytical processes. Since galvanic cells are among the most useful tools available for scientific investigation it is important that the principles underlying their operation are adequately understood.

In this book the following conventions will be used in representing galvanic cells: (a) a semicolon (;) will be used to indicate a metal electrolyte boundary, such as Zn; $ZnSO_4$, (b) a liquid junction or electrolyte-electrolyte boundary will be shown by a colon (:), thus, for instance, hydrochloric acid at the concentrations C_1 and C_2 may form the liquid junction

$$HCl\ (C_1)\ :\ HCl\ (C_2)$$

(c) two solutes in the same solution will be indicated by a comma; thus $ZnSO_4$, H_2SO_4, means that the two substances are both dissolved in the same medium, (d) bold face parentheses **( )** indicate that the substance enclosed is chemically inert, (e) if it is desired to indicate the physical state of a component it will be followed by (s), (l), or (g), indicating, respectively, solid, liquid, and gas. For instance, a cell represented by

$$Zn; ZnSO_4, H_2SO_4 : HNO_3; (C)$$

is one in which the zinc electrode is bathed by a solution containing zinc sulphate and sulphuric acid. This solution is in contact with a solution of nitric acid, into which the chemically inert carbon electrode is inserted. To complete the list of conventions, a double colon (: :) indicates that the potential of the indicated liquid junction has not been included in the potential of the cell as given.

Every galvanic cell contains two *electrodes*, which must be metallic conductors in the general sense outlined on page 16. At each of these electrodes an electrochemical reaction takes place. One of these reactions yields electrons to an electrode and one absorbs electrons from the other electrode. In certain important cases these electrochemical reactions may be the same process acting in reverse directions. We have already considered the reactions

$$Zn = Zn^{++} + 2e^- \quad \text{and} \quad 2H^+ + 2e^- = H_2$$

which occur in the cell shown in Fig. 1. That cell may be represented by

$$Zn; ZnSO_4, H_2SO_4; H_2(Pt) \qquad (35)$$

Another typical galvanic cell may be represented by

$$(Pt)H_2; HCl, AgCl; Ag$$

for which the electrode reactions are

$$H_2 = 2H^+ + 2e^- \quad \text{and} \quad 2AgCl + 2e^- = 2Ag + 2Cl^-$$

The "cell reaction" in this case is

$$H_2 + 2AgCl = 2HCl + 2Ag$$

Such a cell, to which we shall refer a number of times in following chapters, is shown schematically in Fig. 2. Here the electrode *A* consists of an inert noble metal, platinum for instance, which is "platinized," *i. e.*, covered with a coating of finely divided platinum and immersed in the electrolyte, hydrochloric acid in this case, through which a stream of bubbles of hydrogen is passing. The other electrode *B* consists of metallic silver coated with silver chloride and is also immersed in the electrolyte. In this case it is once more seen that the cell is formed by separating the cell reaction into electron yielding and electron absorbing electrochemical reactions. It has not always been sufficiently recognized that success in dealing with galvanic cells consists in a recognition of (a) the fact that the cell reaction should be known and that a single reaction or process should take place during the operation of the cell, and (b) that the reaction should be separable into two electrochemical reactions occurring at the electrodes. The electron yielding reaction is frequently called the *oxidation* reaction

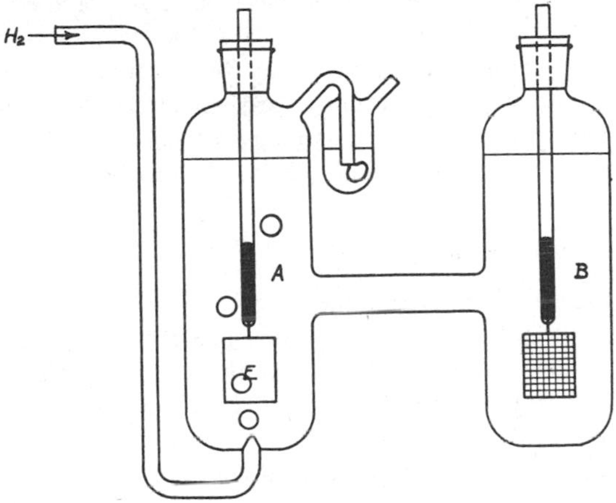

Fig. 2. A Galvanic Cell Containing Hydrogen and Silver-Silver Chloride Electrodes.

and the electron absorbing one a *reduction* reaction. Although well established, the first of these two expressions is misleading, since in many cases oxygen has nothing to do with the electrode reaction. Furthermore the reduction reaction of one cell may be, when taking place in the reverse direction, the oxidation reaction of another cell. It is sometimes necessary to work with galvanic cells for which the cell reaction cannot be definitely stated. For instance, useful information may sometimes be obtained from cells involving complex liquid junctions, which are discussed in Chapter 13. These junctions should, however, be avoided whenever possible.

The electromotive force, E, of the cell indicated by equation (35) will depend somewhat upon the concentrations of zinc sulphate and sulphuric acid, but will have a value of about 0.76 volt. It is necessary, however, to decide upon a convention for the *sign* of the potential of the cell as represented. It is unfortunate that it is about equally usual to regard the potential of such a cell as positive and negative. The convention adopted in this book will be that the electromotive force of the cell will be regarded as positive if the "direction of the current," *i.e.*, the direction of motion of positive ions, is from left to right during the operation of the cell. Thus for the case in question

$Zn; ZnSO_4\ (C_1),\ H_2SO_4\ (C_2);\ H_2\ (Pt)$ $\qquad E = +0.76$ approx.

As already stated, the measured electromotive force of a galvanic cell does not necessarily correspond to a single definite reaction. For example, a powerful primary cell may be constructed by placing a porous cup containing a platinum or carbon electrode, surrounded by nitric acid, into another vessel containing an amalgamated zinc electrode in an electrolyte of sulphuric acid. This cell may be represented by

$$Zn; H_2SO_4 : HNO_3; (Pt)$$

When such a cell is in action the zinc enters the electrolyte as zinc sulphate, and the nitric acid is reduced. The reduction products, however, depend upon the concentration of the acid, the nature and condition of the electrode and other factors. They may be any of the oxides of nitrogen, nitrogen itself, or even ammonia. Under these conditions it is evidently not possible to consider the measured electromotive force of such a cell as a measure of the decrease of the Gibbs free energy of any particular reaction.

The Effect of Temperature on Gibbs Free Energy and Electromotive Force. When the entropy S is eliminated from equations (18) and (22) we obtain

$$Z - T\left(\frac{\partial Z}{\partial T}\right)_P = H \tag{36}$$

If this equation is applied to the reacting substances and to the products of a chemical reaction taking place at constant pressure, and one equation subtracted from the other, the result is

$$\Delta Z - T\left(\frac{\partial \Delta Z}{\partial T}\right)_P = \Delta H \tag{37}$$

Recalling that $-\Delta Z = En\mathbf{F}$ from equation (31) this becomes

$$En\mathbf{F} - Tn\mathbf{F}\left(\frac{\partial E}{\partial T}\right)_P = -\Delta H \tag{38}$$

This is the well known *Gibbs-Helmholtz equation.* From it, for instance, the change of heat content ΔH of a chemical reaction can be computed from the potential of a galvanic cell in which the reaction takes place, and the temperature coefficient of this potential. Since the change of heat content is the heat of reaction as ordinarily measured in a calorimeter (with a change of algebraic sign) it is of interest to demonstrate or test the thermodynamic principles that have been discussed by comparing the values obtained by the two methods. This will be done in a forthcoming paragraph.

It was for a long time believed that the available energy for external work from a chemical reaction could be obtained by measuring the heat of reaction. In terms of the symbols used in this book this is equivalent to

$$\Delta Z = \Delta H \quad \text{or} \quad En\mathbf{F} = -\Delta H \tag{38a}$$

This served as the guiding principle of extended researches by Berthelot (1827-1907) and by Thomsen (1826-1909), to whom the greater part of our data on thermochemistry is due. However, it can be seen from equation (38) that the available work (not including mechanical work which is usually very small) can only be equal to the heat of reaction (a) at absolute zero or (b) if the Gibbs free energy does not change with the temperature. This latter statement also means, of course, that the electromotive force of a cell in which the reaction takes place does not change with the temperature, *i.e.,*

$$(\partial E/\partial T)_P = 0.$$

If we use equation (22) for the reacting substances and the end products and take the difference, it becomes

$$\left(\frac{\partial \Delta Z}{\partial T}\right)_P = -\Delta S = -\frac{\Delta Q}{T} \tag{39}$$

Substituting this in equation (37) we obtain

$$En\mathbf{F} + \Delta Q = \Delta H \tag{40}$$

Here ΔQ is the heat absorbed when the galvanic cell operates reversibly, yielding the electrical energy. A description of a possible experiment to test these thermodynamic relations will probably be useful in clarifying the meanings of the various terms. Experiments of this type have as a matter of fact been carried out by Jahn.[3] A diagram of an

[3] H. Jahn, *Z. physik. Chem.*, **18**, 399 (1895).

apparatus is shown in Fig. 3. A galvanic cell C with a very small internal resistance is placed in a calorimeter A and is connected by heavy metallic leads to the high resistance R in calorimeter B. If the resistance R is sufficiently great the cell C will discharge under approximately reversible conditions. Now as the reaction in cell C takes place the heat evolved in calorimeter B is equal (except for a small heat effect due to the internal resistance of the cell) to the decrease in the Gibbs free energy $(-\Delta Z)$. This energy has, of course, been carried from

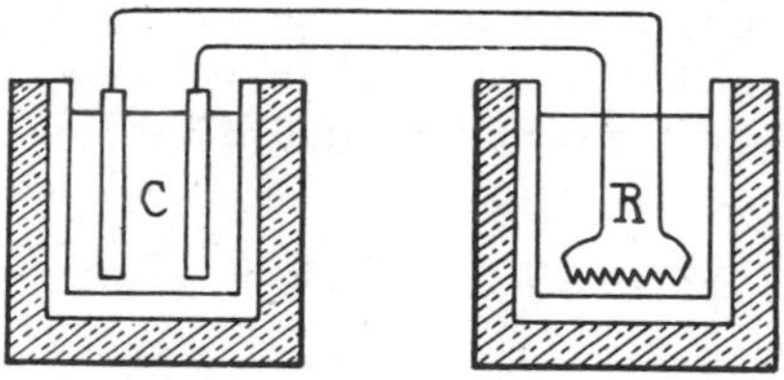

Fig. 3. Diagram of Arrangement for Testing Equation (40).

the galvanic cell to the calorimeter B as electrical energy, and could have served as useful work instead of being converted into heat. The heat *absorbed* in calorimeter A is equal to the term ΔQ in equation (40). correction being made for the small heat effect just mentioned. If the assumption made by Thomsen and Berthelot were correct, no heat would be absorbed or liberated in calorimeter A during the discharge of a cell. However, in practically every case a heat evolution or absorption would be observed. Haber[4] calls ΔQ the "latent heat" of the reaction taking place in the cell, since, in close analogy to the latent heat of vaporization of a liquid, it is the heat absorbed when the maximum work is done. The work done during a vaporization is, however, all mechanical, instead of nearly all electrical as is the case with most galvanic cells.

During the reversible operation of the greater number of types of galvanic cells heat is evolved, *i. e.*, ΔQ has a negative value. In these cases the available electrical energy, $En\mathbf{F}$, is less than the change of the heat content, $-\Delta H$. In some cases, however, heat is absorbed during the operation of the cell, and the electrical energy obtainable from the reaction is greater than would be predicted from the heat of the reaction. As an extreme case, an endothermic reaction, one that is accompanied by the absorption of heat during its progress, may be used as a source of electrical energy. An example will be presently considered.

[4] F. Haber, "Thermodynamics of Gas Reactions," Longmans, Green and Co., New York, 1908.

Instead of making calorimetric experiments such as those described above it is usually far more convenient and accurate to determine the "latent heat" ΔQ of a reaction taking place in a cell by means of the equation

$$\Delta Q = Tn\mathbf{F}\left(\frac{\partial E}{\partial T}\right)_P \tag{41}$$

which follows from equations (31) and (39). However, if the temperature range is not too great this formula may be replaced by

$$\Delta Q = Tn\mathbf{F}\left(\frac{\Delta E}{\Delta T}\right)_P \tag{42}$$

in which ΔE is the increase of the potential of the cell produced by the temperature change ΔT.

By means of the Gibbs-Helmholtz equation (38) it is obviously possible to compute the heats of the reaction, $-\Delta H$, from the electromotive force of a cell in which the reaction takes place, and the temperature coefficient of the electromotive force of the cell. If the temperature range is not too great equation (38) can be replaced by

$$En\mathbf{F} - Tn\mathbf{F}\left(\frac{\Delta E}{\Delta T}\right)_P = -\Delta H \tag{43}$$

in which ΔE is the change of the electromotive force in a finite temperature interval ΔT. Some examples of the use of the equation for that purpose are given in Table II, which is largely self explanatory.

TABLE II. THE USE OF THE GIBBS-HELMHOLTZ EQUATION TO DETERMINE HEATS OF REACTION, $-\Delta H$.

	Cell and reaction	E, volts at 25° C.	$\Delta E/\Delta T$	ΔZ calories	ΔH computed calories	ΔH observed calories
1.	Pb; $PbCl_2$, HCl soln., AgCl; Ag $Pb + 2AgCl = PbCl_2 + 2Ag$	0.4900	− 0.000186	− 22,610	− 25,170	− 24,170
2.	Pb; $PbCl_2$, HCl soln., HgCl; Hg $Pb + 2HgCl = PbCl_2 + 2Hg$	0.5356	+ 0.000145	− 24,720	− 22,720	− 20,100
3.	Tl; TlCl, NaCl soln., AgCl; Ag $Tl + AgCl = TlCl + Ag$	0.7790	− 0.000047	− 17,980	− 18,296	− 18,200
4.	Pb; PbI_2, KI soln., AgI; Ag $Pb + 2AgI = PbI_2 + 2Ag$	0.2135	− 0.000173	− 9,853	− 12,231	−12,200
5.	Ag; AgCl, HCl, HgCl; Hg $Ag + HgCl = AgCl + Hg$	0.0455	+ 0.000338	− 1,050	+ 1,275	+ 1,900

The electromotive force data are all from the work of Gerke.[5] The figures in the column headed "ΔH computed" were obtained from that data and equation (38), whereas those in the column "ΔH observed"

[5] R. H. Gerke, *J. Am. Chem. Soc.*. 44, 1684 (1922).

were compiled from data on heats of reaction in the Landolt and Börnstein "Tabellen" and are based mainly on the researches of Thomsen and Berthelot. It will be seen that the observed and computed values for ΔH for the various reactions agree fairly well. It is altogether probable that the computed values of this quantity are more accurate than those observed, since the early calorimetric determinations carried out by different experimenters often differ in the second significant figure.

A particularly interesting case is that of reaction No. 5 in the table. Here the change of the Gibbs free energy, $\Delta Z = -1050$ calories, has the opposite sign to that of the change of the heat content, ΔH, which has the value of $+1275$ calories. In this case the Thomsen-Berthelot assumption (38a) is not only inaccurate but predicts a result of the opposite sign to that observed. The sign of the computed heat content change is confirmed by the direct calorimetric determinations, though the numerical agreement is not very good, as these latter determinations are subject to a large experimental error.

The Effect of Pressure Changes on the Electromotive Force of Galvanic Cells. If we apply equation (23) to the reacting substances and the products of a chemical reaction and subtract one from the other we obtain

$$\left(\frac{\partial Z_p}{\partial P}\right)_T - \left(\frac{\partial Z_r}{\partial P}\right)_T = V_p - V_r \qquad (44)$$

in which Z_r, Z_p, V_r, and V_p are the Gibbs free energies and the volumes of the reacting substances and the products respectively. This equation may be written

$$\left(\frac{\partial \Delta Z}{\partial P}\right)_T = \Delta V \qquad (45)$$

in which ΔV is the change of volume during the chemical reaction. Furthermore, if the reaction can be caused to take place in a galvanic cell, use may be made of equation (31), and equation (45) becomes

$$\left(\frac{\partial E}{\partial P}\right)_T = -\frac{\Delta V}{n\mathbf{F}} \qquad (46)$$

This equation was first derived, as part of a more general expression, by Gibbs [6] and later by Duhem.[7] To test the equation, Cohen and

[6] J. W. Gibbs, "The Collected Works of J. Willard Gibbs," 1, 338, Longmans, Green and Co., New York, 1928.

[7] P. Duhem, "Le Potential Thermodynamique et ses Applications," A. Hermann, Paris, 1886.

Piepenbroek [8] made measurements on the effect of high pressures on the emf of the cell

Tl amalgam; TlCNS (s), KCNS : KCl, TlCl (s); Tl amalgam

The volume change ΔV when one faraday of electricity is passed through the cell was computed, from density measurements on the reacting substances, to be $-2.66_6 \pm 0.08$ cc. Assuming that this is independent of pressure we have

$$\left(\frac{\Delta E}{\Delta P}\right)_T = \frac{2.66_6 \times 0.10134}{96{,}500} = 2.80 \times 10^{-6} \frac{\text{volt}}{\text{atm.}}$$

(0.10134 is the factor which converts cubic centimeter-atmospheres into joules, *i.e.*, volts × coulombs.) The directly determined value for $(\Delta E/\Delta P)_T$ was found to be 2.84×10^{-6} in very good agreement with the computed value.

However, the greatest volume changes during the operation of a cell and thus the greatest changes of emf with pressure can occur with reactions involving gases. Measurements on the effect of hydrogen pressure on the cell

(Pt) H_2; HCl (*0.1 N*) HgCl; Hg

have been made by various experimenters [9] at low pressures (up to 1.5 atmospheres) and by Hainsworth, Rowley and MacInnes [10] to over 1000 atmospheres.

The reaction occurring in this cell is

$$\tfrac{1}{2} H_2 + HgCl = HCl \text{ (solution)} + Hg$$

which is separated in the galvanic cell into the two electrochemical reactions:

$$\tfrac{1}{2} H_2 = H^+ + e^-$$

which takes place reversibly at the platinized platinum electrode, and the reaction

$$e^- + HgCl = Hg + Cl^-$$

which occurs at the surface of mercury.

[8] E. Cohen and K. Piepenbroek, *Z. physik. Chem.*, **170A,** 145 (1934).
[9] G. N. Lewis and M. Randall, *J. Am. Chem. Soc.*, **36,** 1969 (1914); N. E. Loomis and S. F. Acree, *ibid.*, **38,** 2391 (1916); J. H. Ellis, *ibid.*, **38,** 737 (1916).
[10] W. R. Hainsworth, H. J. Rowley, and D. A. MacInnes, *J. Am. Chem. Soc.*, **46,** 1437 (1924).

The high pressure measurements were made in a steel bomb shown, with the cell in position, in Fig. 4. The hydrogen entered and left the bomb by the tubes *E* and *A*. The platinized platinum electrode is

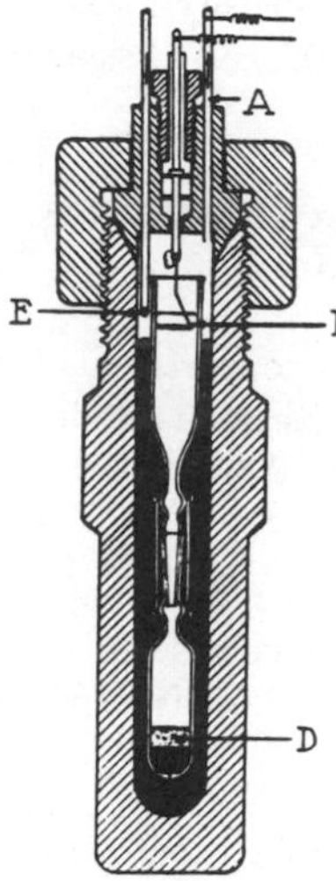

Fig. 4. Apparatus for Measuring the Potential of the Galvanic Cell
(Pt) H_2; HCl, HgCl ; Hg
under High Pressures.

represented by *B* and the calomel electrode by *D*. The first was connected to an insulated conductor and the second to the bomb. The effect of the pressure on the electromotive force of the cell is given in Table III and is shown in curve *A* of Fig. 5, in which the electromotive

TABLE III. MEASUREMENTS OF THE CELL: H_2; HCl (0.1*N*), HgCl; Hg, AT 25° AT DIFFERENT PRESSURES

Pressure atmospheres	emf volts	Pressure atmospheres	emf volts
1.0	0.3990	701.8	0.4891
37.8	.4456	717.8	.4899
51.6	.4496	731.8	.4893
110.2	.4596	754.4	.4903
204.7	.4683	862.2	.4932
386.6	.4784	893.9	.4938
439.3	.4804	974.5	.4963
556.8	.4844	1035.2	.4975
568.8	.4850	.....	.....

forces are plotted as ordinates and the logarithms of the pressures as abscissæ. Curve A represents the measured potentials. It is important to determine how closely the observed change of the electromotive

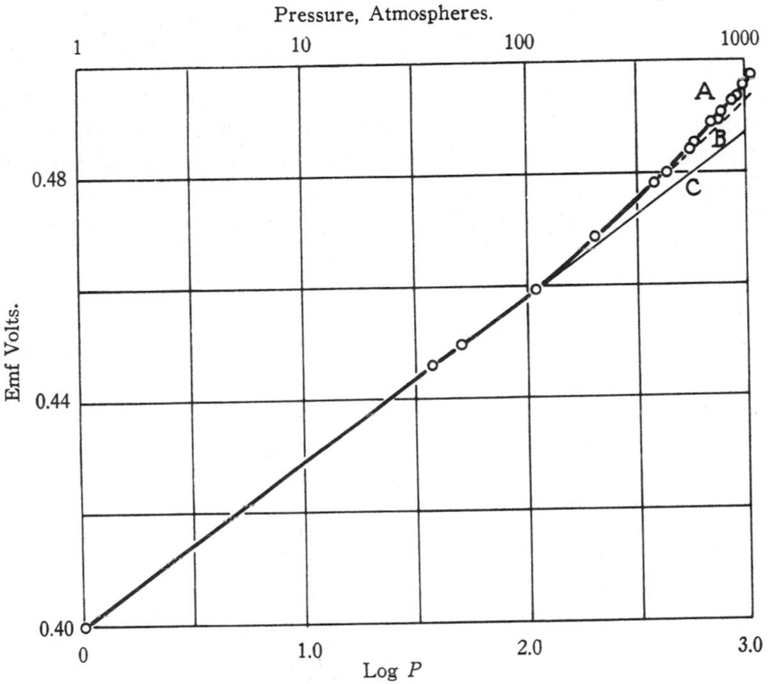

Fig. 5. The Relation Between the emf of the Galvanic Cell (Pt) H_2; HCl, HgCl; Hg and the Pressure.

force with pressure can be computed from an integrated form of equation (46). In order to perform this integration it is necessary to know ΔV, the volume change during the cell reaction, as a function of the pressure P. In this case

$$\Delta V = (\bar{V}_{HCl} + V_{Hg}) - (\tfrac{1}{2}V_{H_2} + V_{HgCl}) \tag{47}$$

in which V_{H_2}, V_{Hg} and V_{HgCl} are, respectively, the molar volumes of the substances represented by the subscripts and $\bar{V}_{HCl}$ is the partial molal volume [11] of HCl. Of these terms the only one we need consider is V_{H_2}

[11] See footnote p. 82.

as the other terms are relatively small, nearly independent of pressure, and compensating. (The computation involving all the terms is, however, given in the original paper.) Since hydrogen is nearly but not quite a perfect gas, the perfect gas law $PV = RT$ may be replaced at 25° by the empirical equation

$$PV = RT\,(1 + 0.000537P + 3.5 \times 10^{-8}\,P^2) \tag{48}$$

which represents accurately the direct measurements on the gas. The terms involving P on the right-hand side of the equation have an appreciable influence only at the higher pressures. On substituting this expression in equations (46) and (47) we obtain

$$dE = \frac{RT}{2\mathbf{F}}\left(\frac{1}{P} + 0.000537 + 3.5 \times 10^{-8}\,P\right)dP \tag{49}$$

Integration yields

$$E_P = E_1 + \frac{RT}{2\mathbf{F}}\,(\ln P + 0.000537\,(P - 1) + 1.7 \times 10^{-8}\,(P^2 - 1)) \tag{50}$$

in which E_P is the electromotive force of the cell at the pressure P and E_1 is its potential at a hydrogen pressure of one atmosphere. In Fig. 5 the dotted curve B represents a plot of equation (50). For the pressures up to 600 atmospheres it is identical with curve A which is drawn through the experimental points. The slight deviation between curves A and B above that pressure is probably due to the fact that hydrogen has an appreciable solubility which affects the term $\bar{V}_{HCl}$ of equation (47). It is evident that the thermodynamic theory as outlined accurately accounts for the change with pressure of the electromotive force of the cell under consideration, at least up to 600 atmospheres. The slight deviations above that pressure can with confidence be accounted for by our lack of knowledge of the various molal volume terms in equation (47) under these unusual conditions.

The straight line C in Fig. 5 is a plot of the equation

$$E_P = E_1 + \frac{RT}{2\mathbf{F}}\ln P \tag{51}$$

which includes only the first two terms on the right-hand side of equation (50). This simple equation is valid up to a pressure of 100 atmospheres of hydrogen, and will be found to be of use later on.

Entropy and the Third Law of Thermodynamics. It has been, for a long time, the ambition of physical chemists to compute Gibbs free energies from calorimetric data. This is due to the fact that direct free energy measurements are frequently difficult to carry out, whereas calorimetric measurements are relatively easy. The problem has also attracted attention because the relations between the changes of free and total energies are of fundamental importance. The following paragraphs will outline the progress that has been made in this field, restricting the discussion, however, mainly to the portion of the subject which has electrochemical interest.

For a change of state or a chemical reaction at constant temperature, equation (18) gives

$$\Delta Z = \Delta H - T\Delta S \tag{52}$$

which indicates that the change of Gibbs free energy, ΔZ, differs from the change of heat content, ΔH, by the term $T\Delta S$, in which ΔS is the change of the value of the entropy. With the aid of equation (39) this may be put in the form

$$\Delta Z - T\left(\frac{\partial \Delta Z}{\partial T}\right)_P = \Delta H \tag{53}$$

which may be readily transformed into

$$\frac{\partial}{\partial T}\left(\frac{\Delta Z}{T}\right) = -\frac{\Delta H}{T^2} \tag{54}$$

This equation may be integrated if ΔH is known as a function of T. However, the value of the Gibbs free energy obtained in this way will be uncertain to the extent of an unknown integration constant. Early speculations as to the value of this constant for different types of physical changes and chemical reactions were made by Le Chatelier [12] and Haber.[13] T. W. Richards [14] made a considerable advance by observing that, for a number of reactions which could be studied in galvanic cells, the values of ΔZ and ΔH tended to approach each other as the temperature was lowered, apparently reaching the same value at absolute zero, as is shown in Fig. 6. That the values of $\partial(\Delta H)/\partial T$ and $\partial(\Delta Z)/\partial T$ approach each other at absolute zero was later postulated by Nernst,[15] and was the basis for the "Nernst Heat Theorem." It is evident that if the change of entropy, ΔS, is zero for reactions or physical changes near absolute zero it furnishes at least a starting point for computations of Gibbs free energy values, ΔZ, from determinations of heat content, ΔH. Planck [16] has gone a step further and has given

[12] H. Le Chatelier, *Ann. Mines,* **13,** 157 (1888).

[13] F. Haber, "The Thermodynamics of Technical Gas Reactions," English translation by A. B. Lamb, Longmans, Green and Co., London, 1908.

[14] T. W. Richards, *Z. physik. Chem.,* **42,** 129 (1903).

[15] W. Nernst, *Nachr. kgl. Ges. Wiss. Göttingen, Math-Phys. Klasse* 1 (1906).

[16] M. Planck, *Ber. deut. chem. Ges.,* **45,** 5 (1912).

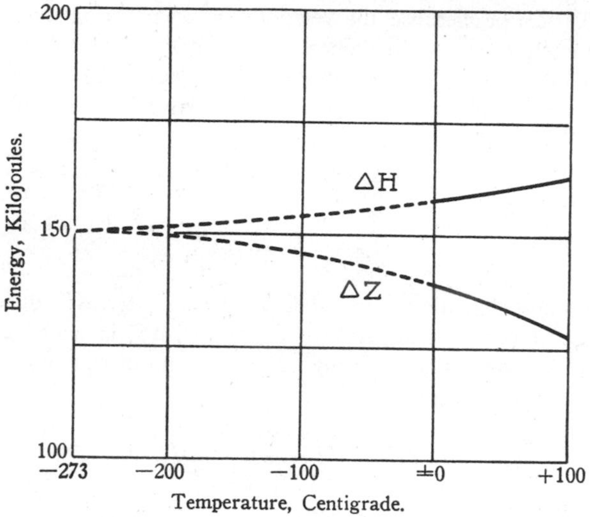

Fig. 6. Temperature Dependence of ΔH and ΔZ for the Reaction $Fe + CuSO_4 = Cu + FeSO_4$

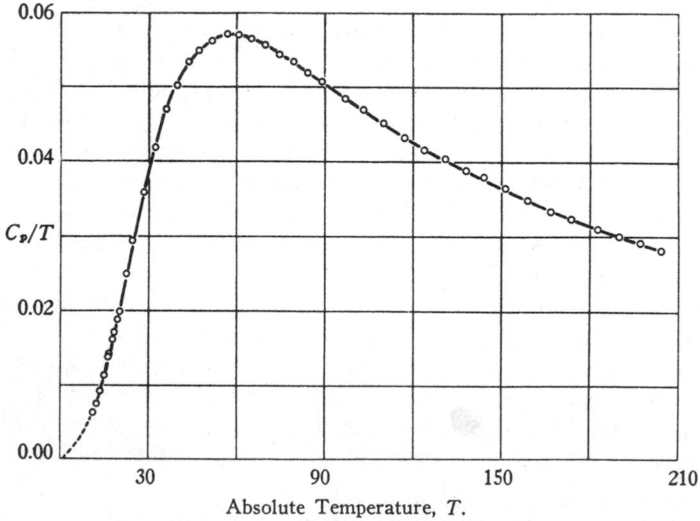

Fig. 7. Heat Capacity-Temperature Diagram for Silver.

theoretical evidence that not only is ΔS zero for a change of state or reaction near the absolute zero, but that the entropies, S, of crystalline solids are zero at that point. Entropies according to his conception may be regarded as a measure of "randomness." A gas or a liquid whose molecules are capable of shifts of position will thus have a large entropy, whereas a crystal at low temperatures, where thermal vibrations are at a minimum, will have very low or zero entropy. However, a supercooled liquid, according to these considerations, may have an appreciable entropy even at absolute zero. The statement that the *entropy of a crystalline substance is zero at absolute zero and is positive at all other temperatures* is called, possibly prematurely, the *third law of thermodynamics*.

According to equation (10) the entropy is defined by

$$dS = Q/T \tag{10}$$

in which Q is the heat absorbed. If a substance is heated at constant pressure through the temperature interval dT, then

$$Q = C_P\, dT \tag{55}$$

in which C_P is the heat capacity at constant pressure. Equation (10) thus becomes

$$dS = C_P \frac{dT}{T} \tag{56}$$

and for a substance which does not undergo any transformations, such as from one crystalline form to another or from solid to liquid

$$S = \int_0^T C_P \frac{dT}{T} \tag{57}$$

For crystals the lower limit of the integral is zero from the considerations just outlined. Equation (57) may be integrated if a relation between C_P and T is known. The available analytical relations are, however, complicated and of limited validity. Fortunately values of S may be obtained from measurements of heat capacity at different temperatures by graphical methods. A convenient method is one proposed by Lewis and Gibson [17] which consists in plotting values of C_P/T against T and determining the area of the enclosed plot. Such a plot is shown in Fig. 7 for the estimation of the entropy of metallic silver, from the work of Eucken, Clusius and Woitinek.[18] Below the lowest

[17] G. N. Lewis and G. E. Gibson, *J. Am. Chem. Soc.*, **39**, 2554 (1917).
[18] A. Eucken, K. Clusius and H. Woitinek, *Z. anorg. Allgem. Chem.*, **203**, 39 (1931).

direct measurement of the heat capacity it is usual to assume the relation

$$C_V = a\,T^3 \tag{58}$$

in which C_V is the heat capacity at constant volume and a is a constant obtained from the data at the lowest temperatures. A relation between C_P and C_V is given by the thermodynamic expression [19]

$$C_P = C_V + \frac{\alpha V T}{\beta} \tag{59}$$

where α is the temperature coefficient of expansion, β the coefficient of compressibility and V the gram molecular volume. The data and plot just given lead to a value of the entropy of silver of 10.0 entropy units. Values of the entropies at 25° C for a number of elements and compounds are given in Table IV, which is from a more complete computation by Latimer, Schultz and Hicks.[20]

TABLE IV. ENTROPY VALUES AT 25° C FOR VARIOUS ELEMENTS AND COMPOUNDS, IN CALORIES PER DEGREE PER MOL

Substance	Entropy	Substance	Entropy	Substance	Entropy
Br_2	36.8	I_2	27.9	AgCl	23.0
Cd	12.3	Pb	15.6	AgI	27.6
C (graphite)	1.4	$PbCl_2$	33.9	S	7.6
Cl_2	53.3	Hg	18.3	Tl	14.9
Cu	7.8	HgCl	22.8	TlBr	28.9
H_2	31.2	O_2	49.0	TlCl	25.8
HCl	44.7	Ag	10.0	TlI	28.6
H_2O	16.9	AgBr	25.6	Zn	9.8

We may thus compute, for instance, the entropy change during the reaction

$$\text{Ag} + \text{HgCl} = \text{AgCl} + \text{Hg} \tag{60}$$

From the figures given in Table IV the increase of entropy during the reaction is 8.5 units. As a test of the third law of thermodynamics this value may be compared with a determination of the entropy change of this reaction made by Gerke,[21] from the potentials of galvanic cells, in a research already referred to. He measured the potential, E, and the change of the potential with temperature, $\Delta E/\Delta T$, of cells of the type

$$\text{Ag; AgCl, MCl (molar), HgCl; Hg} \tag{61}$$

[19] G. N. Lewis, *J. Am. Chem. Soc.*, **29**, 1165, 1516 (1907).

[20] W. M. Latimer, P. W. Schultz, and J. F. G. Hicks, Jr., *J. Chem. Phys.*, **2**, 82 (1934).

[21] R. H. Gerke, *J. Am. Chem. Soc.*, **44**, 1684 (1922).

in which M represents potassium or hydrogen. During the operation of this cell the reaction indicated by equation (60) takes place. From equations (31) and (39) we have the relation

$$n\mathbf{F}\left(\frac{\partial E}{\partial T}\right)_P = \Delta S \tag{62}$$

which, for a not too large temperature interval, may be replaced by

$$\frac{n\mathbf{F}}{4.1835}\left(\frac{\Delta E}{\Delta T}\right)_P = \Delta S \tag{63}$$

in which n is the number of faradays passing through the cell during the reaction and the constant, 4.1835, reduces joules to calories. Gerke's measurements give $(33.7 \pm 1) \times 10^{-6}$ for $(\Delta E/\Delta T)$ for cell (61), which leads to a value of $\Delta S = 7.8$ entropy units as compared to $\Delta S = 8.5$ from the heat capacity measurements. Some other tests of the third law of thermodynamics from Gerke's work are given in Table V. It

TABLE V. COMPARISON OF ENTROPY CHANGES, ΔS, COMPUTED FROM HEAT AND ELECTROMOTIVE FORCE MEASUREMENTS FOR VARIOUS REACTIONS

Reaction	$\Delta E/\Delta T$	ΔS heat	ΔS emf
$Ag + HgCl = AgCl + Hg$	0.000337	8.5	7.8
$\frac{1}{2}Pb + AgCl = \frac{1}{2}PbCl_2 + Ag$	0.0000186	−3.9	−4.3
$Tl + AgCl = TlCl + Ag$	0.000047	−2.1	−1.1
$Ag + \frac{1}{2}Br_2 = AgBr$	0.0001335	−2.8	−3.1

will be seen that the agreement between the entropy values determined from heat capacity and electromotive force determinations average well within one entropy unit. This is probably within the experimental error of the two types of measurement.

There is much more experimental evidence than that given above for the validity of the "third law of thermodynamics." Most of it, however, does not belong in a book devoted to electrochemistry. Although the limits of the validity of the third law have yet to be fully outlined it is a useful working hypothesis in a number of fields of investigation.

[*Comment to 1961 Edition*] More complete and revised values of entropies are given in Wendell L. Latimer's *Oxidation Potentials*, 2nd ed., Prentice-Hall, Inc, 1952, page 359 and following.

Chapter 6

Chemical Potential, Activity and Related Quantities

So far it has been assumed that although heat and work may pass through the boundaries of thermodynamic systems, no transport of matter takes place. Thermodynamics can, however be extended to take account of exchanges of chemical substances as well as energy between a system and its surroundings. For instance, equation (15), Chapter 5, can be modified by the addition of terms as follows

$$dU = T\,dS - P\,dV + \mu_A\,dn_A + \mu_B dn_B + \cdots\cdots + \mu_K dn_K \quad (1)$$

in which n_A, n_B, ... n_K are the number of mols of the components of the system indicated by the subscripts and μ_A, μ_B, ... μ_K are the corresponding "chemical potentials." These useful functions were first used by Willard Gibbs.[1] From equation (1) we may obtain

$$\mu_A = \left(\frac{\partial U}{\partial n_A}\right)_{S,\,V,\,n} \qquad \mu_B = \left(\frac{\partial U}{\partial n_B}\right)_{S,\,V,\,n} \quad \text{etc.} \quad (2)$$

the subscript n representing constancy of the number of mols of all the components except the one in the denominator. The chemical potential of a component is evidently the increase of total energy of the thermodynamic system produced by the reversible addition of one mol of the component to the system under the conditions of constant entropy, temperature and composition. By differentiating equation (18), Chapter 5, and adding to equation (1) we obtain the expression

$$dZ = -\,S\,dT + V\,dP + \mu_A\,dn_A + \mu_B dn_B + \cdots\cdots$$
$$+ \mu_K dn_K + \mu_L dn_L + \cdots \quad (3)$$

from which it appears that the chemical potentials can also be defined by

$$\mu_A = \left(\frac{\partial Z}{\partial n_A}\right)_{T,\,P,\,n} \qquad \mu_B = \left(\frac{\partial Z}{\partial n_B}\right)_{T,\,P,\,n} \quad \text{etc.} \quad (4)$$

[1] J. W. Gibbs, "The Collected Works of J. Willard Gibbs," Vol. 1, 63 and following, Longmans, Green and Co., New York, 1928.

By a similar manipulation with the functions F and H the following expressions may be obtained:

$$\mu_A = \left(\frac{\partial F}{\partial n_A}\right)_{T,\,V,\,n} \text{ etc.} \quad \text{and} \quad \mu_A = \left(\frac{\partial H}{\partial n_A}\right)_{S,\,P,\,n} \text{ etc.} \tag{5}$$

On account of equation (4) chemical potentials are called "partial molal (Gibbs) free energies" by Lewis and Randall.[2] This expresses possibly the most important, but not the only, property of chemical potentials, as is clear from equations (2) and (5).

If the different portions of a thermodynamic system are in true equilibrium the chemical potentials of the components of the system will be the same throughout, whatever the physical state. Thus the chemical potential of a vapor of a component is the same as that of the same component in the solid or liquid state, or of the component in solution.

Another important property of chemical potentials is related to chemical equilibrium in a thermodynamic system. Such an equilibrium may be represented by

$$a\text{A} + b\text{B} + \cdots\cdots \leftrightarrows k\text{K} + l\text{L} + \cdots\cdots \tag{6}$$

in which A, B, $\cdots$ K, L signify chemical components and a, b, $\cdots k$, l the integral proportions in which they react. For an equilibrium at constant temperature and pressure[3]

$$dZ = 0 \tag{6a}$$

so that from equation (3)

$$\mu_A dn_A + \mu_B dn_B + \cdots\cdots + \mu_K dn_K + \mu_L dn_L + \cdots\cdots = 0 \tag{7}$$

Now if the chemical reaction takes place under these conditions it is necessary that

$$\pm\frac{dn_A}{a} = \pm\frac{dn_B}{b} = \cdots\cdots = \mp\frac{dn_K}{k} = \mp\frac{dn_L}{l} = \cdots\cdots \tag{8}$$

From equations (7) and (8) we have therefore

$$a\mu_A + b\mu_B + \cdots\cdots = k\mu_K + l\mu_L + \cdots\cdots \tag{9}$$

[2] G. N. Lewis and M. Randall, "Thermodynamics and the Free Energy of Chemical Substances," McGraw-Hill Book Co., Inc., New York, 1923.

[3] Gibbs gives equation (6a) without proof, "Collected Works," 1, 91, equation (116). A proof is furnished by E. A. Milne, "Commentary on the Writings of J. Willard Gibbs," 213, Yale University Press, New Haven, 1936.

For instance, consider the chemical equilibrium involved in the dissociation of acetic acid into its ions:

$$HAc = H^+ + Ac^-$$

for which, from equation (9)

$$\mu_{HAc} = \mu_{H^+} + \mu_{Ac^-}$$

that is to say, the chemical potential of the undissociated acid is equal to the sum of the chemical potentials of the ions with which it is in equilibrium. Much use of equation (9) will be made later in this book.

If equilibrium does not exist in the system any natural change that takes place will be in the direction of a decrease of the Gibbs free energy, *i.e.*, $dZ < 0$. Therefore a chemical reaction naturally occurring in the system must be accompanied by a decrease of the Gibbs free energy of the system. If the reaction takes place under conditions such that the chemical potentials remain constant, then

$$-\Delta Z = a\mu_A + b\mu_B + \cdots - k\mu_K - l\mu_L - \cdots \quad \textbf{(10)}$$

Chemical Potentials from Galvanic Cells. It has already been shown that Gibbs free energies of certain chemical reactions may be obtained by measuring the potentials of galvanic cells in which the reactions take place. By appropriate selection and design galvanic cells can also be used to measure differences of chemical potentials of single (molecular) components taking part in the cell reaction. To do this the emf values are compared of cells which are identical except for the condition of the component whose chemical potential is under investigation. Thus, for instance, the effect on the chemical potential of hydrogen produced by diluting it with nitrogen may be studied by comparing the emf of the cell

$$\underset{P\,=\,1\text{ atm.}}{(\text{Pt})\ H_2};\ \text{HCl}\ (C),\ \text{HgCl};\ \text{Hg} \quad \textbf{(11)}$$

with one in which the pure hydrogen has been replaced by a mixture of hydrogen and nitrogen and which may be represented by

$$\underset{P\,=\,1\text{ atm.}}{(\text{Pt})\ H_2(N_{H_2}),\ N_2(1 - N_{H_2})};\ \text{HCl}\ (C),\ \text{HgCl};\ \text{Hg} \quad \textbf{(12)}$$

in which N_{H_2} is the mol fraction of the hydrogen in the mixture.[4] If

[4] The mol fraction of a constituent is the number of mols of that constituent present in a phase divided by the total number of mols present. In the case under consideration

$$N_{H_2} = \frac{n_{H_2}}{n_{H_2} + n_{N_2}}$$

and in general

$$N_A = \frac{n_N}{n_A + n_B + \cdots + n_K}$$

in which n represents the number of mols and the subscripts A, B, etc. the different components.

the two cells are arranged so that the difference of their potentials can be measured, we have the combination cell

$$\underset{P\,=\,1\text{ atm.}}{(\mathrm{Pt})\ H_2};\ HCl\ (C),\ HgCl;\ Hg\ -$$
$$-Hg;\ HgCl,\ HCl\ (C);\ \underset{P\,=\,1\text{ atm.}}{H_2(N_{H_2}),\ N_2(1-N_{H_2})}\ (\mathrm{Pt}) \qquad (14)$$

In the left-hand half of this combination cell the reaction

$$H_2 + 2HgCl = 2HCl + 2Hg$$

takes place, and for each two faradays of current passing through the cell one mol of H_2 reacts at one atmosphere pressure and a mol fraction $N_{H_2} = 1$. In the right half cell the reverse reaction

$$2Hg + 2HCl = 2HgCl + H_2$$

occurs, and for the same amount of current one mol of H_2 reappears in the mixture of that gas with nitrogen, and a mol fraction less than unity, but at the same total pressure of one atmosphere. Since the two half cells are identical except for the dilution of the hydrogen, the potential ΔE of the combination cell (14) is a measure of the difference of the chemical potentials of hydrogen in the two half cells, since the chemical potentials of all the other reacting components remain constant, *i.e.*, with equation (31), Chapter 5.

$$\Delta Z = \Delta\mu_{H_2} = -\ 2\mathbf{F}\Delta E \qquad (15)$$

The potential of cell (14) with different mol fractions N_{H_2} of the hydrogen gas is shown in Table I, which is from the work of Romann

TABLE I. THE CHANGE OF CHEMICAL POTENTIAL OF HYDROGEN GAS ON DILUTION WITH NITROGEN

Mol fraction of hydrogen, N_{H_2} in mixture	Potential of cell (14), ΔE —volts— Observed	Computed	Chemical Potential $\Delta\mu_{H_2} = -\ 2\mathbf{F}\Delta E$ joules
1.000	0	0	0
0.496	0.0090	0.0090	− 1737
0.0944	0.0303	0.0303	− 5847
0.00983	0.0594	0.0593	− 11464
0.00517	0.0686	0.0676	− 13240

and Chang.[5] In the fourth column of the table it will be seen that the chemical potential of the hydrogen has an increasing negative value as it is diluted with nitrogen. As a basis for comparison this chemical potential has been assigned a value of zero when the mol fraction is

[5] R. Romann and W. Chang, *Bull. soc. chim.*, **51**, 932 (1932). The cells actually measured in this work were slightly more complicated than represented in cells (11) and (12).

unity. It is of interest to discover what relation exists between the change of emf, ΔE, produced by the dilution of the hydrogen, and the mol fraction N_{H_2} of the hydrogen. If the gas mixture consists of perfect gases the partial pressure of each component of the mixture may be computed from the relation

$$p_i = PN_i \tag{16}$$

in which p_i is the partial pressure of the component i, P is the total pressure, and N_i is the mol fraction of component i. We may *define* the chemical potential of a component i in a mixture of perfect gases by the relation

$$\mu_i = \mu_i^\circ + RT \ln p_i \tag{17}$$

in which μ_i° is a constant which depends upon an arbitrarily chosen "standard state" of the component and upon the temperature. Thus at two different partial pressures, p_i and p_i', the change of chemical potential of component i will be from equations (16) and (17)

$$\Delta\mu_i = RT \ln \frac{p_i}{p_i'} = RT \ln \frac{N_i}{N_i'} \tag{18}$$

If hydrogen is a perfect gas in its mixtures with nitrogen it should follow from equations (15) and (18) that

$$-2\mathbf{F}\Delta E = \Delta\mu_{H_2} = RT \ln \frac{N_{H_2}}{1} \tag{19}$$

since the mol fraction of the pure hydrogen is unity. (A small effect due to the partial pressure of the aqueous solution has been neglected.)

In column 3 of Table I values are given of ΔE computed from equation (19) which may be compared with the directly observed values in column 2 of the same table. It will be seen that the agreement is excellent. This indicates that for the hydrogen gas in the mixtures with nitrogen the chemical potential may be represented within the experimental error by an equation of the form of equation (17). However, it must be emphasized that equation (17) is strictly true only for mixtures of perfect gases. At higher total pressures than one atmosphere, and with gases less nearly perfect than hydrogen and nitrogen, deviations from equation (18) would be observed. A convenient function to use in such cases is the *activity*, *a*, which may be defined by

$$\mu_i = \mu_i^\circ + RT \ln a_i \tag{20}$$

Here again the constant μ_i° depends upon the standard state chosen and the temperature.[6]

Chemical Potentials in Ionic Solutions. It is also possible by emf measurements on galvanic cells to obtain differences of the chemical potentials of the components of the solutions contained in the cells. For instance, the potential of the now familiar cell

$$\underset{P\,=\,1\text{ atm.}}{(\text{Pt})\ H_2};\ HCl\ (C'),\ AgCl;\ Ag \tag{21}$$

may be compared with another cell

$$(\text{Pt})\ H_2;\ HCl\ (C''),\ AgCl;\ Ag$$

identical except for the concentration of hydrochloric acid. If the two cells are arranged thus

$$Ag;\ AgCl,\ HCl\ (C');\ H_2(\text{Pt})\ -\ (\text{Pt})\ H_2;\ HCl\ (C''),\ AgCl;\ Ag \tag{22}$$

we have a *concentration cell.* When a faraday of electric current passes through the cell from left to right the reaction

$$HCl + Ag = \tfrac{1}{2}H_2 + AgCl$$

takes place reversibly in the left-hand half cell, and the opposite reaction

$$AgCl + \tfrac{1}{2}H_2 = Ag + HCl$$

occurs in the right-hand half cell. Since all the components except the hydrogen chloride are constant during the process, the operation of the cell consists essentially of the disappearance of a mol of hydrogen chloride at the concentration C' and its reappearance at the concentration C''. The electrical energy of the process is a measure of the difference of the chemical potentials of the hydrogen chloride at the two concentrations, thus

$$\left(\overset{C'}{\mu}_{HCl} - \overset{C''}{\mu}_{HCl}\right) = \mathbf{F}\Delta E \tag{23}$$

[6] The conception of activity is due to G. N. Lewis, *Proc. Am. Acad.*, **13**, 359 (1907) who also defines the *fugacity* as

$$\mu_i = B + RT \ln \text{F}_i$$

(B is a constant at a given temperature) and the activity as the relative fugacity (Lewis and Randall, "Thermodynamics")

$$a_i = \frac{\text{F}_i}{\text{F}_i^\circ}$$

where F_i° is the fugacity in a chosen standard state. The fugacity is conceived to have the dimensions of pressure and for a gas, to approach the pressure in value as the latter is progressively made smaller. There does not seem to be a need of both fugacity and activity. Although the activity may, in different connections, appear to have the dimensions of pressure, concentration, molality, or mol fraction, it is always used as a ratio, or allowance for the various dimensions is made in the choice of the standard state.

in which ΔE is the potential of cell (22), and $\mu_{\text{HCl}}^{C'}$ and $\mu_{\text{HCl}}^{C''}$ are respectively the chemical potentials of HCl at the concentrations indicated.

Activities and Activity Coefficients. An ideal solution is one in which all the components follow Raoult's law:

$$p_i = p_i^\circ \mathrm{N}_i \tag{24}$$

in which p_i is the partial pressure of the component i from a solution in which N_i is the mol fraction of the same component and p_i° is the vapor pressure of the pure substance i. The relation has been found to hold true for solutions in which the solvent and solute are similar in nature. It has been found valid, for instance, for mixtures of benzene and toluene. It is generally used as the basic relation for any theory of solutions.

Since a solution can be considered to be in equilibrium with its vapor, and dilute vapors can be considered to be perfect gases, equation (24) may be substituted in equation (17) giving, for the chemical potential μ_i of the component i, of an ideal solution:

$$\mu_i = \mu_i^\circ + RT \ln \mathrm{N}_i \tag{25}$$

in which μ_i° has a new value.

If, as is usually the case, the solution is not ideal, equation (24) may be replaced by

$$p_i = p_i^\circ a_i = p_i^\circ \mathrm{N}_i \mathbf{f} \tag{26}$$

in which a_i is the *activity* and $\mathbf{f}$ an *activity coefficient*. And with equation (17) we obtain the expression

$$\mu_i = \mu_i^\circ + RT \ln a_i = \mu_i^\circ + RT \ln \mathrm{N}_i \mathbf{f} \tag{26a}$$

The utility of the conception of activity and activity coefficient is largely mathematical. As the state of infinite dilution is approached the chemical potential tends toward a value of $-\infty$, whereas the activity, by a suitable selection of μ_i° approaches zero, and the activity coefficient a value of unity. The activity may be considered to be an "effective" concentration and the activity coefficient as a measure of the extent an actual solute differs from an ideal solute.

It is, however, more usual to express the composition of a solution in terms of concentrations, C, *i.e.*, mols of solute per liter, or as molalities, m, mols of solute per 1000 grams of solvent. In very dilute solutions the concentrations and molalities are both proportional to the mol fraction, so that for convenience the chemical potentials may also be defined as follows

$$\mu_i = \mu_i^\circ + RT \ln a_i = \mu_i^\circ + RT \ln C_i f_i \tag{26b}$$

$$\mu_i = \mu_i^\circ + RT \ln a_i = \mu_i^\circ + RT \ln m_i \gamma_i \tag{26c}$$

in which f_i and γ_i are activity coefficients according to the corresponding concentration scales. The quantities μ_i° are again functions of the temperature and the standard state. The terms μ_i° and a_i will have different values in equations (26a), (26b) and (26c). The three numerical values of the activities are related to each other by constants which depend upon the concentration scales used.[7] Thus we have the relation:

$$\mathbf{f}N : \gamma m : fC = a_N : a_m : a_C$$

in which the activities are on the concentration scales represented by their subscripts. With the additional convention that $\mathbf{f}$, γ and f all approach unity at infinite dilution, activity coefficients on one scale may be readily computed from those on another scale.[8] The activity coefficients f and γ are measures of the deviations of actual solutions from the ideal solutions defined according to the relations:

$$p_i = k_C C_i \tag{24a}$$

$$p_i = k_m m_i \tag{24b}$$

in which k_C and k_m are constants. Equations (24a) and (24b) follow from equation (24) if C and m are very small.

The Law of Mass Action. Any chemical equilibrium, as has already been shown, may be represented by an equation of the form of

$$a\mathrm{A} + b\mathrm{B} + \cdots \rightleftarrows k\mathrm{K} + l\mathrm{L} + \cdots \tag{6}$$

in which A, B, ... K, L, represent chemical components and $a, b, \ldots k, l$ the integral proportions in which they react. The relation between the chemical potentials, μ, of the components for such an equilibrium is given by equation (9). Substituting an equation of the form of (20) for each component into that equation gives

$$\ln \frac{a_K^k\, a_L^l \cdots}{a_A^a\, a_B^b \cdots} = \frac{a\mu_A^\circ + b\mu_B^\circ + \cdots - k\mu_K^\circ - l\mu_L^\circ - \cdots}{RT} \tag{28}$$

[7] The reader is asked to be patient with this multitude of definitions, which are mostly of a formal nature. They are imposed, not by the author, who regrets the necessity of considering them, but by the manner in which the experimental data have been obtained and discussed in the chemical literature.

[8] The relations between these activity coefficients may be readily shown to be

$$\mathbf{f} = f(\rho - 0.001\mathrm{M}C + 0.001\mathrm{M}_s C\nu)/\rho_0 \tag{27a}$$

$$\mathbf{f} = \gamma(1 + 0.001\nu m \mathrm{M}_s) \quad \text{and} \quad fC = \rho_0 \gamma m \tag{27b-c}$$

in which M_s and M are the molecular weight of the solvent and solute, ρ and ρ_0 the densities of the solution and pure solvent, and ν the number of ions into which one molecule of the solute dissociates. The activity coefficient, $\mathbf{f}$, is sometimes called the "rational" activity coefficient.

Since, at a given temperature, all the terms on the right-hand side of this equation are constant this may be put into the form

$$\frac{a_K^k\, a_L^l \cdots\cdots}{a_A^a\, a_B^b \cdots\cdots} = \mathbf{K} \tag{29}$$

in which **K** is a constant. This relation holds whether the chemical equilibrium is in a gas, liquid or solid phase, or if the system consists of several phases. If the components of the chemical reaction are ideal solutes following equation (24a) then, with equation (17)

$$\frac{C_K^k\, C_L^l \cdots\cdots}{C_A^a\, C_B^b \cdots\cdots} = \mathbf{K} \tag{29a}$$

which is the form in which the *law of mass action* is usually stated. An alternative statement of the law of mass action is from equations (9), (17) and (24)

$$\frac{\mathrm{N}_K^k\, \mathrm{N}_L^l \cdots\cdots}{\mathrm{N}_A^a\, \mathrm{N}_B^b \cdots\cdots} = \mathbf{K} \tag{29b}$$

for ideal solutes following Raoult's law, and, from (17) and (26a)

$$\frac{\mathrm{N}_K^k\, \mathrm{N}_L^l \cdots\cdots}{\mathrm{N}_A^a\, \mathrm{N}_B^b \cdots\cdots} \times \frac{\mathbf{f}_K^k\, \mathbf{f}_L^l \cdots}{\mathbf{f}_A^a\, \mathbf{f}_B^b \cdots} = \mathbf{K} \tag{30}$$

for non-ideal solutes. Equation (30) may also be stated in terms of concentrations or molalities and the corresponding activity coefficients.

Activity Coefficient and Mean Activity Coefficient Ratios of Ion Constituents. It is usual to regard the ion constituents as separate solutes, and utilizing equation (9) to write, for instance,

$$\mu_{\mathrm{HCl}} = \mu_{\mathrm{H^+}} + \mu_{\mathrm{Cl^-}} \tag{31}$$

and in accord with equation (26b)

$$\begin{aligned} \mu_{\mathrm{H^+}} &= \mu^\circ_{\mathrm{H^+}} + RT \ln a_{\mathrm{H^+}} = \mu^\circ_{\mathrm{H^+}} + RT \ln C_{\mathrm{H^+}} f_{\mathrm{H^+}} \\ \mu_{\mathrm{Cl^-}} &= \mu^\circ_{\mathrm{Cl^-}} + RT \ln a_{\mathrm{Cl^-}} = \mu^\circ_{\mathrm{Cl^-}} + RT \ln C_{\mathrm{Cl^-}} f_{\mathrm{Cl^-}} \end{aligned} \tag{32}$$

in which $C_{\mathrm{H^+}}$ and $C_{\mathrm{Cl^-}}$ are the concentrations of the ion constituents indicated, *i. e.*, $C_{\mathrm{H^+}} = C_{\mathrm{Cl^-}} = C_{\mathrm{HCl}}$. Thus from equations (23), (31) and (32) we have for the emf, ΔE, of a concentration cell of type (22) the formula

$$\Delta E = \frac{RT}{\mathbf{F}} \ln \frac{C'^2_{\mathrm{HCl}} f'_{\mathrm{H^+}} f'_{\mathrm{Cl^-}}}{C''^2_{\mathrm{HCl}} f''_{\mathrm{H^+}} f''_{\mathrm{Cl^-}}} \tag{33}$$

in which $f'_{H^+}, f''_{H^+}, \ldots$ represent the values of these coefficients at the concentrations C' and C'' respectively. Since, however we have no methods for determining the activity coefficients of a single ion species, if indeed such activity coefficients have any physical meaning, it is customary to use a mean ion activity coefficient defined by the relation[9]

$$f_{\pm}^{2} = f_{H^+}\, f_{Cl^-} \tag{34}$$

Thus equation (33) becomes

$$\Delta E = \frac{RT}{\mathbf{F}} \ln \frac{C'^2 f'^2_{\pm}}{C''^2 f''^2_{\pm}} \tag{35}$$

A plot of some typical activity coefficient ratios as a function of the square root of the concentration is shown in Fig. 1. The emf data for hydrochloric acid are from the work of Harned and Ehlers[10] on cells of the type represented by equation (21). The mean activity coefficient ratios were computed by arbitrarily assuming that the most dilute solution, 0.003205 normal (which is the most nearly ideal) has the activity coefficient of unity. It will be observed that as the concentration increases the mean activity coefficient, on this arbitrary basis, steadily decreases; whereas, if the ions were perfect solutes the activity coefficient ratios would be unity throughout the concentration range. The shift of activity coefficient ratios in the range of concentration included in this plot is roughly typical of the behavior of uni-univalent electrolytes. Fig. 1 also shows the activity coefficient ratios of zinc sulphate from the work of Cowperthwaite and LaMer,[11] based on emf measurements at 25°, of the cell

$$\begin{aligned} &\mathrm{Zn\,(Hg);\ ZnSO_4\ (}C'\mathrm{),\ PbSO_4;\ Pb(Hg)\ -} \\ &\qquad\qquad\mathrm{Pb(Hg);\ PbSO_4,\ ZnSO_4\ (}C''\mathrm{),\ Zn\,(Hg)} \end{aligned}$$

The expression corresponding to equation (35) for the hydrochloric acid concentration cell, is

$$\Delta E = \frac{RT}{2\mathbf{F}} \ln \frac{C'^2_{ZnSO_4} f'^2_{\pm}}{C''^2_{ZnSO_4} f''^2_{\pm}}$$

[9] If a binary electrolyte dissociates giving ν ions of which ν_+ are positive and ν_- are negative, then the mean ion activity coefficient may be defined by

$$f_{\pm}^{\nu} = f_{+}^{\nu_+} f_{-}^{\nu_-} \tag{34a}$$

If for instance, the electrolyte is lead chloride, then $\nu = 3$, $\nu_+ = 1$, and $\nu_- = 2$, and

$$f^{3}_{\pm PbCl_2} = f_{Pb^{++}} f^{2}_{Cl^-}$$

[10] H. S. Harned and R. W. Ehlers, *J. Am. Chem. Soc.*, **54**, 1350 (1932); **55**, 2179 (1933).
[11] I. A. Cowperthwaite and V. K. LaMer, *J. Am. Chem. Soc.*, **53**, 4333 (1931).

the factor 2 occurring in the denominator since two faradays are necessary for the transport of one mol of zinc sulphate. (The relation of the cell reaction to the potential is discussed in more detail in Chapters

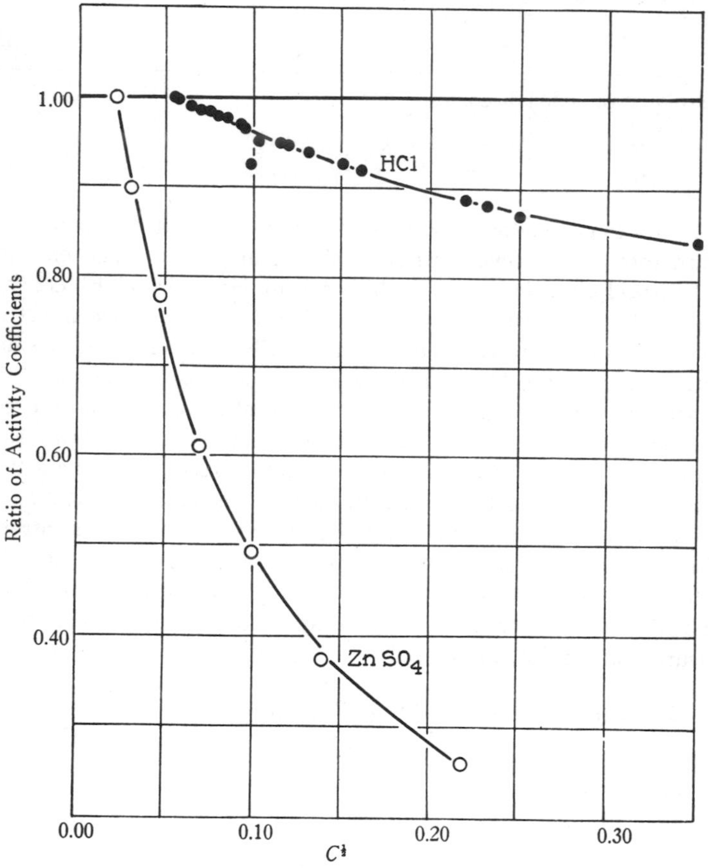

Fig. 1. Concentration Dependence of the Activity Coefficient Ratios of Hydrochloric Acid and Zinc Sulphate.

8 and 10). In the plot the activity coefficient at the lowest concentration, 0.0005 mol per liter, has been arbitrarily assigned the value unity. This plot indicates that the ions of zinc sulphate are much farther from being perfect solutes than the ions of hydrochloric acid. The computation of mean ion activity coefficient *ratios* is, however, as far as it is possible to proceed using pure thermodynamics.

Since the activities of ions and other solutes approach their concentrations as the solutions are made more and more dilute, activity coefficients must approach unity as a limit. However, to adjust the scale of activity coefficients so that they approach unity at infinite dilution, it is necessary to make non-thermodynamic assumptions as to the trend below the concentrations at which they can be measured. A basis for such assumptions is given by the Debye-Hückel theory of interionic attractions which will be discussed in the next chapter.

Summary and Generalization. Since for the sake of clearness most of the formulas derived in this chapter have dealt with special cases, some generalization and recapitulation seems desirable.

For the relation between the chemical potential of a binary electrolyte, μ_s, and the corresponding potentials of the ions into which it dissociates, μ_+, μ_-, we have, from equation (9),

$$\mu_s = \nu_+\mu_+ + \nu_-\mu_- \tag{36}$$

in which ν_+ and ν_- are the numbers of positive and negative ions produced by the dissociation of the electrolyte. Thus for sodium sulphate

$$\mu_{Na_2SO_4} = 2\mu_{Na^+} + \mu_{SO_4^{--}}$$

From equation (26a)

$$\mu_s = \mu_s^\circ + RT \ln a_s = \nu_+\mu_+^\circ + \nu_-\mu_-^\circ + RT \ln a_+^{\nu_+}a_-^{\nu_-} \tag{37}$$

in which a_+ a_- are the activities of the ion constituents. If we make the formal assumption,[12] that for a strong electrolyte

$$\mu_s^\circ - \nu_+\mu_+^\circ - \nu_-\mu_-^\circ = 0$$

then

$$a_s = a_+^{\nu_+}a_-^{\nu_-} \tag{38}$$

The mean ion activity $a_\pm$ is defined by

$$a_s = a_\pm^\nu \tag{39}$$

in which $\nu = \nu_+ + \nu_-$. A mean ion activity coefficient, $f_\pm$, may be obtained as follows. In equation (38) let $a_+ = \nu_+Cf_+$ and $a_- = \nu_-Cf_-$ in which C is the salt concentration and f_+ and f_- are the activity coefficients of the positive and negative ion constituents respectively.

[12] This can be readily shown to be equivalent to assigning a value of unity to the thermodynamic ionization constant

$$\mathbf{K} = \frac{a_1^{\nu_1}, a_2^{\nu_2}, \cdots}{a_s}$$

Thus

$$a_s = (\nu_+ C f_+)^{\nu_+} (\nu_- C f_-)^{\nu_-} \tag{40}$$

$$a_s = (\nu_+^{\nu_+} \nu_-^{\nu_-})\, C^{\nu} \cdot f_+^{\nu_+} f_-^{\nu_-} \tag{41}$$

A mean ion activity coefficient may be defined by

$$f_\pm^{\nu} = f_+^{\nu_+} f_-^{\nu_-} \text{ [13]} \tag{42}$$

from which with equations (39) and (41)

$$f_\pm = \frac{a_s^{\frac{1}{\nu}}}{C(\nu_+^{\nu_+} \nu_-^{\nu_-})^{\frac{1}{\nu}}} = \frac{a_\pm}{C(\nu_+^{\nu_+} \nu_-^{\nu_-})^{\frac{1}{\nu}}} \tag{43}$$

Thus the relation of these mean ion activities and mean ion activity coefficients to the chemical potential of the electrolyte, μ_s, is from equation (37)

$$\begin{aligned} \mu_s &= \mu_s^{\circ} + RT \ln a_\pm^{\nu} \\ \mu_s &= \mu_s^{\circ} + RT \ln (\nu_+^{\nu_+} \nu_-^{\nu_-}) C^{\nu} f_\pm^{\nu} \end{aligned} \tag{44}$$

[13] A similar development leads to the equations

$$\gamma_\pm^{\nu} = \gamma_+^{\nu_+} \gamma_-^{\nu_-} \tag{42a}$$

$$\mathbf{f}_\pm^{\nu} = \mathbf{f}_+^{\nu_+} \mathbf{f}_-^{\nu_-} \tag{42b}$$

of which use will be made later.

Chapter 7

The Debye-Hückel Method for the Theoretical Calculation of Activity Coefficients

Various attempts have been made to account theoretically for the variation of activity coefficients of ion constituents from unity. Such coefficients are, as we have just seen, always less than unity for dilute or moderately concentrated salt solutions. Milner [1] arrived at a partial solution of the problem on the assumption that the deviations are due to interionic attractions and repulsions. However, he encountered mathematical difficulties. Later Debye and Hückel [2] had far greater success, using the same assumption. Their theory has revived interest in the study of electrolytes and has been the basis for all the recent researches on the subject.

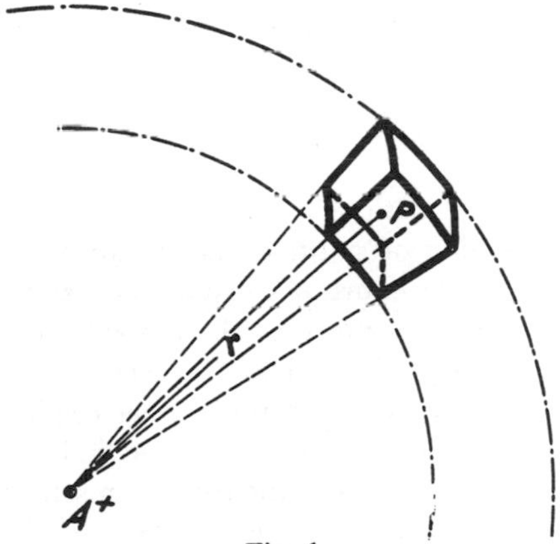

Fig. 1.

The fundamental reasoning underlying the Debye-Hückel theory is as follows. Due to the attraction between electrical charges of unlike sign, and repulsions of charges of like sign, both of which follow Coulomb's law, a given ion (a positive one for instance) will be surrounded by an "ionic atmosphere" in which there are on the average more negative than positive ions. In a small element of volume dV

[1] S. R. Milner, *Phil. Mag.*, [6], 23, 551 (1912); 25, 742 (1913).
[2] P. Debye and E. Hückel, *Physik. Z.*, 24, 185 (1923).

around the point P, Fig. 1, a distance r from the positive ion A^+, there will be, over an interval of time, more negative than positive ions, arising from the attractions of the positive charge for the negative ions and the repulsions of positive ions for positive ions. This tendency of the ions to arrange themselves in some kind of order in the solution is, however, partly but not completely overcome by the thermal vibrations of the ions. On account of these opposing effects, around a given ion there will be a field of potentials ψ whose average values over a time interval will depend on the distance r from the ion. In this chapter the theory of the effect of this ion atmosphere on the activities of ions will be considered, first for the case of electrolytes yielding singly charged ions, and then for the more complex types of salts. In the following chapter the comparison of the theory will be made with the available experimental results.

The Theory for Solutions of Uni-univalent Electrolytes. According to the Boltzmann principle the ionic distribution is a function of the ratio of the electrical energy to the thermal energy, such that in a volume, dV, around a selected ion there will be

$$dn^+ = n^+ e^{-\frac{\epsilon\psi}{kT}} dV \tag{1a}$$

positive and

$$dn^- = n^- e^{+\frac{\epsilon\psi}{kT}} dV \tag{1b}$$

negative ions. In these equations $n^+ = N^+/V$ and $n^- = N^-/V$, N^+ and N^- are the total number of positive and negative ions, and V is the total volume, k is the Boltzmann constant (equal to $R/\mathbf{N}$ where R is the molar gas constant) and T is the absolute temperature.[3] It is evident that dn^+/dV and dn^-/dV become respectively n^+ and n^- if the temperature becomes infinite, since under that condition the exponent becomes zero. This takes account, mathematically, of the fact that the tendency of the ions to arrange themselves in solution is increasingly opposed by thermal vibration as the temperature is raised. If the number of positive and negative ions is the same, *i.e.*, $n^+ = n^- = n$, the net charge per unit volume at a point where the potential is ψ will be

$$\rho = \epsilon(dn^+ - dn^-)/dV = n\epsilon\left(e^{-\frac{\epsilon\psi}{kT}} - e^{+\frac{\epsilon\psi}{kT}}\right) \tag{3}$$

which, after expanding the two exponential expressions into series, becomes

$$\rho = -2n\epsilon\left[\frac{\epsilon\psi}{kT} + \frac{1}{\underline{|3}}\left(\frac{\epsilon\psi}{kT}\right)^3 + \frac{1}{\underline{|5}}\left(\frac{\epsilon\psi}{kT}\right)^5 + \cdots\cdots\right] \tag{4}$$

However, as a close approximation the higher terms may be neglected, which gives the relation [4]

$$\rho = -\frac{2n\epsilon^2\psi}{kT} \tag{5}$$

Another relation between ρ and ψ is given by the Poisson equation [5] which for the special case of spherical symmetry of the values of ψ is

$$\frac{1}{r^2}\frac{d}{dr}\left(r^2\frac{d\psi}{dr}\right) = -\frac{4\pi}{D}\rho \tag{6}$$

in which r is the distance from the center of a chosen ion and D is the dielectric constant of the medium, in this case of the solvent. Now eliminating ρ between (5) and (6)

$$\frac{1}{r^2}\frac{d}{dr}\left(r^2\frac{d\psi}{dr}\right) = \frac{8\pi n\epsilon^2}{DkT}\psi \tag{7}$$

This expression is simplified by Debye and Hückel by setting

$$\kappa = \sqrt{\frac{8\pi n\epsilon^2}{DkT}} \tag{8}$$

[3] A more general form of the Boltzmann equation is

$$dn = A\,e^{-\frac{W}{kT}}\,dV \tag{2}$$

in which n is the number of molecules per unit of volume, A is a constant and W is the work necessary to introduce a molecule into the element of volume dV. An elementary derivation of this equation does not appear to be available. However, its use may be made clearer by consideration of a special case, the change of pressure of a gas with the height, h. Consider a vertical tube of unit cross-sectional area filled with the gas. The volume, dV, may then be set equal to dh. The work necessary to lift a molecule from the zero of height to a height h will be equal to mgh, in which m is the mass of the molecule and g is the acceleration due to gravity. Thus equation (2) becomes, for this case,

$$dn = A\,e^{-\frac{mgh}{kT}}\,dh$$

Integrating we obtain

$$n = -A\frac{kT}{mg}e^{-\frac{mgh}{kT}}$$

If n_0 is the value of n when $h = 0$ then $n_0 = -AkT/mg$ and

$$n = n_0\,e^{-\frac{mgh}{kT}}$$

Since for perfect gases $p = nRT$ this equation may be put in the more familiar form

$$\ln\frac{p}{p_0} = -\frac{mgh}{kT} = -\frac{Mgh}{RT}$$

in which p_0 is the pressure at $h = 0$ and M is the molecular weight of the gas. In the case just considered the product $W = mgh$ corresponds to $\epsilon\psi$ in equations (1a) and (1b).

[4] The effect of the higher terms has been investigated by T. H. Gronwall, V. K. LaMer and K. Sandved, *Physik. Z.*, **29**, 358 (1928), and is discussed later in this book, page 148.

[5] For a derivation of this equation see R. A. Houstoun, "Introduction to Mathematical Physics," 21, Longmans, Green and Co., New York, 1912.

The term κ has the dimensions of reciprocal length, and is a very important quantity in the development and use of the theory. Equation (7) thus becomes

$$\frac{1}{r^2}\frac{d}{dr}\left(r^2\frac{d\psi}{dr}\right) = \kappa^2\psi \tag{9}$$

which can be integrated to give [6]

$$\psi = A\frac{e^{-\kappa r}}{r} + A'\frac{e^{\kappa r}}{r} \tag{10}$$

Of the two constants of integration A' must equal zero since the value of ψ approaches zero as r increases, and the term $e^{\kappa r}/r$ becomes indefinitely large under these conditions. The relation desired between ψ and r is therefore

$$\psi = A\frac{e^{-\kappa r}}{r} \tag{11}$$

The constant A may be evaluated as follows. The charge ϵ on an ion must be equal and opposite to the charge located in all the field surrounding it. This field commences at the "distance of closest approach," a_i, see Fig. 2, equal to the sum of the radii of oppositely charged ions in

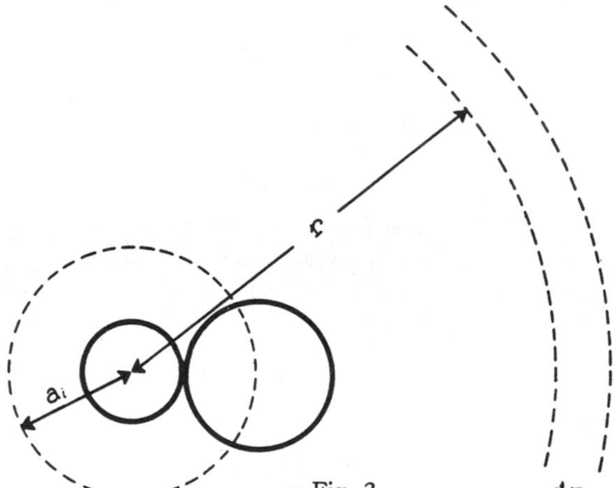

Fig. 2.

[6] Equation (9) can be put in the form

$$r\frac{d^2\psi}{dr^2} + 2\frac{d\psi}{dr} = r\kappa^2\psi$$

Now

$$\frac{d^2(r\psi)}{dr^2} = \frac{d}{dr}\left(r\frac{d\psi}{dr} + \psi\right) = r\frac{d^2\psi}{dr^2} + 2\frac{d\psi}{dr}$$

Therefore

$$\frac{d^2(r\psi)}{dr^2} = \kappa^2(r\psi)$$

It can be seen that a general solution will have the form

$$\psi r = Ae^{-\kappa r} + A'e^{\kappa r}$$

in which A and A' are constants.

contact. Since the field is spherically symmetrical the variable charge density ρ around an ion has the same value at a distance r from the ion in any direction. Thus the contribution to the total charge of a spherical shell of area $4\pi r^2$ and thickness dr will be $4\pi r^2 \rho\, dr$ and the integration from the distance a_i to infinity will be

$$-\epsilon = \int_{a_i}^{\infty} 4\pi r^2 \rho dr \tag{12}$$

Since, from equations (6), (7), and (8), $-4\pi\rho/D = \kappa^2\psi$ we have

$$\epsilon = \int_{a_i}^{\infty} D\kappa^2 \psi r^2 dr = \int_{a_i}^{\infty} D\kappa^2 r\, A\, e^{-\kappa r} dr$$

from equation (11). Integrating by parts,

$$\epsilon = \Big[-DAe^{-\kappa r}(\kappa r + 1)\Big]_{a_i}^{\infty} = DAe^{-\kappa a_i}(\kappa a_i + 1)$$

Therefore

$$A = \frac{\epsilon\, e^{\kappa a_i}}{D(1 + \kappa a_i)} \tag{13}$$

Substituting this value of A in equation (11)

$$\psi = \frac{\epsilon}{Dr} \frac{e^{\kappa(a_i - r)}}{1 + \kappa a_i} \tag{14}$$

At the distance of closest approach, $r = a_i$ and

$$\psi = \frac{\epsilon}{Da_i} \frac{1}{1 + \kappa a_i} = \frac{\epsilon}{Da_i} - \frac{\epsilon}{D} \frac{\kappa}{1 + \kappa a_i} \tag{15}$$

The first term on the right-hand side of equation (15) is the potential Ψ^0 at the surface of the ion due solely to the charge on the ion itself and is independent of the concentration. The second term is the portion Ψ_i of the total potential that is due to the arrangement of the surrounding ions in the vicinity of the given ion, that is to say, to the "ionic atmosphere." This portion is of particular interest since it is a function of the concentration.

Now according to equations (35) of Chapter 6 and (31) of Chapter 5, the relation between the change of Gibbs free energy ΔZ, the concentrations C_1, C_2, and the mean activity coefficients f_1, f_2 is

$$\Delta Z = 2RT \ln \frac{f_1 C_1}{f_2 C_2} \tag{16}$$

for an electrolyte of the type of potassium chloride. This may be rearranged to give

$$\Delta Z = 2RT \ln \frac{C_1}{C_2} + 2RT \ln \frac{f_1}{f_2} \tag{17}$$

The first term on the right-hand side is the change of Gibbs free energy produced by dilution of the ions as ideal solutes. The second term takes care of the effect on the Gibbs free energy of the deviations of the ions from ideal solutes. Let us take C_2 so small that the corresponding value of f_2 is unity, then

$$\Delta Z = 2RT \ln \frac{C_1}{C_2} + 2RT \ln f_1 \tag{18}$$

The distinguishing feature of the theory we are discussing is that the deviation term

$$2RT \ln f_1$$

which is negative because f_1, for dilute solutions, is less than one, is considered to be equal to the energy per mol of solute necessary to charge the ions in the interionic atmosphere prevailing in the more concentrated solution. For a single ion this energy will be

$$Z_{(\text{electrical})} = \frac{RT}{\mathbf{N}} \ln f_1 = kT \ln f_1 \tag{18a}$$

To be concrete let us consider the concentration cell

$$(\text{Pt})\ H_2;\ \underset{\text{infinitely dilute, } C_0}{H^+ + Cl^-},\ AgCl;\ Ag - Ag;\ AgCl,\ \underset{C_1}{H^+ + Cl^-};\ H_2\ (\text{Pt}) \tag{19}$$

If this cell is operated reversibly so that one mol each of hydrogen and chloride ions is gradually discharged in the dilute solution and gradually charged in the more concentrated one, the "ideal" [7] portion of the necessary energy will be

$$\Delta Z_{(\text{ideal})} = 2RT \ln \frac{C_1}{C_0} \tag{20}$$

The energy associated with the transfer of the ions will, however, be diminished in charging the ions in the more concentrated solution, in which the excess potential Ψ_i of the ions is given by the expression

$$\Psi_i = -\frac{\epsilon}{D} \frac{\kappa}{1 + \kappa a_i} \tag{21}$$

Since κ depends, as is shown in equation (8), on the square root of the

[7] For the concentration range in which the Debye-Hückel theory would be expected to hold, the upper limit being possibly 0.05 molar, this equation based on equation (24a), Chapter 6, as an ideal solution is not appreciably different from

$$\Delta Z_{(\text{ideal})} = 2RT \ln \frac{N_1}{N_0} \tag{20a}$$

based on Raoult's law, equation (24), Chapter 6. For the semi-empirical extensions of the theory, however, it is desirable to base the computations on Raoult's law, which can be readily done by introducing another term in the final equation, as is shown on page 166.

number of ions per unit volume, the value of Ψ_i at infinite dilution is zero. On the other hand, the portion of the potential due to the ion itself, $\Psi^0 = \epsilon/Da_i$, is the same in both the dilute and concentrated solution. The change in electrical energy corresponding to this potential, accompanying the discharging process, is therefore equal and opposite to that accompanying the charging process and so we need consider only the excess potential Ψ_i, due to the ion atmosphere.

According to Müller [8] and Güntelberg [9] the Gibbs free energy accompanying the charging of a single ion may be obtained, using equation (21), from the following expression

$$Z_{(\text{electrical})} = kT \ln f = \int_0^\epsilon \Psi_i d\epsilon = -\int_0^\epsilon \frac{\epsilon}{D} \cdot \frac{\kappa}{1 + \kappa a_i} d\epsilon$$

$$= -\frac{\epsilon^2 \kappa}{2D(1 + \kappa a_i)}$$

This corresponds to a process of charging an ion, which has a potential Ψ_i, due to an already existing ion atmosphere, by successive increments of $d\epsilon$. With the value of κ from equation (8) we obtain the important equation

$$-\ln f = \frac{\epsilon^3 \sqrt{2\pi n}}{(DkT)^{3/2}(1 + a_i\sqrt{8\pi n\epsilon^2/(DkT)}} \tag{22}$$

The number, n, of positive or negative ions per cubic centimeter is equal to

$$n = \frac{C}{1000} \mathbf{N} = \frac{C}{1000} \times 6.064 \times 10^{23} \tag{23}$$

in which C is the number of mols per liter, so that

$$-\log f = 1.8123 \times 10^6 \frac{\sqrt{C}}{(DT)^{3/2}} \cdot \frac{1}{1 + 50.288 \times 10^8 (DT)^{-1/2} a_i \sqrt{C}} \tag{24}$$

With any given solvent and temperature therefore the equation takes the form

$$-\log f = \frac{A\sqrt{C}}{1 + \beta a_i \sqrt{C}} \tag{25}$$

in which the parameters A and β are constants for a given solvent. This is the Debye-Hückel equation for uni-univalent electrolytes. The

[8] H. Müller, *Physik. Z.*, **28**, 324 (1927).
[9] E. Güntelberg, *Z. physik. Chem.*, **123**, 199 (1926).

values of A and β for aqueous solutions at several different temperatures are given in the following table.

TABLE I. DEBYE-HÜCKEL CONSTANTS FOR VARIOUS TEMPERATURES

Temp. °C.	A	$\beta \times 10^{-8}$
0	0.4863	0.3243
18	.4992	.3272
25	.5056	.3286
38	.5186	.3314

In computing the values of A and β in this table the measurements of the dielectric constant of water were averaged from the data obtained by Wyman [10] which can be represented by the expression

$$D = 78.54\ [1 - 0.00460\ (t - 25) + 0.0000088\ (t - 25)^2] \quad (25a)$$

and that of Drake, Pierce and Dow [11] which is given by

$$D = 78.57\ [1 - 0.00461\ (t - 25) + 0.0000155\ (t - 25)^2] \quad (25b)$$

The experimental methods of these authors are described in Chapter 22. For very dilute solutions the second term in the denominator of equation (25) becomes negligible and the Debye-Hückel equation approaches its limiting form,

$$-\log f = A\sqrt{C} \quad (26)$$

an equation to which frequent reference will be made in the following chapters.

The Theory for Multiply Charged Ions. Let us consider a solution with

$$n_1,\ n_2,\ \ldots\ldots\ n_i,\ \ldots\ldots\ldots\ n_s$$

ions of different kinds per cubic centimeter, with the charges

$$z_1\epsilon,\ z_2\epsilon,\ \ldots\ldots\ z_i\epsilon,\ \ldots\ldots\ z_s\epsilon$$

which can be either positive or negative, in which ϵ is the charge on a positron. Since such a solution must be electrically neutral

$$\sum_s n_i z_i \epsilon = 0 \quad (27)$$

The Boltzmann equation gives the density of ions of the ith kind as

$$n_i e^{-z_i \frac{\epsilon\psi}{kT}} \quad (28)$$

[10] J. Wyman, Jr., *Phys. Rev.*, [II] **35**, 623 (1930).
[11] F. H. Drake, G. W. Pierce and M. T. Dow, *ibid.*, [II] **35**, 613 (1930).

so that (corresponding to equation (3) for the case of uni-univalent electrolytes) the electrical density, ρ, is given by

$$\rho = \epsilon \sum n_i z_i \, e^{-z_i \frac{\epsilon\psi}{kT}} \tag{29}$$

Expanding the exponentials and retaining only the first two terms

$$e^{-z_i \frac{\epsilon\psi}{kT}} = 1 - z_i \frac{\epsilon\psi}{kT}$$

the electrical density becomes

$$\rho = \epsilon \sum n_i z_i \left(1 - z_i \frac{\epsilon\psi}{kT}\right) \tag{30}$$

and with equation (27)

$$\rho = -\epsilon \sum n_i z_i^2 \frac{\epsilon\psi}{kT} \tag{31}$$

As in equations (6) and (7)

$$\frac{1}{r^2}\frac{d}{dr}\left(r^2 \frac{d\psi}{dr}\right) = -\frac{4\pi}{D}\rho = \frac{4\pi\epsilon^2}{DkT} \Sigma n_i z_i^2 \psi \tag{32}$$

Now giving κ the general value

$$\kappa = \sqrt{\frac{4\pi\epsilon^2}{DkT} \Sigma n_i z_i^2} \tag{33}$$

we again have

$$\frac{1}{r^2}\frac{d}{dr}\left(r^2 \frac{d\psi}{dr}\right) = \kappa^2\psi \tag{9}$$

As we have seen this integrates to

$$\psi = A \frac{e^{-\kappa r}}{r} \tag{11}$$

The value of A can be obtained again by means of the same considerations as were used for the case of singly charged ions. Thus

$$-z_i\epsilon = \int_{a_i}^{\infty} 4\pi r^2 \rho dr = -\int_{a_i}^{\infty} D\kappa^2 r A e^{-\kappa r} dr$$

Integration yields

$$z_i\epsilon = DAe^{-\kappa a_i}(\kappa a_i + 1)$$

giving for A the expression

$$A = \frac{z_i\epsilon}{D} \frac{e^{\kappa a_i}}{1 + \kappa a_i}$$

and substituting this value of A in equation (11) we have

$$\psi = \frac{z_i\epsilon}{Dr}\frac{e^{\kappa(a_i - r)}}{1 + \kappa a_i} \tag{33a}$$

Therefore, at the distance of closest approach where $r = a_i$

$$\psi = \frac{z_i\epsilon}{Da_i} - \frac{z_i\epsilon}{D}\frac{\kappa}{1 + \kappa a_i} \tag{34}$$

Again, ψ can be divided into two portions, one due to the charge on the ion and independent of concentration, and the other arising from the ionic atmosphere and consequently dependent on concentration. Thus

$$\psi = \Psi^0 + \Psi_i \quad \text{where} \quad \Psi^0 = \frac{z_i\epsilon}{Da}$$

and

$$\Psi_i = -\frac{z_i\epsilon}{D}\frac{\kappa}{1 + \kappa a_i} \tag{35}$$

The Gibbs free energy per ion, $Z_{(\text{electrical})}$, resulting from the interionic attractions will be the electrical energy required to charge the ion in the presence of the ionic atmosphere. Utilizing the Müller-Güntelberg charging process once more we obtain [12]

$$Z_{(\text{electrical})} = -\frac{z_i^2\epsilon^2}{2D}\frac{\kappa}{1 + \kappa a_i} \tag{36}$$

and with equation (33)

$$Z_{(\text{electrical})} = -\frac{z_i^2\epsilon^2}{2D}\sqrt{\frac{4\pi\epsilon^2}{DkT}\Sigma n_i z_i^2}\;\frac{1}{1 + \kappa a_i} \tag{37}$$

and with equation (18a)

$$\ln f_i = -\frac{z_i^2\epsilon^3}{(DkT)^{3/2}}\sqrt{\pi\Sigma n_i z_i^2}\;\frac{1}{1 + \kappa a_i} \tag{38}$$

The stoichiometric concentration, C_i, in mols per liter, corresponding to n_i of the ions per cm.^{3} is given by the formula

$$\frac{C_i}{1000} = \frac{n_i}{\mathbf{N}}$$

and the summation $\sum n_i z_i^2$ becomes

$$\sum n_i z_i^2 = \frac{\mathbf{N}}{1000}\sum C_i z_i^2 \tag{39}$$

[12] Another and possibly more convincing charging process has been used by P. Debye, *Physik. Z.*, **25**, 97 (1924), and consists in charging all the ions simultaneously by a given fraction of their total charge.

Lewis and Randall[13] have however defined the "ionic strength," ω, by means of the relation

$$\omega = \tfrac{1}{2}\sum C_i z_i^2 \tag{40}$$

Thus for instance a solution of magnesium chloride containing 0.05 mol per liter has an ionic strength of

$$\omega = \frac{0.05 \times 4 + 0.05 \times 2}{2} = 0.15$$

With equations (38), (39) and (40) in (38) we obtain

$$-\ln f_i = \frac{\epsilon^3 z_i^2}{(DkT)^{3/2}} \sqrt{\frac{2\pi \mathbf{N} \omega}{1000}} \frac{1}{1 + a_i \sqrt{\dfrac{8\pi \mathbf{N} \epsilon^2 \omega}{1000 DkT}}} \tag{41}$$

Thus the Debye-Hückel equation for multivalent ions becomes

$$\log f_i = \frac{z_i^2 A \sqrt{\omega}}{1 + \beta a_i \sqrt{\omega}} \tag{42}$$

This f_i is the ion activity coefficient of the ion constituent i. To obtain a mean ion activity coefficient, $f_\pm$, for a *binary* electrolyte we may use the relation, (34a), Chapter 6,

$$f_\pm^\nu = f_+^{\nu_+} f_-^{\nu_-} \tag{43}$$

in which f_+ and f_- are the activity coefficients of the positive and negative ions respectively and ν, ν_+, and ν_-, are the total number and the number of positive and negative ions given by the dissociation of one molecule of the electrolyte. Equations (42) and (43) give

$$-\nu \log f_\pm = (\nu_+ z_+^2 + \nu_- z_-^2) A \sqrt{\omega} \frac{1}{1 + \beta a_i \sqrt{\omega}} \tag{44}$$

Since electrical neutrality requires

$$\nu_+ z_+ = \nu_- z_-$$

equation (44) may be rearranged to give

$$-\log f_\pm = z_+ z_- A \sqrt{\omega} \frac{1}{1 + \beta a_i \sqrt{\omega}} \tag{45}$$

which is the Debye-Hückel relation for computing the mean ion activity coefficient of a binary electrolyte having ions with valences z_+ and z

[13] G. N. Lewis and M. Randall, *J. Am. Chem. Soc.*, **43**, 1112 (1921).

General Remarks Concerning the Debye-Hückel Theory. Experimental studies concerning the validity of the Debye-Hückel theory have been made using freezing-point data, measurements of the solubilities of electrolytes, and the results of determinations of the potentials of concentration cells. The discussion in this book will be limited to the last of these methods since the interpretation of the first two types of data will be considered, somewhat arbitrarily, to be outside the field of electrochemistry. Although the interionic attractions postulated in the theory would be expected to exist at all concentrations in soluions of ions, and to be even more effective in determining the properties of such solutions at high concentrations than in dilute solutions, the validity of the theory as developed in the preceding pages is limited to dilute solutions. At higher concentrations various complicating effects must be considered. For instance there is probably a change of the dielectric constant of the solvent, due to the presence of the charged ions. Also the mathematical approximations that were made in the derivation as given must be considered in connection with the range of applicability of the theory to actual solutions.

There is no detail of the derivation of the equations of the Debye-Hückel theory that has not been criticized. Its incompleteness mathematically is evident, since only the first term of the expansion of equations (3) and (29) is used. The extensions of the theory to overcome this deficiency are, however, briefly considered below. A possibly more serious deficiency of the theory as given is that it does not take account of "fluctuation terms." This amounts to the statement that the Boltzmann equation does not yield a correct average potential ψ, this being subject to wide variations for which allowance should be made in the theory. This part of the criticism is still in active progress, and cannot be briefly summarized. In Chapters 8 and 12 an attempt will be made to show both the successes and the limitations of the theory in its present state from the experimental point of view.

The "Extended" Debye-Hückel Theory. The Debye-Hückel theory is successful in accounting for the experimental results when its application is limited to solutions in which the ratio of the electrical to the thermal energy of the ions is very small, *i.e.*,

$$\frac{\epsilon\psi}{kT} \ll 1$$

If this is not true, which will be the case if the ions are highly charged, if the dielectric constant of the medium is low or if the ions are small, the higher terms in the expansions of equations (3) and (29) cannot be neglected. Modifications of the theory to include these higher terms have been made by Müller,[14] by Gronwall, LaMer and Sandved,[15] and by Bjerrum.[16] In the first two papers mentioned no additional physical concepts are introduced, the attempt being made only to complete the theory mathematically. Müller dispenses entirely with the series expansion of equation (29) and obtains the values of the integral by graphical means. Gronwall, LaMer and Sandved, on the other hand, expand equation (29) and retain further terms in the series. The manipulation of the resulting equation involves a complex mathematical development for the details of which the original paper must be consulted. The final equation, for a symmetrical valence type salt, written in the abbreviated form usually employed, is

$$\ln f = -\frac{\epsilon^2 z^2}{DkTa_i}\frac{1}{2}\frac{x}{1+x} + \sum_{m=1}^{\infty}\left(\frac{\epsilon^2 z^2}{DkTa_i}\right)^{2m+1}\left[\frac{1}{2}X_{2m+1}(x) - 2mY_{2m+1}(x)\right] \qquad (46)$$

Where $x = \kappa a_i$, m is a running number and X and Y are complicated functions of x. The mathematical difficulties are still further increased if the extension deals with electrolytes which are "unsymmetrical," *i.e.*, have ions of different charges. This case has been investigated by LaMer, Gronwall and Grieff.[17] On substituting numerical values for water at 25° equation (46) becomes

$$-\log f = 1.53917\left(\frac{z^2}{10^8 a_i}\right)\frac{x}{1+x} - 0.15466\left(\frac{z^2}{10^8 a_i}\right)^3 \cdot 10^3 \left[\frac{1}{2}X_3(x) - 2Y_3(x)\right] - 0.077706\left(\frac{z^2}{10^8 a_i}\right)^5 \cdot 10^5 \left[\frac{1}{2}X_5(x) - 4Y_5(x)\right] - \cdots\cdots \qquad (47)$$

Here a_i is in centimeters and thus $x = 10^{-8}\kappa \cdot 10^8 a_i$, and $10^{-8}\kappa = 0.3286\, z\sqrt{C}$. This equation is useful only if the quantities in brackets are evaluated. This has been done by Gronwall, LaMer and Sandved for $m = 1$ and $m = 2$. The values of

$$10^3 \cdot \left[\frac{1}{2}X_3(x) - 2Y_3(x)\right] \quad \text{and} \quad 10^5 \cdot \left[\frac{1}{2}X_5(x) - 4Y_5(x)\right]$$

[14] H. Müller, *Physik. Z.*, **28**, 324 (1927); **29**, 78 (1928).
[15] T. H. Gronwall, V. K. LaMer and K. Sandved, *Physik. Z.*, **29**, 358 (1928).
[16] N. Bjerrum, *Det. Kgl. Danske vidensk. Selsk. Math. fysik. Medd.*, **VII**, No. 9, (1926).
[17] V. K. LaMer, H. T. Gronwall and L. J. Grieff, *J. Phys. Chem.*, **35**, 2245 (1931).

TABLE II. CONSTANTS FOR THE THIRD AND FIFTH APPROXIMATIONS OF THE GRONWALL, LAMER AND SANDVED EXTENSION OF THE DEBYE-HÜCKEL THEORY

x	$10^3[\frac{1}{2}X_3 - 2Y_3]$	$10^5[\frac{1}{2}X_5 - 4Y_5]$
0.05	− 0.05711	− 0.06564
0.06	− 0.07522	− 0.08394
0.07	− 0.09403	− 0.10138
0.08	− 0.11316	− 0.11737
0.09	− 0.13231	− 0.13153
0.10	− 0.15130	− 0.14363
0.11	− 0.16992	− 0.15356
0.12	− 0.18802	− 0.16126
0.13	− 0.20555	− 0.16680
0.14	− 0.22240	− 0.17023
0.15	− 0.23853	− 0.17166
0.16	− 0.25391	− 0.17123
0.17	− 0.26851	− 0.16910
0.18	− 0.28231	− 0.16543
0.19	− 0.29530	− 0.16037
0.20	− 0.30750	− 0.15409
0.21	− 0.31892	− 0.14674
0.22	− 0.32955	− 0.13847
0.23	− 0.33943	− 0.12942
0.24	− 0.34859	− 0.11973

as calculated by these authors are given in Table II for values of x from 0.05 to 0.24. The original tables cover a much greater range. It may be useful, at least as a generalization, to bear in mind Müller's conclusion that the minimum a_i value at which equation (45) can be expected to be applicable is

$$a_i = \frac{z^2_{max}\epsilon^2}{4DkT}$$

where z_{max} is the numerical value of the valence of the most highly charged ions present. For a uni-univalent salt in water at 25° this would make the lowest value of a_i at which equation (25) would be expected to be valid about 2Å. The usefulness of the "extended" theory for accounting for the experimental results on solutions of highly charged, and relatively small ions is discussed in Chapter 10. Much work on the higher valence types of electrolytes remains to be done.

It will be observed that in the preceding treatment the assumption has been made that the ion concentration is in each case the same as the total concentration of the electrolyte. That is to say, a "degree of dissociation" has not been introduced. For this reason the Debye-Hückel theory has been frequently referred to as a "theory of complete dissociation." Although the evidence, Chapters 8 and 12, appears to support the assumption that at least in dilute aqueous solutions strong electrolytes are substantially completely dissociated, it will be shown later that there are many solutions of electrolyte for which this is not the case. However, it will be found that the Debye-Hückel theory is equally useful in dealing with the *ionic portion* of these incompletely dissociated substances.

[*Comment to 1961 Edition*] In recent years, a vast amount of scientific research has been stimulated by the Debye-Hückel theory. Useful summaries of this work are given in:

Harned, Herbert S. and Owen, B. B. *Physical Chemistry of Electrolytic Solutions.* Reinhold Publishing Corp., 3rd ed., 1958.

Robinson, R. A. and Stokes, R. H. *Electrolyte Solutions.* Academic Press, 2nd ed., 1959.

Hamer, Walter J. (ed.). *Structure of Electrolytic Solutions.* John Wiley & Sons, 1959.

Chapter 8

Concentration Cells and the Validity of the Debye-Hückel Theory

This chapter is concerned with the determination of activity coefficients with the aid of various types of concentration cells, and with the comparison of such activity coefficients with the predictions of the Debye-Hückel theory, developed in the previous chapter. The types of cells discussed are (a) cells without transference, including those containing amalgam electrodes, (b) cells with transference, and (c) cells without transference containing mixtures of electrolytes.

Amalgam Cells. For the determination of activity coefficients of electrolytes the method depending upon the measurement of the potentials of concentration cells "without transference" has already been discussed in Chapter 6. Two examples of such cells were mentioned. These were the following:

$$\text{Ag; AgCl, HCl } (C'); \text{ H}_2\text{(Pt)} - \text{(Pt) H}_2\text{; HCl } (C''), \text{ AgCl; Ag} \quad (1)$$

and

$$\text{Zn(Hg); ZnSO}_4\ (C'), \text{ PbSO}_4\text{; Pb(Hg)} - \text{Pb(Hg); PbSO}_4\ \text{ZnSO}_4\ (C''); \text{ Zn(Hg)} \quad (2)$$

It is distinctive of both of these examples and of this type of cell in general that two types of electrodes are involved, each of which must be reversible to an ion constituent in the solutions contained in the cells. Since the mechanism of operation of such cells is relatively simple they should be used whenever possible. However, experimental difficulties which may make the construction of such cells impossible or limit their range of usefulness frequently arise when measurements on such cells are attempted. For instance a cell which may be used to obtain data on the activity coefficients of the ions of potassium chloride is as follows:

$$\text{Ag; AgCl, KCl } (C'); \text{ K(Hg)}_x - \text{K(Hg)}_x\text{; KCl } (C'') \text{ AgCl; Ag} \quad (3)$$

Because of the relatively rapid reaction of the potassium amalgam electrodes, (represented by $K(Hg)_x$), with the potassium chloride solutions, and the disturbing effects of traces of oxygen, it requires elaborate experimental technique to produce cells having the reversible electromotive forces and this is possible only at relatively high concentrations. Similar cells can obviously be set up in which the potassium is replaced by any alkali metal, and the chloride ion by any halogen ion. The first determinations of the potentials of cells of this type were carried out by MacInnes and Parker.[1] Concentration cells involving amal-

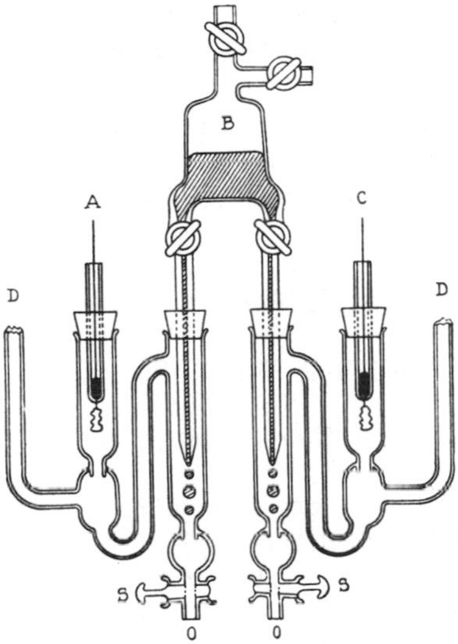

Fig. 1. Cell for use with Liquid Amalgam Electrodes.

gam electrodes have also been studied by other workers,[2, 3, 4] and particularly by Harned and associates.[5, 6, 7] The apparatus used in this type of work is shown in Fig. 1. The two silver-silver chloride electrodes are at *A* and *C*. A reservoir of dilute amalgam is represented at *B*. Solutions of the salt under investigation, at the desired concentrations, are run into the cell through the tubes *D,D*. These solutions are boiled under vacuum to remove the last traces of oxygen. In operation amalgam is allowed to flow through the capillary tubes leading from the reservoir *B* into the solutions in the two half cells and out through

[1] D A. MacInnes and K. Parker, *J. Am. Chem. Soc.*, **37**, 1445 (1915).
[2] D. A. MacInnes and J. A. Beattie, *ibid.*, **42**, 1117 (1920).
[3] A. J. Allmand and W. G. Polack, *J. Chem. Soc.*, **115**, 1020 (1919).
[4] J. N. Pearce and H. B. Hart, *J. Am. Chem. Soc.*, **43**, 2483 (1921).
[5] H. S. Harned, *ibid.*, **47**, 930 (1925); **47**, 676 (1925); **51**, 416 (1929).
[6] H. S. Harned and F. E. Swindells, *ibid.*, **48**, 126 (1926).
[7] H. S. Harned and L. F. Nims, *ibid.*, **54**, 423 (1932).

the reject tubes *O,O*. A current of solution may also flow at the same time from solution reservoirs attached to tubes *D,D*. By constantly replacing the amalgam and solution the effect of attack of one of these on the other may be minimized, and is, possibly, negligible at concentrations of alkali halides 0.1 normal and above. Since the results of the method can be trusted only at these relatively high concentrations they cannot be readily interpreted by the Debye-Hückel theory as outlined in Chapter 7. However, such results are valuable in connection with the results obtained by another method about to be described, for the purpose of computing activity coefficients at the higher concentrations.

With solutions of sodium hydroxide, however, the attack of the amalgam on the electrolyte appears to be less rapid and the experimental data reliable to a much lower concentration. Harned [8] has studied concentration cells of the form

$$\text{Na(Hg); NaOH } (C'); \text{ H}_2 \text{ (Pt)} - \text{(Pt) H}_2; \text{ NaOH } (C'') \text{ Na(Hg)} \quad (4)$$

The reaction occurring in one-half of this concentration cell is

$$\text{Na} + \text{H}_2\text{O} = \text{NaOH} + \tfrac{1}{2}\text{H}_2$$

and the reverse reaction occurs in the other half. Therefore the result of the operation of the complete cell will be, per faraday of current passing, the transfer of an ion equivalent each of sodium and hydroxyl from the more concentrated to the more dilute solution and the transfer of a mol of water in the reverse direction. Therefore for a reversible operation of the cell

$$\Delta Z = -En\mathbf{F} = (\mu'_{\text{Na}^+} - \mu''_{\text{Na}^+}) + (\mu'_{\text{OH}^-} - \mu''_{\text{OH}^-}) - (\mu'_{\text{H}_2\text{O}} - \mu''_{\text{H}_2\text{O}}) \quad (5)$$

in which, for example, μ'_{Na^+} is the chemical potential of the sodium ion constituent in one of the solutions and $\mu'_{\text{H}_2\text{O}}$ is the corresponding potential of the water. The hydrogen pressure is kept one atmosphere so that its chemical potential, being constant, does not enter into the equation. This concentration cell differs from that for hydrochloric acid, already considered, page 129, in that we must deal with the term $\mu'_{\text{H}_2\text{O}} - \mu''_{\text{H}_2\text{O}}$. However this is of relatively small magnitude. Since the solvent water is in equilibrium with its vapor, which to a sufficient approximation, is a perfect gas, the chemical potential of the water is given by equation (17), Chapter 6

$$\mu_{\text{H}_2\text{O}} = \mu^{\circ}_{\text{H}_2\text{O}} + RT \ln p_{\text{H}_2\text{O}}$$

[8] H. S. Harned, *J. Am. Chem. Soc.*, **47**, 676 (1925).

in which p_{H_2O} is the vapor pressure of water from the solution. Using in addition equations (26b) and (34a) of Chapter 6 we have

$$\Delta E = \frac{2RT}{\mathbf{F}} \ln \frac{C'_{NaOH} f'}{C''_{NaOH} f''} - \frac{RT}{\mathbf{F}} \ln \frac{p'_{H_2O}}{p''_{H_2O}} \tag{6}$$

in which f' and f'' are mean ion activity coefficients. Harned's data are given in Table I, which is mostly self-explanatory. The vapor pressures of the solution of concentration, C', are given in the third column and have been interpolated from data in the International Critical

TABLE I. THE COMPUTATION OF THE ACTIVITY COEFFICIENTS OF SODIUM HYDROXIDE FROM CELLS WITHOUT TRANSFERENCE AT 25°

Concentration mols per liter C'	Potential of cell, volts E	Vapor pressure of solution mm of Hg p'	Log of activity coefficient ratio $\Delta \log f$	activity coefficient f
0.00997	0.0000	23.75	0.0000	0.899
.0201	.0338	23.74	.0186	.858
.0525	.0795	23.71	.0490	.803
.1047	.1116	23.68	.0786	.750
.1097	.1142	23.67	.0769	.752

Tables.[9] Using these data the values of $\Delta \log f = \log (f'/f'')$ were obtained with equation (6) and are given in the fourth column of the table. To obtain, from this function of the ratios f'/f'', activity coefficients which approach unity as the solution is progressively diluted, it is necessary to supplement the thermodynamics which so far has been sufficient. To do this we may use the Debye-Hückel theory in a manner suggested by Hitchcock.[10] With sufficient accuracy for the present purpose equation (25), Chapter 7, may be put into the approximate form

$$-\log f = A\sqrt{C} - B'C \tag{7}$$

With the relation

$$\log f = G - \Delta \log f \tag{8}$$

in which G is a constant, equation (7) becomes

$$\Delta \log f - A\sqrt{C} = G - B'C \tag{9}$$

Thus if equation (7) is valid a plot of values of $(\Delta \log f - A\sqrt{C})$ as a function of C should be a straight line with a slope equal to B' and an intercept equal to G. Values of $\log f$ may then be found with the aid of equation (8) and the value of G (which is the logarithm of

[9] *International Critical Tables,* McGraw-Hill Book Co., Vol. **III**, 370, New York (1928).
[10] D. I. Hitchcock, *J. Am. Chem. Soc.,* **50,** 2076 (1928).

the activity coefficient of the ion constituents of the reference solution). A plot of the kind just described is given in Fig. 2. A more accurate method for extrapolation, which may be used when the precision of the data justifies it, is described later in this chapter. The mean activity coefficients, f, in the last column of the table have been obtained in the manner just described.

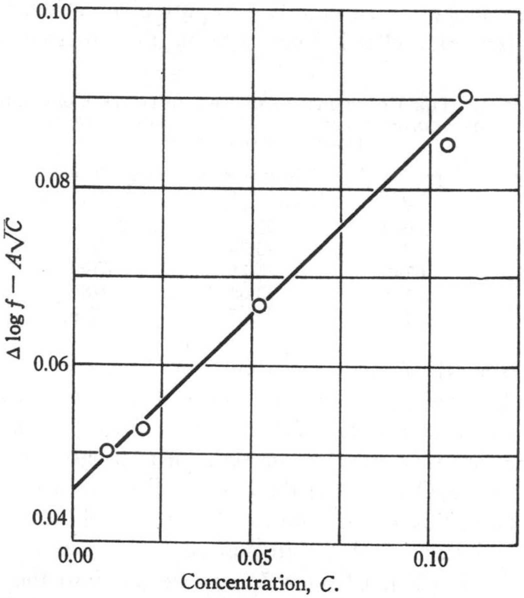

Fig. 2. Hitchcock Plot for the Extrapolation of Activity Coefficient Ratios for Sodium Hydroxide.

Concentration Cells with Transference. While, as has already been stated, it is, in many cases, difficult or impossible to find electrodes that are reversible to both ion constituents of a binary electrolyte, it is much more frequently possible to find electrodes which are reversible to *one* of the constituents. The silver-silver chloride electrode, for instance, will serve in most chloride solutions. Another example is the hydrogen electrode, which will work in most solutions of acid or alkali. Using two electrodes reversible to the same ion constituent it is possible to set up and measure potentials of *concentration cells with transference.* Typical cells of this kind are the following:

$$\text{Ag; AgCl, NaCl } (C_1) : \text{NaCl } (C_2), \text{ AgCl; Ag} \tag{10}$$

and

$$\text{Ag; AgNO}_3 \ (C_1) : \text{AgNO}_3 \ (C_2) \text{; Ag} \tag{11}$$

These cells have an electrolyte-electrolyte boundary, more commonly referred to as a "liquid junction" between two solutions of the same salt, which differ only in their concentrations.[11]

The theory of the operation of such cells is as follows. If the cell

$$\text{Ag; AgCl, NaCl }(\mu) : \text{NaCl }(\mu + d\mu), \text{ AgCl; Ag} \qquad (12)$$

is set up its potential dE is

$$-\mathbf{F}dE = t_{\text{Na}}d\mu \qquad (13)$$

Here NaCl (μ) represents a solution of sodium chloride, the chemical potential of the solute of which is μ, and t_{Na} is the number of equivalents per faraday transferred from the higher to the lower concentration during the reversible operation of the cell. It will be noted that the operation of cell (12) is, differentially and reversibly, the opposite of the Hittorf method for determining the Hittorf transference number t_{Na} as described in Chapter 4. If now a second cell is made as follows

$$\text{Ag; AgCl, NaCl }(\mu + d\mu) : \text{NaCl }(\mu + d\mu + d\mu'), \text{ AgCl; Ag} \qquad (14)$$

its potential dE' can be obtained from

$$-\mathbf{F}dE' = t'_{\text{Na}}d\mu' \qquad (15)$$

in which t'_{Na} may have a slightly different value from t_{Na}. When cells (12) and (14) are put in series, the potentials of the electrodes in contact with solutions of like chemical potential will cancel, and the resulting cell will have the same emf as one having the composition

$$\text{Ag; AgCl, NaCl }(\mu), \text{ NaCl }(\mu + d\mu + d\mu'), \text{ AgCl; Ag} \qquad (16)$$

Such a series of cells could, of course, be continued indefinitely, from which it follows that

$$-E = -\int dE = \frac{1}{\mathbf{F}}\int t_{\text{Na}}d\mu \qquad (17)$$

Since $\mu_{\text{NaCl}} = \mu_{\text{Na}} + \mu_{\text{Cl}}$, this may, with the aid of equations (9), (26b) and (34a) of Chapter 6, be put in the form

$$-E = \frac{2RT}{\mathbf{F}}\int_{C_1}^{C_2} t_{\text{Na}}\, d\ln Cf_{\pm} \qquad (18)$$

[11] There is some confusion in the literature concerning the expression "cells with transference" and "cells with liquid junction." In this book the first of these will be reserved for cells of the type represented by (10) and (11) which contain only one solute, and the second expression will include these and also cells containing more complicated types of liquid junctions.

In general for univalent electrolytes

$$-E = \frac{2RT}{\mathbf{F}} \int_{C_1}^{C_2} t \, d \ln Cf_{\pm} \tag{19}$$

in which t is the transference number of the ion constituent to which the electrodes are *not* reversible.

If the transference number, t, is constant in the concentration range C_1 to C_2 this equation takes the form

$$-E = \frac{2tRT}{\mathbf{F}} \ln \frac{C_1 f_1}{C_2 f_2} \tag{20}$$

in which f_1 and f_2 are the mean ion activities at the concentrations C_1 and C_2. In nearly all cases, as can be seen by a reference to Table IV of Chapter 4, the transference number is not constant, even over small concentration ranges, so that equation (19) must be used to obtain activity coefficients from the potentials of cells with transference. However, before going into the details of that computation, the experimental technique involved in obtaining potentials of cells of type (10) and (11) will be discussed.

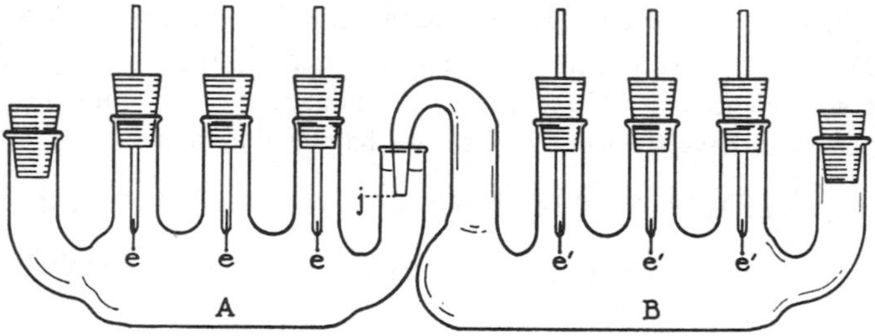

Fig. 3. A Cell for Measuring the Potentials of Concentration Cells with Transference.

A simple type of cell with which measurements of the potentials of concentration cells with transference may be carried out is shown in Fig. 3. The vessel A holds the three electrodes, e, e, e and contains the more concentrated solution. The vessel B is filled with the dilute solution into which dip the electrodes e', e', e'. A liquid junction between the two solutions is formed at the point j. Due to the slowness of the diffusion process the solution in the region of the electrodes is not disturbed until quite a long period has elapsed and the potential of such a cell remains constant. The potentials of cells containing most types of liquid junctions vary with time. The concentration cell with transference, of which cells (10) and (11) are examples, is a special case for reasons to be discussed in Chapter 13.

A more recent form of the apparatus used in making measurements on cells with transference is shown in Fig. 4.[12] This apparatus has the decided advantage, over the one described in the previous paragraph, that it can be filled with solutions that have not come into contact with

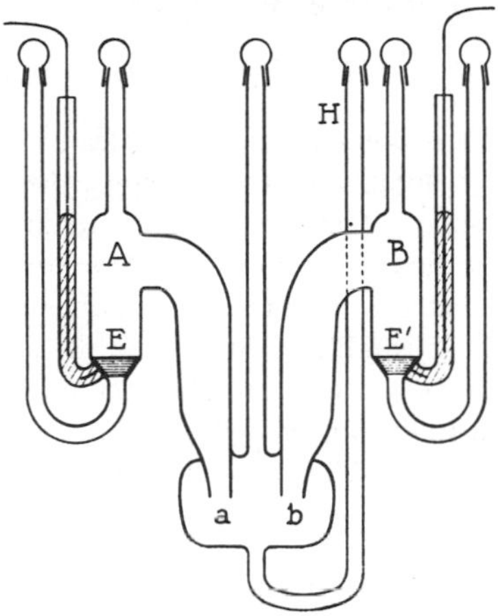

Fig. 4. A Cell for Measuring the Potentials of Concentration Cells with Transference.

laboratory air. So far the measurements have been carried out only with chlorides, using the silver-silver chloride electrodes. The electrodes E and E' are hollow truncated cones of platinum foil and are sealed into the glass walls of the apparatus. They are plated with silver and coated with chloride electrolytically. The half cells A and B are filled, in an inverted position, with the two chloride solutions under investigation. The apparatus is then turned to its normal position and the heavier of the two solutions is run through the tube H forming a liquid junction at the point a or b. The potential of such a cell is found to be constant, within a few thousandths of a millivolt, for several hours. As will be seen in Chapter 13 such constancy is obtained, without some special device such as a flowing junction, only if the same electrolyte is present on both sides of the junction. Corrections for

[12] Other types of cell used in measuring the potentials of cells with transference have been described by D. A. MacInnes and K. Parker, *J. Am. Chem. Soc.*, **37**, 1445 (1915) and D. A. MacInnes and J. A. Beattie, *ibid.*, **42**, 1117 (1920) and by A. S. Brown and D. A. MacInnes, *ibid.*, **57**, 1356 (1935).

the difference, which is usually very slight, between the potentials of the electrodes may be made by placing the same chloride solution throughout the cell, and also by making duplicate determinations with each pair of solutions, but with the positions of the solutions reversed in the apparatus.

The computation involved in obtaining activity coefficients from data on the potentials of concentration cells with transference is as follows. According to equation (19) the potential of a cell with transference of the type, for instance, of (10) or (11) is the integral of

$$dE = -(2RT/\mathbf{F})\, t\,(d \ln C + d \ln f) \tag{21}$$

in which C is the salt concentration, f the mean ion activity coefficient, and t the transference number of the ion constituent to which the electrodes are not reversible. The use of this equation for the determination of the activity coefficients from the electromotive force data is complicated by the fact that the transference number t usually varies with the concentration C. It would, of course, be possible to express both t and E as analytic functions of C, but the resulting equations are likely to be unwieldy. The following procedure which avoids this difficulty is due to L. G. Longsworth. The transference number at any concentration may be expressed by

$$t = t_1 + \Delta t \tag{22}$$

t_1 being the transference number at some reference concentration. Values of the transference number at concentrations so low that direct determinations are difficult may be obtained by interpolation between measured values and the limiting value of the transference number, t_0, given by Kohlrausch's law, as discussed in Chapter 18. If the temperature is 25° equation (21) becomes

$$dE = -2 \times 0.05915\,(t_1 + \Delta t)\,(d \log C + d \log f) \tag{23}$$

Expanding and rearranging we obtain

$$\frac{-dE}{0.1183 t_1} = d \log C + \frac{\Delta t}{t_1} d \log C + d \log f + \frac{\Delta t}{t_1} d \log f \tag{24}$$

Integrating and again rearranging yields

$$-\Delta \log f = \log f_2 - \log f_1 = \frac{-E}{0.1183\, t_1} - (\log C_2 - \log C_1) - \frac{1}{t_1}\int_1^2 \Delta t\, d \log C + \frac{1}{t_1}\int_1^2 \Delta t\, d\,\Delta \log f \tag{25}$$

Of the four terms on the right-hand side of this equation the first two are computed directly from the data. The third term is obtained by graphical integration, using a plot of Δt values against values of $\log C$. The fourth term, of relatively small magnitude, is obtained by graphical integration using preliminary values of $\Delta \log f$, obtained by adding the first three terms of the equation, and plotting against Δt. This process may be repeated with more accurate values of $\Delta \log f$ but a further approximation is usually not found necessary.

The computation of activity coefficients from $\Delta \log f$ values will be illustrated for hydrochloric acid since in that case direct comparison can be made with the results of measurements on concentration cells without transference of the type described in Chapter 6. The relevant data are given in Table II and are from the work of Shedlovsky and MacInnes.[13] The emf data in the second column were obtained from a cell of the type illustrated in Fig. 4. The transference numbers in the third column were interpolated from the measurements of Longsworth given in Table IV of Chapter 4. The $\Delta \log f$ values in the fourth column were computed as described in the last paragraph.

To provide a basis for the activity coefficients, f, such that they will approach unity as the concentration is progressively decreased, use can be made of the more accurate equation, (25), Chapter 7, of the Debye-Hückel theory

$$- \log f = \frac{A\sqrt{C}}{(1 + \beta a_i \sqrt{C})} \tag{26}$$

since the extrapolation, suggested by Hitchcock, and used on the data for sodium hydroxide is not, for these data, and for much of the recent work, sufficiently precise. To use equation (26) with the data in Table II the following procedure used by Brown and MacInnes [14] was found to be convenient. We may set, once more,

$$\log f = G - \Delta \log f \tag{27}$$

in which G is a constant. Combining equations (26) and (27) and rearranging terms

$$\Delta \log f - A\sqrt{C} = G + \beta a_i (G - \Delta \log f)\sqrt{C} \tag{28}$$

Thus, through the range of validity of equation (26) a plot of $\Delta \log f - A\sqrt{C}$ against $(G - \Delta \log f)\sqrt{C}$ should be a straight line with intercept G and slope βa_i. Such a plot is shown in Fig. 5. The constant G

[13] T. Shedlovsky and D. A. MacInnes, *J. Am. Chem. Soc.*, **58**, 1970 (1936).
[14] A. S. Brown and D. A. MacInnes, *J. Am. Chem. Soc.*, **57**, 1356 (1935).

TABLE II. COMPUTATION OF ACTIVITY COEFFICIENTS FROM THE POTENTIALS OF THE CONCENTRATION CELL Ag; AgCl, HCl (0.1*N*): HCl (C_2), AgCl; Ag AT 25°

Concentration mols per liter at 25° C_2	emf volt	Transference number t_H	Log activity coefficient ratio $-\Delta \log f$	Activity Coefficients f (observed)	f(computed) equation (26)	f (computed) equation (31)
0.0034468	0.136264	0.8234	0.07065	0.9405	0.9402	0.9400
.0052590	.118815	.8239	.06486	.9280	.9283	.9280
.010017	.092529	.8251	.05453	.9062	.9063	.9061
.010029	.092480	.8251	.05450	.9061	.9063	.9060
.019914	.064730	.8266	.04068	.8778	.8777	.8778
.020037	.064464	.8266	.04072	.8778	.8775	.8775
.020132	.064282	.8266	.04051	.8774	.8773	.8773
.040492	.036214	.8286	.02372	.8441	.8430	.8442
.059826	.020600	.8297	.01344	.8244	.8219	.8246
.078076	.009948	.8306	.00630	.8110	.8070	.8113
.100000	.000000	.8314	.00000	.7993	.7926	.7993

is obtained by means of a short series of approximations consisting of adjusting the value of G used in computing the abscissae until it agrees with the intercept. Using a value of $A = 0.5056$ at 25° this computation for the data on hydrochloric acid yields a value of G of -0.0973 and of βa_i of 1.847. This value of βa_i corresponds to a "distance of

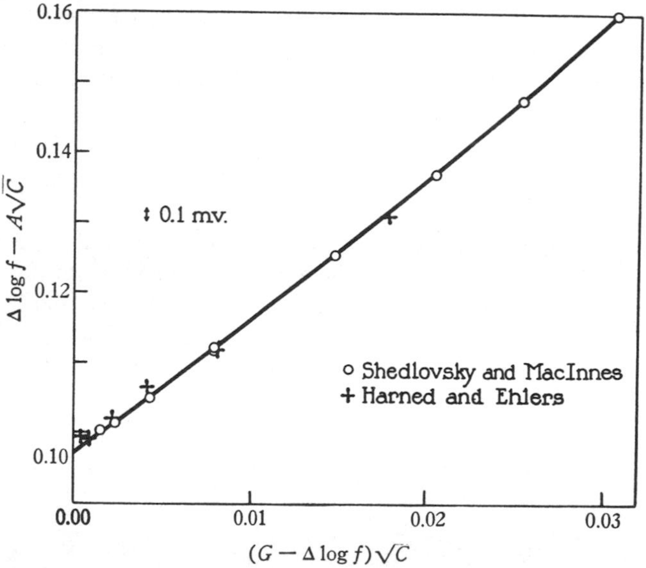

Fig. 5. An Extrapolation Plot Used for Obtaining a Basis of Reference for the Activity Coefficients of Hydrochloric Acid.

closest approach" of 5.62 Å, which is sufficiently large for the higher terms of the extended theory of Gronwall, LaMer and Sandved, discussed in Chapter 7, to be negligible. In columns 5 and 6 of Table II are given the observed activity coefficients and those computed from equation (26). It will be observed that there is agreement of the corresponding values up to a concentration of about 0.04 normal.

The most extensive study of cells "without liquid junction" involving hydrochloric acid has been made by Harned and Ehlers.[15] From a critical summary of their work they have obtained potentials at rounded molalities from which emf values of cells of the type

$$\text{Ag; AgCl, HCl } (C_1)\text{; } H_2\text{(Pt)} \; - \; \text{(Pt) } H_2\text{; HCl } (C_2)\text{, AgCl; Ag}$$

may be obtained. These results are also plotted in Fig. 5, using the G value given above. It will be seen that there is excellent agreement in the results from the cells with and without liquid junction. That is to say, the measurements with cells with and without transference yield substantially the same activity coefficients. Less complete agreement, though probably within the limits of error of the available data, is found with the earlier critical summary of results from cells without liquid junction by Scatchard.[16]

In addition to hydrochloric acid, the results for which have just been described in detail, the method utilizing concentration cells with transference has been used in obtaining the activity coefficients of potassium chloride,[17] sodium chloride,[18] silver nitrate,[19] and calcium chloride.[17] The resulting activity coefficients, f, and comparisons with equation (45), Chapter 7, of the Debye-Hückel theory,

$$-\log f = \frac{A z_1 z_2 \sqrt{\omega}}{1 + \beta a_i \sqrt{\omega}} \qquad (29)$$

are given in Tables III and IV. This equation reduces to equation (25), Chapter 7, for uni-univalent solutes and to

$$-\log f = \frac{1.7515\sqrt{C}}{1 + 0.5692 a_i \sqrt{C}} \qquad (30)$$

in which C is in mols per liter, for a bi-univalent solute such as calcium chloride. It will be seen by comparing the "observed" and "computed" columns that the two correspond for potassium chloride within a very small experimental error for the range of measurements, up to $C = 0.1$

[15] H. S. Harned and R. W. Ehlers, *J. Am. Chem. Soc.*, **54**, 1350 (1932); **55**, 652 (1933); **55**, 2179 (1933).
[16] G. Scatchard, *ibid.*, **47**, 641 (1925).
[17] T. Shedlovsky and D. A. MacInnes, *J. Am. Chem. Soc.*, **59**, 503 (1937).
[18] A. S. Brown and D. A. MacInnes, *ibid.*, **57**, 1356 (1935).
[19] D. A. MacInnes and A. S. Brown, *Chem. Rev.*, **18**, 335 (1936).

TABLE III. ACTIVITY COEFFICIENTS ON THE CONCENTRATION, C, SCALE OF SOME TYPICAL ELECTROLYTES FROM CONCENTRATION CELLS WITH TRANSFERENCE AT 25°

Concentration mols per liter C	potassium chloride f(observed)	potassium chloride f(computed) equation (29) $a_i = 4.07$ A°	sodium chloride f(observed)	sodium chloride f(computed) equation (29) $a_i = 4.45$ A°	silver nitrate f(observed)	silver nitrate f(computed) equation (47) chapter 7 $a_i = 2.3$ A°
0.005	0.9274	0.9276	0.9283	0.9281	0.922	0.922
.01	.9024	.9024	.9034	.9036	.892	.894
.02	.8702	.8707	.8726	.8726	.858	.857
.03	.8492	.8490	.8513	.8515	....	....
.04	.8320	.8322	.8354	.8354	....	....
.05	.8191	.8183	.8220	.8220	.795	.794
.06	.8070	.8067	.8117	.8108	....	....
.08	.7872	.7874	.7938	.7923	....	....
.10	.7718	.7720	.7793	.7776	.733	.735

mol per liter, and to about $C = 0.05$ mol per liter for sodium and calcium chlorides. This agreement is of particular interest in the case of calcium chloride since it is of a higher valence type than the others here considered and is "unsymmetrical." The "distances of closest approach," a_i, are at least of the order of magnitude expected from X-ray studies on crystals. The case of silver nitrate requires additional comment. The data may be fitted fairly well to an equation of the form just given with a value of $a_i = 2.0$ Å. However this is so low that the higher terms of the Debye-Hückel theory discussed at the end of Chapter 7 begin to have an influence. Using the method of computation described

TABLE IV. ACTIVITY COEFFICIENTS OF CALCIUM CHLORIDE FROM CELLS WITH TRANSFERENCE AT 25°

Concentration mols per liter C	f (observed)	f (computed) equation (29) $a_i = 5.24$	f (computed) equation (32) $B = 0.147$ $a_i = 4.944$
0.0018153	0.8588	0.8586	0.8586
.0060915	.7745	.7748	.7743
.0095837	.7361	.7366	.7361
.024167	.6514	.6515	.6513
.037526	.6097	.6092	.6099
.050000	.5834	.5819	.5836
.096540	.5275	.5214	.5276

a somewhat higher but still very small value, 2.3 Å, of a_i was found, and somewhat better agreement of the observed and computed activity coefficients. However, silver nitrate as well as other nitrates show abnormal behavior, as will be shown when the conductances of their aqueous solutions are considered. Fig. 6 shows the relations of the different activity coefficients to each other and to the limiting equations

$$-\log f = 0.5056\sqrt{C}$$

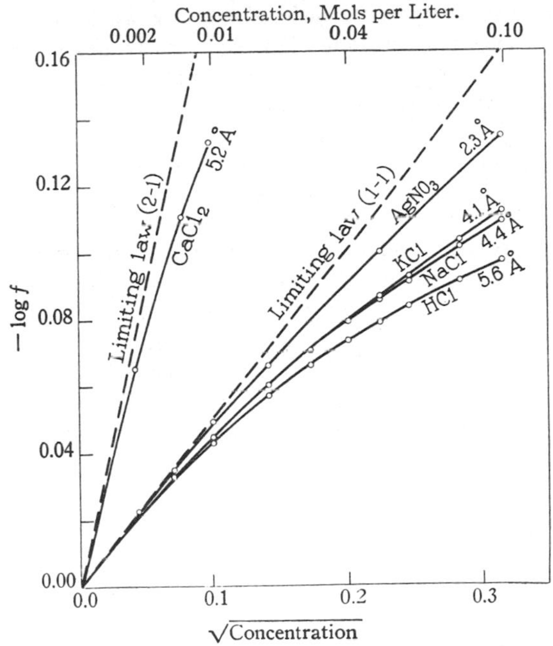

Fig. 6. The Concentration Dependence of the Logarithm of the Activity Coefficients of Some Typical Electrolytes at 25°.

for a uni-univalent, (1–1), electrolyte and

$$- \log f = 1.7515\sqrt{C}$$

for a bi-univalent (2–1) electrolyte.

Extensions of the Debye-Hückel Equations to Higher Concentrations. For solutions of higher concentrations Hückel[20] has proposed an equation of the form

$$- \log \mathbf{f} = \frac{Az_1z_2\sqrt{\omega}}{1 + \beta a_i\sqrt{\omega}} - B\omega \tag{31}$$

in which **f** is the activity coefficient based on Raoult's law. The additional term $B\omega$ in which B is a constant was intended in Hückel's derivation to take account of the change of the dielectric constant D with the concentration. The actual use of the extra term has, however, been empirical. As a matter of fact Harned[21] has shown that in certain

[20] E. Hückel, *Physik, Z.*, 26, 93 (1925).
[21] H. S. Harned, *J. Am. Chem. Soc.*, 48, 326 (1926).

cases the original interpretation of the equation would lead to negative values of the dielectric constant. The equation is, nevertheless, of service for interpolation purposes. Using equation (27a) of Chapter 6, equation (31) may also be stated in the form:

$$-\log f = \frac{Az_1z_2\sqrt{\omega}}{1 + \beta a_i\sqrt{\omega}} - B\omega + \log\left(\frac{\rho - 0.001\mathrm{M}C + 0.001\mathrm{M}_s\nu C}{\rho_0}\right) \quad (32)$$

in which the last term is added to take account of the difference between the activity coefficient f and the coefficient f which comes from equation (24a), Chapter 6. In this term ρ and ρ_0 are the densities of the solution and the pure solvent respectively, M_s and M are molecular weight of the solvent and solute, C is the concentration, in mols per liter, of the solute, and ν is the number of ions into which one molecule of the solute dissociates. Another frequently used modification of the same equation is:

$$-\log \gamma = \frac{Az_1z_2\sqrt{\omega}}{1 + \beta a_i\sqrt{\omega}} - B\omega + \log(1 + 0.001\mathrm{M}_s\nu m) \quad (33)$$

in which the last term, from equation (27b) accounts for the difference between equations (26a) and (26c) of Chapter 6. For solutions of sodium chloride and hydrochloric acid Brown and MacInnes [22] and Shedlovsky and MacInnes [23] have shown that the activity coefficients of these substances given in Tables II and III may be accurately represented by equations of the form of (31) or (32) up to $C = 0.1$ mol per liter, with the constants $a_i = 4.00$, $B = 0.047$ for sodium chloride and $a_i = 4.65$, $B = 0.105$ for hydrochloric acid. The value of the last term of equation (32) is practically zero in these two cases. To obtain this agreement it has been necessary to adjust the values of the distance of closest approach, a_i, somewhat from the values found using equations (26) and (30). Harned and associates [24] have made extensive use of equation (33) and have found that it is useful in accounting for the results of their measurements, which were mostly at concentrations above 0.1 mol per liter. However, the a_i values found were, in general, different from those that hold in the more dilute solutions in which the assumptions made in obtaining the Debye-Hückel equations would be expected to be valid.

[22] A. S. Brown and D. A. MacInnes, *J. Am. Chem. Soc.*, **57**, 1356 (1935).
[23] T. Shedlovsky and D. A. MacInnes, *ibid.*, **58**, 1970 (1936).
[24] H. S. Harned and G. Akerlof, *Physik. Z.*, **27**, 411 (1926).
H. S. Harned, *J. Am. Chem. Soc.*, **51**, 416 (1929).
H. S. Harned and O. E. Schupp, *ibid.*, **52**, 3886 (1930).
H. S. Harned and C. M. Mason, *ibid.*, **54**, 1439 (1932).
H. S. Harned and R. W. Ehlers, *ibid.*, **55**, 2179 (1933).
H. S. Harned and J. C. Hecker, *ibid.*, **56**, 650 (1934).

The results of the more accurate determination of the activity coefficient, γ, at 25° over a wide range of molalities, m, are given in Table V, and typical values from these data are plotted as functions of

TABLE V. ACTIVITY COEFFICIENTS γ, AT 25° ON THE MOLALITY SCALE OVER A WIDE RANGE OF MOLALITIES, m

Concentration, mols per 1000g of H_2O, m	NaCl	KCl	HCl	HBr	NaOH	$CaCl_2$	$ZnCl_2$	H_2SO_4	$ZnSO_4$	$CdSO_4$
0.005	0.928	0.927	0.930	0.930	....	0.789	0.767	0.643	0.477	0.476
.01	.903	.902	.906	.906	0.89_9	.732	.708	.545	.387	.383
.02	.872	.869	.878	.879	$.86_0$	.669	.642	.455	.298	....
.05	.821	.817	.833	.838	$.80_5$	.584	.556	.341	.202	.199
.10	.778	.770	.798	.805	$.75_9$	.524	.502	.266	.148	.137
.20	.732	.719	.768	.782	$.71_9$	.491	.448	.210	.104	....
.50	.680	.652	.769	.790	$.68_1$	.510	.376	.155	.063	.061
1.00	.656	.607	.811	.871	$.66_7$	.725	.325	.131	.044	.042
1.50	.655	.587	.898	....	$.67_1$	....	.290	....	.037	.039
2.00	.670	.578	1.011	....	$.68_5$	1.554	....	.125	.035	.030
3.00	.719	.574	1.31	....	....	3.384	....	.142	.041	.026
4.00	.791	....	1.74	....	....	....	....	.172	....	....
Reference	*1,2*	*3*	*4,5, 6,21*	*7*	*8,9*	*3,10, 11*	*12*	*13,14, 15*	*16,17, 18*	*19,20*

1. A. S. Brown and D. A. MacInnes, *J. Am. Chem. Soc.*, **57**, 1356 (1935).
2. H. S. Harned and L. F. Nims, *ibid.*, **54**, 423 (1932).
3. T. Shedlovsky and D. A. MacInnes, *ibid.*, **59**, 503 (1937).
4. T. Shedlovsky and D. A. MacInnes, *ibid.*, **58**, 1970 (1936).
5. H. S. Harned and R. W. Ehlers, *ibid.*, **54**, 1350 (1932).
6. H. S. Harned and R. W. Ehlers, *ibid.*, **55**, 2179 (1933).
7. H. S. Harned, A. S. Keston and J. G. Donelson, *ibid.*, **58**, 989 (1936).
8. H. S. Harned, *ibid.*, **47**, 677 (1925).
9. H. S. Harned and J. C. Hecker, *ibid.*, **55**, 4838 (1933).
10. W. W. Lucasse, *ibid.*, **47**, 743 (1925).
11. H. S. Harned and G. Åkerlof, *Physik. Z.*, **27**, 411 (1926).
12. G. Scatchard and R. F. Tefft, *J. Am. Chem. Soc.*, **52**, 2272 (1930).
13. H. S. Harned and W. J. Hamer, *ibid.*, **57**, 27 (1935).
14. G. Baumstark, Dissertation, Catholic Univ. of Amer., Washington, D. C. (1932).
15. J. Shrawder and I. A. Cowperthwaite, *J. Am. Chem. Soc.*, **56**, 2340 (1934).
16. I. A. Cowperthwaite and V. K. LaMer, *ibid.*, **53**, 4333 (1931).
17. U. B. Bray, *ibid.*, **49**, 2372 (1927).
18. J. Kielland, *ibid.*, **58**, 1855 (1936).
19. V. K. LaMer and W. G. Parks, *ibid.*, **53**, 2040 (1931).
20. V. K. LaMer and W. G. Parks, *ibid.*, **55**, 4343 (1933).
21. G. Åkerlof and J. W. Teare, *ibid.*, **59**, 1855 (1937).

the square root of the ionic strength, $\omega^{\frac{1}{2}}$, in Fig. 7. It will be observed from this plot that the deviation of values of γ from unity are greater the more complex the valence type of electrolyte. However there can be quite wide variation of activity coefficients for electrolytes of the same valence type as can be seen from the curves for hydrochloric acid and potassium chloride. In a number of cases the activity coefficients pass through a minimum and may increase to values greater than unity.

Quite a number of attempts have been made to account theoretically for the activity coefficients at higher concentrations by introducing, in addition to the variation of the dielectric constant already mentioned, the ideas of partial dissociation, complex ions, etc. However, the discussions in this field are still highly speculative.

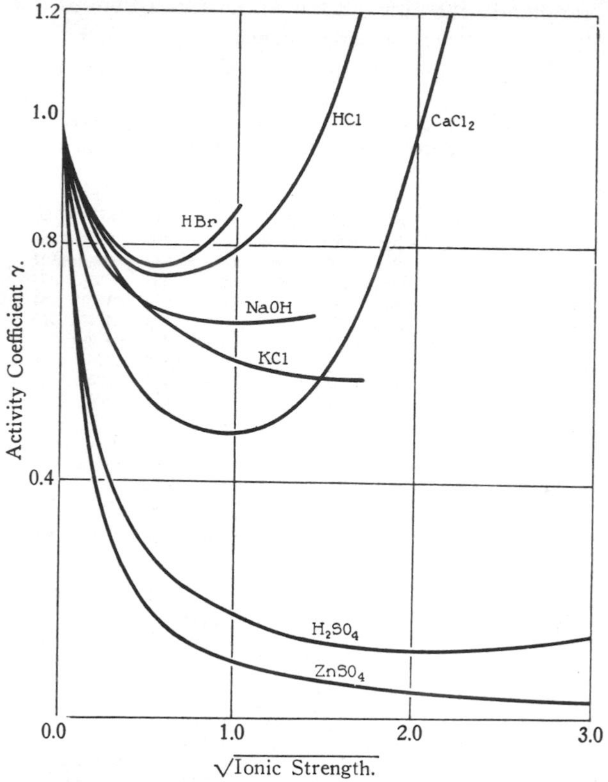

Fig. 7. Change in the Activity Coefficients of Different Valence Type Salts with the Square Root of the Ionic Strength.

The Determination of Transference Numbers from the Potentials of Concentration Cells. Another use of concentration cells, which involves the principles already discussed in this chapter, is that of the determination of transference numbers. Since a cell without liquid junctions of the type

$$\text{Ag; AgCl, NaCl } (C_1)\text{; Na(Hg)} - \text{Na(Hg); NaCl } (C_2)\text{, AgCl; Ag} \quad (34)$$

involves the reversible transport of one equivalent of sodium chloride from the concentration C_1 to C_2 per faraday of current passed through the cell and the operation of a cell of the form

$$\text{Ag; AgCl, NaCl } (C_1) : \text{NaCl } (C_2)\text{, AgCl; Ag} \quad (35)$$

is accompanied by the transfer of t_{Na} equivalents through the same concentration range, the transference number of the sodium ion constituent may be obtained from the simple relation

$$t_{Na} = \frac{E_t}{E} \tag{36}$$

in which E and E_t represent the potentials of cells of the types (34) and (35), respectively, containing solutions of the same concentrations, C_1 and C_2. However as this can hold only if the transference number is constant in the range C_1 to C_2, a more detailed analysis of the method is as follows. We have already seen that the differential equation for the relation between the chemical potential of the solute and the emf, E_t of a cell of type (35) is given by

$$-\mathbf{F}\,dE_t = t_{Na}d\mu \tag{37}$$

Here E_t represents the potential of a cell "with transference." Since a whole equivalent is involved in the operation of cell (34) the corresponding differential equation for a cell of that type is

$$-\mathbf{F}\,dE = d\mu \tag{38}$$

Therefore the transference number may be obtained by the relation

$$dE_t/dE = t_{Na} \tag{39}$$

or in general

$$dE_t/dE = t \tag{40}$$

in which E_t and E refer to the potentials of cells with transference and without liquid junction, and t is the transference number of the ion constituent to which the electrodes in the first-named type of cell are not reversible.

Though thermodynamically sound this method for obtaining transference numbers has not attained the accuracy of the recent moving boundary of Hittorf methods, and no illustrations of its use will be given here. One difficulty is that of obtaining reversible potentials with amalgam electrodes, which are usually necessary. If in addition the transference number changes at all rapidly with concentration the evaluation of dE_t/dE may present difficulties. It may be done graphically or by fitting analytical expressions to the data and differentiating. The final results are apparently too dependent, up to the present at least, on the means employed, for adequate accuracy. The method has been studied with varying success by MacInnes and Parker,[25] MacInnes and Beattie,[26] Jones and Dole,[27] Hamer [28] and others.

[25] D. A. MacInnes and K. Parker, *J. Am. Chem. Soc.*, **37**, 1445 (1915).
[26] D. A. MacInnes and J. A. Beattie, *ibid.*, **42**, 1117 (1920).
[27] G. Jones and M. Dole, *ibid.*, **51**, 1073 (1929).
[28] W. J. Hamer, *ibid.*, **57**, 662 (1935).

Concentration Cells Involving Mixtures of Electrolytes. There is a type of concentration cell, in addition to those already discussed, the study of which yields data of value. The data are useful, for instance, in the computation of liquid junction potentials as described in Chapter 13. These can be represented by the example:

$$\begin{aligned}&Ag;\ AgCl,\ HCl\ (m_1),\ KCl\ (m_2);\ H_2(Pt)\\&\qquad -\ (Pt)H_2;\ KCl\ (m_3),\ HCl\ (m_4),\ AgCl;\ Ag\end{aligned} \tag{41}$$

As indicated they are concentration cells, without liquid junctions, containing mixtures of electrolytes. Since the electrodes of the two half-cells are reversible to the chloride and hydrogen ions, the operation of such a cell involves per faraday of current the removal of one mol of hydrogen chloride from one of the two solutions and the reappearance of the same amount in the other, the mechanism being exactly that described for cell (22) of Chapter 6. The potential ΔE of the cell is therefore a measure of the difference of the chemical potential of hydrogen chloride in the two solutions, according to the equation:

$$(\mu'_{HCl} - \mu''_{HCl}) = \mathbf{F}\Delta E$$

With the aid of equations (9), (26c) and (42a), Chapter 6, this may be given the form

$$\Delta E = \frac{RT}{\mathbf{F}} \ln \frac{m'_H m'_{Cl} \gamma'^2_{HCl}}{m''_H m''_{Cl} \gamma''^2_{HCl}} \tag{42}$$

An important investigation in this field has been carried out by Güntelberg,[29] who measured, at 20°, the potentials of cells of type

$$\begin{aligned}&Ag;\ AgCl,\ HCl\ (0.1m);\ H_2(Pt)\\&\qquad -\ (Pt)\ H_2;\ HCl\ (0.1 - m_1),\ KCl\ m_1,\ AgCl;\ Ag\end{aligned}$$

in which the total molality, m, was kept at 0.1, but the proportion of potassium and hydrogen chloride in the solution contained in the half-cell on the right hand side was varied. Similar measurements were made in which the potassium ion was replaced by lithium, sodium and cesium. The measurements are apparently of unusual accuracy. The results are perhaps best shown in Fig. 8, in which the logarithms of the ratios of activity coefficients, $\gamma_{HCl(MCl)}/\gamma_{HCl}$, in which $\gamma_{HCl(MCl)}$ represents the mean ion activity coefficient of hydrochloric acid in a solution which also contains MCl, are plotted against the proportion, $x = m_1/0.1$, of the added salt. It will be seen that for the four added salts studied the logarithms of the activity coefficients decrease linearly

[29] E. Güntelberg, *Z. physik. Chem.*, 123, 199 (1926).

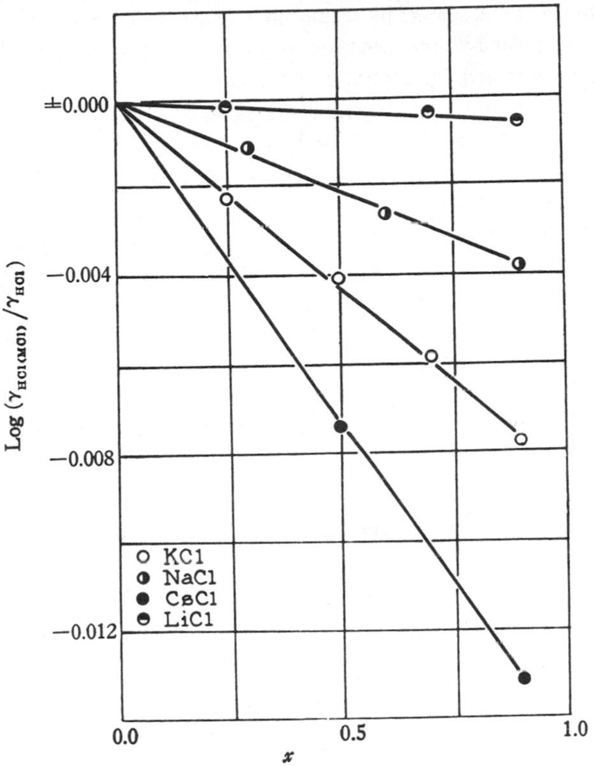

Fig. 8. The Linear Variation of the Logarithm of the Activity Coefficient of Hydrochloric Acid in Alkali Halide Solutions at a Constant Total Molality of 0.10 Molal at 20°.

with x, the effect being greatest for cesium and least for lithium as the added positive ion constituent. The relation may be put in the form

$$\log \left(\frac{\gamma_{HCl(MCl)}}{\gamma_{HCl}}\right)_m = - A_M x \tag{43}$$

in which $\gamma_{HCl(MCl)}$ represents the activity coefficient of hydrochloric acid, with the salt MCl at the molality m_1; γ_{HCl} is the activity coefficient of the acid alone at the molality m. A_M is a constant depending upon the ion constituent M and the total molality m. The values of the constant A_M for the different ion constituents at a total molality, m, of 0.1, are:

$$A_{Li} = 0.0007, \quad A_{Na} = 0.0042, \quad A_K = 0.0084 \quad A_{Cs} = 0.0146$$

The relation, represented by equation (43), was independently demonstrated by Harned[30] who worked at 25° and at higher molalities. The A_M values obtained by Harned at $0.1m$ for the effect of the lithium, sodium and potassium ions were the same as those found by Güntelberg. The relationship was found to hold at still higher molalities, up to six molal, by Hawkins.[31] The linear variation of the logarithm of activity coefficients at constant concentration represented by equation (43) was, however, first observed by Brønsted[32] who obtained activity coefficients from solubility measurements. The change of A_M values with the total molality has been discussed by Harned.[33] No simple relation appears to be valid. Harned and Cook[34] have found that equation (43) does not represent the change of activity coefficient with the proportion, x, for mixtures of alkali chlorides and hydroxides, a quadratic function being necessary to account for the results in these cases.

It is also possible to study the effect on the activity coefficient of, for instance, 0.01 molal hydrochloric acid, produced by increasing molalities of an added chloride, MCl, by means of the cell

$$\begin{aligned}&\text{Ag; AgCl, HCl }(0.01m)\text{; }H_2(\text{Pt})\\&\qquad - (\text{Pt})\ H_2\text{; HCl }(0.01m)\text{, MCl }m\text{, AgCl; Ag}\end{aligned} \qquad (44)$$

and the resulting activity coefficients may be computed from equation (42). The results of several series of such determinations of the activity coefficients, γ, are shown in Fig. 9. It is interesting to note that the activity coefficient of 0.01 molal hydrochloric acid in increasing molalities of lithium chloride is practically identical with that of hydrochloric acid itself at the same total molality. The family of curves given are all of the same general form and can all be expressed by means of Hückel's semi-empirical equation (33). Curves of the same shape as those in Fig. 9 have been obtained using chlorides of higher valence cations for the substance MCl in cell (44).[35, 36, 37, 38]

[30] H. S. Harned, *J. Am. Chem.*, **48**, 326 (1926).
[31] J. E. Hawkins, *J. Am. Chem. Soc.*, **54**, 4480 (1932).
[32] N. Brønsted, *ibid.*, **45**, 2898 (1923).
[33] H. S. Harned, *ibid.*, **57**, 1865 (1935).
[34] H. S. Harned and M. A. Cook, *ibid.*, **59**, 1890 (1937).
[35] H. S. Harned and N. J. Brumbaugh, *J. Am. Chem. Soc.*, **44**, 2729 (1922) ($BaCl_2$, $SrCl_2$, $CaCl_2$).
[36] M. Randall and G. F. Breckenridge, *ibid.*, **49**, 1435 (1927) ($BaCl_2$, $LaCl_3$).
[37] C. M. Mason and D. B. Kellam, *J. Phys. Chem.*, **38**, 689 (1934) ($CeCl_3$).
[38] H. S. Harned and C. M. Mason, *J. Am. Chem. Soc.*, **53**, 3377 (1931) ($AlCl_3$).

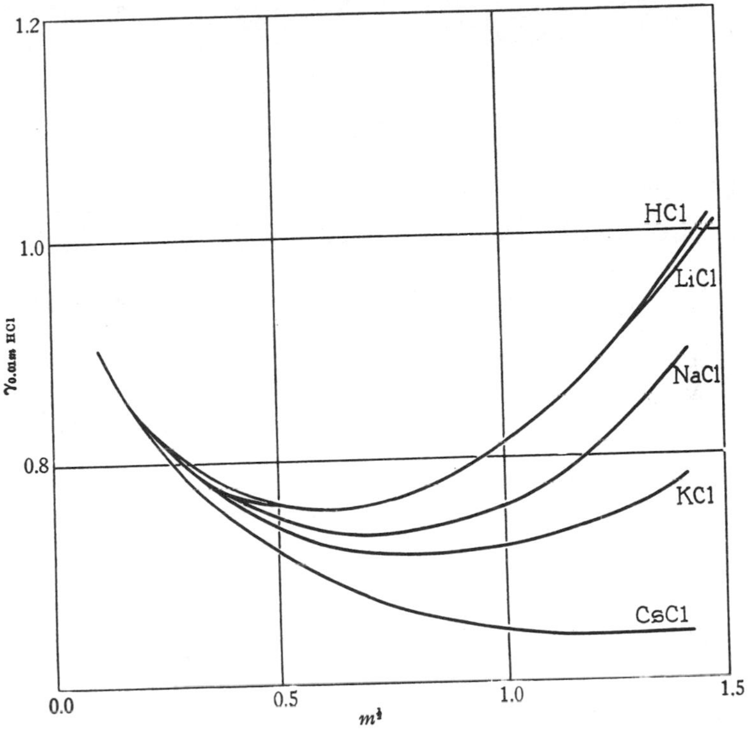

Fig. 9. The Activity Coefficients of 0.01 Molal Hydrochloric Acid in Solutions of the Alkali Halides at 25°.

KCl, NaCl—H. S. Harned and G. Murphy, *J. Am. Chem. Soc.*, **53**, 8 (1931); LiCl—H. S. Harned and F. E. Swindells, *ibid.*, **48**, 126 (1926); LiCl—H. S. Harned and H. R. Copson, *ibid.*, **55**, 2206 (1933); KCl—H. S. Harned and W. J. Hamer, *ibid.*, **55**, 2194 (1933); CsCl—H. S. Harned and O. E. Schupp, *ibid.*, **52**, 3892 (1930); HCl—H. S. Harned and R. W. Ehlers, *ibid.*, **55**, 2179 (1933).

[*Comment to 1961 Edition*] Values of the activities of HCl, NaCl, KCl, $CaCl_2$ and $LaCl_3$ in aqueous solution are given by Shedlovsky.* All these electrolytes, including the highly unsymmetrical salt $LaCl_3$, were found to follow Debye-Hückel equations.

Much new data on activity coefficients is given in:

Harned, Herbert S. and Owen, B. B. *Physical Chemistry of Electrolytic Solutions.* Reinhold Publishing Corp., 3rd ed., 1958.

* T. Shedlovsky, *J. Am. Chem. Soc.*, **72**, 3680 (1950).

Chapter 9

The Effects of Gravity and Centrifugal Force on the Electromotive Force of Galvanic Cells

In our discussion up to this point the effect of the force of gravity on potentials of galvanic cells has been neglected. In general this neglect is justified, but there are cases, as will be demonstrated, in which the effect of gravity is appreciable. Such effects can, as we shall also see, be greatly increased by substituting a centrifugal field of force for a gravitational field.

The differential equation relating the chemical potential, μ_i, of the substance i with its position in a gravitational field is

$$\left(\frac{\partial \mu_i}{\partial h}\right)_{P,\,T} = \mathrm{M}_i g \tag{1}$$

in which M_i is the molecular weight of the substance, g is the gravitational force per gram, and h is the height. If, however, the substance, i, is in solution two more variables must be considered. These are the pressure, P, on the substance and its mol fraction, N_i, thus

$$d\mu_i = \frac{\partial \mu_i}{\partial h} dh + \frac{\partial \mu_i}{\partial P} dP + \frac{\partial \mu_i}{\partial \mathrm{N}_i} d\mathrm{N}_i \tag{2}$$

Referring to equation (3), of Chapter 6:

$$dZ = -\,sdT + VdP + \mu_A dn_A + \mu_B dn_B + \cdots\cdots$$

it can be seen that if

$$n_A + n_B + \cdots\cdots + n_i + \cdots\cdots = 1$$

then n_i, etc. are mol fractions, *i.e.*, $dn_i = d\mathrm{N}_i$. By cross-differentiation of that equation

$$\frac{\partial \mu_i}{\partial P} = \frac{\partial V}{\partial \mathrm{N}_i} \tag{3}$$

the term on the right-hand side being, by definition, the partial molal volume, $\overline{V}_i$.[1] Since

$$-\,dP = \rho g dh \tag{4}$$

[1] See footnote, page 82.

in which ρ is the density of the solution, with equation (1), equation (2) becomes

$$d\mu_i = (M_i - \bar{V}_i\rho)g\,dh + \frac{\partial \mu_i}{\partial N_i}\,dN_i \tag{5}$$

Under equilibrium conditions $d\mu_i = 0$ throughout the system, so that the mol fraction of the substance will vary with the height according to

$$\frac{dN_i}{dh} = \frac{-(M_i - \bar{V}_i\rho)g}{\partial \mu_i/\partial N_i} \tag{6}$$

However, due to the slowness with which equilibrium is established, the variation of composition with height indicated by this equation may not take place immediately. If the composition of the solution can be assumed to be constant, *i.e.*, $dN_i = 0$, then from equation (5)

$$d\mu_i = (M_i - \bar{V}_i\rho)g\,dh \tag{7}$$

The electrical potentials of galvanic cells resulting from differences in chemical potentials are already familiar from numerous examples discussed in this book. In particular the potential of cells of the general type

Hg; HgCl, MCl : MCl, HgCl; Hg

has been considered at length in Chapter 8. Up to the present, however, the variations of the chemical potentials, μ_i, have been attained by variations of the concentration of the electrolyte, MCl. Such variations can also arise, as we have seen, from the different positions of the components in a gravitational field. However, such a field also produces changes in the chemical potentials of the electrode materials which must be considered in the discussion of the mechanism of the cell. The experimental results on the effect of gravity and of centrifugal force will be discussed in that order.

The Effect of Gravity. Experiments undertaken to study the effect of differences of position in a gravitational field on the emf of a galvanic cell have been carried out by Des Coudres.[2] His apparatus is shown diagrammatically in Fig. 1. Two reversible calomel electrodes *A* and *B* were connected by means of the rubber tube *a* to *b* which could be filled with a chloride solution, MCl. The electrodes were suspended by means of a rope and pulley as shown, so that the height, *h* (the maximum value of which was 375 centimeters) could be altered at will. The effective height could be doubled by reversing the positions of the electrodes and adding the resulting electromotive forces. Table I gives the results of Des Coudres' measurements. The emf, per

[2] Th. Des Coudres, *Ann. der Physik u. Chemie*, **49**, 284 (1893); **57**, 232 (1896).

centimeter difference in height of electrodes, is given in the third column for each of the different chlorides listed in the first column.

The operation of this cell, which may be represented by:

$$\overrightarrow{\underset{A}{\text{Hg; HgCl, MCl}} \underset{h=0}{:} \underset{h=h}{\text{MCl, HgCl; Hg,}}}\quad B$$

per faraday of current passing through the cell in the direction indicated, is as follows: (a) the transfer of the Hittorf transference number, t_M, of equivalents of the salt, MCl, from the region of electrode A

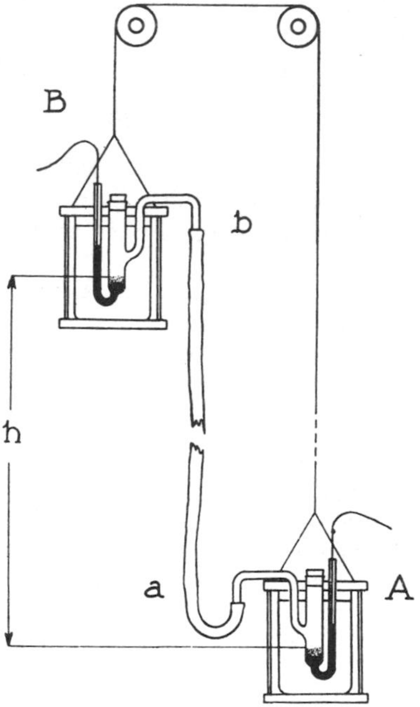

Fig. 1. Arrangement for Determining the Effect of Gravity on the Potential of a Galvanic Cell.

and to electrode B and (b) the reaction with mercury of an equivalent of chloride ion at electrode A and the reduction of an equivalent of HgCl at electrode B, *i. e.*, the transfer of an equivalent of combined chlorine from electrode B to electrode A. Since the change of Gibbs free energy is equal to the algebraic sum of the changes in the chemical potentials

$$-\Delta Z = E\mathbf{F} = t_M \Delta\mu_{MCl} - \Delta\mu_{Cl},$$

in which μ_{MCl} and μ_{Cl_s} are the chemical potentials of the solute, MCl and of combined chloride respectively, we have, using equation (7)

$$EF = gh\,[t_M(M_{MCl} - \bar{V}_{MCl}\rho) - (M_{Cl} - V_{Cl_s}\rho)] \qquad (8)$$

in which V_{Cl_s} is the molal volume of the combined chloride, *i.e.*, the difference between the molal volumes of the calomel and mercury. From equation (8) it is evidently possible to compute the transference number, t_M, which is given by the formula

$$t_M = \frac{EF/gh + M_{Cl} - \bar{V}_{Cl_s}\rho}{M_{MCl} - \bar{V}_{MCl}\rho} \qquad (8a)$$

An illustrative computation from Des Coudres' results is as follows. He found 0.510×10^{-8} volt for the ratio E/h for 2.71 molal potassium chloride. Multiplying the product EF/h by 10^7 to reduce it to absolute units, and dividing by the value 980.7, also in absolute units, for g, the acceleration due to gravity, yields 5.02 grams. The values of the other terms in equation (8a) are $V_{Cl_s} = 18.36$ cc., $\bar{V}_{KCl} = 31.3$ cc. and $\rho = 1.111$. From these figures and the molecular weights equation (8a) gives the value 0.50 for the cation transference number of potassium chloride. If the same values of the transference number are obtained as are found by other methods it is evident that a correct analysis of the mechanism of the process has been made. Such a comparison is made in Table I. The values of the

TABLE I. COMPARISON OF TRANSFERENCE NUMBERS OBTAINED FROM THE EFFECT OF GRAVITY ON EMF WITH THE HITTORF TRANSFERENCE NUMBERS

Solution	Molality of solution m	Emf, volts, per centimeter height, $\times 10^8$	Cation Transference Number: Gravity method	Cation Transference Number: Hittorf method	Reference
NaCl	4.25	− 0.315	0.34	0.365	*1*
$BaCl_2$	0.98	+ 1.70	0.36	0.379	*2*
LiCl	4.93	− 1.09	0.23	0.245	*3*
KCl	2.71	+ 0.510	0.50	0.486	*4*
HCl	1.01	− 0.218	0.85	0.835	*1*

1. International Critical Tables (1926).
2. Chapter 4.
3. G. Jones and B. C. Bradshaw, *J. Am. Chem. Soc.*, **54**, 138 (1932).
4. D. A. MacInnes and M. Dole, *ibid.*, **53**, 1357 (1931).

transference numbers determined by the Hittorf method, at as nearly the same molality as possible from the available data, are given in the last column. Considering the experimental difficulties of the method, and the small values of the electromotive force measured, the agreement between the transference numbers by the two methods is surprisingly good.

The Effect of Centrifugal Force. Des Coudres and more particularly Tolman [3] have sought to produce an increase of the effects described in the previous section by substituting a centrifugal field of force for a gravitational field.

The arrangement used by Des Coudres for this purpose is shown in Fig. 2. The cells *A* and *B* were mounted on a horizontal platform which could be rotated about the axle *D, D* by means of the belt *e, e*

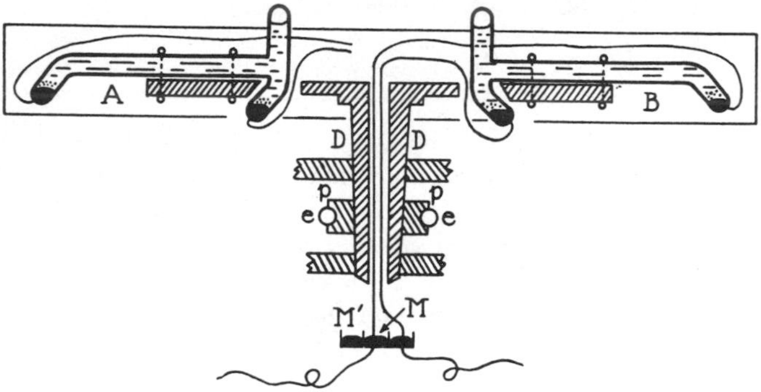

Fig. 2. Arrangement for Determining the Effect of Centrifugal Force on the Potential of a Galvanic Cell.

and pulley *p, p*. The connecting wires from one of the cells were run as shown through the center of the axle to mercury contacts, *M, M'*, the latter being arranged as an annular ring around the former. By this means the potential of a cell could be measured while the apparatus was in motion.

Des Coudres used comparatively feeble centrifugal forces. Tolman, however, utilized a steam turbine in his experiments and his apparatus attained relatively high velocities. The effect of centrifugal force may be obtained as follows. The centrifugal force acting on one gram at a radius, r, rotating n times per second is $4\pi^2 n^2 r$, so the work done in carrying one gram from radius r_2 to r_1 will evidently be

$$\int_{r_1}^{r_2} 4\pi^2 n^2 r dr = 2\pi^2 n^2 (r_2^2 - r_1^2) \tag{9}$$

Tolman used cells of the type

$$(Pt)\ I_2;\ MI\ (C_1);\ I_2\ (Pt)$$

[3] R. C. Tolman, *Proc. Am. Acad. Arts Sci.*, **46**, 109 (1910); *J. Am. Chem. Soc.*, **33**, 121 (1911).

in which M again represents hydrogen or one of the alkali metals, so that the equation corresponding to equation (8) is

$$EF = 2\pi^2 n^2 (r_2^2 - r_1^2) [t_M (M_{MI} - \bar{V}_{MI}\rho) - (M_I - \bar{V}_{I_s}\rho)] \qquad (10)$$

in which work in the centrifugal force field from equation (9) has been substituted for the product gh and the molecular weight and partial molal volume of iodine replace those values for combined chlorine. Tolman's apparatus attained speeds of 80 revolutions per second giving values of $2\pi^2 n^2 (r_2^2 - r_1^2)$ reaching 114,000,000 dyne-cms., whereas in Des Coudres' work on gravity cells the corresponding term gh did not rise above 36,000.

From equation (10) it is evident that if the theory is correct the values of the potential, E, for a given cell should vary directly with the square of the number of revolutions per second, n^2, or in other words E/n^2 should be a constant. That this is substantially true is shown in Table II which gives the data for a cell containing molal lithium iodide and 0.01 molal iodine.

TABLE II. THE EFFECT OF SPEED OF ROTATION ON THE EMF OF THE CELL

(Pt) I_2; LiI, I_2 (Pt)

$r_2 = 29.5$ cm. $r_1 = 4.2$ cm.

Revolutions per second n	$E \times 10^3$ volts	$E/n^2 \times 10^6$
52.3	3.23	1.181
56.2	3.72	1.178
59.2	4.16	1.187
63.8	4.80	1.179
68.3	5.51	1.181
72.4	6.22	1.187

Just as with gravity cells, the measurements involving centrifugal force may be used with the aid of equation (10) to compute transference numbers. Table III contains the results of Tolman's measure-

TABLE III. COMPARISON OF TRANSFERENCE NUMBERS OBTAINED FROM THE EFFECT OF CENTRIFUGAL FORCE ON EMF WITH THOSE BY OTHER METHODS

	Anion Transference Numbers			
Substance	Centrifugal method in 1.0 molal solution	At infinite dilution	Hittorf or moving boundary method in 0.1 normal solution	Reference
KI	0.514	0.509	0.512	*1*
NaI	0.615	0.603	0.624	*2*
LiI	0.732	0.663	.....	..
HI	0.184	0.179	0.174	*3*

1. Table IV, Chapter 4.
2. A. A. Noyes and K. G. Falk, *J. Am. Chem. Soc.*, **33**, 1436 (1911).
3. E. K. Strachan and V. G. Chu, *ibid.*, **36**, 810 (1914).

ments which were all carried out on molal solutions of the electrolytes. For comparison the transference numbers at infinite dilution, computed as described in Chapter 18, and also a few direct measurements at 0.1 normal by the Hittorf and moving boundary methods are given. The values by the different methods are not strictly comparable, being at widely different concentrations, but they at least show that the centrifugal method gives very nearly correct results in spite of the large number of possible sources of experimental error.

[*Comment to 1961 Edition*] Grinell and Koenig* have investigated the effect of the gravitational field on the cell

$$(Pt)I_2\,;\,KI\,;\,I_2(Pt)$$

using excellent experimental technique. The result was a confirmation of the theory underlying equation (10) (with the term gh replacing $2\pi n^2(r_2^2 - r_1^2)$).

The effect of centrifugal fields on this and closely related cells has been studied by MacInnes and associates.† These researches indicate that equation (10) is a limiting expression for very low ratios of iodine to iodide in solution. For higher ratios, allowance must be made for the formation of the ion I_3^-, incidentally confirming the composition of that complex.

* S. W. Grinell and F. O. Koenig, *J. Am. Chem. Soc.*, **64**, 682 (1942).

† D. A. MacInnes and B. Roger Ray, *J. Am. Chem. Soc.*, **71**, 2987 (1949); *Rev. Sci. Ins.*, **20**, 52 (1949). D. A. MacInnes, *Proc. Am. Phil. Soc.*, **97**, 51 (1953). D. A. MacInnes and M. O. Dayhoff, *J. Chem. Phys.*, **20**, 1034 (1952).

Chapter 10

Standard Electrode Potentials from Galvanic Cells without Liquid Junctions

The potential of an ordinary dry cell (Leclanché cell) is about 1.5 volts. According to the scheme we have adopted such a cell can be represented by

$$Zn; ZnCl_2, NH_4Cl, MnO_2\ (s); (C) \tag{1}$$

It is natural to ask the question: How much of this potential should be assigned to the zinc electrode and how much to the carbon electrode? Unfortunately no conclusive or satisfying answer can be given to this question, in spite of much research and speculation. It is evident that if an electrode with an absolute zero of potential were available, the zinc and carbon electrodes could be measured against it separately, and the question answered. However, no such zero reference electrode has been found. There is the more fundamental question as to how such an absolute zero of potential would be recognized if, by some means, it were attained. As a matter of fact, there is doubt whether the concept of an absolute zero of potential has any meaning.

Some writers, particularly physicists, consider that the potentials do not originate at the metal-electrolyte boundaries of a galvanic cell, but rather at the metal-metal junctions, such as, for instance, the contacts Cu-Zn and C-Cu of the connecting wires with the electrodes. That such a fundamental question should have remained open indicates that we are dealing with a subject of real difficulty.

Fortunately all practical purposes are served if we choose an arbitrary zero of potential and refer all other potentials to it. The arbitrary electrochemical zero of potential which will be adopted is that the electrode:

$$(Pt)\ H_2; H^+ \tag{2}$$

has a zero potential *at any temperature,* when the partial pressure of the hydrogen gas is one atmosphere, and the mean ion activity of the hydrogen ion constituent is unity. The qualification concerning the hydrogen ion activity is necessary, and is due to the fact that there is much the same kind of difficulty in deciding upon activities of single ionic components as there is in evaluating the potentials of single electrodes. The two conceptions are, as a matter of fact, intimately connected. This question is discussed in more detail in Chapter 13.

This chapter is concerned with the definition and determination of *standard potentials*. Such standard potentials are of use in obtaining Gibbs free energies of electrochemical reactions under definite "standard" conditions, and for obtaining activity coefficients of solutes, as will be discussed in detail below. However, in order to deal with such potentials it is first necessary to discuss the matter of *standard states*.

Standard States. According to equation (10), Chapter 6, the change of the Gibbs free energy, ΔZ, of a reaction occurring at constant temperature and pressure is

$$\Delta Z = k\mu_K + l\mu_L + \cdots - a\mu_A - b\mu_B - \cdots$$

and if the reaction takes place in a galvanic cell

$$\Delta Z = -En\mathbf{F}$$

As has already been pointed out, the Gibbs free energy and the chemical potentials are each indefinite to the extent of an arbitrary constant. It is therefore desirable to adopt reference or standard states of each component at which the chemical potentials may be assumed to have definite known values. Some of the conventions which have been adopted concerning standard states are as follows.

If the component A is a solid its chemical potential under equilibrium conditions is, at a given temperature and pressure, a constant, *i.e.*,

$$\mu_A = \mu_A^\circ \tag{3}$$

in which μ_A° is the chemical potential of A in the standard state. Thus a pure, one-component, solid is in its standard state. In a few cases, notably those of mercury and bromine, the chemical potential of the liquid may be assumed to be constant, and the liquid component may be taken as the standard of such a substance.

If the component A is a gas then from equation (20), Chapter 6,

$$\mu_A = \mu_A^\circ + RT \ln a_A \tag{4}$$

It will be recalled that the activity a_A approaches the partial pressure p_A as the latter is reduced, also that the term $RT \ln a$ equals zero when $a = 1$. The gas A is defined to be in its standard state when it is at one atmosphere partial pressure but with the properties (those of a perfect gas) of the component at very low pressures. The standard state is evidently unrealizable experimentally.

If the component A is a solute then from equation (26b), Chapter 6,

$$\mu_A = \mu_A^\circ + RT \ln a_A = \mu_A^\circ + RT \ln C_A f_A \quad (5)$$

As explained in Chapter 8 the scale of activities is so chosen that a_A approaches C_A as the solution of A is made more dilute. In other words $C_A = a_A$ when $f_A = 1$. The standard state of the solute A is therefore defined as the hypothetical state in which it has a concentration of unity but with the properties associated with infinite dilution. Under these conditions

$$\mu_A = \mu_A^\circ$$

Since the results of much of the experimental work are expressed in terms of molalities, m, (mols per 1000 grams of solvent, instead of volume concentrations) it is frequently convenient to give chemical potentials the form, from equation (26c) of Chapter 6, of

$$\mu_A = \mu_A^\circ + RT \ln m_A \gamma_A \quad (6)$$

in which γ_A is an activity coefficient which, if μ_A° is properly chosen, approaches the same limit, unity, as the activity coefficient f, but deviates from it as the m and C take on larger values. The quantity μ_A° will, in general, have a different value from that defined by equation (5).

If a chemical reaction takes place at constant temperature and pressure under conditions such that all the components are in their standard state, then the value of the change of the Gibbs free energy, ΔZ_0, will be

$$\Delta Z_0 = k\mu_K^\circ + l\mu_L^\circ + \cdots - a\mu_A^\circ - b\mu_B^\circ - \cdots \quad (7)$$

and if this takes place in a galvanic cell from equation (31), Chapter 5,

$$\Delta Z_0 = -E_0 n\mathbf{F} \quad (8)$$

in which E_0 is the *standard potential of the galvanic cell* in which the reaction takes place. From equations (10), (20), Chapter 6, and (31), Chapter 5, we have

$$-En\mathbf{F} = k\mu_K^\circ + l\mu_L^\circ + \cdots - a\mu_A^\circ - b\mu_B^\circ - \cdots + RT \ln \frac{a_K^k \times a_L^l \cdots}{a_A^a \times a_B^b \cdots} \quad (9)$$

With equations (7) and (8) this becomes

$$En\mathbf{F} = E_0 n\mathbf{F} - RT \ln \frac{a_K^k \times a_L^l \cdots\cdots}{a_A^a \times a_B^b \cdots\cdots} \quad (10)$$

In general, only the components which are present as solutes or as gases have factors which appear in the last term of this equation.

The *standard potential of an electrode* is defined as the standard potential of a cell in which the other (reference) electrode is the arbitrary zero of potential (equation (2)) as described above. In this chapter the methods for obtaining standard potentials from emf measurements of cells without liquid junctions will be discussed, and the available data will be used for computing such potentials. The order adopted will be, more or less, that of the increasing complexity of the methods employed. Later chapters will deal with liquid junctions and the less accurate standard potentials that can be obtained from emf values of cells containing such junctions.

The Standard Potentials of the Silver-Silver Halide Electrodes. The determination of the standard potential of the silver-silver chloride electrode will be dealt with in detail because it has been carefully studied by various authors and because such electrodes have been much used, as secondary reference electrodes, in recent researches. We have already considered the galvanic cell

$$\text{(Pt) } H_2;\ HCl\ (m),\ AgCl\ (s);\ Ag \qquad (11)$$

in a number of connections.[1] It will be recalled that the operation of this cell involves the reaction

$$\tfrac{1}{2}H_2 + AgCl = H^+ + Cl^- + Ag$$

and the passage through the cell of one faraday.* Of the substances entering into this reaction silver chloride and silver are solids, so that their chemical potentials are constant at a given temperature. Using equation (10) we have

$$E = E_0 - \frac{RT}{\mathbf{F}} \ln \frac{(H^+)(Cl^-)}{(H_2)^{1/2}} \qquad (12)$$

Since, as has been shown, on pages 118 and 128, unless the pressure is very high, the activity of hydrogen is proportional to its partial pressure p_{H_2} equation (12) may be put in the form

$$E = E_0 + \frac{RT}{2\mathbf{F}} \ln p_{H_2} - \frac{RT}{\mathbf{F}} \ln m^2_{HCl}\, \gamma^2_{HCl} \qquad (13)$$

* The convention will be adopted in this book that the "cell reaction" is the one that will take place when current flows, through the cell as written, from left to right. If the potential of the cell is positive the reaction will tend to take place spontaneously in that direction, since, from the equation $En\mathbf{F} = -\Delta Z$, such a reaction will occur with a decrease of Gibbs free energy. If the cell potential is negative the reaction will occur spontaneously in the reverse direction.

[1] Methods for the preparation of hydrogen electrodes are given by S. Popoff, A. H. Kunz and R. D. Snow, *J. Phys. Chem.*, **32**, 1056 (1928) and of silver-silver chloride electrodes by A. S. Brown, *J. Am. Chem. Soc.*, **56**, 646 (1934).

Making the (usually) small correction for the variation of p_{H_2} from one atmosphere and calling the corrected emf of the cell E' we have

$$E' = E_0 - \frac{RT}{\mathbf{F}} \ln m^2 \gamma^2 \tag{14}$$

In accordance with almost universal usage E_0 values will be determined in terms of the molality, m, rather than the concentration, C. The reason for this is that the molality scale has the distinct advantage of being independent of the temperature. The relation between E_0 values on the two concentration scales will be discussed in Chapter 12. The method used by Brown and MacInnes [2] for obtaining E_0 values from data on the potentials of cells of type (11) is as follows. If we define

$$E'' = E' + \frac{2RT}{\mathbf{F}} \ln m \tag{15}$$

and use equation (25), Chapter 7, which may be put into the sufficiently accurate form, for very dilute solutions and extrapolation, of [3]

$$-\log \gamma = \frac{A'\sqrt{m}}{1 + \beta' a_i \sqrt{m}} \tag{16}$$

then equation (14) may be stated as

$$E'' - E_0 = -\frac{2RT}{\mathbf{F}} \ln \gamma = \frac{2R'T}{\mathbf{F}} \frac{A'\sqrt{m}}{1 + \beta' a_i \sqrt{m}} \tag{17}$$

in which R' is equal to the gas constant, R, multiplied by 2.3026. This equation may be rearranged to give

$$E'' - \frac{2R'T}{\mathbf{F}} A'\sqrt{m} = E_0 - (E'' - E_0)\beta' a_i \sqrt{m} \tag{18}$$

Thus if the Debye-Hückel theory as given in equation (16) holds, a plot of values of $(E'' - 2A'R'T\sqrt{m}/\mathbf{F})$ against $(E'' - E_0)\sqrt{m}$ as abscissae should be a straight line with a slope of $\beta' a_i$ and an intercept of E_0. In using this method a short series of approximations is made until the value of E_0 used in computing values of the abscissae agrees with that of the intercept.[4] Such a plot is shown in Fig. 1, based on the emf data of Harned and Ehlers,[5] and Roberts [6] for the cell repre-

[2] A. S. Brown and D. A. MacInnes, *J. Am. Chem. Soc.*, **57**, 1356 (1935).

[3] A' and β' differ from A and β of equation (25), Chapter 7, by the factor $\sqrt{\rho_0}$ in which ρ_0 is the density of the solvent.

[4] A preliminary estimation of E_0 may be made by the method of D. I. Hitchcock, *J. Am. Chem. Soc.*, **50**, 2076 (1928). In this case it consists in plotting $(E'' - 2A'R'T\sqrt{m}/\mathbf{F})$ as a function of m, the intercept giving a value of E_0. With accurate data such a plot usually shows a slight curvature.

[5] H. S. Harned and R. W. Ehlers, *J. Am. Chem. Soc.*, **54**, 1350 (1932).

[6] E. J. Roberts, *ibid.*, **52**, 3877 (1930).

sented by equation (11). This extrapolation leads clearly to a value of 0.2225 (to the nearest tenth of a millivolt) for the E_0 value of the cell. This plot is based on the most recent series of measurements on this type of cell. Similar measurements have been carried out by Linhart,[7] Nonhebel,[8] Scatchard,[9] Carmody [10] and other workers, and

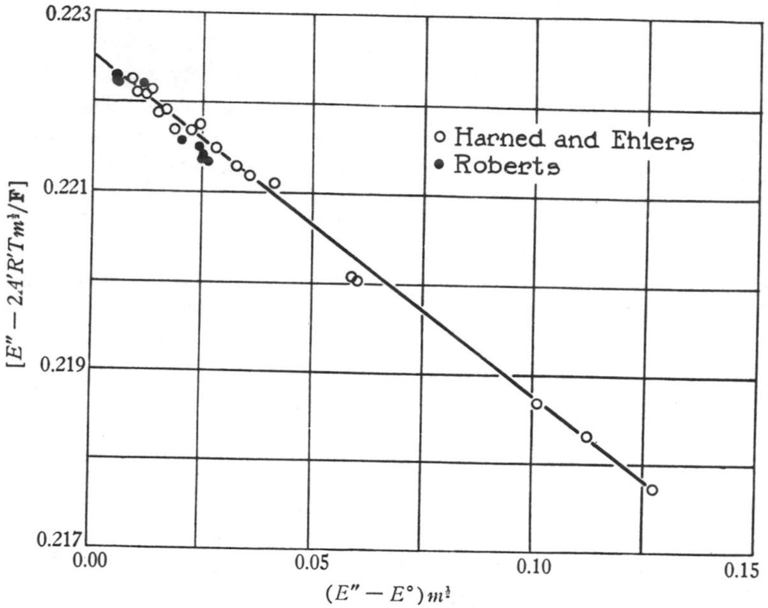

Fig. 1. Plot for Obtaining the Standard Potential of the Silver-Silver Chloride Electrode at 25°.

their results which have been carefully analyzed by Prentiss and Scatchard [11] lead to values about ± 0.1 millivolt of the one given.

Since the method just outlined gives possibly undue weight to the data obtained in very dilute solutions where the accuracy is lowest, part at least of this difficulty may be overcome by utilizing the activity coefficients of hydrochloric acid, obtained by Shedlovsky and MacInnes,[12] and given in Table II of Chapter 8. These latter data do not involve the hydrogen electrode, the variations of which are probably the chief cause of the scattering of the points in Fig. 1. By interpolating with

[7] G. A. Linhart, *J. Am. Chem. Soc.*, **41**, 1175 (1919).
[8] G. Nonhebel, *Phil. Mag.*, [7] **2**, 1085 (1926).
[9] G. Scatchard, *J. Am. Chem. Soc.*, **47**, 641 (1925).
[10] W. R. Carmody, *ibid.*, **54**, 188 (1932).
[11] S. S. Prentiss and G. Scatchard, *Chem. Rev.*, **13**, 139 (1933).
[12] T. Shedlovsky and D. A. MacInnes, *J. Am. Chem. Soc.*, **58**, 1970 (1936).

the aid of equation (33), Chapter 8, γ values were obtained corresponding to the molalities, m, at which Harned and Ehlers' measurements were made. Substituting these quantities in equation (14), gives the series of E_0 values in Table I. The average of the figures given in Table I, column 4, is 0.2225 volt in agreement with the value given

TABLE I. THE COMPUTATION OF THE STANDARD POTENTIAL OF THE CELL (Pt)H_2: HCl, AgCl(s), Ag

Molality of HCl m	Activity coefficient γ	Emf in volts, at 1 atm. pressure, E'	Standard potential volts E_0'
0.003215	0.9418	0.52053	0.22255
.004488	.9328	.50384	.22251
.005619	.9259	.49257	.22241
.007311	.9173	.47948	.22236
.009138	.9094	.46860	.22250
.011195	.9031	.45861	.22258
.013407	.8946	.44974	.22248
.01710	.8843	.43783	.22247
.02563	.8660	.41824	.22260
.05391	.8293	.38222	.22256
.1238	.7877	.34199	.22244
		mean value	0.22250

above. Thus the standard potential of the cell

$$\text{Ag; AgCl (s), HCl ; } H_2 \text{ (Pt)}$$

is $-$ 0.2225 volt and, recalling the definition of the arbitrary zero of potential, this yields for the standard potential of the silver-silver chloride electrode at 25°,

$$\text{Ag; AgCl (s), Cl}^- \qquad E_0 = -\ 0.2225 \tag{19}$$

Values at other temperatures are given in Table V. Knowing E_0 it is evidently possible to compute the activity coefficients, γ, of hydrochloric acid at the molality, m, from the potential E of the cell of equation (11). Such activity coefficients, as a matter of fact, agree very closely with those obtained from concentration cells with transference. This is indicated indirectly by Fig. 5, Chapter 8.

Recently, very accurate measurements have been made on the potentials of the cell

$$\text{Ag; AgBr (s), HBr } (m) \text{ ; } H_2 \text{ (Pt)} \tag{20}$$

by Keston [13] in dilute solutions, and by Harned, Keston and Donelson[14] in more concentrated solutions. Keston has computed the standard

[13] A. S. Keston, *J. Am. Chem. Soc.*, **57**, 1671 (1935).

[14] H. S. Harned, A. S. Keston and J. G. Donelson, *ibid.*, **58**, 989 (1936); see also H. S. Harned and J. G. Donelson, *ibid.*, **59**, 1280 (1937).

potential of the cell substantially as described for the cell containing the silver-silver chloride electrode and has obtained, at 25°

$$\text{Ag ; AgBr (s), Br}^- \qquad E_0 = -0.0711 \tag{21}$$

The activity coefficients of hydrobromic acid, in Table V, Chapter 8, are based on these measurements, the constant just given and equation (14).

The Standard Potential of Zinc. The determination of the standard potential of zinc will also be considered in detail since it can be obtained from measurements on several different types of galvanic cells. In addition, a discussion of the computations from the different data brings up most of the questions involved in this field. The types of cell on which the measurements have been made are:

$$\text{(Hg)Zn; ZnCl}_2\text{, AgCl (s); Ag} \tag{23}$$

$$\text{Zn; ZnCl}_2\text{, HgCl (s); Hg} \tag{24}$$

$$\text{(Hg)Zn; ZnSO}_4\text{, PbSO}_4\text{ (s); Pb(Hg)} \tag{25}$$

in which, in these cases, the symbol (Hg)Zn stands for the two phase zinc amalgam. Scatchard and Tefft [15] obtained potentials of cell (23) and used them for an extrapolation of the type suggested by Hitchcock [16] and described in principle in Chapter 8. The equation for the potential of cell (23), in which the reaction

$$\text{Zn} + 2\text{AgCl} = \text{Ag} + \text{ZnCl}_2$$

occurs, from equations (10) of this chapter and (42a) of Chapter 6, is

$$E = E_0 - \frac{RT}{2\mathbf{F}} \ln\,[\text{Zn}]\;[\text{Cl}]^2\,\gamma^3 \tag{26}$$

Here [Zn] and [Cl] represent the molalities of the constituents indicated and γ is the mean ion activity coefficient. This equation can be rearranged to

$$E = E_0 - \frac{RT}{2\mathbf{F}} \ln m\,(2m)^2 - \frac{RT}{2\mathbf{F}} \ln \gamma^3 \tag{27}$$

or

$$E = E_0 - \frac{3RT}{2\mathbf{F}} \ln 4^{1/3}m - \frac{3RT}{2\mathbf{F}} \ln \gamma \tag{28}$$

in which m is the molality. For the purposes of this extrapolation equation (45), Chapter 7, is given, for a uni-bivalent salt, the approximate form

$$-\log \gamma = 2A'\sqrt{\omega} - B\omega \tag{29}$$

[15] G. Scatchard and R. F. Tefft, *J. Am. Chem. Soc.*, **52**, 2272 (1930).
[16] D. I. Hitchcock, *ibid.*, **50**, 2076 (1928).

Substituting equation (29) in equation (28), and recalling that R' equals $2.3026R$

$$E = E_0 - \frac{3RT}{2\mathbf{F}} \ln 4^{1/3}m + \frac{3R'T}{\mathbf{F}} A'\sqrt{\omega} - \frac{3R'T}{2\mathbf{F}} B\omega \qquad (30)$$

If we define E_0'' as

$$E + \frac{3RT}{2\mathbf{F}} \ln 4^{1/3}m - \frac{3R'T}{\mathbf{F}} A'\sqrt{\omega} \qquad (31)$$

and plot it as a function of ω then E_0 will be the intercept and the slope, if equation (29) is valid, will be $3R'TB/2\mathbf{F}$. The E_0 value for a cell of type (23) obtained by this means by Scatchard and Tefft is 0.9834 volt.[17] With the aid of this standard potential, together with the potentials of cell (23) and equation (26) a series of values of the activity coefficients of zinc chloride may be computed. These are given in Table V, Chapter 8.

The difference of potential between metallic zinc and two phase zinc amalgam, *i.e.*, for instance, the potential of the cell

$$\text{Zn; ZnSO}_4\text{: Zn(Hg)}$$

has a value of zero[18] so that for the combination

$$\text{Zn; ZnCl}_2\text{, AgCl (s) ; Ag} \qquad E_0 = 0.9834 \qquad (32)$$

This potential is the algebraic sum of the standard potentials of the electrodes Zn; Zn^{++} and Ag; AgCl (s), Cl^-. The value of the latter we have decided to be -0.2225 volt from which

$$\text{Zn ; Zn}^{++} \qquad E_0 = 0.7609 \qquad (33)$$

Another series of measurements has been made by Getman[19] on cells of the type shown by equation (24), using, instead of zinc amalgam, large crystals of the metal. An extrapolation of his results by the method of Hitchcock, as just described, leads to a value of 1.029_2 for the E_0 value of the cell. Accurate measurements at low concentrations of the potentials of cells of the form

$$\text{(Pt) H}_2\text{; HCl, HgCl (s); Hg} \qquad (34)$$

are not available so that there is no directly determined value of the E_0

[17] Using the extended Debye-Hückel theory for non-symmetrical electrolytes V. K. LaMer, T. H. Gronwall and L. J. Greiff, *J. Phys. Chem.*, **35**, 2245 (1931), have obtained a slightly higher value, $E_0 = 0.9837$ volt, from these same data.

[18] W. J. Clayton and W. C. Vosburg, *J. Am. Chem. Soc.*, **58**, 2093 (1936).

[19] F. H. Getman, *J. Phys. Chem.*, **35**, 2749 (1931).

for the reference "calomel electrode," Hg; HgCl (s), Cl^-. However, the potential of the combination

$$Hg;\ HgCl\ (s),\ MCl,\ AgCl\ (s);\ Ag$$

has been measured at 25° by various workers. Using hydrochloric acid for the solute, MCl, Randall and Young[20] have shown that the potential is constant over a wide range of concentration, and equal to −0.0456 volt at 25°. This value with the E_0 of the silver-silver chloride electrode yields, at 25°

$$Hg;\ HgCl\ (s),\ Cl^- \qquad E_0 = -\ 0.2681 \qquad (35)$$

Together with 1.029_2 volts for the E_0 value for cell (24) we have another value for the standard potential of zinc

$$Zn;\ Zn^{++} \qquad E_0 = 0.7611 \qquad (36)$$

When applicable, the extended Debye-Hückel theory of Gronwall, LaMer and Sandved,[21] although rather laborious, furnishes a means for the determination of E_0 values. Its use may be illustrated by the work of Cowperthwaite and LaMer[22] on zinc sulphate, using a cell of type (25). Their data are given in Table II. Due to the fact that this

TABLE II. STANDARD POTENTIAL OF THE CELL, Zn; $ZnSO_4$, $PbSO_4$(s); Pb(Hg) AT 25° AS COMPUTED BY THE GRONWALL, LAMER AND SANDVED EXTENSION OF THE DEBYE-HÜCKEL THEORY

Molality of $ZnSO_4$ m	Emf observed volts E	E'_0	Standard potential volts E_0	$(0.41089 - E_0)$ millivolts
0.0005	0.61144	0.41725	0.41088	− 0.01
.001	.59714	.42002	.41093	+ 0.04
.002	.58319	.42365	.41089	± 0.00
.005	.56598	.42989	.41094	+ 0.05
.01	.55353	.43524	.41084	− 0.05
.02	.54252	.44203	(.41176)	+ 0.87
.05	.52867	.45171	(.41296)	+ 2.07

salt is of a higher valence type than any thus far considered the effect of the higher terms, neglected in the simple Debye-Hückel theory, is greater. One effect of these higher terms is to produce a bend or hump in a curve corresponding to Fig. 1 making an accurate extrapolation by the Hitchcock or Brown and MacInnes methods impossible. For the cell represented by equation (25) the reaction is

$$Zn + PbSO_4 = ZnSO_4 + Pb$$

[20] M. Randall and L. E. Young, *J. Am. Chem. Soc.*, **50**, 989 (1928).
[21] T. H. Gronwall, V. K. LaMer and K. Sandved, *Physik. Z.*, **29**, 358 (1928).
[22] I. A. Cowperthwaite and V. K. LaMer, *J. Am. Chem. Soc.*, **53**, 4333 (1931).

and for this case equation (10) takes the form

$$E = E_0 - \frac{RT}{\mathbf{F}} \ln (m\gamma) \tag{37}$$

If

$$E_0' = E + \frac{RT}{\mathbf{F}} \ln m$$

then

$$E_0' = E_0 - \frac{RT}{\mathbf{F}} \ln \gamma \tag{38}$$

Values of E_0' are given in the third column of Table II. They include a small correction for the appreciable solubility of lead sulphate. Equation (47), Chapter 7, and the measured values of E_0' give relations from which the unknown quantities E_0 and the distance of closest approach, a_i may be computed.[23] Trial a_i values may therefore be substituted until one is found which yields the most nearly constant E_0 values. In this case a value of 3.64Å has been found to give the figures for E_0 in the fourth column of the table. They will be observed to be very close to $E_0 = 0.4109$ volt for the measurements with the more dilute solutions. Thus at 25° for the cell

$$\text{Zn(Hg); ZnSO}_4\text{, PbSO}_4\text{ (s); Pb(Hg)} \qquad E_0 = 0.4109 \tag{39}$$

This standard potential has been used for computing the activity coefficients of zinc sulphate given in Table V, Chapter 8. From this value the standard potential of zinc can obviously be obtained if an E_0 value of the cell

$$\underset{1\text{ atm.}}{\text{(Pt) H}_2}\text{; H}_2\text{SO}_4\text{, PbSO}_4\text{ (s); Pb(Hg)} \tag{40}$$

is available. Data, given in Table III, on the emf of this cell have been obtained by Baumstark [24] and by Shrawder and Cowperthwaite.[25] The extrapolation to obtain an E_0 value involves a variable of a different kind from those so far discussed. This variable is the partial dissociation of the sulphuric acid according to the equations

$$H_2SO_4 \rightleftarrows H^+ + HSO_4^- \rightleftarrows 2H^+ + SO_4^{--}$$

The first of these two steps goes to substantial completion. The ionization of HSO_4^- ion is, however, only partial, and this fact must be

[23] In this computation f, calculated by equation (57), Chapter 7, has been substituted for γ in equation (38) since the extended theory has meaning only when concentrations are expressed on a volume scale. Cowperthwaite and LaMer have, however, pointed out that if m is substituted for C in equation (47), Chapter 7, the resulting E_0 for cell (25) at 25, 37.5 and 50° will be lowered by only 0.01, 0.04 and 0.11 millivolt, respectively.

[24] G. Baumstark, *Dissertation,* Catholic University of America, Washington, D. C. (1932).

[25] J. Shrawder and I. A. Cowperthwaite, *J. Am. Chem. Soc.*, **56,** 2340 (1934).

considered in making the extrapolation. The ionization constant **K** for the dissociation

$$HSO_4^- \rightleftarrows H^+ + SO_4^{--}$$

has been determined by various authors, the most recent being Hamer [26] whose method and results will be considered in Chapter 11. From that constant, degrees of dissociation, α, of the HSO_4^- ion may be estimated. Applying equation (10) to the process for the cell, represented by equation (40), gives

$$E = E_0 - \frac{3RT}{2\mathbf{F}} \ln 4^{1/3} (m\gamma) \tag{41}$$

In close analogy to the consideration of the cell represented by equation (23) the extrapolation by the Hitchcock method for this case can be readily seen to consist in plotting E_0'' against the ionic strength, ω, where

$$E_0'' = E + \frac{3R'T}{2\mathbf{F}} (\log m + \log 4^{1/3} - 2A'\sqrt{\omega})$$

In computing the ionic strength the ionic molalities are, however, obtained as follows:

$$m_{H^+} = m(1 + \alpha), m_{HSO_4} = m(1 - \alpha) \text{ and } m_{SO_4} = m\alpha$$

from which, with equation (40), Chapter 7,

$$\omega = m(1 + 2\alpha)$$

TABLE III. STANDARD POTENTIAL OF THE CELL[1]
(Pt)H_2; H_2SO_4, $PbSO_4$(s); Pb(Hg) at 25°

Molality of H_2SO_4 m	Measured potential volts E	Degree of dissociation, α	Ionic strength ω	E_0''
0.0004704	– 0.07023	0.944	0.001358	– 0.34961
0.0007269	– 0.08496	0.918	0.002063	– 0.34900
0.0008120	– 0.08867	0.910	0.002289	– 0.34877
0.00100*	– 0.09589	0.895	0.002790	– 0.34840
0.001008	– 0.09603	0.894	0.002810	– 0.34844
0.00200*	– 0.11895	0.836	0.005342	– 0.34688
0.002664	– 0.12802	0.805	0.006950	– 0.34603
0.004080	– 0.14108	0.752	0.010219	– 0.34429
0.004938	– 0.14683	0.727	0.012116	– 0.34350
0.00500*	– 0.14745	0.726	0.012257	– 0.34360
0.005297	– 0.14871	0.717	0.01290	– 0.34300
0.007181	– 0.15737	0.676	0.01689	– 0.34142
0.009711	– 0.16802	0.629	0.02192	– 0.34208

[1] Starred measurements are from Shrawder and Cowperthwaite, the others from Baumstark.

Here, the ionic strength is computed, to quite sufficient approximation, in terms of molalities, m instead of molar concentrations, C. The data are given in Table III, which is self-explanatory, and the resulting plot,

[26] W. J. Hamer, *J. Am. Chem. Soc.*, **56**, 860 (1934).

according to the Hitchcock method, is shown in Fig. 2, which extrapolates to an E_0 value, -0.3505 volt, and thus yields

$$Pb(Hg);\ PbSO_4\ (s),\ SO_4^{--} \qquad E_0 = 0.3505 \qquad (42)$$

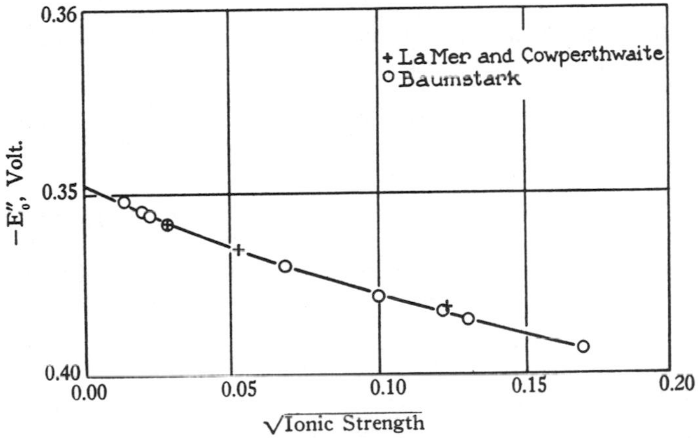

Fig. 2. Extrapolation Plot for Obtaining the Standard Potential of the Cell $Pb\ (Hg)\ ;\ PbSO_4,\ H_2SO_4;\ H_2\ (Pt)$ at 25°.

from which with the aid of the E_0 value for cell (25) we obtain another value for the standard potential of zinc at 25° of

$$Zn(Hg);\ Zn^{++} \qquad E_0 = 0.7614 \qquad (43)$$

and recalling that the potential between the metal and the two phase amalgam has been found to be zero;

$$Zn;\ Zn^{++} \qquad E_0 = 0.7614$$

agreeing excellently with the other determinations of this potential as described above. We will adopt, at 25° the average value

$$Zn;\ Zn^{++} \qquad E_0 = 0.7611 \qquad (44)$$

The Standard Potential of Cadmium. LaMer and Parks[27] have measured the potentials of cells of the type

$$Cd(Hg);\ CdSO_4\ (m),\ PbSO_4\ (s);\ Pb(Hg) \qquad (45)$$

at a series of temperatures and have interpreted their results in terms of the extension of the Debye-Hückel theory, the computations being

[27] V. K. LaMer and W. G. Parks, *J. Am. Chem. Soc.*, **55**, 4343 (1933).

essentially the same as those for the analogous zinc sulphate concentration cells. At 25° they compute a value of 0.0013 volt for the E_0 value. They have also measured [28] the potentials of cells of the type

$$Cd;\ CdSO_4\ (0.05m);\ Cd(Hg) \qquad (46)$$

for which they obtain $E = 0.0505$ volt at 25°. These two results together with the E_0 value for

$$Pb(Hg);\ PbSO_4\ (s),\ SO_4^{--}$$

discussed above yield, at 25°

$$Cd;\ Cd^{++} \qquad E_0 = 0.4023 \qquad (47)$$

The standard potential of cadmium has also been studied by Harned and Fitzgerald,[29] using the cell

$$Cd\ (Hg);\ CdCl_2,\ AgCl\ (s);\ Ag \qquad (47a)$$

Harned and Fitzgerald interpreted their results on the assumption that cadmium chloride is incompletely ionized and dissociates according to the equations

$$CdCl_2 \rightleftarrows CdCl^+ + Cl^-$$
$$CdCl^+ \rightleftarrows Cd^{++} + Cl^-$$

Their treatment of the data was, therefore, closely analogous to that for the cell represented by equation (40), since the sulphuric acid contained in that cell is also considered to ionize incompletely, forming intermediate ions. In this manner they arrived at the E_0 value, at 25°, for the cell (47a) of 0.5739 volt which with equations (46) and (19) gives

$$Cd;\ Cd^{++} \qquad E_0 = 0.4019 \qquad (47b)$$

in good agreement with the value given above.

The Standard Potential of Lead. Carmody [30] has studied cells of the type

$$Pb(Hg);\ PbCl_2\ (m),\ AgCl\ (s);\ Ag$$

at 25° throughout the range of molalities 0.0002 to 0.04. This cell is evidently analogous to that shown in equation (23) and computations of the standard potential may be made with equation (28). Using

[28] W. G. Parks and V. K. LaMer, *ibid.*, **56**, 90 (1934).
[29] H. S. Harned and M. Fitzgerald, *J. Am. Chem. Soc.*, **58**, 2624 (1936).
[30] W. R. Carmody, *ibid.*, **51**, 2905 (1929).

either the Hitchcock or Brown and MacInnes procedure the E_0 value of this cell is 0.3432 volt, which with equation (19) yields, at 25°

$$Pb\,(Hg);\,Pb^{++} \qquad E_0 = 0.1207 \tag{48}$$

Carmody[31] has measured the potential between lead amalgam and lead crystals and found it to be 0.0058 volt from which, at 25°

$$Pb;\,Pb^{++} \qquad E_0 = 0.1265 \tag{49}$$

The Standard Potentials of Nickel, Cobalt and Copper. Data on galvanic cells containing dilute solutions of nickel, cobalt and copper salts are not available. A number of workers have, however, measured the potentials of the combination

$$M;\, MSO_4\,(m),\, Hg_2SO_4\,(s);\, Hg \tag{50}$$

where M is one of the three metals mentioned and the molality, m, was usually above 0.05 molal. The relations holding in dilute solutions cannot, therefore, be used in interpreting the data. However on making a plot according to the Hitchcock procedure, the resulting lines through the experimental points were found to be parallel to each other and to a similar plot for the emf data at those concentrations from the zinc sulphate cells of Bray[32] and Kielland.[33] This is an indication that the activity coefficients of the four sulphates are very nearly the same. Using the activity coefficients of zinc sulphate from Table V, Chapter 8, which are based on the data of the authors just mentioned and upon those of Cowperthwaite and LaMer[34] for very dilute solutions, a series of E_0 values was computed, using equation (10), which for this case takes the form

$$E = E_0 - \frac{RT}{\mathbf{F}} \ln m\gamma \tag{51}$$

Here E represents a measured potential of a cell of the type shown in equation (50). The resulting E_0 values for copper,[35] cobalt[36] and nickel[37] were constant within the experimental accuracy of the original data, which in the case of cobalt at least, was not very good. However, reproducibility in emf measurements is difficult to obtain when hard

[31] W. R. Carmody, *loc. cit.*
[32] U. B. Bray, *J. Am. Chem. Soc.*, **49,** 2372 (1927).
[33] J. Kielland, *ibid.*, **58,** 1855 (1936).
[34] I. A. Cowperthwaite and V. K. LaMer, *ibid.*, **53,** 4333 (1931).
[35] R. F. Neilson and D. J. Brown, *J. Am. Chem. Soc.*, **49,** 2423 (1927).
G. N. Lewis and W. N. Lacey, *ibid.*, **36,** 804 (1914).
F. H. Getman, *J. Phys. Chem.*, **34,** 1454 (1930).
F. E. W. Wetmore and A. R. Gordon, *J. Chem. Phys.*, **5,** 60 (1937).
[36] M. M. Haring and B. B. Westfall, *Trans. Electrochem. Soc.*, **55,** 235 (1934).
[37] M. M. Haring and V. Bosche, *J. Phys. Chem.*, **33,** 161 (1929).

metals are used as electrodes, variations being observed due to strains arising from mechanical and heat treatment of the metals, to absorption of gases, etc. Much greater reproducibility is obtained if the metals are amalgamated, but the resulting potential usually differs, as we have already seen, markedly from that of the pure metal. Averaging the E values computed as just described yields, at 25°, the following:

$$\text{Cu (Hg); CuSO}_4\text{, Hg}_2\text{SO}_4\text{ (s); Hg} \qquad E_0 = 0.270$$
$$\text{Co; CoSO}_4\text{ Hg}_2\text{SO}_4\text{ (s); Hg} \qquad E_0 = 0.897$$
$$\text{Ni; NiSO}_4\text{, Hg}_2\text{SO}_4\text{ (s); Hg} \qquad E_0 = 0.850$$

To obtain the standard potentials of the metals from these values it is necessary to have a value of the standard potential of the reference electrode

$$\text{Hg; Hg}_2\text{SO}_4\text{ (s), SO}_4^{--}$$

This can be obtained from the E_0 value of the cell given in equation (42) and the potential of the combination

$$\text{Pb (Hg); PbSO}_4\text{ (s), Na}_2\text{SO}_4\text{, Hg}_2\text{SO}_4\text{ (s); Hg} \qquad (52)$$

which has been measured by Henderson and Stegeman,[38] Harned and Hamer [39] and LaMer and Carpenter [40] who agree upon the potential of 0.9646 at 25°. We thus obtain, with the aid of equation (42)

$$\text{Hg; Hg}_2\text{SO}_4\text{ (s), SO}_4^{--} \qquad E_0 = -\ 0.6141 \qquad (53)$$

The potential of the cell

$$\text{Cu (Hg); CuSO}_4\text{; Cu} \qquad E = 0.0051 \qquad (54)$$

which is also needed, has been measured by Oku.[41] Thus

$$\text{Cu; Cu}^{++} \qquad E_0 = -\ 0.339 \qquad (55)$$
$$\text{Co; Co}^{++} \qquad E_0 = +\ 0.283 \qquad (56)$$
$$\text{Ni; Ni}^{++} \qquad E_0 = +\ 0.236 \qquad (57)$$

Standard Potentials of the Alkali Metals. Since the alkali metals: sodium, potassium, etc., react violently with water their standard potentials cannot be determined directly. However, by using an ingenious device, due to G. N. Lewis, these constants can also be obtained. The method will be illustrated by the determination of the standard potential of potassium. The potentials of two cells are measured. One of them has as electrodes potassium metal and dilute

[38] W. E. Henderson and G. Stegeman, *J. Am. Chem. Soc.*, **40**, 84 (1918).
[39] H. S. Harned and W. J. Hamer, *ibid.*, **57**, 33 (1935).
[40] V. K. LaMer and E. L. Carpenter, *J. Phys. Chem.*, **40**, 287 (1936).
[41] M. Oku, *Sci. Repts. Tôhoku Imp. Univ.*, [1] **22**, 288 (1933).

potassium amalgam, with a non-aqueous solution of a potassium salt as electrolyte. The solvent chosen is one which does not react with the metal or amalgam, but dissolves a potassium salt to yield a conducting solution. The actual cell measured may be represented by

K (metal); KI in ethylamine; K (amalgam, 0.2216%) (58)

for which the potential of 1.0478 volts was found at 25°, by Lewis and Keyes.[42] The other cell measured may be represented by:

K (amalgam); KCl (1.0168*m*), AgCl (s); Ag (59)

This was found by Armbruster and Crenshaw [43] to have an emf of 2.0704 volts at 25°. These authors also obtained data from which they obtained the difference (0.0532 volt) between the amalgam used by them and that used by Lewis and Keyes. The measurements of cell (59) involved the use of flowing amalgam electrodes much as described in Chapter 8, page 153. Adding these three potentials we have for the cell

K (metal); KCl (1.0168 *m*), AgCl (s); Ag $E = 3.1714$ (60)

From the equation

$$E = E_0 - \frac{RT}{\mathbf{F}} \ln m^2\gamma^2$$

and interpolating a value of 0.605 for the mean ion activity for potassium chloride at the concentration given, we obtain 3.1464 volts for this value of E_0, which, together with the value of E_0 for the silver-silver chloride electrode yields, at 25°

K; K^+ $E_0 = 2.9239$ (61)

An analogous computation, leading to the standard potential of sodium, may be made by combining the measurements of Lewis and Kraus [44] who measured the potential, at 25°, of the cell:

Na (metal); NaI in ethylamine; Na (amalgam, 0.206%) (62)

obtaining a potential of 0.8453 volt, with the determination of Allmand and Polack [45] from whose work the emf of the cell:

Na (amalgam, 0.206%) ; NaCl (0.1005*m*) , HgCl(s) ; Hg (63)

may be interpolated to be 2.2676 volts. These values, together with a value of 0.778 for γ for 0.1005 molal sodium chloride from Table V,

[42] G. N. Lewis and F. G. Keyes, *J. Am. Chem. Soc.*, **34,** 119 (1912).
[43] M. H. Armbruster and J. L. Crenshaw, *ibid.*, **56,** 2525 (1934).
[44] G. N. Lewis and C. A. Kraus, *J. Am. Chem. Soc.*, **32,** 1459 (1910).
[45] A. J. Allmand and W. G. Polack, *J. Chem. Soc.*, **115,** 1020 (1919).

Chapter 8 and the E_0 value of the Hg; HgCl(s), Cl^- electrode given on page 190 yield

$$\text{Na; Na}^+ \qquad E_0 = 2.7139 \qquad (64)$$

These values are in fair agreement with the earlier work by Lewis and associates involving cells with liquid junctions, which have been used in all the available determinations of the standard potentials of the other alkali metals, and will be discussed in Chapter 14.

The Standard Potential of Chlorine. Measurements of the potentials of galvanic cells without liquid junctions from which the standard potential of chlorine may be deduced have been made by Lewis and Ruppert [46] who used, as one electrode, platinum over which a mixture of chlorine and nitrogen was bubbled, and, as reference, a calomel electrode and hydrochloric acid as electrolyte. The arrangement may be represented by

$$(\text{Pt})\ \underset{p_1}{\text{Cl}_2},\ \underset{p_2}{\text{N}_2};\ \text{HCl}\ (m),\ \text{HgCl (s); Hg} \qquad (65)$$

The partial pressure p_1 of chlorine was kept low to diminish the effects of hydrolysis of the dissolved gas, but later workers have apparently found that precaution unnecessary. Similar measurements were made by Kameyma, Yamamoto, and Oku,[47] except that they substituted saturated potassium chloride for the hydrochloric acid. Since both electrodes used are presumably reversible to the chloride ion constituent the potentials of such cells should be independent of the nature of the positive ion constituent and of the concentration. In both cells the reaction is

$$2\text{HgCl} = \text{Cl}_2 + 2\text{Hg}$$

and the corresponding expression according to equation (10) is

$$E = E_0 - \frac{RT}{2\mathbf{F}} \ln a_{\text{Cl}_2} = E_0 - \frac{RT}{2\mathbf{F}} \ln p_1 f \qquad (66)$$

in which p_1 is the partial pressure of chlorine and f an activity coefficient. The data obtained by the Japanese workers are given in Table IV, together with values of E_0 computed from equation (66) assuming f to be unity, which can be shown to be well within the experimental error. It will be observed that the values of E_0 are quite constant over a wide range of partial pressures of chlorine, with some deviations at the lower pressures. Taking an average value from

[46] G. N. Lewis and F. F. Ruppert, *J. Am. Chem. Soc.*, **33**, 299 (1911).
[47] N. Kameyma, H. Yamamoto, and S. Oku, *Proc. Imp. Acad. (Japan)*, **3**, 41 (1927).

TABLE IV. ELECTROMOTIVE FORCE AND STANDARD POTENTIAL OF THE CELL $(Pt)Cl_2(N_2)$; KCl(sat'd.), HgCl(s), Hg

Partial Pressure of chlorine P_{Cl_2}	Emf volts E	Standard potential volts E_0
0.0675	− 1.0578	(− 1.0924)
.0712	− 1.0583	(− 1.0922)
.1137	− 1.0626	− 1.0905
.1495	− 1.0662	− 1.0906
.228	− 1.0726	− 1.0916
.283	− 1.0746	− 1.0903
.616	− 1.0838	− 1.0900
.833	− 1.0885	− 1.0908
.956	− 1.0899	− 1.0905
	mean value	− 1.0906

these figures, and the E_0 value (page 190) for the Hg; HgCl(s), Cl^- electrode we obtain − 1.3587 volts for the standard potential of chlorine.

Measurements on cell (65) have also been made by Gerke[48] who obtained the potential, corrected as described to one atmosphere partial pressure of chlorine, −1.0903 volts at 25°, corresponding to a standard potential of chlorine of −1.3584 volts. He also determined the potential of the combination

$$(Pt)\ Cl_2;\ HCl\ (m),\ AgCl\ (s);\ Ag$$

which yielded − 1.1364 volts at one atmosphere partial pressure of chlorine. With equation (19) this gives −1.3589 volts for the E_0 value of chlorine. We will therefore adopt, at 25°, the average value

$$(Pt)\ Cl_2\ (1.\ atm.);\ Cl^- \qquad E_0 = -\ 1.3587 \qquad (67)$$

The Standard Potential of the Quinhydrone Electrode. The quinhydrone electrode is of interest and importance as a method for the determination of pH values and because the oxidation-reduction relations of quinone and hydroquinone have been extensively studied. It will however receive consideration here because it is an excellent example of the use of cells without[49] liquid junctions for the determination of the standard potential of a galvanic cell of a somewhat more complex type than those so far considered.

Quinhydrone is a compound of one molecule each of quinone ($C_6H_4O_2$) and hydroquinone ($C_6H_6O_2$). The substance dissociates when dissolved in aqueous solution. At an electrode the reaction

$$\text{hydroquinone} = \text{quinone} + 2H^+ + 2e^-$$

[48] R. H. Gerke, *J. Am. Chem. Soc.*, **44**, 1684 (1922).
[49] See, however, the second paragraph of Chapter 14.

takes place. If therefore the cell

$$\text{(Au); quinone, hydroquinone, HCl } (m) : \text{HCl } (m);\ H_2\text{(Pt)} \quad (68)$$

is set up the cell reaction will be

$$\text{hydroquinone} = \text{quinone} + H_2$$

to which we may apply equation (10) in the form

$$E = E_0 - \frac{RT}{2\mathbf{F}} \ln \frac{a_Q}{a_{Hy}} = E_0 - \frac{RT}{2\mathbf{F}} \ln \frac{m\gamma_Q}{m\gamma_{Hy}} \quad (69)$$

if the pressure of hydrogen has its standard value of one atmosphere. In equation (69) a_{Hy} and a_Q are respectively the activities of hydroquinone and quinone. Since the quinhydrone is present as solid, it maintains equal molalities, m, of quinone and hydroquinone in solution, but not constant activities of these substances unless the activity coefficients γ_Q and γ_{Hy} happen to be the same. It is evident that the directly measured potential E differs from E_0 by a quantity which depends on the ratio γ_{Hy}/γ_Q, which would be expected to be very nearly unity. Hovorka and Dearing[50] have made an extended and careful study of cells of the type represented by equation (68) and of similar cells in which the hydrochloric acid was replaced by other acids, and by mixtures of salts and acids. It was found that the effect of even large increases of the concentration of the electrolyte in the cell was small, and very nearly linear, so that the effect of the various electrolytes on the values of γ_{Hy}/γ_Q and on the potential could be eliminated by a short extrapolation. Fourteen series of experiments, at 25°, all gave lines that converge to a value of:

$$\text{(Au); quinhydrone (s), } H^+ \qquad E_0 = -0.6994 \quad (70)$$

Hovorka and Dearing's work will be mentioned once more in Chapter 15 in connection with the "salt error" of the quinhydrone electrode.

The Variation of the Standard Potentials of Some Electrodes with the Temperature. In a number of cases the standard potentials of galvanic cells without liquid junctions have been determined over a range of temperatures. From these determinations it has been possible to prepare Table V, which gives the standard potentials of a number of electrodes at intervals of 12.5° from 0° to 50°. Some slight adjustments, of the order of ±0.2 millivolt, of the original data have been necessary to bring the figures into accord with the E_0 values at 25° adopted in this book. A more complete table of standard potentials of the elements at 25° will be found at the end of Chapter 14.

[50] F. Hovorka and W. C. Dearing, *J. Am. Chem. Soc.*, **57**, 446 (1935).

TABLE V. STANDARD POTENTIALS IN VOLTS OF SOME ELECTRODES FROM 0° TO 50° FROM GALVANIC CELLS WITHOUT LIQUID JUNCTIONS

Electrode	Temperature C° 0	12.5	25	37.5	50	References
Zn; Zn^{++}	+0.7640	+0.7533	+0.7611	+0.7582	+0.7557	*1,2*
Cd; Cd^{++}	+0.4004	+0.4014	+0.4021	+0.4024	....	*3,14*
Cd(Hg) ; Cd^{++}	+0.3450	+0.3483	+0.3516	+0.3544	....	*3,14*
Pb(Hg) ; $PbSO_4$(s), SO_4^{--}	+0.3281	+0.3392	+0.3505	+0.3619	+0.3738	*4*
Tl; Tl^+	+0.3031	+0.3194	+0.3385	+0.3521	+0.3688	*5*
Ag; AgI(s), I^-	....	+0.1486	+0.1522	+0.1568	....	*6*
(Pt)H_2; H^+	±0.0000	±0.0000	±0.0000	±0.0000	±0.0000	
Ag; AgBr(s), Br^-	−0.0813	−0.0766	−0.0711	−0.0644	−0.0569	*7,8*
Ag; AgCl(s), Cl^-	−0.2364	−0.2300	−0.2225	−0.2140	−0.2045	*9*
Hg; $HgSO_4$(s), SO_4^{--}	−0.6339	−0.6241	−0.6141	−0.6040	−0.5938	*10*
(Au) quinhydrone(s), H^+	−0.7176	−0.7086	−0.6994	−0.6902	....	*11,12*
Pb(Hg); PbO_2(s), $PbSO_4$(s), SO_4^{--}	−1.6769	−1.6808	−1.6849	−1.6894	−1.6944	*13*

1. I. A. Cowperthwaite and V. K. LaMer, *J. Am. Chem. Soc.*, **53**, 4333.
2. W. J. Clayton and W. C. Vosberg, *ibid.*, **58**, 2093 (1936).
3. V. K. LaMer and W. G. Parks, *ibid.*, **55**, 4343 (1933); **56**, 90 (1934).
4. J. Shrawder and I. A. Cowperthwaite, *ibid.*, **56**, 2340 (1934).
5. I. A. Cowperthwaite, V. K. LaMer and J. Barksdale, *ibid.*, **56**, 544.
6. B. B. Owen, *ibid.*, **57**, 1526 (1935).
7. A. S. Keston, *ibid.*, **57**, 1671 (1935).
8. H. S. Harned, A. S. Keston and J. G. Donelson, *ibid.*, **58**, 989 (1936).
9. H. S. Harned and R. W. Ehlers, *ibid.*, **55**, 2179 (1933).
10. H. S. Harned and W. J. Hamer, *ibid.*, **57**, 33 (1935); and ref. 4.
11. F. Hovorka and W. C. Dearing, *ibid.*, **57**, 446 (1935).
12. H. S. Harned and D. D. Wright, *ibid.*, **55**, 4849 (1933).
13. W. J. Hamer, *ibid.*, **57**, 9 (1935).
14. H. S. Harned and M. Fitzgerald, *ibid.*, **58**, 2624 (1936).

Chapter 11

Thermodynamic Ionization Constants from the Potentials of Cells without Liquid Junctions

In recent years much work has been carried out, particularly by H. S. Harned and his associates, on concentration cells without liquid junction for the purpose of obtaining ionization constants of weak electrolytes. The principle involved in these investigations is as follows. Galvanic cells are set up of the form:

$$\underset{p\,=\,1\text{ atm.}}{(\text{Pt})\ H_2};\ HA\ (m_1),\ NaA\ (m_2),\ NaCl\ (m_3),\ AgCl;\ Ag \tag{1}$$

in which HA represents a weak acid and NaA its sodium salt. (Different alkali metal ion constituents and other halogen ions were used. Cell (1) may, however, be considered to be typical of them all.) The actual cell is sufficiently represented by Fig. 2, Chapter 5. The experimental results have, in the original papers, usually been expressed in terms of molalities (m) rather than concentrations, and that convention will be followed in this chapter. The reaction of cell (1) is, per faraday,

$$\tfrac{1}{2}H_2 + AgCl = Ag + H^+ + Cl^-$$

for which equation (10) of Chapter 10 yields, at one atmosphere pressure of hydrogen,

$$E = E_0 - \frac{RT}{\mathbf{F}} \ln (H^+)(Cl^-) = E_0 - \frac{RT}{\mathbf{F}} \ln m_H\gamma_H m_{Cl}\gamma_{Cl} \tag{2}$$

in which the E_0 value is that for the silver-silver chloride electrode with a reversed sign. The hydrogen ion activity arises from the ionization of the weak acid HA, an expression for the thermodynamic ionization constant of which may be obtained as follows. For the equilibrium

$$HA \leftrightarrows H^+ + A^- \tag{3}$$

we have the relation

$$\mu_{HA} = \mu_H + \mu_A \tag{4}$$

and using equation (26c) of Chapter 6, this yields

$$RT \ln \frac{m_H m_A}{m_{HA}} \frac{\gamma_H \gamma_A}{\gamma_{HA}} = \mu^\circ_{HA} - \mu^\circ_H - \mu^\circ_A \quad (5)$$

The terms μ°_{HA}, etc. are constants at a given temperature, so that we may therefore use the relation

$$\mu^\circ_{HA} - \mu^\circ_H - \mu^\circ_A = RT \ln \mathbf{K} \quad (6)$$

from which

$$\frac{m_H m_A}{m_{HA}} \frac{\gamma_H \gamma_A}{\gamma_{HA}} = \mathbf{K} \quad (7)$$

in which **K** is the thermodynamic ionization constant. Now if equation (7) is solved for m_H and substituted in equation (2) there results after rearrangement

$$E - E_0 + \frac{RT}{\mathbf{F}} \ln \frac{m_{HA} m_{Cl}}{m_A} = -\frac{RT}{\mathbf{F}} \ln \frac{\gamma_H \gamma_{Cl} \gamma_{HA}}{\gamma_H \gamma_A} - \frac{RT}{\mathbf{F}} \ln \mathbf{K} \quad (8)$$

In practice the potentials of a series of cells of type (1) are measured in which the molalities of the acid, its salt, and of sodium chloride are known. In the more dilute solutions however there is an appreciable change of the molalities from the stoichiometric values due to the reaction represented by equation (3); but a correction can be made for this effect using a preliminary value of the ionization constant in evaluating the third term on the left-hand side of equation (8) and in computing the ionic strength, ω, which is used as described below. The method may thus involve a series of approximations, but a single approximation is usually sufficient.

If the right-hand side of the equation is represented by $-(RT/\mathbf{F}) \ln \mathbf{K}'$, then $\mathbf{K}'$ should approach **K** as the ionic strength, ω, approaches zero, since the activity coefficients in the first term on the right-hand side of the equation approach unity as ω decreases. Some typical results from the work of Harned and Ehlers[1] on acetic acid are given in Table I. A plot of $\log \mathbf{K}'$ against ω, obtained from those data is shown in Fig. 1. This procedure is adopted because the products $\gamma_H \gamma_{Cl}$ and $\gamma_H \gamma_A$ tend to cancel each other and the activity coefficient of a non-electrolyte, such as γ_{HA}, has been found to be roughly a linear

[1] H. S. Harned and R. W. Ehlers, *J. Am. Chem. Soc.*, **54**, 1350 (1932).

TABLE I. DATA FOR OBTAINING THE IONIZATION CONSTANT OF ACETIC ACID AT 25° FROM THE POTENTIALS OF CONCENTRATION CELLS WITHOUT LIQUID JUNCTIONS

HAc m_1	Molalities NaAc m_2	NaCl m_3	Ionic Strength ω	Emf of cell at 25° volts E	$K' \times 10^5$
0.004779	0.004599	0.004896	0.00951	0.63959	1.752
0.012035	0.011582	0.012426	0.02403	0.61583	1.747
0.021006	0.020216	0.21516	0.04175	0.60154	1.743
0.04922	0.04737	0.05042	0.09781	0.57977	1.734
0.08101	0.07796	0.08297	0.16095	0.56712	1.724
0.09056	0.08716	0.09276	0.17994	0.56423	1.726
				Limiting value **K** =	1.754

function of the ionic strength.[2, 3] The value of the ionization constant, $\mathbf{K} = 1.754 \times 10^{-5}$ for acetic acid at 25° obtained by this method is in close agreement with the result[4] 1.758×10^{-5} obtained by MacInnes and Shedlovsky[5] by an electrical conductance method, to be discussed in Chapter 18.

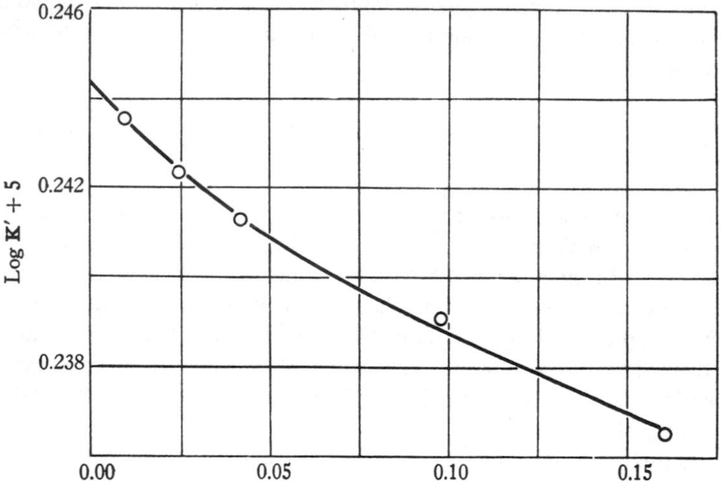

Fig. 1. Plot for the Extrapolation to Zero Ionic Strength for the Ionization Constant of Acetic Acid.

A summary of the extensive results on ionization constants obtained by Harned and associates in the Yale laboratories, using the emf method as just described, is given in Table II.

[2] See for instance M. Randall and C. F. Failey, *Chem. Rev.*, **4**, 285 (1927) for a discussion of this matter.

[3] The E_0 value for the silver-silver chloride electrode at 25° of −0.2224 volt obtained by Harned and associates has been retained in discussing their work. It is only slightly different from the value −0.2225 volt adopted in Chapter 10, and is possibly within experimental error.

[4] A small correction to take account of the different concentration scales has been made using equation (54) of Chapter 18.

[5] D. A. MacInnes and T. Shedlovsky, *J. Am. Chem. Soc.*, **54**, 1429 (1932).

TABLE II. DISSOCIATION CONSTANTS FROM CELLS WITHOUT LIQUID JUNCTIONS FROM THE WORK OF HARNED AND ASSOCIATES

Substance		Temperature, t — 0	10	20	25	30	40	50	60	Reference
Formic Acid	$K \times 10^4$	1.638	1.728	1.765	1.772	1.768	1.716	1.650	1.551	*1*
Acetic Acid	$K \times 10^5$	1.657	1.729	1.753	1.754	1.750	1.703	1.633	1.542	*2*
in 10% MeOH	$K \times 10^5$	1.138	1.200	1.242	1.247	1.237	1.214	.....	.....	*3*
in 20% MeOH	$K \times 10^6$	7.38	7.94	8.24	8.34	8.30	8.19	.....	.....	*3*
Chloroacetic Acid	$K \times 10^3$	1.528	1.491	.....	1.378	.....	1.229	.....	.....	*4*
Propionic Acid	$K \times 10^5$	1.274	1.326	1.338	1.336	1.326	1.284	1.229	1.160	*5*
n-Butyric Acid	$K \times 10^5$	1.563	1.576	1.542	1.515	1.484	1.395	1.302	1.199	*6*
Phosphoric Acid	$K_1 \times 10^3$	8.983	8.519	7.896	7.537	7.152	6.330	5.475	.....	*7*
Phosphoric Acid	$K_2 \times 10^8$	.....	.....	6.056	6.226	6.349	6.471	6.439	.....	*8*
Bisulphate ion	$K_2 \times 10^3$	14.8	13.9	12.7	12.0	11.3	9.73	7.94	5.96	*9*
Boric Acid	$K \times 10^{10}$	.....	4.17	5.25	5.80	6.35	7.39	8.33	.....	*10*
Glycolic Acid	$K \times 10^4$	1.334	1.413	1.463	1.475	1.481	1.480	1.449	.....	*11*
Lactic Acid	$K \times 10^4$	1.290	1.346	1.372	1.374	1.367	1.332	1.267	.....	*12*
Glycine	$K_A \times 10^3$	.....	3.94	4.31	4.47	4.59	4.81	.....	.....	*13*
Glycine	$K_B \times 10^5$	.....	4.68	5.57	6.07	6.52	7.43	.....	.....	*13*
dl-Alanine	$K_1 \times 10^3$	.....	.....	7.30	6.54	6.68	6.76	.....	.....	*14*
dl-Alanine	$K_2 \times 10^{10}$	.....	.....	0.986	1.31	2.29	3.19	.....	.....	*14*
Water	$K \times 10^{14}$	0.115	0.293	0.681	1.008	1.471	2.916	5.476	9.614	*15*

1. H. S. Harned and N. D. Embree, *J. Am. Chem. Soc.*, **56**, 1042 (1934).
2. H. S. Harned and R. W. Ehlers, *ibid.*, **54**, 1350 (1932); **55**, 652 (1933).
3. H. S. Harned and N. D. Embree, *ibid.*, **57**, 1669 (1935).
4. D. D. Wright, *ibid.*, **56**, 314 (1934).
5. H. S. Harned and R. W. Ehlers, *ibid.*, **55**, 2379 (1933).
6. H. S. Harned and R. O. Sutherland, *ibid.*, **56**, 2039 (1934).
7. L. F. Nims, *ibid.*, **56**, 1110 (1934).
8. L. F. Nims, *ibid.*, **55**, 1946 (1933).
9. W. J. Hamer, *ibid.*, **56**, 860 (1934).
10. B. B. Owen, *ibid.*, **56**, 1695 (1934).
11. L. F. Nims, *ibid.*, **58**, 987 (1936).
12. L. F. Nims and P. K. Smith, *J. Biol. Chem.*, **113**, 145 (1936).
13. B. B. Owen, *ibid.*, **56**, 24 (1934).
14. L. F. Nims and P. K. Smith, *J. Biol. Chem.*, **101**, 401 (1933).
15. H. S. Harned and W. J. Hamer, *J. Am. Chem. Soc.*, **55**, 2194 (1933).

The Temperature Dependence of Ionization Constants. It is an interesting fact, pointed out by Harned and Embree,[6] that if values of the logarithm of the ionization constant $\mathbf{K}_t$, are plotted against the temperature, t, the resulting curves for each substance are superposable. This is shown by the curve in Fig. 2 in which values of ($\log \mathbf{K}_t$

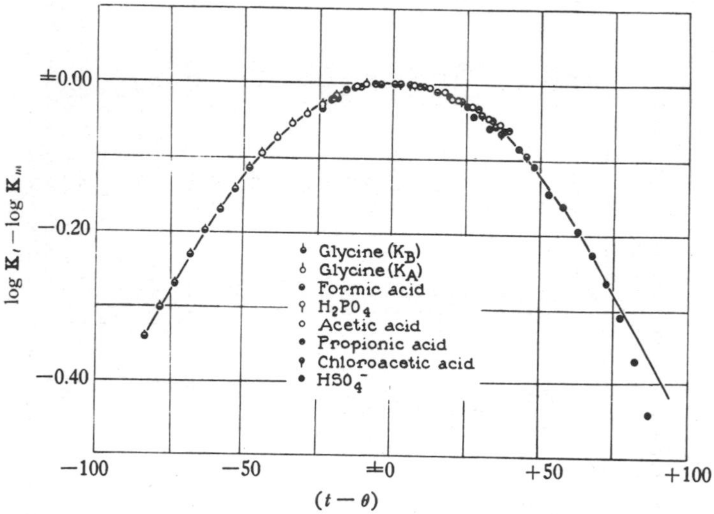

Fig. 2. Plot of ($\log \mathbf{K}_t - \log \mathbf{K}_m$) against ($t-\theta$) for a Series of Acids. The smooth curve represents equation (9).

$- \log \mathbf{K}_m$) are the ordinates and values of ($t - \theta$) are the abscissae. $\mathbf{K}_m$ is the maximum value of the ionization constant $\mathbf{K}_t$ and θ is the temperature corresponding to this maximum. Within a small experimental error the data can also be represented by means of the relation

$$\log \mathbf{K}_t = \log \mathbf{K}_m - 5.0 \times 10^{-5}(t - \theta)^2 \tag{9}$$

Some of the electrolytes, for example chloroacetic acid, do not exhibit maxima in the temperature range studied. For these cases $\mathbf{K}_m$ and θ are of course empirical constants.

The First Ionization Constant of Carbonic Acid. Using a method similar in principle but differing in details from that just described, MacInnes and Belcher[7] have determined the first and second ionization

[6] H. S. Harned and N. D. Embree, *J. Am. Chem. Soc.*, **56**, 1050 (1934).
[7] D. A. MacInnes and D. Belcher, *J. Am. Chem. Soc.*, **55**, 2630 (1933); **57**, 1683 (1935).

constants of carbonic acid at 25° and 38°. Their method for the first of these constants was to measure the potential of the arrangement

$$\textit{glass electrode};\ KHCO_3\ (C),\ CO_2(\text{dissolved}),\ KCl,\ AgCl;\ Ag \qquad (10)$$

The "glass electrode," or more accurately, "glass half cell," is discussed fully in Chapter 15. For the present purpose it can be regarded as a hydrogen electrode with the difference, which was convenient for the experiments under discussion, that hydrogen gas does not have to be used.

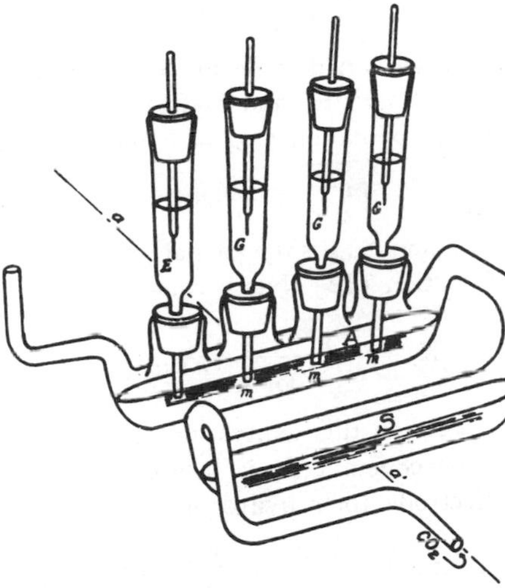

Fig. 3. Apparatus for Measuring the Ionization Constant of Carbonic Acid.

The apparatus used in these measurements is shown in Fig. 3. The solution was contained in the vessel *A*, into which was inserted the tube *E*, open at the lower end, and containing the reference silver-silver chloride electrode. Three glass electrodes *G*, were also inserted into the vessel, the thin glass membranes being represented at the points *m*. The whole apparatus was gently rocked around the axis *a* to *a′* and a slow stream of carbon dioxide was passed first over the solution in the saturator *S* and then over the solution, of the same concentration, in the vessel *A*.

The computations are the same as for the results of measurements on the cell

$$(Pt)\ H_2;\ KHCO_3,\ CO_2,\ KCl,\ AgCl;\ Ag$$

except that allowance must be made for the difference in potential, E_g, between the glass and the hydrogen electrodes. This was a constant for each series of the measurements under discussion. The primary dissociation of carbonic acid follows the reaction

$$CO_2 + H_2O \rightleftarrows H_2CO_3 \rightleftarrows H^+ + HCO_3^- \tag{11}$$

and has the thermodynamic dissociation constant

$$\mathbf{K}_1 = \frac{m_H m_{HCO_3}}{m_{CO_2}} \cdot \frac{\gamma_H \gamma_{HCO_3}}{\gamma_{CO_2}} \tag{12}$$

The terms referring to carbon dioxide are considered to refer to that substance in solution both as CO_2 and H_2CO_3.[8]

To a very close approximation $\gamma_{CO_2} = 1$ and from Henry's law

$$m_{CO_2} = Sp_{CO_2} \tag{13}$$

in which S is a constant and p_{CO_2} is the partial pressure of carbon dioxide in the gas phase. Equation (12) thus becomes

$$\mathbf{K}_1 = \frac{m_H m_{HCO_3}}{Sp_{CO_2}} \gamma_H \gamma_{HCO_3} \tag{14}$$

If m_H is eliminated between this expression and equation (2), and if, as explained above, the constant E_g is inserted to allow for the substitution of a glass electrode for a hydrogen electrode, we obtain

$$E - E_g - E_0 + \frac{RT}{\mathbf{F}} \ln \frac{m_{Cl}}{m_{HCO_3}} + \frac{RT}{\mathbf{F}} \ln S + \frac{RT}{\mathbf{F}} \ln p_{CO_2} = -\frac{RT}{\mathbf{F}} \ln \mathbf{K}_1 - \frac{RT}{\mathbf{F}} \ln \frac{\gamma_H \gamma_{Cl}}{\gamma_H \gamma_{HCO_3}} \tag{15}$$

In the experiments by MacInnes and Belcher a series of measurements of the potentials of cells of the form (10) was made in which the concentrations of KCl and $KHCO_3$ were kept equal, *i.e.*, $m_{Cl} = m_{HCO_3}$ but with changing values of the ionic strength ω. Thus the term $(RT/\mathbf{F})$ $\ln (m_{Cl}/m_{HCO_3})$ was zero except for the very dilute solutions when

[8] The constant $\mathbf{K}_1$, which is strictly equal to $(H^+)(HCO_3^-)/(CO_2 + H_2CO_3)$, has been shown by C. Faurholt (*J. chim. phys.*, **21**, 400 (1924)) to be equal to $\mathbf{K}_1 = \mathbf{K}_j/(1/\mathbf{K}_h + 1)$ in which $\mathbf{K}_j = (H^+)(HCO_3^-)/(H_2CO_3)$, the "true" ionization constant of H_2CO_3, and $\mathbf{K}_h = (H_2CO_3)/(CO_2)$. Since $\mathbf{K}_h$ is a constant if the activity of the solvent water does not change, $\mathbf{K}_1$ will be a constant with this same limitation. A. Thiel and R. Strohecker (*Ber.*, **47**, 945 (1914)) and L. Pusch (*Z. Elektrochem.*, **22**, 206 (1916)) have shown that less than 1% of the dissolved carbon dioxide is in the hydrated form H_2CO_3.

it had a low value, due to the fact that small concentrations of bicarbonate ion can originate from the dissolved carbon dioxide. This involves the ionization constant $\mathbf{K}_1$, but a preliminary value is sufficient for making the correction. Carbon dioxide, usually at atmospheric pressure, was passed over the gently agitated solution in the cell shown in Fig. 3. A few typical data are given in Table III. The values of E

TABLE III. DATA FOR THE COMPUTATION OF THE FIRST DISSOCIATION CONSTANT OF CARBONIC ACID. SOLUTION: EQUAL CONCENTRATIONS OF $KHCO_3$ AND KCl. CO_2 = 99.54 MOL PERCENT, $E_g = +0.6326$ AND HENRY'S LAW CONSTANT, S, = 0.03372

Ionic Strength, ω	Observed emf, volts, E	Barometric Pressure mms. of mercury	Log $\mathbf{K}'$ observed	Log $\mathbf{K}'$ equation (16)
0.002181	0.0530	762.5	6.341	6.345
0.003620	0.0530	761.3	6.343	6.345
0.004281	0.0533	756.0	6.246	6.344
0.005058	0.0531	761.5	6.347	6.344
0.006814	0.0532	754.0	6.344	6.344
0.01109	0.0530	751.9	6.340	6.344
0.02139	0.0530	756.2	6.342	6.342
0.04020	0.0529	758.0	6.342	6.340
0.1006	0.0523	760.4	6.333	6.333
0.2486	0.0512	754.3	6.311	6.316

given in the second column of the table are the results of measurements of cell (10) at the ionic strengths listed in the first column.[9]

As described in the discussion of the computations to determine the ionization constant of acetic acid, if the right-hand side of equation (15) is set equal to $(RT/\mathbf{F}) \ln \mathbf{K}_1'$, then the value of $\mathbf{K}_1'$ should approach the thermodynamic ionization constant $\mathbf{K}_1$ as the activity coefficients, which nearly cancel each other, approach unity. The values of $\mathbf{K}_1'$ given in the fifth column of the table can be expressed by the equation

$$\log \mathbf{K}_1' = 6.345 - 0.119\omega \tag{16}$$

For comparison the values computed from this equation are given in the last column of the table. The constant 6.345 in equation (16) is evidently equal to the value of $\log \mathbf{K}_1'$ when $\omega = 0$, and is thus the logarithm of the thermodynamic ionization constant $\mathbf{K}_1$, which has, therefore, the value 4.52×10^{-7}.

In the computations outlined in Table III the partial pressure of carbon dioxide, p_{CO_2}, was obtained from the relation

$$p_{CO_2} = \frac{(p_{Bar} - p_{H_2O})}{760}\, x \tag{17}$$

[9] The values of the ionic strength, ω, in this table are in concentrations, C, rather than in molalities, m. The equations used are of the same form if the appropriate changes are made of γ for f values, and the differences in the constants involved are just within the experimental error in the present case.

in which p_{Bar} and p_{H_2O} are respectively the barometric pressure and the vapor pressure of water in millimeters of mercury and x is the mol per cent of carbon dioxide in the gas used in the experiments. The effect on the potential of cell (10) of wide variation of the mol per cent x is shown in Table IV, also from the work of MacInnes and Belcher,

TABLE IV. THE EFFECT OF THE PARTIAL PRESSURE OF CARBON DIOXIDE ON THE EMF OF CELL (10)

Mol percent of CO_2, x	Barometric pressure mms. of mercury	Partial Pressure CO_2 (atm)	emf of Cell (10), volts Observed	emf of Cell (10), volts Computed
99.59	726.4	0.9674	− 0.0528	(− 0.0528)
30.50	761.9	0.2962	− 0.0832	− 0.0832
12.03	763.1	0.1170	− 0.1069	− 0.1071
0.54	758.0	0.0052	− 0.1872	− 0.1870

in which the fourth column contains the emf of the cell corresponding to the partial pressures of carbon dioxide given in the third column. The fifth column contains computed values of the emf obtained with the equation

$$E = E^* + \frac{RT}{\mathbf{F}} \ln p_{CO_2} \tag{18}$$

in which E^* is the measured potential of cell (10) when p_{CO_2} is one atmosphere. It will be seen that the computed potentials agree with those observed within a small experimental error. Partial pressures obtained from equation (17) are thus valid for use in equation (15). There is evidence,[10] however, that at higher total pressures equation (17) would not be adequate.

The Second Ionization Constant of Carbonic Acid. MacInnes and Belcher also determined the second ionization constant, $\mathbf{K}_2$, of carbonic acid, corresponding to the equilibrium

$$HCO_3^- \leftrightarrows H^+ + CO_3^{--} \tag{19}$$

using the cell

$$H_2;\ KHCO_3,\ K_2CO_3,\ KCl,\ AgCl;\ Ag \tag{20}$$

the principle involved being much the same as in the cases already discussed. However, since the mixture of carbonate and bicarbonate has an appreciable equilibrium pressure of carbon dioxide it was necessary, by preliminary saturation, to supply that partial pressure of the gas to the hydrogen used for the hydrogen electrode. The two ionization constants of carbon dioxide obtained at 25° and 38° from measure-

[10] E. Lurie and L. J. Gillespie, *J. Am. Chem. Soc.*, **49**, 1146 (1927).

ments on cells without liquid junctions are given in Table V. Other determinations of the first ionization constant from conductance measurements are given in Chapter 18.

TABLE V. THERMODYNAMIC DISSOCIATION CONSTANTS OF CARBONIC ACID

Temperature, t	$K_1 \times 10^7$	$K_2 \times 10^{11}$
25	4.52	5.59
38	4.91	6.25

The Thermodynamic Ionization Constant of Water. Harned and his associates have used the methods outlined in this chapter in studying the ionization of water,

$$H_2O \leftrightarrows H^+ + OH^-$$

and in determining its limiting, or thermodynamic, ionization constant. Of several methods utilized, the simplest and possibly the most effective is as follows. This constant is defined as

$$\mathbf{K}_w = m_H m_{OH} \frac{\gamma_H \gamma_{OH}}{\gamma_{H_2O}} \tag{21}$$

the molality of 1000 grams of pure water being taken for this purpose as unity. The galvanic cells measured were of the form

$$(\text{Pt})\ H_2;\ KOH\ (m_1),\ KCl\ (m_2),\ AgCl;\ Ag \tag{22}$$

Cells were also used in which the potassium was replaced by other alkali metals and the chloride by other halogen ions. On solving equation (21) for m_H substituting in equation (2) and rearranging, the resulting expression is

$$E - E^\circ + \frac{RT}{\mathbf{F}} \ln \frac{m_{Cl}}{m_{OH}} = -\frac{RT}{\mathbf{F}} \ln \mathbf{K}_w - \frac{RT}{\mathbf{F}} \ln \frac{\gamma_H \gamma_{Cl} \gamma_{H_2O}}{\gamma_H \gamma_{OH}} \tag{23}$$

As in the cases already discussed, values of the right-hand side of the equation may be called $(RT/\mathbf{F}) \ln \mathbf{K}'_w$. To obtain $\mathbf{K}_w$ values of $\log \mathbf{K}'_w$ are plotted against the ionic strength ω and the resulting line is extrapolated to zero abscissa as shown in Fig. 4. Some typical results are given in Table VI and are taken from a paper by Harned and Hamer.[11] The values of $\mathbf{K}_w$ at various temperatures obtained by this and closely related methods are given in Table II.

[11] H. S. Harned and W. J. Hamer, *J. Am. Chem. Soc.*, **55**, 2194 (1933).

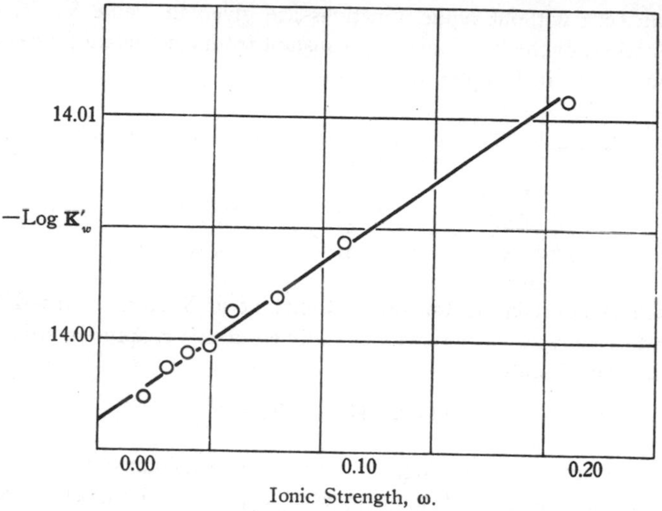

Fig. 4. Logarithm of the Apparent Ionization Constant of Water in Mixtures of 0.01 Molal Potassium Hydroxide with Potassium Chloride.

As already mentioned, there are other methods for obtaining ionization constants than that just discussed. Of these the conductance method briefly outlined in Chapter 3 and more fully considered in Chapter 18 is responsible for by far the greatest portion of the available data.

TABLE VI. THE DETERMINATION OF THE IONIZATION CONSTANT OF WATER FROM MEASUREMENTS ON THE GALVANIC CELLS: H_2; KOH(0.01m), KCl(m), AgCl; Ag

Ionic Strength ω	Emf, volts at 25°	$K'_w \times 10^{14}$
0.21	0.97418	0.9750
0.11	0.99169	0.9899
0.08	1.00061	0.9958
0.06	1.00922	0.9971
0.05	1.01486	1.000_8
0.04	1.02223	1.001_6
0.03	1.03260	1.003_1
0.02	1.05033	1.006_2
	Limiting Value, K_w	1.008_3

Such conclusions as are possible as to the meaning of ionization constants with respect to molecular structure, etc. will be deferred until the conductance method has been considered.

Chapter 12

Thermodynamic Studies of Non-Aqueous Solutions

While by far the greater portion of the electrochemical studies on solutions have been made using water as solvent, researches have also been carried out in which the water has been replaced by other solvents, or mixtures of non-aqueous solvents with water. Until very recently such studies yielded little of value because of their scattered nature and, usually, lack of accuracy, to which was added a lack of an adequate theory for their interpretation. It will be shown in the following chapter that some headway has been made in the difficult field of study of the thermodynamic properties of such solutions of electrolytes. The Debye-Hückel theory is, if anything, more valuable in the interpretation of the results in this field than in that of aqueous solutions.

Because of the necessity, with galvanic cells without liquid junctions, of finding two reversible electrodes, one for a positive and one for a negative ion constituent, most of the recent work in this field has been carried out with cells of the type:

$$\text{(Pt) } H_2;\ HCl\ (m),\ AgCl;\ Ag \tag{1}$$

in which the hydrogen chloride is dissolved in various non-aqueous solvents, or mixtures of such solvents with water. Fortunately hydrogen chloride is a fairly typical strong electrolyte, and is probably roughly at least representative of other electrolytes of the same valence type. It is to be hoped that transference numbers of salts in non-aqueous solutions will soon be available so that measurements of the potentials of concentration cells with liquid junctions can be interpreted. As shown in Chapter 8 this will make it possible to study solutions containing solutes for which a reversible electrode is available for only one of the ion constituents.

The Activity Coefficients of Hydrogen Chloride, and the Standard Potential of the Silver-Silver Chloride Electrode, in Methyl and Ethyl Alcohol Solutions. The potentials, E, of cells of the type shown in equation (1) with ethyl alcohol as solvent for the dissolved hydrogen chloride have been measured by Woolcock and Hartley,[1] and with methyl alcohol as solvent by Nonhebel and Hartley.[2] It is of particular interest to see whether the activity coefficients for hydrogen chloride in these solutions follow the predictions of the Debye-Hückel theory, since the dielectric constants, D, are less than half those of water, and the interionic attractions and repulsions of the ions must therefore be considerably greater. From equations (24) and (25), Chapter 7, it will be seen that the constants A and β of the Debye-Hückel equations are, respectively, proportional to $D^{-\frac{3}{2}}$ and $D^{-\frac{1}{2}}$. The dielectric constants[3]

TABLE I. DEBYE-HÜCKEL CONSTANTS AND STANDARD POTENTIALS OF THE SILVER-SILVER CHLORIDE ELECTRODES IN WATER AND IN METHYL AND ETHYL ALCOHOL

Solvent	Dielectric Constant, D	Debye-Hückel constants at 25°		Standard Potential, E_0, of Ag; AgCl, Cl⁻ volts
		A	β	
Ethyl Alcohol	24.3	2.94	0.591	+ 0.0740
Methyl Alcohol	31.5	1.99	0.519	+ 0.0100
Water	78.6	0.506	0.329	− 0.2225

and the corresponding values of A and β at 25° are given in Table I together, for comparison, with those of water.

In order to obtain activity coefficients or standard potentials from measurements on cells of the type shown in equation (1) it is necessary, as described in Chapters 8 and 10, to make some form of extrapolation. Furthermore in the test of the Debye-Hückel relations for these solutions it has been found that, because of the higher molecular weight of the non-aqueous solvents, the difference between the activity coefficient, f, and the "rational" coefficient, $\mathbf{f}$, based on Raoult's law, cannot be neglected, even below a concentration of 0.1 normal, as it usually can with aqueous solutions. This difference was overlooked by the workers just mentioned, who found only partial agreement of their results with the Debye-Hückel theory. Their data have therefore been recomputed as follows. For the ethyl alcohol solutions, for instance, a reference solution with a molality, m_1, of 0.09501, was chosen. The relation

$$\Delta E' = \frac{2RT}{\mathbf{F}} \ln \frac{m\gamma}{m_1\gamma_1} \tag{2}$$

together with equation (27b), Chapter 6, may be, at 25°, put in the form

$$\Delta E' = 0.1183\left(\log \frac{m(1 + 0.002\mathrm{M}_s m)}{m_1(1 + 0.002\mathrm{M}_s m_1)} + \Delta \log \mathbf{f}\right) \tag{3}$$

[1] J. W. Woolcock and H. Hartley, *Phil. Mag.*, [7] **5**, 1133 (1928).
[2] G. Nonhebel and H. Hartley, *ibid.*, [6] **50**, 729 (1925).
[3] The dielectric constants of the two alcohols are from the work of G. Åkerlof, *J. Am. Chem. Soc.*, **54**, 4125 (1932).

In this equation $\Delta E'$ is the difference in the potentials of two cells of the type given in equation (1), containing hydrochloric acid at the molalities m and m_1, and M_s is the molecular weight of the solvent. Using $\Delta \log \mathbf{f}$, obtained from equation (3), instead of $\Delta \log f$, in an extrapolation of the type described on page 161, and an example of which is plotted in Fig. 5 of Chapter 8, values of **f** may be obtained. It is of considerable interest that plots of this type for the data for methyl and ethyl alcohol solutions are straight lines, up to concentrations of 0.07 and 0.05 normal respectively, within the limits of the experimental error, indicating that the Debye-Hückel relation

$$-\log \mathbf{f} = \frac{A\sqrt{C}}{1 + \beta a_i \sqrt{C}} \tag{4}$$

holds to those concentrations. The relations of observed activity coefficients for the different solvents, obtained as just described, to the limiting law

$$-\log \mathbf{f} = A\sqrt{C} \tag{5}$$

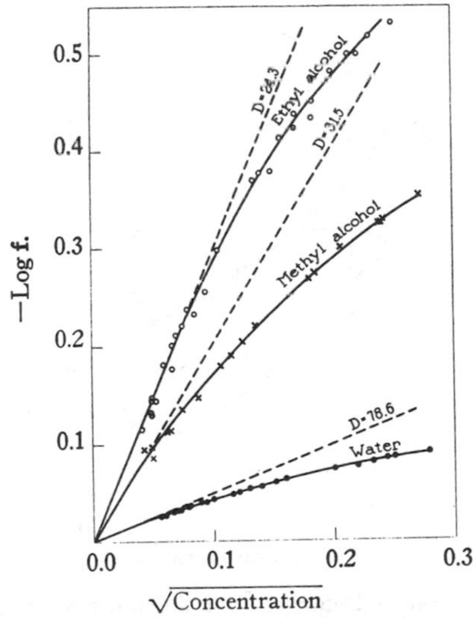

Fig. 1. Change in the Logarithm of the Rational Activity Coefficient, **f**, of Hydrogen Chloride with the Square Root of the Concentration in Water, Methyl and Ethyl Alcohol.

are shown graphically in Fig. 1, in which values of $-\log \mathbf{f}$ are plotted against $\sqrt{C}$. The data for aqueous solutions are given for comparison.

In this plot the straight lines have slopes equal to the values of *A* given in Table I. The deviations from the limiting law are such as are predicted from distances of closest approach of 5.9Å and 4.2Å respectively for ethyl and methyl alcohol solutions. These values are at least of the same order of magnitude as the value 5.6Å for aqueous solutions. It is quite evident from the figure that the activity coefficients for hydrogen chloride in the non-aqueous solvents vary from unity considerably more than do the values for aqueous solutions. This is illustrated by Table II in which the activity coefficients at three different

TABLE II. COMPARISON OF THE ACTIVITY COEFFICIENT, **f**, OF HYDROGEN CHLORIDE IN THREE SOLVENTS AT 25°

Solvent	Concentrations, mols per liter 0.01	0.05	0.10
Ethyl Alcohol	0.603	0.423	0.241
Methyl Alcohol	0.678	0.489	0.410
Water	0.906	0.833	0.799

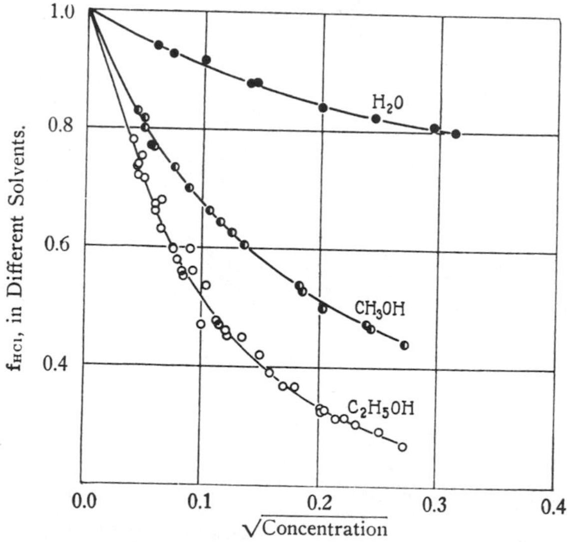

Fig. 2. The Concentration Dependence of the Rational Activity Coefficient, **f**, of Hydrogen Chloride in Water, Methyl and Ethyl Alcohol.

concentrations are compared, and from Fig. 2, in which the activity coefficients of hydrochloric acid in these solvents are plotted as functions of the concentration. The low values of these activity coefficients for the non-aqueous solutions indicate why the tacit assumption, of early investigators, that the activity is equal to the concentration worked fairly well for water solutions but failed for other solvents. The neglect of the effects of interionic attractions and repulsions becomes more serious as the dielectric constant takes lower and lower values. It is also evident that it is not necessary to assume incomplete dissociation for the solutions under consideration. It will be noted that the three solvents considered in this discussion all contain the hydroxyl group.

It is also possible from potential measurements on the cell given in equation (1) to obtain standard potentials of the silver-silver chloride electrode in other solvents than water. From the activity coefficients just discussed, values of the standard potential, E_0, may be computed. For that purpose the activity coefficient, γ, may be computed from the value of **f**, at the molality m, by means of equation (27b) of Chapter 6. E_0 may then be obtained from the equation

$$E' = E_0 - \frac{2RT}{\mathbf{F}} \ln m\gamma \tag{6}$$

when corresponding values of E' for the cell indicated in equation (1) are substituted. The resulting E_0 values for the silver-silver chloride electrode in ethyl and methyl alcohol at 25°, on the molality basis, to correspond to those listed elsewhere in this book, are given in Table I. These E_0 values are referred to a mean ion activity, of the hydrogen ion component, of unity *in the solvent in question.*

Although the E_0 values on the molality, m, and the concentration, C, bases are very nearly the same when dealing with aqueous solutions, the difference between the two values may be quite important when other solvents are used in the measurements. For uni-univalent electrolytes the difference may be readily shown to be

$$E_0^C - E_0^m = \frac{2RT}{\mathbf{F}} \ln \rho_0$$

in which ρ_0 is the density of the solvent. For ethyl alcohol solutions at 25°, $E_0^C - E_0^m$ is -0.0124 volt.

The Activity Coefficients of Hydrogen Chloride and the Standard Potential of the Silver-Silver Chloride Electrode in Mixed Solvents. In addition to the measurements described in the previous paragraphs determinations of the potential of the cell

$$(\text{Pt})\ H_2;\ HCl\ (m),\ AgCl;\ Ag$$

have been made in which the hydrogen chloride was dissolved in various mixed solvents. The most complete series of measurements are those which have been carried out by Harned and Morrison[4] on cells using,

[4] H. S. Harned and J. O. Morrison, *J. Am. Chem. Soc.*, **58**, 1908 (1936).

as solvent for hydrochloric acid, dioxane-water mixtures containing 20, 45 and 70 weight per cent of dioxane. With the Hitchcock method of extrapolation they obtained the standard potentials, E_0, on the molar basis, given in Table III.

TABLE III. STANDARD POTENTIALS OF THE SILVER-SILVER CHLORIDE ELECTRODE, AT 25°, IN WATER-DIOXANE MIXTURES

Percentage by Weight of Dioxane	Debye-Hückel Coefficient A	Dielectric Constant D	Standard Potentials E_0, volts
70	4.738	17.69	− 0.0662
45	1.477	38.48	− 0.1634
20	0.7437	60.79	− 0.2032
0	0.5056	78.56	− 0.2225

The product, $-\Delta E_0\mathbf{F}$, in which ΔE_0 is the difference of any two of these standard potentials, is the Gibbs free energy of transfer of hydrogen chloride, at infinite dilution, from one solvent to another.

With the aid of equation (6) the activity coefficients of hydrogen chloride may be computed and from equation (27b), of Chapter 6,

$$\mathbf{f} = \gamma(1 + 0.002m\text{M}_s) \tag{7}$$

the rational coefficient may be obtained. In this expression M_s is the "mean molecular weight" of the solvent and is equal to

$$\text{M}_s = \frac{100}{\left(\dfrac{X}{\text{M}_d} + \dfrac{Y}{\text{M}_w}\right)} \tag{8}$$

in which X and Y are the weight percentages of dioxane and water in the mixtures, and M_d and M_w are the corresponding molecular weights. A plot of $-\log \mathbf{f}$ against $\sqrt{C}$ for hydrogen chloride in the three mixtures is given in Fig. 3, in which the straight line is again in each case a plot of the limiting law. For comparison the values of the activity coefficient of hydrochloric acid in pure water are also included in the plot. It will be seen that the curves passing through the observed values merge into the line representing the limiting law. However, it is an interesting and significant fact that it is not possible to assign a value for the distance of closest approach, a_i, to hydrogen chloride in these mixtures, or to put it in another way, the Debye-Hückel equation (4) does not hold, as it does for water and pure methyl and ethyl alcohol solutions. A rather speculative explanation of this observation may lie in the fact that water has a dipole moment, and dioxane has not. As will be made clear in Chapter 22, this means that the centers of positive and negative charges are not at the same point in the water molecule, whereas these centers coincide in the dioxane molecule. A result of the presence of the dipole moment of water is that the molecules will be attracted by, and oriented around, the charged ions. If this preferential attraction and orientation occurs it will have at least two effects. It will

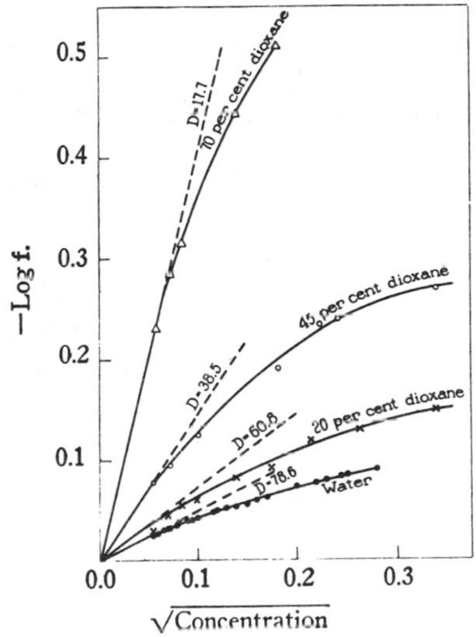

Fig. 3. The Concentration Dependence of the Logarithm of the Rational Activity Coefficient, f, of Hydrogen Chloride in a Series of Dioxane-Water Mixtures.

change the composition of the bulk of the solvent, and also affect the dielectric constant of the medium in close proximity to the ions. Both these effects could produce apparent disagreements with the simple Debye-Hückel theory.

Chapter 13

Galvanic Cells with Liquid Junction Potentials

In Chapter 10 standard potentials were obtained from measurements on galvanic cells involving only one electrolyte. These "cells without transference" thus do not involve surfaces between solutions of electrolytes, more commonly called "liquid junctions." Although the measurements on cells without liquid junctions can be much more readily interpreted than the results from cells containing such boundaries most of the earlier work was carried out with the latter type of cell. Thus for instance instead of measuring the potential of the cell:

$$H_2;\ HCl,\ HgCl;\ Hg$$

a more complicated arrangement such as

$$H_2;\ HCl : KCl,\ HgCl;\ Hg$$

was used, making it necessary in the computation of standard potentials, etc., to evaluate potentials of junctions of which

$$HCl : KCl$$

is typical.

As we shall see, much experimental and theoretical ingenuity has been utilized in dealing with liquid junctions, since, in addition to being the reason for necessary corrections in a number of types of electrochemical researches, they are of considerable interest in themselves.

In this chapter a general differential equation for liquid junctions will first be derived, after which the application of the equation to various types of junction will be discussed.

The General Differential Equation for Liquid Junction Potentials. If two solutions of electrolytes are brought into contact there will be a region in which the composition varies from that of one of the two solutions to that of the other. The "make up" of this region will depend upon such factors as diffusion, convection and mixing. However, such a region can be divided into layers that are so thin that the variation of the composition of the solution can be assumed to be indefinitely small.

Consider such a thin layer a to b of Fig. 1, on one A side of which is a solution containing ions[1] $1, 2, \ldots i, \ldots n$ at the concentrations

$$C_1,\ C_2,\ \ldots\ C_i,\ \ldots\ C_n$$

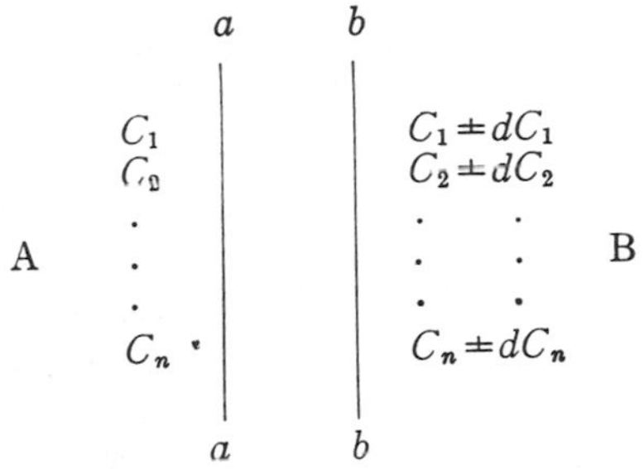

Fig. 1.

On the other B side of this layer the concentrations are

$$C_1 \pm dC_1,\ C_2 \pm dC_2,\ \ldots\ C_i \pm dC_i,\ \ldots\ C_n \pm dC_n$$

Let

$$z_1,\ z_2,\ \ldots\ z_i,\ \ldots\ z_n$$

be the valences of these ions (the sign + or − of these valences being retained) and

$$t_1,\ t_2,\ \ldots\ t_i,\ \ldots\ t_n$$

the transference numbers of the corresponding ions. A potential dE_L will result from the processes occurring in this layer.

If now a faraday of current is passed reversibly, from solution A to solution B, t_i/z_i mol of each ion will pass through the layer. This passage will be from left to right if the valence z_i is positive and from right to left if this valence is negative. The total change of Gibbs free energy, dZ, of this process will be

$$dZ = \sum_n \frac{t_i}{z_i}\, d\mu_i \tag{1}$$

in which μ_i is the chemical potential of the ion i. Recalling that

$$t_1 + t_2 + \cdots + t_i + \cdots + t_n = 1 \tag{2}$$

and using equations (31), Chapter 5, and (20), Chapter 6, we have

$$-dE_L = \frac{RT}{\mathbf{F}} \sum_n \frac{t_i}{z_i}\, d \ln a_i \tag{3}$$

[1] Throughout this chapter the terms "ion" and "ion constituent" are regarded as synonymous.

Any liquid junction between two solutions, I and II, will consist of an indefinite number of such layers as we have been considering and its potential will be the integral of equation (3), thus

$$-E_L = \frac{RT}{\mathbf{F}} \int_{\mathrm{I}}^{\mathrm{II}} \sum_n \frac{t_i}{z_i} d \ln a_i \tag{4}$$

It is important to note that this equation is not rigorously thermodynamic since neither the single potential of a liquid junction nor the ion activities, a_i, can be measured. However, as will be seen in what follows, correct thermodynamic equations are obtained if equation (3) is combined with equations for electrode processes in such a manner as to include the process for a complete galvanic cell. In such cases it will always be possible to combine the single ion activities into physically measurable mean ion activities. Some examples will be discussed in detail.

Liquid Junctions between Two Solutions of the Same Electrolyte. A type of cell involving liquid junctions has already been considered, in Chapter 8, and the measurements have been shown to yield potentials which can be interpreted thermodynamically. A cell of the type is

$$\underset{E_1}{\mathrm{Ag}\ ;\ \mathrm{AgCl},\ \mathrm{NaCl}\ (C_1)} : \underset{E_L}{\mathrm{NaCl}\ (C_2),} \ \underset{E_2}{\mathrm{AgCl}\ ;\ \mathrm{Ag}} \tag{5}$$

the liquid junction being between two solutions, at different concentrations, of the same salt. It is of interest to analyze the mechanism of such a cell in somewhat more detail than has already been done. For this purpose it is useful to consider that the total potential E of the cell is divided into three portions, E_1 and E_2 at the two electrodes and E_L at the liquid junction. The potential E_1 will be:

$$E_1 = E_0 + \frac{RT}{\mathbf{F}} \ln a_{\mathrm{Cl}}^{\mathrm{I}} \tag{6}$$

in which $a_{\mathrm{Cl}}^{\mathrm{I}}$ is the activity of the chloride ion to which the electrode is reversible in solution I. Similarly the potential E_2 will be

$$E_2 = -E_0 - \frac{RT}{\mathbf{F}} \ln a_{\mathrm{Cl}}^{\mathrm{II}} \tag{7}$$

Now if the transference number of the sodium ion can be assumed to be constant between C_1 and C_2, which are the concentrations of solutions I and II, the operation of the cell involves the transport of t_{Na} mol per faraday of sodium ion in the direction of the current and $(1 - t_{\mathrm{Na}})$

mol of chloride ion in the reverse direction. The liquid junction potential E_L will therefore be

$$E_L = \frac{RT}{\mathbf{F}}\left[t_{Na} \ln \frac{a^{I}_{Na}}{a^{II}_{Na}} - (1 - t_{Na}) \ln \frac{a^{I}_{Cl}}{a^{II}_{Cl}}\right] \quad (8)$$

This is, of course, an application of equation (4) to this specific case. If the transference number is not constant in the range C_1 to C_2 equation (4) gives

$$-E_L = \frac{RT}{\mathbf{F}}\int_{I}^{II} t_{Na}\, d\ln a_{Na} - \frac{RT}{\mathbf{F}}\int_{I}^{II} (1 - t_{Na})\, d\ln a_{Cl} \quad (9)$$

Adding (6), (7) and (8) yields

$$E = E_1 + E_2 + E_L = \frac{t_{Na}RT}{\mathbf{F}} \ln \frac{a^{I}_{Na}\, a^{I}_{Cl}}{a^{II}_{Na}\, a^{II}_{Cl}} \quad (10)$$

for the total potential of cell A, for the case in which the transference number is constant. The sum of equations (6), (7) and (9) is

$$E_1 + E_2 + E_L = -\frac{RT}{\mathbf{F}}\int_{I}^{II} t_{Na}\, d\ln a_{Na}a_{Cl} \quad (11)$$

which may be used if the transference number is not constant between C_1 and C_2.

However, as has already been mentioned,[2] we have no means for evaluating the individual ion activities such as a_{Na} and a_{Cl}. There is, on the other hand, always an equation connecting them with the mean ion activity, which can be obtained thermodynamically. In this case the equation is, from equations (38) and (39) of Chapter 6,

$$a_{Na}\, a_{Cl} = a^{2}_{NaCl} \quad (12)$$

a_{NaCl} being the mean ion activity for sodium chloride. Substituting this expression in equation (11) gives

$$-E = \frac{2RT}{\mathbf{F}}\int_{I}^{II} t_{Na}\, d\ln a_{NaCl} \quad (13)$$

which is equivalent to the thermodynamic equation (18) of Chapter 8 for the same type of cell since $a_{NaCl} = Cf_{\pm}$

[2] p. 133.

It is important to note that to apportion the total potential E of the cell between the electrode potentials E_1 and E_2 and the liquid junction potential E_L with the aid of equations (6), (7), and (8) it is necessary to use individual ion activities, and thus involve a non-thermodynamic assumption, by which is meant an assumption that, at least in the present state of our knowledge, cannot be given an unambiguous test in the laboratory. For the case under discussion, any assumption as to the single ion activities which is in agreement with equation (12) will also yield equation (13) for the whole cell, but different assumptions will give different values for potentials E_1, E_2 and E_L and there is no means at present available for deciding which of the infinite number of possible values is the correct one. Similar difficulties always arise when attempts are made to evaluate any single potentials. Guggenheim[3] goes so far as to say that since we have no method for measuring them, single electrode potentials and single ion activities have no physical meaning. This may be an extreme position. The conceptions of single potentials and single ion activities are useful as mathematical concepts, and in making mental pictures of the operation of different portions of cell mechanisms. In what follows it will be found desirable to make non-thermodynamic assumptions but an attempt will be made to outline clearly what they are.

As already mentioned in Chapter 8, cells having liquid junctions of the type

$$\text{NaCl}\ (C_1) : \text{NaCl}\ (C_2)$$

have potentials that are independent of the manner in which the liquid junctions are formed. This may be accounted for by the fact that in the integration

$$-E_L = \frac{RT}{\mathbf{F}} \int_{\text{I}}^{\text{II}} \sum_n \frac{t_i}{z_i} a \ln a_i \tag{4}$$

t_i and a_i are both single valued functions of the concentration.

Although it has just been emphasized that no proof is available as to the validity of the computations for any single potential, *reasonable* values of such potentials can be obtained in certain cases. For a cell of the type

$$\text{Ag; AgCl, MCl}\ (C_1) : \text{MCl}\ (C_2),\ \text{AgCl; Ag} \tag{14}$$

in which M is any univalent ion, equations (8) and (10) (which are, of course, valid if the univalent cation, M, replaces sodium) have been

[3] E. A. Guggenheim, *J. Phys. Chem.*, 33, 842 (1929).

derived for the liquid junction E_L and the potential of the whole cell E for the case in which the transference number t_M is constant. If further it is assumed that the ion activities of the positive and negative ions are both equal to the mean ion activity, *i.e.*, in this case that

$$a_M = a_{Cl} = a_{MCl}$$

equation (8) may be put in the form

$$E_L = (2t_M - 1)\frac{RT}{\mathbf{F}}\ln\frac{a^{I}_{Cl}}{a^{II}_{Cl}} \tag{15}$$

and the potential, equation (10), of the whole cell by

$$E = 2t_M\frac{RT}{\mathbf{F}}\ln\frac{a^{I}_{Cl}}{a^{II}_{Cl}} \tag{16}$$

Dividing equation (15) by equation (16) we obtain [4]

$$E_L = E\frac{(2t_M - 1)}{2t_M} \tag{17}$$

an equation which gives the potential of the liquid junction in terms of that of the whole cell and the transference number only. Jahn [5] and more recently Brown and MacInnes [6] and Shedlovsky and MacInnes [7] have made measurements on the potentials of the cells of the type just mentioned which furnish data that may be used for the purpose of testing equation (17). The basis of the test is as follows. Since silver-silver chloride electrodes are reversible to the chloride ion, it seems reasonable to suppose that, for dilute solutions at least, for a given pair of concentrations C_1 and C_2 the differences of the electrode potentials will be nearly the same whatever the nature of the (univalent) positive ion. The data for the potentials of cells of type (14) and the computed values of the liquid junctions E_L are given in Table I. The transference numbers given in column four of the table are from the work of Longsworth given in Table IV of Chapter 4. It will be seen that although the liquid junction potentials E_L computed from equation (17) and given in column six vary both in magnitude and sign, the difference of the electrode potentials, $E_1 - E_2$, recorded in the last column, obtained by subtracting the liquid junction potential

[4] D. A. MacInnes, *J. Am. Chem. Soc.*, **37**, 2301 (1915).
[5] H. Jahn, *Z. physik. Chem.*, **33**, 545 (1900).
[6] A. S. Brown and D. A. MacInnes, *J. Am. Chem. Soc.*, **57**, 1356 (1935).
[7] T. Shedlovsky and D. A. MacInnes, *ibid.*, **59**, 503 (1937).

from that of the whole cell, E, is very nearly constant for any pair of concentrations C_1 and C_2. This appears to be evidence that equation (17) yields at least reasonable values for the liquid junction for the cases under consideration. Since the concentrations C_1 and C_2 do not differ greatly a mean value of the transference number t_M has been taken in these computations.

TABLE I. COMPUTED POTENTIALS OF THE LIQUID JUNCTION IN CELLS OF THE TYPE Ag; AgCl, MCl (C_1): MCl (C_2), AgCl; Ag

Concentrations mols per liter C_1	C_2	Electro-lyte	Transference number t_M	Potentials millivolts Measured E	Liquid Junction E_L	Electrode Potential millivolts $E_1 - E_2$
0.01	0.005	NaCl	0.3924	13.41	− 3.68	17.1
		KCl	0.4903	16.77	− 0.33	17.1
		HCl	0.8245	28.29	+ 11.13	17.2
0.02	0.01	NaCl	0.3910	13.22	− 3.69	16.9
		KCl	0.4900	16.56	− 0.34	16.9
		HCl	0.8258	28.05	+ 11.07	17.0
0.04	0.005	NaCl	0.3907	39.63	− 11.09	50.7
		KCl	0.4902	49.63	− 1.00	50.6
		HCl	0.8261	84.16	+ 33.22	50.9
0.03	0.02	NaCl	0.3893	7.62	− 2.17	9.8
		KCl	0.4902	9.59	− 0.20	9.8
0.04	0.02	NaCl	0.3893	13.00	− 3.70	16.7
		KCl	0.4899	16.31	− 0.34	16.7
		HCl	0.8274	27.82	+ 11.01	16.8
0.06	0.04	NaCl	0.3877	7.50	− 2.18	9.7
		KCl	0.4899	9.44	− 0.20	9.6
		HCl	0.8292	16.25	+ 6.45	9.8

Cells Containing Junctions of Different Electrolytes, Experimental. For junctions of the types

$$HCl : KCl \quad \text{or} \quad HCl : NaBr$$
$$\text{or} \quad CaCl_2 : NaBr$$

in which ions appear on one side of the boundary that are not present on the other, measurements of the potentials reveal that the results depend on the manner in which the junction has been formed. Experimental work has been carried out on two main types of liquid junctions, known respectively as the "static" and "flowing" boundaries. In the formation of the first of these the two solutions are brought together by any convenient means, generally with the boundary in a horizontal plane, and the lighter solution on top to avoid convection currents. Frequently arrangements are made to remake the boundary at intervals. The device used by Lewis, Brighton and Sebastian [8] is shown in Fig. 2.

[8] G. N. Lewis, T. B. Brighton and R. L. Sebastian, *ibid.*, 39, 2245 (1917).

The tubes *O* and *R* are connected with the electrodes and also with reservoirs of the solutions brought into contact at the point *Q*. Excess of both solutions pass out through the tube *P*. The junction can be

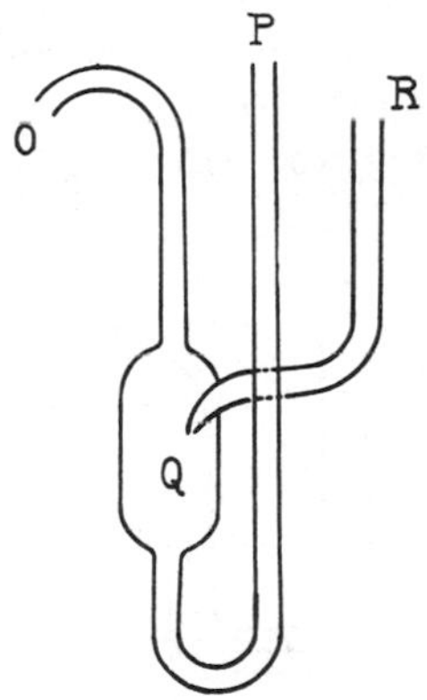

Fig. 2. Device for the Formation of a "Static" Boundary.

renewed as often as desired by running fresh solutions through *O* and *R*. The potentials of cells involving such junctions are usually found to vary more or less with time. This is due to the fact that, for instance, a boundary between 0.1 normal HCl and 0.1 normal KCl arranged as shown in Fig. 3, if initially sharp, will soon consist of a region in

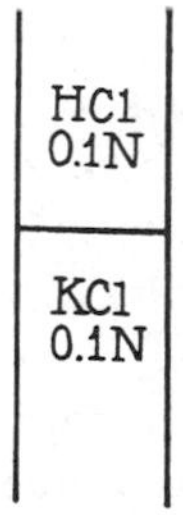

Fig. 3.

which the salt has diffused into the acid and *vice versa,* the composition of the different layers depending, if actual mixing and convection have not taken place, upon the relative rates of diffusion of the different components.

Reproducible and constant potentials of cells involving liquid junctions may be obtained with "flowing" junctions, which have been studied by Lamb and Larson,[9] and MacInnes and Yeh[10] and others. The method of formation of flowing junctions is shown in Fig. 4. The boundary which forms at *A* results from the meeting of two slowly flowing streams of solutions from tubes *B* and *C*. These streams also pass by the tubes containing the electrodes. The heavier solution must obviously enter the boundary from below. The nature of the boundary

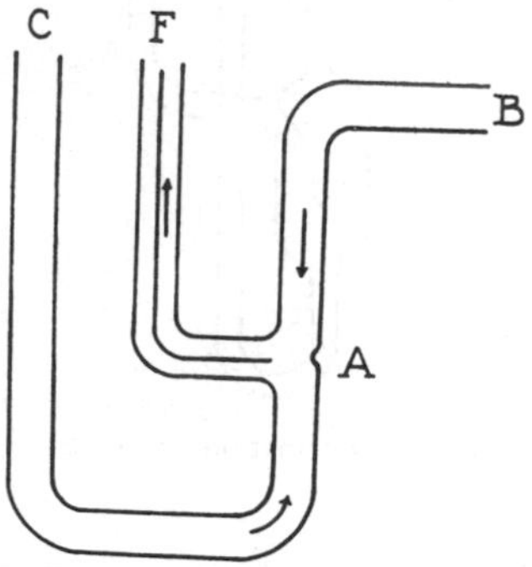

Fig. 4. Device for the Formation of a "Flowing" Junction.

between the two solutions may be made clearly visible by operating the apparatus with two solutions, one of which contains a few drops of phenolphthalein and the other a small amount of sodium hydroxide, the pink color of the indicator showing only at the surface of the solutions and in the region in which they mix or diffuse. A surface of almost microscopic thinness starts at *A* and persists, with a slight thickening, throughout the length of tube *F*. Fig. 5, in which the ordinates are potentials in millivolts and the abscissae times in minutes, summarizes, graphically, the results of the experiments on a cell containing the junction between 0.1 normal hydrochloric acid and 0.1 normal potassium chloride with a flowing junction as just described. Curve I shows the effect of stopping the movement of a flowing junction. The potential rose rapidly through a millivolt or more, and this was followed by a slow decrease of potential. At *b* on this curve the flow was started again and the original potential was quickly regained at *c*. At *d* the flow was again stopped and the potential rose to a different maximum, after which it slowly decreased. This time, at *e*, the

[9] A. B. Lamb and A. T. Larson, *ibid.*, **42**, 229 (1920).
[10] D. A. MacInnes and Y. L. Yeh, *ibid.*, **43**, 2563 (1921).

original potential was almost instantly obtained from the higher value by starting the flowing. Curves II and III show the effect of a low rate of flow, 1 and 2 drops a minute respectively dripping from the outlet tube *F*. The heavy line represents the constant potential of the junction obtained by the flow of from 3 to 7 drops a minute. Above the latter rate (from 8 drops per minute to rapid streaming) there was

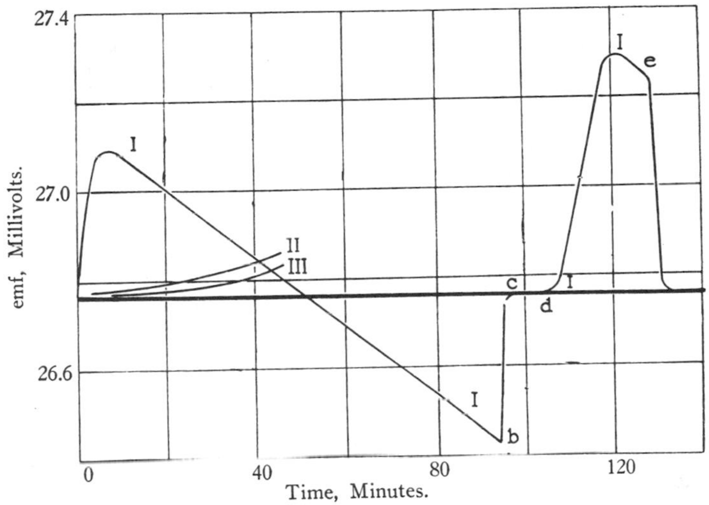

Fig. 5. The Effect of Rate of Flow on the Potential of a Cell Containing a Flowing Junction.

a decrease in the potential of the order of a few hundredths of a millivolt. The results of measurements obtained as just described and with a sufficiently large rate of flow of the solutions were extremely constant and reproducible, duplicate determinations agreeing within 0.02 millivolt. The actual figures obtained will be discussed later.

Modifications of the flowing junction as described have been made by several workers. Roberts and Fenwick [11] use an apparatus shown diagrammatically in Fig. 6a, in which streams of the two solutions from tubes as shown are directed against a piece of mica through which a tiny hole has been pierced, the rejected solution flowing on both sides of the mica. Lekhani [12] has further modified this arrangement, Fig. 6b, by omitting the mica and directing two fine streams of the solutions against each other.

[11] E. J. Roberts and F. Fenwick, *ibid.*, **49**, 2787 (1927).
[12] J. V. Lekhani. *J. Chem. Soc.*, **1932**, 179.

It is of interest to compare the results which are given in Table II of the several types of measurements on cells containing the boundary

$$HCl\ (0.1\ N) : KCl\ (0.1\ N),$$

with silver-silver chloride or calomel electrodes. An extreme variation of 1.5 millivolts is seen in the figures in this table. Some of this may

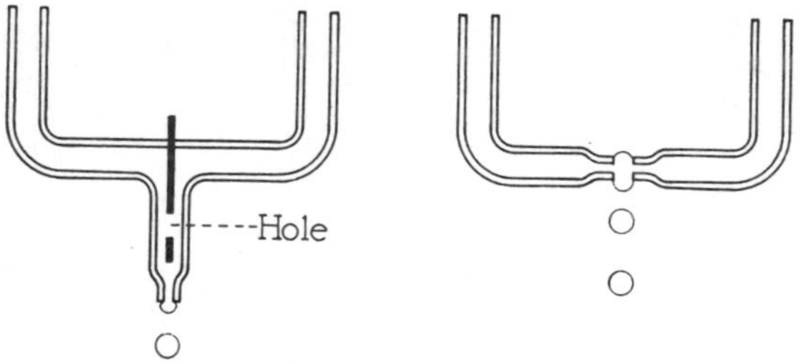

(a) (b)

Fig. 6. Modifications of the Flowing Junction.

be experimental error. However, much of the variation of the values observed must be due to the distribution of the ions in the layers of solution constituting the boundary. Unfortunately in no case has

TABLE II. POTENTIALS OF CELLS CONTAINING THE LIQUID JUNCTION HCl (0.1 *N*): KCl (0.1 *N*) at 25°

Worker	Reference	Type of Junction	Potential millivolts
Roberts and Fenwick	*1*	Flowing	28.00
MacInnes and Yeh	*2*	Flowing	26.78
Chloupek, Daneš and Danešova	*3*	Free Diffusion	27.08
Lewis, Brighton and Sebastian	*4*	Dipping tube	27.8
Ghosh	*5*	Drop Contact	28.27
Bjerrum	*6*	In sand	27.8
Meyers and Acree	*7*	In sand	27.79

1. E. J. Roberts and F. Fenwick, *J. Am. Chem. Soc.*, **49**, 2789 (1927).
2. D. A. MacInnes and Y. L. Yeh, *ibid.*, **43**, 2563 (1921).
3. J. B. Chloupek, V. Z. Daneš and B. A. Danešova, *Coll. Czechoslov. Chem. Comm.*, **5**, 469, 527 (1933).
4. G. N. Lewis, T. B. Brighton and R. L. Sebastian, *J. Am. Chem. Soc.*, **39**, 2245 (1917).
5. D. N. Ghosh, *J. Indian Chem. Soc.*, **12**, 15 (1935).
6. N. Bjerrum, *Z. Elektrochem.*, **17**, 58 (1911).
7. C. N. Meyers and S. F. Acree, *Am. Chem., J.*, **50**, 396 (1913).

this distribution been accurately known. A theoretical discussion of the question is given in the following paragraphs.

The Integration of the Differential Equation for Liquid Junction Potential. To account for the results of such measurements as have been described in the previous section there have been a number of integrations of the fundamental differential equation (3). Of these the following will deal only with the integrations by Henderson and by Planck. In addition a graphical integration method devised by MacInnes and Longsworth will be discussed.

(*a*) *Henderson's Integration for the "Mixture Boundary."* With the exception of the case, already discussed, of a boundary between two solutions of the same salt, it is necessary, in integrating equation (3)

$$-\,dE_L = \frac{RT}{\mathbf{F}} \sum_n \frac{t_i}{z_i}\, d \ln a_i \tag{3}$$

to know or to assume the relations of the concentrations of the ions with respect to each other, and also to make an assumption concerning the single ion activities a_i.

A simple assumption as to the nature of the boundary which leads to an easy integration of the equation is that the junction consists of a continuous series of solutions produced by mixing the two solutions initially brought into contact. If x, the mixing fraction, is the proportion of solution II at a particular point in the boundary, then the proportion of solution I will be $(1 - x)$ and x will vary between zero and unity.

Let $c'_1, c'_2, \ldots c'_i, \ldots c'_n$ be the concentrations expressed in equivalents of the ions in solution I, and $c''_1, c''_2, \ldots c''_i, \ldots c''_n$ be the corresponding concentrations in solution II, and $z_1, z_2, \ldots z_i, \ldots z_n$ the valencies (retaining the signs). At a position in the boundary where the mixing fraction is x the concentration, c_i, of the ion i will obviously be

$$c_i = c'_i + (c''_i - c'_i)\,x \tag{18}$$

The transference number t_i of the ion i will be at that same position

$$t_i = \frac{c_i u_i}{(1 - x)\Sigma_n c'_i u_i + x\Sigma_n c''_i u_i} \tag{19}$$

in which u_i represents the mobility of the ion i and Σ_n represents summation of all the terms indicated. It will be noted that this transference number is the ratio of the conductance by the ion i to that of all the ions at the same point in the boundary. Assuming that the activities a_i are equal to the concentrations, c_i, and that the mobilities are constant in the concentration range c′ to c″, equation (3) becomes, using equations (18) and (19),

$$- E_L = \frac{RT}{\mathbf{F}} \int_0^1 \sum_n \frac{u_i/z_i \cdot (c_i'' - c_i')\, dx}{\Sigma_n c_i' u_i + x \Sigma_n u_i (c_i'' - c_i')} \tag{20}$$

which on integrating between the limits indicated gives

$$E_L = \frac{RT}{\mathbf{F}} \frac{\Sigma_n u_i/z_i \cdot (c_i'' - c_i')}{\Sigma_n u_i (c_i'' - c_i')} \ln \frac{\Sigma_n c_i' u_i}{\Sigma_n c_i'' u_i} \tag{21}$$

This equation is due to Henderson.[14] It is important to note that this derivation requires no particular spatial arrangement of the layers of solution in the boundary, the only requirement being that the solutions composing the boundary should be a series of mixtures of the end solutions I and II.

Three special cases are of interest:

(1) If the *ions are all univalent*, then $z_i = +1$ for positive ions and -1 for negative ions. If in addition the following functions are defined:

$$\begin{aligned} \mathbf{U}_1 &= C_1^{+\prime} u_1^{+\prime} + C_2^{+\prime} u_2^{+\prime} + \cdots\cdots \\ \mathbf{V}_1 &= C_1^{-\prime} u_1^{-\prime} + C_2^{-\prime} u_2^{-\prime} + \cdots\cdots \\ \mathbf{U}_2 &= C_1^{+\prime\prime} u_1^{+\prime\prime} + C_2^{+\prime\prime} u_2^{+\prime\prime} + \cdots\cdots \\ \mathbf{V}_2 &= C_1^{-\prime\prime} u_1^{-\prime\prime} + C_2^{-\prime\prime} u_2^{-\prime\prime} + \cdots\cdots \end{aligned} \tag{22}$$

then the Henderson equation (20) takes the more familiar form[15]

$$E_L = \frac{RT}{\mathbf{F}} \cdot \frac{(\mathbf{U}_1 - \mathbf{V}_1) - (\mathbf{U}_2 - \mathbf{V}_2)}{(\mathbf{U}_1 + \mathbf{V}_1) - (\mathbf{U}_2 + \mathbf{V}_2)} \ln \frac{\mathbf{U}_1 + \mathbf{V}_1}{\mathbf{U}_2 + \mathbf{V}_2} \tag{23}$$

(2) If the solutions I and II are a single univalent salt at two concentrations C' and C'', then equation (21) yields

$$E_L = \frac{RT}{\mathbf{F}} \frac{u^+ - u^-}{u^+ + u^-} \ln \frac{C'}{C''} \tag{24}$$

Since the transference number, t^+, of the positive ion is $u^+/(u^+ + u^-)$ and of the negative ion $t^- = u^-/(u^+ + u^-)$, this equation becomes

$$E_L = (2t^+ - 1) \frac{RT}{\mathbf{F}} \ln \frac{C'}{C''} \tag{25}$$

[14] P. Henderson, *Z. physik. Chem.*, **59**, 118 (1907); **63**, 325 (1908).
[15] P. Henderson, *loc. cit.*

which can be seen to be equivalent to equation (15) for the junction

$$NaCl\ (C_1) : NaCl\ (C_2)$$

if the activities are equal to the concentrations.

(3) *If the solutions are two uni-univalent electrolytes both at the concentration C,* and with one ion in common, such as for instance

$$HCl\ (C) : KCl\ (C)$$

equation (21) takes the form:

$$E_L = \frac{RT}{\mathbf{F}} \ln \frac{\text{U}^{+\prime} + \text{U}^{-\prime}}{\text{U}^{+\prime\prime} + \text{U}^{-\prime\prime}} \tag{26}$$

From equation (14), Chapter 3, we have $\Lambda = \mathbf{F}\alpha(\text{U}^+ + \text{U}^-)$ in which Λ is the equivalent conductance and α is the degree of dissociation. For strong electrolytes we may assume that the dissociation is complete *i.e.,* that α equals unity. Equation (26) may therefore be given the form [16]

$$E_L = \frac{RT}{\mathbf{F}} \ln \frac{\Lambda'}{\Lambda''} \tag{26a}$$

in which Λ' and Λ'' are the equivalent conductances of solutions I and II at the concentration $C' = C''$. Equation (26a) was suggested by Lewis and Sargent.[17] Originally, however, Henderson gave the equation the form

$$E_L = \frac{RT}{\mathbf{F}} \ln \frac{\Lambda_0'}{\Lambda_0''} \tag{26b}$$

in which Λ_0' and Λ_0'' are the limiting equivalent conductances.

(*b*) *Planck's Integration for the "Constrained Diffusion" Boundary.* Another integration of the differential equation for liquid junction potentials which, as a matter of fact, preceded Henderson's integration, was carried out by Planck.[18] He assumed what has been called a "constrained diffusion" boundary. Such a boundary could be produced,

[16] Equations (26) and (26a) are somewhat more general in application than the derivation just given would lead us to suppose, in that they remain valid if for each ion $a_i = f_i C_i$ in which f_i is a *constant* activity coefficient. In that case

$$d \ln a_i = \frac{d(f_i C_i)}{f_i C_i} = \frac{f_i dC_i}{f_i C_i} = \frac{dC_i}{C_i}$$

The activity coefficient f_i can evidently have different values for different ions. (See D. A. MacInnes and Y. L. Yeh, *J. Am. Chem. Soc.*, **43**, 2563 (1921).)

[17] G. N. Lewis and L. W. Sargent, *J. Am. Chem. Soc.*, **31**, 363 (1909).

[18] M. Planck, *Ann. physik.*, [3] **39**, 161 (1890); [3] **40**, 561 (1890).

as shown in Fig. 7, by bathing with solutions I and II two sides of a porous plug (supposed to have no effect on the solution in its pores), the solutions I and II being continuously renewed to keep their composition

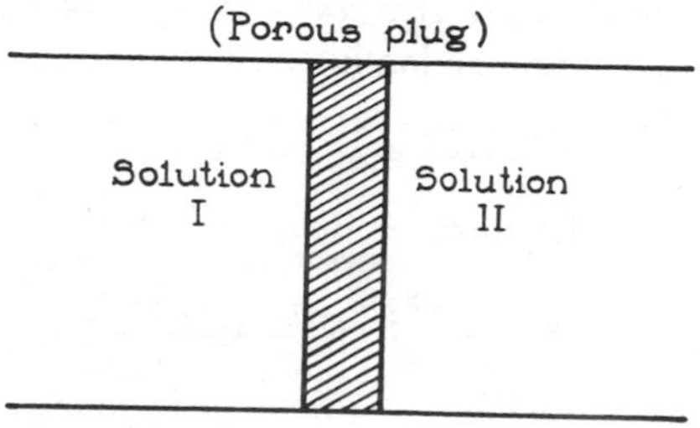

Fig. 7. Diagrammatic Representation of a "Constrained Diffusion" Boundary.

constant. The diffusion through the layer represented by the plug is presumed to take place until a stationary state is reached. In addition the assumptions were made that the ions are normal solutes and that their mobilities are constant. The rather involved integration is given in an appendix, page 461. Planck's integration leads to the equation

$$\frac{\xi \mathbf{U}_2 - \mathbf{U}_1}{\mathbf{V}_2 - \xi \mathbf{V}_1} = \frac{\ln \dfrac{C_2}{C_1} - \ln \xi}{\ln \dfrac{C_2}{C_1} + \ln \xi} \cdot \frac{\xi C_2 - C_1}{C_2 - \xi C_1} \tag{27}$$

$\mathbf{U}_1$, $\mathbf{V}_1$, $\mathbf{U}_2$ and $\mathbf{V}_2$ being defined as for equation (22), the function ξ by the equation

$$E_L = \frac{RT}{\mathbf{F}} \ln \xi$$

and C_1 and C_2 are the total concentrations of solutions I and II.

In two important cases equation (27) reduces to equations that have already been derived by other means. For the junction

$$HCl\ (C) : KCl\ (C)$$

in which the two solutions have a common ion and the same concentration, C, the equation becomes

$$E_L = \frac{RT}{\mathbf{F}} \ln \frac{u_{H^+} + u_{Cl^-}}{u_{K^+} + u_{Cl^-}} = \frac{RT}{\mathbf{F}} \ln \frac{\Lambda_{HCl}}{\Lambda_{KCl}} \tag{28}$$

in agreement with equation (26).

Furthermore for the junction

$$HCl\ (C_1) : HCl\ (C_2)$$

equation (27) yields

$$E_L = \frac{U_{H^+} - U_{Cl^-}}{U_{H^+} + U_{Cl^-}} \frac{RT}{\mathbf{F}} \ln \frac{C_1}{C_2} = \frac{(2t^+ - 1)\,RT}{\mathbf{F}} \ln \frac{C_1}{C_2} \tag{29}$$

in agreement, for normal solutes, with equation (15).

For liquid junctions of the type

$$HCl\ (C_1) : KCl\ (C_2)$$

equation (27) does not reduce to a simple form. The method proposed by Planck to evaluate ξ is as follows.

Let $$\frac{\xi \mathbf{U}_2 - \mathbf{U}_1}{\mathbf{V}_2 - \xi \mathbf{V}_1} = \eta_1 \quad \text{and} \quad \frac{\ln \dfrac{C_2}{C_1} - \ln \xi}{\ln \dfrac{C_2}{C_1} + \ln \xi} \cdot \frac{\xi C_2 - C_1}{C_2 - \xi C_1} = \eta_2$$

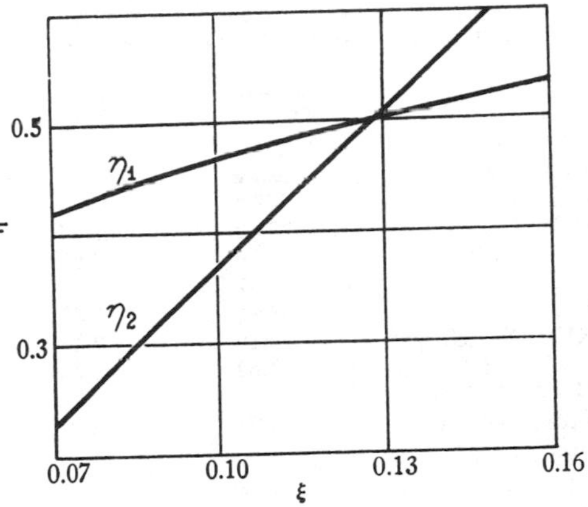

Fig. 8. Plot for the Evaluation of ξ for the Junction HCl (0.1 *N*) : KCl (0.01 *N*)

Since all the terms except ξ are presumably known, values of η_1 and η_2 may be computed for, and plotted as ordinates against, a series of assumed values of ξ. The value of ξ fitting equation (24) will be that corresponding to $\eta_1 = \eta_2$ where the resulting lines cross. The plot used for obtaining ξ for the boundary

$$HCl\ (0.1\ N) : KCl\ (0.01\ N)$$

at 25°, is given in Fig. 8. The value of $\xi = 0.1285$ leads to 0.05271 volt for the potential of this boundary. Henderson's equation, (21) or (23), yields the somewhat higher value of 0.05726 volt. As already stated if $C_1 = C_2$ in the junction HCl (C_1) : KCl (C_2) the two equations give the same computed potential. If $C_1 > C_2$, then the Henderson equation gives a higher potential than the Planck and *vice versa.* The values obtained by the two equations are discussed more fully by Cumming and Gilcrist.[19]

Comparison of the Theory with the Experimental Results. As already mentioned, MacInnes and Yeh [19a] have made studies of the potentials of cells of the form:

$$\text{Ag; AgCl, M'Cl } (C) : \text{M''Cl } (C)\text{, AgCl; Ag,}$$

M′ and M″ being various univalent cations, using a flowing junction. The concentration C was the same throughout the cell and was 0.1 normal in one series of experiments and 0.01 normal in another. The results are given in Table III, in which the measured values are com-

TABLE III. LIQUID JUNCTION POTENTIALS, E_L, CALCULATED BY THE LEWIS AND SARGENT EQUATION AND COMPARED WITH MEASURED VALUES

Concentration mols per liter	Junction	E_L observed millivolts	E_L computed (L. & S.) millivolts	E_L from characteristic potentials
0.1N	HCl : KCl	+ 26.78	+ 28.52	+ 26.78
	HCl : NaCl	33.09	33.38	33.08
	HCl : LiCl	34.86	36.14	35.65
	HCl : NH_4Cl	28.40	28.57	28.78
	KCl : LiCl	8.79	7.62	8.87
	KCl : NaCl	6.42	4.86	6.30
	KCl : LiCl	8.76	7.62	8.86
	KCl : NH_4Cl	2.16	0.046	2.00
	NaCl : LiCl	2.62	2.76	2.57
	NaCl : NH_4Cl	− 4.21	− 4.81	− 4.30
	LiCl : NH_4Cl	− 6.93	− 7.57	− 6.87
0.01N	HCl : KCl	+ 25.73	+ 27.48	+ 25.62
	HCl : NaCl	31.16	32.02	31.19
	HCl : NH_4Cl	27.02	27.50	26.93
	HCl : LiCl	33.75	34.56	33.82
	KCl : NaCl	5.65	4.54	5.57
	KCl : LiCl	8.20	7.08	8.20
	KCl : NH_4Cl	1.31	0.018	1.31
	KCl : CsCl	0.31	− 0.60	0.30
	NaCl : LiCl	2.63	2.53	2.63
	NaCl : NH_4Cl	− 4.26	− 4.52	− 4.26
	NaCl : CsCl	− 5.39	− 5.13	− 5.17
	LiCl : NH_4Cl	− 6.89	− 7.06	− 6.89
	LiCl : CsCl	− 7.80	− 7.67	− 7.80
	CsCl : NH_4Cl	+ 0.95	+ 0.61	− 0.91

[19] A. C. Cumming and E. Gilcrist, *Trans. Farad. Soc.*, 9, 174 (1913).
[19a] D. A. MacInnes and Y. L. Yeh, *J. Am. Chem. Soc.*, 43, 2563 (1921).

pared with values computed from equation

$$E_L = \frac{RT}{\mathbf{F}} \ln \frac{\Lambda'}{\Lambda''} \tag{26a}$$

in which Λ' and Λ'' are the equivalent conductances of the two solutions meeting at the boundary. It will be recalled that this equation is obtained, for electrolytes having the same concentration and one common ion, by both the Henderson and Planck integrations of the fundamental differential equation. In making this comparison the assumption is made that the electrode potentials cancel, or, in other words, that the chloride ion activities are the same in two different chlorides at the same concentration. It will be seen by comparing the figures in the third and fourth columns of the table, that although the agreement between the observed and computed values is surprisingly close in certain cases, there are a number of values in which the agreement is far from satisfactory. It must be recalled however that quite a number of assumptions have been made in obtaining equation (26a), and still another assumption has been made in identifying the potential of the cell measured with the liquid juction. Strangely enough, the difference between the measured and computed values is greatest for boundaries involving potassium chloride.

It is of interest that although the agreement of the measured values with the theory is far from perfect the observed values are quite consistent among themselves. Using differences between the "characteristic potentials" obtained from the data and given in Table IV, the figures

TABLE IV. "CHARACTERISTIC POTENTIALS" IN MILLIVOLTS

Salt	Normality 0.1	0.01
LiCl	0.00	0.00
KCl	8.87	8.20
HCl	35.65	33.87
NaCl	2.57	2.63
NH_4Cl	6.92	6.89
CsCl	.....	7.80

given in the fifth column of Table III are obtained. For instance the potential of the junction

$$\text{HCl } (0.1\ N) : \text{KCl } (0.1\ N)$$

is thus seen to be 35.65 − 8.87, or 26.78 millivolts. The potentials obtained by this additive principle are seen to agree very closely with the measured values.

Graphical Methods for Computing Potentials of Cells with Liquid Junctions. The assumptions made by Planck and Henderson in obtaining their equations for liquid junction potentials were made, not because they were in the closest possible accord with reality, but were compromises necessary in order to effect analytical integrations of the equation

$$E_L = -\frac{RT}{\mathbf{F}}\int_{\mathrm{I}}^{\mathrm{II}}\sum_n \frac{t_i}{z_i}\,d\ln a_i \tag{4}$$

It is however possible to evaluate the liquid junction potential using all the data available for the transference numbers and the ionic activities and any assumed distribution of electrolytes in the boundary with the aid of graphical methods. In the discussion below the procedure followed so far in this chapter will be reversed. The potential for a complete cell will be computed, after which the liquid junction will be obtained by subtracting a computed value of the electrode potentials.

If we consider the cell:

$$\mathrm{Ag;\ AgCl,\ HCl}\ (C_1) : \mathrm{KCl}\ (C_2),\ \mathrm{AgCl;\ Ag} \tag{31}$$

the electrode potentials are given by

$$E_{\mathrm{I}} = E^\circ + \frac{RT}{\mathbf{F}}\ln a_{\mathrm{Cl}}^{\mathrm{I}} \quad \text{and} \quad E_{\mathrm{II}} = -E^\circ - \frac{RT}{\mathbf{F}}\ln a_{\mathrm{Cl}}^{\mathrm{II}}$$

The total potential E of the cell will therefore be, using equation (4) for the liquid junction potential,

$$E = \frac{RT}{\mathbf{F}}\left[\ln a_{\mathrm{Cl}}^{\mathrm{I}} - \int_{\mathrm{I}}^{\mathrm{II}} t_{\mathrm{H}}\,d\ln a_{\mathrm{H}} - \int_{\mathrm{I}}^{\mathrm{II}} t_{\mathrm{K}}\,d\ln a_{\mathrm{K}} + \int_{\mathrm{I}}^{\mathrm{II}} t_{\mathrm{Cl}}\,d\ln a_{\mathrm{Cl}}^{\mathrm{I}} - \ln a_{\mathrm{Cl}}^{\mathrm{II}}\right] \tag{32}$$

Since $t_{\mathrm{Cl}} = 1 - t_{\mathrm{H}} - t_{\mathrm{K}}$ this equation reduces to

$$E = -\frac{RT}{\mathbf{F}}\int_{\mathrm{I}}^{\mathrm{II}} t_{\mathrm{H}}\,d\ln a_{\mathrm{H}}a_{\mathrm{Cl}} - \frac{RT}{\mathbf{F}}\int_{\mathrm{I}}^{\mathrm{II}} t_{\mathrm{K}}\,d\ln a_{\mathrm{K}}a_{\mathrm{Cl}} \tag{33}$$

the non-thermodynamic quantities a_{H}, a_{K} and a_{Cl} in this equation may be replaced by the thermodynamic mean ion activities a_{HCl} and a_{KCl} by means of the relations

$$a_{\mathrm{H}}a_{\mathrm{Cl}} = a_{\mathrm{HCl}}^2 \quad \text{and} \quad a_{\mathrm{K}}a_{\mathrm{Cl}} = a_{\mathrm{KCl}}^2$$

These mean ion activities may be put into the form

$$a^2_{\mathrm{HCl}} = C_{\mathrm{H}} C_{\mathrm{Cl}} f^2_{\mathrm{HCl}} \quad \text{and} \quad a^2_{\mathrm{KCl}} = C_{\mathrm{K}} C_{\mathrm{Cl}} f^2_{\mathrm{KCl}}$$

in which C_{H}, C_{K} and C_{Cl} represent concentrations of the indicated ions and f_{HCl} and f_{KCl} are mean ion activity coefficients. Equation (33) thus becomes

$$-E = \frac{RT}{\mathbf{F}} \int_{\mathrm{I}}^{\mathrm{II}} \frac{t_{\mathrm{H}}}{C_{\mathrm{H}} C_{\mathrm{Cl}} \cdot f^2_{\mathrm{HCl}}} d(C_{\mathrm{H}} C_{\mathrm{Cl}} \cdot f^2_{\mathrm{HCl}}) + \frac{RT}{\mathbf{F}} \int_{\mathrm{I}}^{\mathrm{II}} \frac{t_{\mathrm{K}}}{C_{\mathrm{K}} C_{\mathrm{Cl}} \cdot f^2_{\mathrm{HCl}}} d(C_{\mathrm{K}} C_{\mathrm{Cl}} \cdot f^2_{\mathrm{KCl}}) \qquad (34)$$

This equation contains only measurable quantities. However in order to integrate it, information concerning the point to point variation of the concentrations in the boundary is necessary, since the values of the transference numbers and the activity coefficients depend both upon the total concentrations of the solutions I and II and upon the proportions in which these solutions are mixed. The distribution of electrolytes in the boundary assumed by Planck and by Henderson have already been discussed. These were chosen, it is well to repeat, not because of their inherent probability, but because with them analytical integrations could be carried out.

A graphical method for integrating equation (33) is as follows.[19b] To evaluate the first integral of the equation, values of $t_{\mathrm{H}}/(C_{\mathrm{H}} C_{\mathrm{Cl}} f^2_{\mathrm{HCl}})$ are plotted against values of $(C_{\mathrm{H}} C_{\mathrm{Cl}} f^2_{\mathrm{HCl}})$ and the area is conveniently obtained with a planimeter. The second integral is obtained in like manner. It is evident that using this method it is not necessary to obtain functional relationships between the variables. Such functional relationships would be very complex, and the integration difficult if not impossible, if use were made of the experimental results for the transference numbers and activity coefficients and any physically possible distribution of the ion concentrations in tne boundary.

Relatively complete data for transference numbers and activity coefficients on which such a graphical integration may be based are available only for the cell:

$$\text{Ag; AgCl, HCl } (0.1\,N) : \text{KCl } (0.1\,N), \text{ AgCl; Ag} \qquad (35)$$

Transference numbers for mixtures of HCl and KCl have been

[19b] D. A. MacInnes and L. G. Longsworth, *Cold Spring Harbor Symposia on Quantitative Biology,* **4,** 18 (1936).

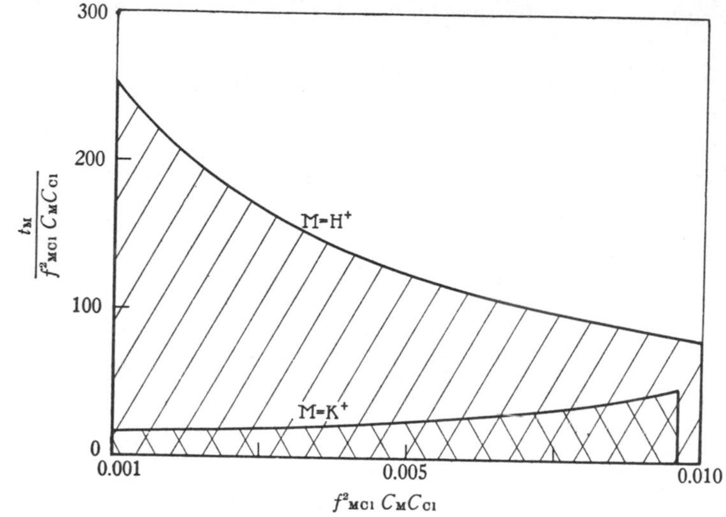

Fig. 9. Plot for the Graphical Integration of Equation (34).

obtained by Longsworth and are given in Table VI of Chapter 4, and the mean ion activity coefficients $f_{HCl(KCl)}$ and $f_{KCl(HCl)}$ in such mixtures have been measured by Güntelberg as described in Chapter 8. (For the purpose of this computation γ and f values are not sufficiently

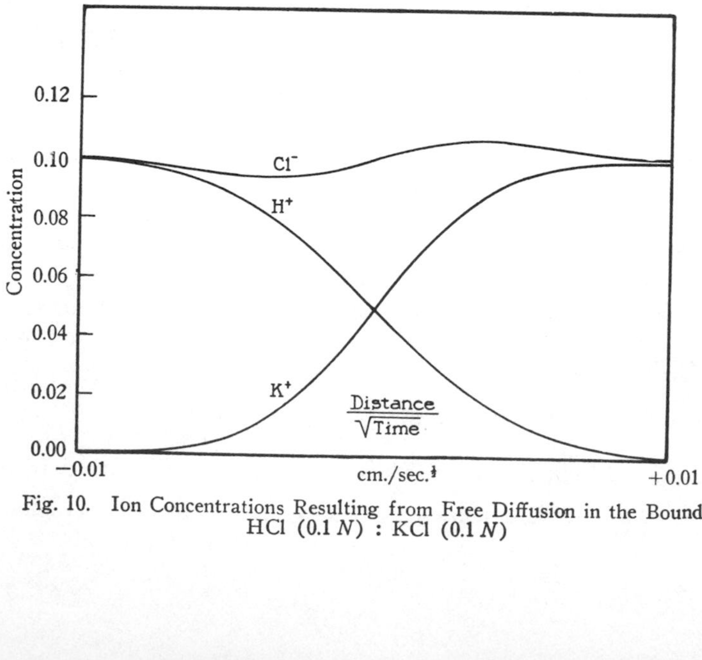

Fig. 10. Ion Concentrations Resulting from Free Diffusion in the Boundary HCl (0.1 N) : KCl (0.1 N)

different to introduce an appreciable error). Utilizing these data, and assuming a Henderson or mixture boundary, plots of the type described in the previous paragraph are given in Fig. 9. Since in this case $d(C_H C_{Cl} f^2_{HCl})$ is negative and $d(C_K C_{Cl} f^2_{KCl})$ positive, the two areas indicated in the plot have different signs.[20] Estimation of the two areas in Fig. 9 and substitution in equation (34) yielded a value for the potential of cell (35) of 28.57 millivolts. This may be compared with the directly measured potentials for the cell as given in Table II which average a little less than a millivolt lower, but differ about 1.5 millivolts among themselves.

Actually, unless disturbed by convection or mixing, the contact between two solutions of electrolytes, such as that indicated in cell (35), is a "free diffusion" boundary. Experimental determinations of the distribution in such boundaries are not available, and a complete mathematical solution of the problem is very difficult, though progress has been made by Taylor.[21] A close approximation to such a distribution for a boundary initially 0.1 N HCl : 0.1 N KCl is however given in Fig. 10 in which the concentration of the ions is shown as a function of the distance x at the time s. The computation was made with the aid of a suggestion of Guggenheim[22] that the electrolytes be assumed to diffuse independently. It will be seen that the free diffusion boundary differs from the mixture boundary for the case under consideration in that the total chloride ion concentration may be greater than 0.1 normal in one portion of the boundary and less in another. This results from the relatively greater rate of diffusion of HCl.

[20] The limiting values of $t_H/(C_H C_{Cl} f^2_{HCl})$ and $t_K/(C_K C_{Cl} f^2_{KCl})$ are indeterminate but they may be evaluated as follows. In terms of ionic conductances the transference number t_H is equal to

$$t_H = \frac{\Lambda_H C_H}{\Lambda_H C_H + \Lambda_K C_K + \Lambda_{Cl} C_{Cl}}$$

for which we may obtain

$$\lim_{C_H \to 0} \frac{t_H}{C_H} = \frac{\Lambda_H}{\Lambda_K C_K + \Lambda_{Cl} C_{Cl}}$$

which for the present case is

$$\lim_{C_H \to 0} \frac{t_H}{C_H} = \frac{\Lambda_H^{0.1\,N\,\mathrm{KCl}}}{0.1 \Lambda_{0.1\,N\,\mathrm{KCl}}}$$

With the series of transference numbers for mixtures of HCl and KCl already referred to and conductance values for the same solutions, values of Λ_H for the various mixtures were obtained by Longsworth. These were found to vary linearly with the mixing fraction, and could be extrapolated without error to yield $\Lambda_H^{0.1 N \mathrm{KCl}}$, *i.e.*, the equivalent conductance of vanishingly small amounts of hydrogen ion in 0.1 normal potassium chloride solution. The limiting values of f_{HCl} were obtained from the extrapolation of Guntelberg's data given in Fig. 6 of Chapter 8. Similar limiting values for the potassium ion were obtained in the same manner.

[21] P. B. Taylor, *J. Phys. Chem.*, **31**, 1478 (1927).

[22] E. A. Guggenheim, *J. Am. Chem. Soc.*, **52**, 1315 (1930).

Using the graphical method described in the previous paragraphs the potential 28.19 millivolts was obtained for cell (35) when it includes a free diffusion boundary. A slight approximation was necessarily made in this case since the relevant transference numbers and activity coefficients for the mixtures have been obtained only at the total concentration of 0.1 normal. The influence of this approximation is however negligible. Here again comparison may be made with the experimental results given in Table II. How much of the remaining discordance is due to experimental error and how much to unknown factors connected with the distribution in the boundary there is at present no means of telling. Carefully organized experimental work appears to be desirable in this field. The experiments might well include the fixing of the distribution in the boundary along the lines suggested by the work of Plettig [23] and Guggenheim.[24]

The two examples given above indicate that the graphical method may be used in computing the potentials of cells including liquid junctions if the necessary thermodynamic data are available. Any distribution of the electrolytes in the boundary may be used. Such distributions may be experimentally determined, or based on assumptions, and may be of any complexity.

Having computed values for the total potential of cell (35) it is of interest to discuss the matter of the distribution of this potential between the differences of the electrode potentials $E_1 - E_2$ and the liquid junction potential E_L. To do this we must again resort to non-thermodynamic assumptions. The assumption made by MacInnes [25] and frequently used in computations of this sort is that at any concentration

$$a_{\mathrm{Cl(MCl)}} = a_{\mathrm{Cl(KCl)}} = a_{\mathrm{KCl}}$$

i.e., that the chloride ion activity in a solution of any chloride is the same as in a solution of potassium chloride at the same total concentration. Such an assumption may be considered reasonable because of the very similar electronic structures of potassium and chloride ions. Another assumption, which will be referred to as the "Guggenheim assumption," [26] is for the case under discussion

$$a_{\mathrm{M}} = a_{\mathrm{Cl}} = a_{\mathrm{MC.}}$$

[23] V. Plettig, *Ann. Physik.*, **5**, 735 (1930).

[24] E. A. Guggenheim, *J. Am. Chem. Soc.*, **52**, 1315 (1930).

[25] D. A. MacInnes, *ibid.*, **41**, 1086 (1919).

[26] It has been used by others, G. Scatchard, (*J. Am. Chem. Soc.*, **47**, 696 (1925)) for instance, but was clearly stated and adopted by Guggenheim.

which means that the activity of the chloride ion of any univalent chloride is equal to the mean ion activity.

The computed values of the liquid junction

$$\text{HCl } (0.1\ N) : \text{KCl } (0.1\ N)$$

according to the various methods discussed in this chapter are collected in Table V. It will be seen that if the MacInnes assumption is

TABLE V. COMPUTED VALUES OF THE POTENTIAL, E, OF THE CELL
Ag; AgCl, HCl (0.1 N) : KCl (0.1 N), AgCl; Ag
AND OF THE LIQUID JUNCTION HCl (0.1 N) : KCl (0.1 N)

Equation or Method	Author	Cell Potential millivolts E	Liquid Junction Potential, millivolts E_L
26*b*	Henderson or Planck	.....	26.85
26*a*	Lewis and Sargent	.....	28.52
Graphical (mixture boundary)	MacInnes and Longsworth	28.57	28.57[1] 28.65[2]
Graphical (diffusion boundary)	MacInnes and Longsworth	28.19	28.19[1] 28.27[2]

[1] MacInnes assumption.
[2] Guggenheim assumption.

adopted the electrode potentials cancel each other and the potential of the liquid junction is equal to the potential of the whole cell. Using the Guggenheim assumption the difference of the electrode potentials may be computed from the relation

$$E_1 - E_2 = \frac{RT}{\mathbf{F}} \ln \frac{a_{\text{HCl}}}{a_{\text{KCl}}}$$

and is equal in this instance to 0.92 millivolt, which must be subtracted from the total potential to give that of the liquid junction. For this computation the values of a_{HCl} and a_{KCl} have been obtained from data in Table V of Chapter 8.

Attempts to Minimize Liquid Junction Potentials by Use of Salt Bridges. Since in many researches involving galvanic cells liquid junctions have appeared, not as interesting subjects for study, but as troublesome variables, attempts have been made to eliminate or minimize them. The most usual device for attaining this end is to introduce a strong solution of some electrolyte, known as a "salt bridge," between the solutions otherwise in direct contact. Thus the liquid junction

$$\text{HCl } (C_1) : \text{NaCl } (C_2)$$

may for instance be replaced by

$$\text{HCl } (C_1) : \text{KCl (saturated)} : \text{NaCl } (C_2)$$

The reason for the choice of saturated potassium chloride for the salt bridge is evident from a consideration of equation (23). Take, for example, the junction

$$\text{HCl } (C_1) : \text{KCl (saturated)}$$

Since saturated potassium chloride is about 4.2 normal at 25° the potential of the liquid junction is given (on the assumptions with which the equation is derived) by

$$E_L = \frac{RT}{\mathbf{F}} \cdot \frac{4.2\,(\text{U}_\text{K} - \text{U}_\text{Cl}) - C_1(\text{U}_\text{H} - \text{U}_\text{Cl})}{4.2\,(\text{U}_\text{K} + \text{U}_\text{Cl}) - C_1(\text{U}_\text{H} + \text{U}_\text{Cl})} \ln \frac{4.2\,(\text{U}_\text{K} + \text{U}_\text{Cl})}{C_1(\text{U}_\text{H} + \text{U}_\text{Cl})}$$

If the concentration, C_1, is small in comparison with 4.2 normal, the coefficient of the logarithm will be

$$\frac{4.2\,(\text{U}_\text{K} - \text{U}_\text{Cl})}{4.2\,(\text{U}_\text{K} + \text{U}_\text{Cl})}$$

and as U_K is very nearly equal to U_Cl this coefficient will be small.

A short review of the assumptions made in obtaining equation (23) is sufficient to indicate the uncertainty involved in this type of computation. It has been assumed (a) that the boundary is of a mixture type, (b) that the mobilities are constant throughout the concentration range to 4.2 normal and (c) that the ions are normal solutes. All these assumptions, as has been shown in preceding pages, are contrary to fact. Furthermore the difference between the mobilities of potassium and chloride ions is not as small as it was thought to be until recently. This is indicated by the fact that the transference number of potassium chloride,

$$t_\text{K} = \frac{\text{U}_\text{K}}{\text{U}_\text{K} + \text{U}_\text{Cl}}$$

found by early workers to be 0.498, is now known (page 85) to be 0.490. If the potassium and chloride ions had the same mobility the transference number would, of course, be 0.50. It has, as a matter of fact, long been recognized that the use of a salt bridge is a makeshift introducing considerable uncertainty. Other salts than potassium chloride and mixtures of salts have been proposed, but have not come into general use, probably because they do not help to minimize the fundamental difficulties with salt bridges.

There is another procedure, proposed by Bjerrum,[27] which must be mentioned since it has been considerably used. This is to make potential measurements with two salt bridges in succession, one of 3.5 normal potassium chloride and another of the same salt at 1.75 normal. The difference between these two potentials is then subtracted from that obtained with the more concentrated solution. This worker regarded the device as satisfactory only if the difference measured was found to be small.

In spite of the uncertainties involved in dealing with salt bridges they have been much used. In some cases in which they have been used in the past they are not necessary, and the information desired could have been obtained with less ambiguity using cells without liquid junctions, such as have been described in Chapter 10. In certain types of measurement, however, they appear, at present at least, to be necessary. In this category belong the very important pH determinations discussed in Chapter 15, certain types of oxidation-reduction investigations, and some others.

[27] N. Bjerrum, *Z. Elektrochem.*, **17**, 389 (1911).

Chapter 14

Standard Potential Determinations Involving Cells with Liquid Junctions, or Equilibrium Measurements

The most accurate method for the determination of standard potentials has been outlined in Chapter 10. That method, which avoids the use of cells containing liquid junctions, is, however, of comparatively recent origin. Most of the early work in this field involved cells in which liquid junctions contributed to the measured potential. In addition there are standard potentials that cannot, at least by methods at present available, be measured without the use of these additional sources of potential. So far it has not been found possible to obtain the standard potentials for pH measurements, or for many of the oxidation-reduction processes involving organic compounds, without the use of liquid junctions.

It is important to note that the difference between cells with and without liquid junctions is one of degree and not of kind. Thus the familiar cell:

$$\text{(Pt)};\ H_2,\ HCl,\ AgCl\ (s);\ Ag$$

really contains two (or more) solutions, and should, strictly, be represented as follows:

$$\text{(Pt)};\ H_2,\ HCl\ (\text{satd. with } H_2) : HCl\ (\text{satd. with } AgCl),\ AgCl\ (s);\ Ag$$

In addition a solution of pure hydrochloric acid might be present between the two saturated solutions. In this particular case the considerations outlined in the preceding chapter would lead us to believe that the liquid junctions in such a cell are of negligible magnitude. If however the saturating substances were more soluble the resulting liquid junction potentials could not be neglected in computations involving the potential of such a cell.

In computing a standard potential of silver from, for instance, a potential measurement, E, on a cell of the type

$$Ag;\ AgNO_3\ (0.1\,N) \underset{a}{:} HNO_3\ (0.1\,N) \underset{b}{:} HCl\ (0.1\,N);\ H_2\ \text{(Pt)}$$

it is necessary, in the first place, to correct the potential for the liquid junction potentials at *a* and *b*. The principles involved in that correction are adequately outlined in the foregoing chapter. The result will be the potential, E', of the combination

$$\text{Ag; AgNO}_3\ (0.1\ N) :: \text{HCl}\ (0.1\ N);\ \text{H}_2\ (\text{Pt})$$

for which we have the expression

$$E' = E_0^{\text{Ag}} - \frac{RT}{\mathbf{F}} \ln (\text{Ag}^+) - 0 + \frac{RT}{\mathbf{F}} \ln (\text{H}^+)$$

$$= E_0^{\text{Ag}} - \frac{RT}{\mathbf{F}} \ln m_{\text{Ag}}\gamma_{\text{Ag}} + \frac{RT}{\mathbf{F}} \ln m_{\text{H}}\gamma_{\text{H}}$$

in which m_{Ag}, m_{H}, γ_{Ag}, γ_{H} are respectively the molalities and activity coefficients of the silver and hydrogen ion constituents. It will be recalled that the symbol (::), page 108, means that the liquid junction (or junctions) is not included in the potential of the cell. In all the computations of standard potentials discussed so far it has been possible to replace the activities of single ion constituents by means of ion activities which, except for a constant obtained by extrapolation, are thermodynamic. In the present case, and in all the cases to be considered in this chapter, it is necessary to make a further non-thermodynamic assumption as to the activities of single ion constituents. The assumption adopted will be that called the "Guggenheim assumption" in the foregoing chapter, *i.e.*, that the single ion activity is equal to the mean ion activity. For the case in hand the activity of the silver ion constituent, $m_{\text{Ag}}\gamma_{\text{Ag}}$, will be assumed to be equal to the mean ion activity of silver nitrate and the activity of the hydrogen ion, $m_{\text{H}}\gamma_{\text{H}}$, to be equal to the mean ion activity of hydrochloric acid. This assumption will be adhered to in spite of the fact that in certain cases a better agreement between the results of measurements on cells with and without liquid junction can be obtained using a somewhat more elaborate assumption. However, the uncertainties in the computations, when liquid junctions are involved, do not appear to justify the greater complications.

The Potentials of the Normal and Decinormal "Calomel" Electrodes. A large portion of the studies on the potentials of galvanic cells has been made using calomel electrodes containing normal or decinormal potassium chloride. Such cells, in general, involve liquid junctions. It is therefore important for the interpretation of these results to decide upon the potentials of the combinations:

$$\text{Hg; HgCl (s), KCl}\ (1.0\ N) :: (\text{H}^+);\ \text{H}_2\ (\text{Pt}) \qquad (1)$$

$$\text{Hg; HgCl (s), KCl}\ (0.1\ N) :: (\text{H}^+);\ \text{H}_2\ (\text{Pt}) \qquad (2)$$

It will be recalled that the electrode H^+; $H_2(Pt)$ with (H^+) at unit mean ion activity is our arbitrary zero of potential. The passage of current from left to right through either cell results in the reaction

$$Hg + H^+ + Cl^- = HgCl + \tfrac{1}{2}H_2 \tag{3}$$

for which equation (10), Chapter 10, gives

$$E = E_0 - \frac{RT}{\mathbf{F}} \ln \frac{(H_2)^{1/2}}{(H^+)(Cl^-)} \tag{4}$$

Since for the reference electrode (H_2) and (H^+) are both unity we have [1]

$$E = E_0 + \frac{RT}{\mathbf{F}} \ln (Cl^-) \tag{5}$$

The value, −0.2681, of E_0 for the electrode Hg; HCl(s), Cl^- or for the cell:

$$Hg;\ HgCl\ (s),\ Cl^- :: H^+;\ H_2\ (Pt) \tag{6}$$

is given on page 190. As was made clear in the previous chapter, computations of the potentials of single electrodes such as Hg; HgCl(s), KCl (0.1 *N*) or the activities of single ion constituents cannot be made without non-thermodynamic assumptions. For the purpose of computing the potentials of cells (1) and (2) the assumption will be made that the activity of the chloride ion constituent of potassium chloride is equal to the mean ion activity, $a = m\gamma$, of potassium chloride at the molality in question. From the data in Table V, Chapter 8, the activity coefficients for 1.0 normal (1.0327 molal), and 0.1 normal (0.1006 molal), are respectively 0.607 and 0.770. Using these figures in equation (5) we obtain, at 25°

Hg; HgCl (s), KCl (*1.0 N*)	$E = -0.2801$	(10)
Hg; HgCl (s), KCl (*0.1 N*)	$E = -0.3338$	(11)

[1] Another way of developing equation (5) is as follows. The calomel electrode is essentially a mercury-mercurous ion Hg; Hg^+ electrode and its potential follows the relation

$$E = E^*_0 - \frac{RT}{\mathbf{F}} \ln (Hg^+) \tag{7}$$

in which E^*_0 is the standard potential of the electrode. However in a saturated calomel solution the relation

$$(Hg^+)(Cl^-) = S \tag{8}$$

holds, in which S is the solubility product and the parentheses represent the activities of the included ions. Substituting equation (8) in equation (7)

$$E = E^*_0 - \frac{RT}{\mathbf{F}} \ln S + \frac{RT}{\mathbf{F}} \ln (Cl^-) \tag{9}$$

Since at a given temperature S is a constant the first two terms on the right-hand side of this equation may be combined, yielding equation (5). This explains the positive (+) sign before the term $RT/\mathbf{F} \ln (Cl^-)$. Incidentally equation (9) illustrates a method for obtaining solubilities from emf data, or for finding standard potentials from solubility data, both of which are discussed elsewhere in this book.

The Standard Potentials of the Alkali Metals from Cells with Liquid Junctions. The determination of the standard potentials of sodium and potassium, using cells without liquid junctions, has already been described in Chapter 10. It is of interest to compare the value obtained in that way for potassium with the result of measurements on cells with liquid junctions, especially as the available data for computing the standard potentials of lithium, rubidium and cesium are of the latter type. Lewis and Keyes [2] have found the potentials of the cells:

$$\text{K (amal., } 0.2216\%\text{) ; KOH (}0.2026\,N\text{)} \underset{a}{:} \text{KCl (}0.2\,N\text{)} \underset{b}{:} \text{KCl (}1.0\,N\text{), HgCl(s) ; Hg}$$

$$\text{K (metal) ; KI in ethylamine ; K (amalgam, } 0.2216\%\text{)}$$

to be, respectively, 2.1891 and 1.0481 volts at 25°. Using the formulas (23) and (25), Chapter 13, the potentials of the liquid junctions at *a* and *b* are computed to be − 0.0168 and 0.0008 volt, from which the combination

$$\text{K (metal); KOH (}0.2026\,N\text{) :: KCl (}1.0\,N\text{), HgCl (s); Hg}$$

would have a potential of 3.2532 volts. Assuming that potassium hydroxide has the same activity coefficient as sodium hydroxide and interpolating a value of 0.718 for a 0.2014 molal solution from Table V, Chapter 8, we may compute the potential 3.2035 volts for the cell

$$\text{K (metal); K}^+ (a = 1) \text{ :: KCl (}1.0\,N\text{), HgCl (s); Hg}$$

which together with equation (10) yields, at 25°

$$\text{K (metal); K}^+ \qquad E_0 = 2.9234 \tag{12}$$

in good agreement with the value 2.9239 volts, obtained from cells without liquid junctions.

Lewis and Keyes [3] have found the potentials of the cells

$$\text{Li (amal., } 0.0350\%\text{) ; LiOH (}0.1\,N\text{)} \underset{a}{:} \text{LiCl (}0.1\,N\text{)} \underset{b}{:} \text{KCl (}0.1\,N\text{)} \underset{c}{:} \text{KCl (}1.0\,N\text{), HgCl (s) ; Hg}$$

$$\text{Li (metal) ; LiI in propylamine; Li (amalgam, } 0.0350\%\text{)}$$

to be 2.3952 and 0.9502 volts. Correcting for the liquid junctions, by equations (26a) and (25), Chapter 13, at *a*, *b*, and *c*, and adding, gives 3.3706 volts for the combination

$$\text{Li (metal); LiOH (}0.1\,N\text{) :: KCl (}1.0\,N\text{), HgCl (s); Hg}$$

[2] G. N. Lewis and F. G. Keyes, *J. Am. Chem. Soc.*, **34**, 119 (1912).
[3] G. N. Lewis and F. G. Keyes, *J. Am. Chem. Soc.*, **35**, 340 (1913).

Assuming that the lithium hydroxide solution has the same activity coefficient, 0.759, as sodium hydroxide gives, with equation (10), at 25°

$$\text{Li (metal); Li}^+ \qquad E_0 = 3.0243 \qquad (13)$$

A similar investigation on the standard potential of rubidium has been carried out by Lewis and Argo[4] who measured at 25° the potentials of the cells:

Rb (metal) ; RbI in ethylamine; Rb (amalgam)

$$\text{Rb (amal.) ; RbOH } (0.1\,N) \underset{a}{:} \text{RbCl } (0.1\,N) \underset{b}{:} \text{KCl } (1.0\,N),\ \text{HgCl (s) ; Hg}$$

finding respectively, 1.0745 and 2.1805 volts. Since the rubidium ion has almost exactly the same mobility as the potassium ion, equations (26a) and (25), Chapter 13, give −0.0165 and 0.0013 volt for the potentials of the liquid junctions at *a* and *b*. Assuming that the activity coefficient of rubidium hydroxide is the same as that for sodium hydroxide, and using equation (10) yields, at 25°

$$\text{Rb (metal); Rb}^+ \qquad E = 2.9239 \qquad (14)$$

The Standard Potentials of the Halogens. The standard potential of chlorine, obtained from cells without liquid junctions, has been considered on page 198. The standard potential of *iodine* has been determined by Maitland[5] and more recently by Jones and Schumb[6] and Jones and Kaplan.[7] It is most convenient to take the solid as the standard state of iodine, instead of the gas at one atmosphere pressure as was done with chlorine. Jones and Schumb used cells of the type

$$\text{(Pt); } I_2\text{ (s), KI } (C_1) : \text{KCl } (0.1\,N),\ \text{HgCl (s); Hg} \qquad (15)$$

in which the concentration C_1 ranged from 0.01 to 0.1 normal. The interpretation of the results is complicated by the formation of I_3^- ions according to the equilibrium

$$I^- + I_2 \rightleftarrows I_3^- \qquad (16)$$

With solid iodine present this reaction goes nearly to completion with the result that the concentration of the iodide, I^-, ion could not be esti-

[4] G. N. Lewis and W. L. Argo, *J. Am. Chem. Soc.*, **37**, 1983 (1915).
[5] W. Maitland, *Z. Elektrochem.*, **12**, 263 (1906).
[6] G. Jones and W. C. Schumb, *Proc. Am. Acad.*, **56**, 199 (1921).
[7] G. Jones and B. B. Kaplan, *J. Am. Chem. Soc.*, **50**, 2066 (1928).

mated with accuracy. Jones and Kaplan therefore extended the measurements so that potentials of the combination

$$(\text{Pt});\ I_2\ (C_2),\ \text{KI}\ (C_1) \underset{a}{\vdots} \text{KI}\ (C_1),\ I_2\ (s);\ (\text{Pt}) -$$
$$-(\text{Pt});\ I_2\ (s),\ \text{KI}\ (C_1) \underset{b}{\vdots} \text{KCl}\ (0.1\ N),\ \text{HgCl}\ (s);\ \text{Hg} \qquad (17)$$

were available. The concentration C_2 of the iodine was kept at relatively low values and was determined by equilibrium measurements. Using the value of 1.40×10^{-3} for the equilibrium constant of equation (16) the compositions of the solutions in contact at the junctions *a* and *b* were found and the liquid junction potentials computed by means of the Henderson equation (23), Chapter 13. Since the chosen standard state for iodine is that of the solid, a correction must be made for the difference between the iodine concentration C_2 and the concentration of saturated iodine by means of the relation

$$\frac{RT}{2\mathbf{F}} \ln \frac{0.00132}{C_2}$$

in which 0.00132 mol per liter is the solubility of iodine in water at 25°. Another correction is necessary for the concentration and activity of the iodide, I^-, ion as actually present in the solution around the electrode at the left of the cell as represented. Jones and Kaplan,[8] and Gelbach[9] give a limited amount of data on the concentration cells

$$\text{Ag};\ \text{AgI},\ \text{KI}\ (C_1) : \text{KI}\ (C_2),\ \text{AgI};\ \text{Ag} \qquad (18)$$

from which, using the transference data given in Table IV, Chapter 4, the activity coefficients of potassium iodide may be obtained. These prove to be very close to those of potassium chloride given in Table V, Chapter 8. The result of these measurements and computations is the potential, *E*, of the cell

$$(\text{Pt});\ I_2\ (s),\ I^- :: \text{KCl}\ (0.1\ N);\ \text{HgCl}\ (s);\ \text{Hg} \qquad (19)$$

which at the corresponding potassium iodide concentrations, C_1, is as follows:

Concentration of KI mols per liter C_1	Potential of Cell (19) volts E
0.1	− 0.2012
0.05	− 0.2010
0.02	− 0.2011
0.01	− 0.2015

[8] G. Jones and B. B. Kaplan, *loc. cit.*
[9] R. W. Gelbach, *J. Am. Chem. Soc.*, 55, 4858 (1933).

The average of these potentials together with equation (11) yields

$$(Pt);\ I_2\ (s),\ I^- \qquad E_0 = -\ 0.5350 \qquad (20)$$

The standard potential of *bromine* has been determined by Lewis and Storch[10] and quite recently by Jones and Baekström.[11] The latter workers determined the potentials of cells which may be represented by

$$(Pt)\cdot\ Br_2\ (C_2),\ KBr\ (C_1) \underset{a}{:} KBr,\ AgBr\ (s);\ Ag \qquad (21)$$

The interpretation of the results of the measurements is complicated by the fact that bromine reacts with water according to the equation

$$H_2O + Br_2 = HOBr + HBr$$

and with the bromide ion according to

$$Br_2 + Br^- \rightleftarrows Br_3^- \quad \text{and} \quad 2Br_2 + Br^- \rightleftarrows Br_5^-$$

As the standard state of bromine we are free to choose the liquid or the gaseous state, of which the former will be selected. Correcting for the liquid junction at a by the Henderson equation, (23), Chapter 13, and for incomplete saturation with bromine of the solutions used in the cell, Jones and Baekström obtain -0.9940 volt at 25° for the E_0 value of the combination

$$(Pt);\ Br_2\ (l),\ (Br^-) :: (Br^-),\ AgBr\ (s);\ Ag \qquad (22)$$

which, together with the standard potential, -0.0711 volt, of the electrode Ag; AgBr(s), Br^- from Table V, Chapter 10, yields, at 25°

$$(Pt);\ Br_2\ (l),\ Br^- \qquad E_0 = -\ 1.0651 \qquad (23)$$

The Standard Potential of the Hydroxyl Electrode. The electrode

$$(Pt)\ H_2;\ H_2O\ (l),\ OH^- \qquad (24)$$

may be considered to be a hydrogen electrode at which the hydrogen ion activity is controlled by the equilibrium

$$H^+ + OH^- \rightleftarrows H_2O$$

for which the mass law constant is

$$\mathbf{K}_w = (H^+)\ (OH^-) = 1.0083 \times 10^{-14} \text{ at } 25°$$

[10] G. N. Lewis and H. Storch, *J. Am. Chem. Soc.*, **39**, 2544 (1917).
[11] G. Jones and S. Baekström, *ibid.*, **56**, 1524 (1934).

from Chapter 11. The electrode will have its standard potential when the hydroxyl ion activity, (OH^-) is unity. Thus

$$E = 0 - \frac{RT}{\mathbf{F}} \ln (H^+) = - \frac{RT}{\mathbf{F}} \ln \frac{\mathbf{K}_w}{(OH^-)} \qquad (25)$$

If the activity of the hydroxyl ion constituent, (OH), is unity then the value of the potential, E, is equal to the standard potential, E_0, of the hydroxyl electrode, at 25° ; therefore

$$\text{(Pt) } H_2;\ H_2O\ (l),\ OH^- \qquad E_0 = 0.8279 \qquad (26)$$

Lorenz and Böhi [12] and Lewis and Randall [13] have obtained values not far from this, using cells of the type

$$\text{(Pt) } H_2;\ KOH : KCl : HCl;\ H_2 \text{ (Pt)}$$

with the three solutions all of the same concentration so that the liquid junctions could be computed by equation (26a), Chapter 13. However the assumptions necessary in computing the standard potential on this basis make the result rather uncertain.

The Mercuric Oxide Electrode. The mercuric oxide electrode has been extensively studied since it is one of the few electrodes which are of service in alkaline solutions. Early work was done by Brønsted,[14] Donnan and Allmand,[15] and Knobel.[16] The more recent measurements on the cell:

$$Hg;\ HgO\ (s),\ NaOH\ (m);\ H_2 \text{ (Pt)} \qquad (27)$$

have been carried out by Fried,[17] Kobayashi and Wang [18] and Shibata and Murata,[19] using various molalities of the alkali solution. The cell reaction is

$$Hg + H_2O = H_2 + HgO$$

so that at a constant pressure of hydrogen the potential of the cell should depend only upon the activity of the water in the solution of electrolyte. Since that activity changes slowly with the concentration of alkali the potential of the cell should change but little with the electrolyte concentration. Some of Kobayashi and Wang's results are given in Table I, and illustrate this slow change with molality, m, of the alkali

[12] R. Lorenz and A. Böhi, *Z. physik. Chem.*, **66,** 733 (1909).
[13] G. N. Lewis and M. Randall, *J. Am. Chem. Soc.*, **36,** 1969 (1914).
[14] N. Brønsted, *Z. physik. Chem.*, **65,** 84 (1908).
[15] F. G. Donnan and Allmand, *J. Chem. Soc.*, **99,** 845 (1911).
[16] M. Knobel, *J. Am. Chem. Soc.*, **45,** 70 (1923).
[17] F. Fried, *Z. physik. Chem.*, **123A,** 406 (1926).
[18] Y. Kobayashi and H. L. Wang, *J. Sci. Hiroshima Univ.*, **5A,** 71 (1934).
[19] F. L. E. Shibata and M. Murata, *J. Chem. Soc.* (Japan), **52,** 399 (1931).

TABLE I. POTENTIAL OF THE CELL Hg; HgO (s), NaOH (m); H_2 (Pt) at 25°

Molality of NaOH m	Potential volt
0.04873	− 0.92550
.20001	− 0.92550
.27366	− 0.92552
.50013	− 0.92556
.89951	− 0.92581

in the electrolyte. These workers also obtained results at 22.5 and 27.5° which are in close accord with Fried's work at 23.5°. The value of this potential at the lowest concentrations, −0.9255 volt, may thus be regarded as the potential of

$$\text{Hg; HgO (s), (OH}^-\text{) : (OH}^-\text{); H}_2\text{ (Pt)}$$

which with the value of the standard potential of (Pt) H_2; $H_2O(l)$, OH^- just obtained gives, at 25°

$$\text{Hg; HgO (s), OH}^- \qquad E_0 = -0.0976 \tag{28}$$

The Standard Potential of Silver. The standard potential of silver has been measured by Lewis [20] and by Noyes and Brann [21] who used the cell

$$\text{Ag; AgNO}_3\ (0.1\,N) : \text{KNO}_3\ (0.1\,N) : \text{KCl}\ (0.1\,N),\ \text{HgCl (s); Hg}$$

obtaining −0.3992 volt. Correcting for the liquid junction potential using equation (26a), Chapter 13, gives −0.3985 volt for the potential of

$$\text{Ag; AgNO}_3\ (0.1\,N) :: \text{KCl}\ (0.1\,N),\ \text{HgCl (s); Hg}$$

This value, together with the activity coefficient 0.733 for 0.1 normal silver nitrate from Table III, Chapter 8, and the value, page 248, of the potential of the tenth normal calomel cell gives, at 25°

$$\text{Ag; Ag}^+ \qquad E_0 = -0.7994$$

The uncertainty of computations involving liquid junctions is however indicated by the fact that if use is made of equation (26b) instead of (26a), Chapter 13, a potential is obtained that is almost one millivolt lower, numerically, than the one just given.[22]

[20] G. N. Lewis, *J. Am. Chem. Soc.*, **28**, 158 (1906).

[21] A. A. Noyes and B. F. Brann, *ibid.*, **34**, 1016 (1912).

[22] A. Brester, *Rec. Trav.*, **46**, 328 (1927), has recently investigated the standard potential of the silver electrode and obtained an average value of $E_0 = -0.7992$ from measurements on cells using three different silver salts. However, his cells involve liquid junctions the potentials of which are difficult to estimate.

The standard potential for silver can however, be readily computed from that of the silver-silver chloride electrode and the solubility product, s, of silver chloride in terms of the activities of the ions for which Brown and MacInnes [23] have recently obtained $(1.309 \times 10^{-5}$ mol per liter$)^2$ at 25°. The silver-silver chloride electrode is essentially a silver, Ag; Ag^+, electrode at which the silver ion activity is controlled by the solubility product

$$(Ag^+)\,(Cl^-) = s \tag{29}$$

Thus at a silver-silver chloride electrode the potential E is

$$E = E_0^{Ag} - \frac{RT}{\mathbf{F}} \ln (Ag^+) = E_0^{Ag} - \frac{RT}{\mathbf{F}} \ln \frac{s}{(Cl^-)}$$

$$E = E_0^{Ag} - \frac{RT}{\mathbf{F}} \ln s + \frac{RT}{\mathbf{F}} \ln (Cl^-) = E^{Ag;\,AgCl} + \frac{RT}{\mathbf{F}} \ln (Cl^-) \tag{30}$$

in which E_0^{Ag} and $E_0^{Ag;\,AgCl}$ are the indicated standard potentials. It is evident therefore that

$$E_0^{Ag} = E_0^{Ag;\,AgCl;} + \frac{RT}{\mathbf{F}} \ln s \tag{34}$$

from which, since the standard potential for the silver-silver chloride electrode is -0.2225 volt, we obtain, at 25°

Ag; Ag^+ $\qquad E_0 = -\ 0.8002$ (32)

a value probably more accurate than the one given above since it does not involve the computation of liquid junction potentials. It does, however, depend upon the accuracy of the determination of the solubility of silver chloride. The five most recent determinations of the solubility of silver chloride at 25° by nephelometric methods [24] yield an average value of 1.33×10^{-5} mol per liter, in terms of activity. Use of this value would change the potential given above to Ag;Ag^+, $E_0 = 0.7993$. As a result of these computations we shall adopt, at 25°, the rounded value

Ag; Ag^+ $\qquad E_0 = -\ 0.799$ (33)

The Standard Potential of Tin. Direct determinations of the standard potential of tin are of doubtful accuracy. The potential can however be readily computed from the standard potential of lead and a knowledge of the equilibrium

$$Sn\,(s) + Pb\,(ClO_4)_2 \rightleftarrows Pb\,(s) + Sn\,(ClO_4)_2 \tag{34}$$

[23] Page 318.
[24] Landolt-Börnstein, "Physikalisch-chemische Tabellen," 1, 5th ed., 483, Julius Springer, Berlin, 1935.

When this equilibrium is established the Gibbs free energy of the reaction is zero. Therefore a cell, in which the reaction takes place and in which the solutes are at their equilibrium values, will have a potential of zero. For the case in question the cell would be:

$$\text{Pb; Pb}(\text{ClO}_4)_2\,(m_1) :: \text{Sn}(\text{ClO}_4)_2\,(m_2)\text{; Sn} \tag{35}$$

for which

$$E = 0 = E_0^{\text{Pb}} - \frac{RT}{2\mathbf{F}} \ln m_{\text{Pb}}\gamma_{\text{Pb}} - E_0^{\text{Sn}} + \frac{RT}{2\mathbf{F}} \ln m_{\text{Sn}}\gamma_{\text{Sn}} \tag{36}$$

This equation may be put in the form

$$E_0^{\text{Sn}} - E_0^{\text{Pb}} = \frac{RT}{2\mathbf{F}} \ln \frac{m_{\text{Sn}}\gamma_{\text{Sn}}}{m_{\text{Pb}}\gamma_{\text{Pb}}} = \frac{RT}{2\mathbf{F}} \ln \mathbf{K} \tag{37}$$

in which **K** is the mass law constant of the reaction in equation (34). Thus if any two of the quantities E_0^{Sn}, E_0^{Pb} or **K** are known the other may be computed. Noyes and Toabe[25] have determined the ratio m_2/m_1 at equilibrium for reaction (34) and find it, roughly at least, independent of the concentration and equal to 2.98. Since the two salts are of the same ionic type, γ_1 is nearly equal to γ_2, and thus **K** = 2.98. From equation (37) and the E_0 value for the Pb; Pb^{++} electrode, from equation (49), of Chapter 10, this leads, at 25°, to

$$\text{Sn; Sn}^{++} \qquad E_0 = 0.1405 \tag{38}$$

A Table of the Standard Potentials of the Elements, at 25°. The Electromotive Series. In Table II the standard potentials of the

TABLE II. STANDARD POTENTIALS OF THE ELEMENTS AT 25°

Electrode	Standard Potential volts E_0	Electrode	Standard Potential volts E_0
Li ; Li^+	+ 3.0243	Sn ; Sn^{++}	+ 0.140_5
K ; K^+	+ 2.9239	Pb ; Pb^{++}	+ 0.1265
Rb ; Rb^+	+ 2.9239	(Pt)H_2 ; H^+	± 0.0000
Na ; Na^+	+ 2.7139	Cu ; Cu^{++}	− 0.339
Zn ; Zn^{++}	+ 0.7611	(Pt) ; I_2(s), I^-	− 0.5350
Cd ; Cd^{++}	+ 0.4023	Ag ; Ag^+	− 0.799
Tl ; Tl^+	+ 0.3385	(Pt) ; Br_2(l), Br^-	− 1.0651
Co ; Co^{++}	+ 0.283	(Pt) ; Cl_2(g), Cl^-	− 1.3587
Ni ; Ni^{++}	+ 0.236		

elements at 25°, obtained as described in this chapter and Chapter 10, are listed in the order of their values, starting with the alkali metals, and ending with the halogens. The order is that of the "electromotive

[25] A. A. Noyes and K. Toabe, *J. Am. Chem. Soc.*, 39, 1537 (1917).

series." Early workers considered the electromotive series to be that in which metals displaced one another from their salts. Thus zinc will react with copper sulphate according to the equation

$$Zn + CuSO_4 = Cu + ZnSO_4$$

In general a metal higher in the series tends to displace one lower down. However, the figures given in Table II are for unit activities of the ions. By shifting the relative ion activities of the reacting substances, or by changing the salt concentrations, or by forming ionic complexes, displacements may occur which are in the contrary order to those predicted by the series as given in the table.

Chapter 15

The Determination and Meaning of "pH" Values

A useful concept, and one of very general practical application is that of "pH." The symbol was suggested by Sørensen in 1909 [1] to obviate the necessity of using negative exponents in expressing what were then regarded as hydrogen ion concentrations. A 0.001 normal hydrochloric acid solution was considered to have a hydrogen ion concentration of 9.9×10^{-4}, which is the concentration times the conductance ratio, Λ/Λ_0. The negative Briggsian logarithm of this number, by which Sørensen defined the pH, is 3.015, a number which is evidently more convenient to use, both in writing and conversation, than 9.9×10^{-4}. On this scale a normal solution of a strong acid will have a pH in the neighborhood of zero, the pH of pure water will be about seven, and a normal solution of a strong alkali will have a pH of about fourteen. The term pH is frequently defined by one or the other of the expressions

$$\text{pH} = -\log C_{\text{H}^+} \tag{1}$$

in which C_{H^+} is the hydrogen ion concentration, and

$$\text{pH} = -\log a_{\text{H}^+} \tag{2}$$

where a_{H^+} is the hydrogen ion activity. Neither of these equations, as we shall shortly see, is satisfactory. It seems best to define pH in terms of a method for determining it, which is to measure the potential of a galvanic cell of the form:

$$\text{(Pt) H}_2\text{; solution X} \underset{a}{\vdots} \text{saturated KCl} \underset{b}{\vdots} \text{reference electrode} \tag{3}$$

In this cell "solution X" is the fluid the pH of which is desired, and the reference electrode is usually the tenth normal calomel electrode or the saturated calomel electrode. From the potential, E', of this cell at 1 Atm hydrogen pressure the pH may be computed from the formula, in which $R' = 2.3026\,R$,

$$\text{pH} = \frac{E' - \text{E}_0}{R'T/\mathbf{F}} \tag{4}$$

[1] S. P. L. Sørensen, *Compt. Rend. Lab. Carlsberg*, **8**, 1 (1909).

In this expression, T and $\mathbf{F}$ have their customary significance and E_0 is a constant the evaluation of which will be discussed later in this chapter. The quantity E_0 is not, strictly speaking, a standard potential. This is indicated by the change of type. It is the variable liquid junction at the point *a* in cell (3) that makes the interpretation of pH measurements difficult. For most purposes it is not necessary to consider pH values in terms of hydrogen ion activities or concentrations, the pH *numbers* being sufficient. For such cases it would be necessary only to agree upon a value of E_0 for each reference electrode. Sørensen for instance obtained that constant from the equation

$$E' = \text{E}_0 - \frac{RT}{\mathbf{F}} \ln C_{\text{H}^+} = \text{E}_0 + \frac{R'T}{\mathbf{F}} \text{pH} \tag{5}$$

utilizing for "solution X" dilute hydrochloric acid for which he assumed, as was then usual, $C_{\text{H}^+} = C_{\text{HCl}} \cdot \Lambda/\Lambda_0$ in which Λ and Λ_0 are the equivalent and limiting equivalent conductances of the acid. Although the theoretical basis for that computation is now agreed to be erroneous, Sorensen's values of E_0 for different temperatures are still in general use today. It is a fact, however, that if a series of concentrations of hydrochloric acid is used in cell (3), and the corresponding values of E_0 are computed on the basis just described, they are not found to be constant. The variation from constancy is still greater if it is assumed, in accord with present theories, that the acid is completely dissociated.

Since certain types of galvanic cell yield activities it has been thought that pH measurements should be more nearly in accord with equation (2) *i.e.*, $\text{pH} = -\log a_{\text{H}}$ than with equation (1). However, as has been made clear in previous chapters, the concentration cell determinations yield mean ion activities, and not single ion activities, such as a_{H^+}, single ion activities being unobtainable without non-thermodynamic assumptions. Equation (2) is therefore not susceptible of proof. A value of E_0 for each temperature can however be chosen, as will be shown below, that will bring pH measurements into at least fair accord with the equation

$$\text{pH} = -\log C_{\text{H}^+} f_{\pm} \tag{6}$$

in which C_{H^+} is the hydrogen ion concentration, and $f_{\pm}$ is the mean ion activity coefficient.

The Practical Determination of pH by Potentiometric Methods. There are four potentiometric methods in general use for determining pH values. These depend upon (a) the hydrogen electrode, (b) the quinhydrone electrode, (c) the antimony-antimony trioxide electrode, and (d) the glass "electrode." These will be discussed in the order given. Some other methods used for special purposes will receive brief notice. The indicator and catalytic methods for obtaining pH values are of great utility in certain cases but are not in the province of this book.

(a) *The Hydrogen Elcctrode.* The hydrogen electrode has been mentioned so many times in the preceding chapters that little further comment seems necessary. A type of half cell, including a hydrogen electrode, which has been used in many investigations, is shown in Fig. 1. The solution in the vessel, A, is in contact with the platinized

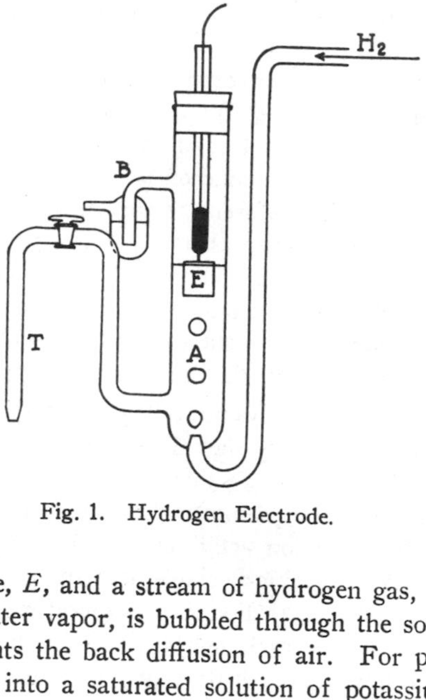

Fig. 1. Hydrogen Electrode.

platinum electrode, E, and a stream of hydrogen gas, oxygen free, and saturated with water vapor, is bubbled through the solution, as shown. A trap, B, prevents the back diffusion of air. For pH measurements the tube, T, dips into a saturated solution of potassium chloride, into which is inserted the connecting tube from a calomel electrode, usually containing either tenth normal or saturated potassium chloride. The potential, E, of such a cell is corrected to the value, E', which it would have if the partial pressure of hydrogen were one atmosphere, by means of the equation

$$E' = E - \frac{R'T}{2\mathbf{F}} \log p \tag{7}$$

in which the partial pressure, p, in atmospheres, is obtained by subtracting the vapor pressure of the solution in the vessel A from the observed barometric pressure. (See equation (51) of Chapter 5.) The values of pH may then be computed from the corrected potential, E', and equation (3). This may be regarded as the standard method for determining pH values, and the accuracy of other methods is judged by the closeness with which they conform to it.

The type of half cell devised by Clark, especially for work with hydrogen electrodes in solutions of biological origin, is shown in Fig. 2. By manipulation of the stopcock, *S*, solution from the vessel, *D*, can be flowed into the chamber, *E*, or hydrogen bubbled in from the tube, *A*,

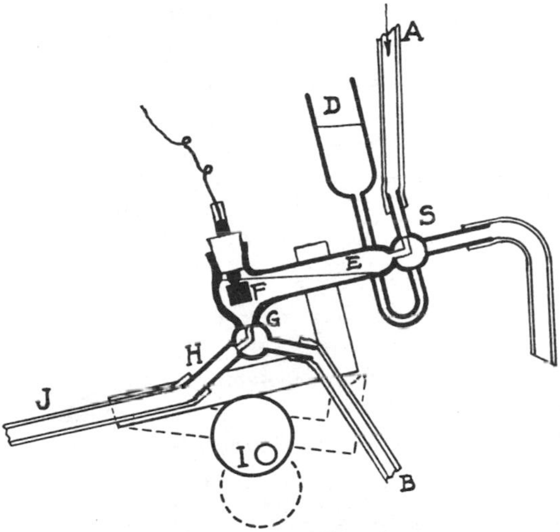

Fig. 2. Clark's Cell for Obtaining pH Values.

and out of the exit tube, *B*. After filling with the gas, the vessel is rocked by the eccentric, *I*, after which, by turning the stopcock *G*, a liquid junction is made with the tube, *J*, which is filled with saturated potassium chloride, and leads to the reference electrode. An advantage of this apparatus is that the presence of oxygen in the hydrogen and the solution does not lead to appreciable error, since the oxygen is "cleaned up" by reaction with hydrogen on the surface of the platinized platinum electrode, *F*. This apparatus is in very general use for measurements of moderate accuracy.[2]

[2] For further details concerning the operation of this cell, and much else on the determination of pH, see W. M. Clark's "The Determination of Hydrogen Ions," 3rd ed., The Williams and Wilkins Co., Baltimore, 1928.

It is evident, however, that the hydrogen electrode method can be used in obtaining pH values only of solutions which do not contain substances which will interfere with the attainment of equilibrium in the electrochemical reaction

$$H_2 = 2H^+ + 2e^- \qquad (8)$$

Solutions which contain strong oxidizing or reducing substances may react with the hydrogen, or establish other potentials than that determined by the equilibrium represented by equation (8), or do both at the same time. The pH values of solutions of chromic or permanganic acids, for instance, cannot be determined by the hydrogen electrode method just described, because these solutions will react with hydrogen, in the presence of platinized platinum, and because such substances and their reduction products tend to establish their own potentials at the electrode, as is discussed in Chapter 16. In addition there are present in many solutions, especially those of biological origin, what are known somewhat vaguely as "poisons" for the hydrogen electrode, which reduce the accuracy of pH determinations. These facts have led to the extensive use of alternative methods for determining pH values.

(b) *The Quinhydrone Electrode.* If to the solution surrounding a smooth platinum or gold electrode some crystals of quinhydrone (a compound of equimolecular proportions of hydroquinone and quinone) are added, a definite reproducible potential, against any convenient reference electrode, will be observed, at least for solutions on the acid side of neutrality, *i. e.*, that have pH values less than seven. The type of cell employed is usually:

(Au); quinhydrone (s), solution X : saturated KCl : KCl, HgCl; Hg (9)

The potential at the gold electrode arises from the electrochemical reaction

$$\underset{\text{hydroquinone}}{C_6H_4O_2H_2} = \underset{\text{quinone}}{C_6H_4O_2} + 2H^+ + 2e^-$$

The careful studies leading to the evaluation of the standard potential of this reaction have been discussed in Chapter 10. It will be observed, however, that cell (9) involves the same liquid junction as in pH measurements with hydrogen electrodes. Ignoring change in the liquid junction potential, a cell of this type will have a potential given by the formula

$$E = E_0^q - \frac{RT}{2F} \ln \frac{(Q)}{(Hy)} - \frac{RT}{F} \ln (H^+) \qquad (10)$$

or by

$$E = E_0^q - \frac{RT}{2\mathbf{F}} \ln \frac{[Q] f_Q}{[Hy] f_{Hy}} + \frac{R'T}{\mathbf{F}} \text{pH} \tag{11}$$

in which E_0^q is a constant. By use of quinhydrone the *concentrations* of quinone [Q] and hydroquinone [Hy] are automatically made equal so that the activity ratio, (Q)/(Hy), will be equivalent to the ratio of the activity coefficients f_Q/f_{Hy}. Since, in dilute salt solutions at least, this ratio will be close to unity, equation (11) can be given the form

$$\text{pH} = \frac{E - E_0^q}{R'T/\mathbf{F}} \tag{12}$$

Values of E_0^q for use, at various temperatures, in equation (12) are given in Table III.

The quinhydrone electrode was first studied by Haber and Russ[3] but its value as an electrode for measuring pH was recognized and developed by Biilmann.[4] It can be used, at low pH values, in the presence of oxygen. This fact, and the ease with which it can be prepared, has led to its extensive use. Obviously, the quinhydrone electrode cannot be used for the pH determination of solutions containing strong oxidizing or reducing agents, due to direct reaction of such agents with the quinhydrone or their effects upon the potential of the electrode. Due both to the acid nature of the hydroquinone and to the tendency of that substance to oxidize in alkaline solutions the electrode is inaccurate if the pH is above 7.0. Above that value LaMer and Parsons[5] have found that the pH values determined by the quinhydrone electrode are low, the difference reaching a maximum of 2.4 pH units (148 millivolts) at pH = 11. The electrode is further subject to a "salt error," due apparently to the different effects of dissolved substances upon the activity coefficients f_Q and f_{Hy} in equation (11). This is, fortunately, not large for solutions of moderate concentrations, but must not be overlooked for accurate work. It has been studied directly by Hovorka and Dearing[6] whose work has already been discussed in Chapter 10. These workers have found that the salt error, ΔpH, can be computed from the simple relation

$$\Delta\text{pH} = Bc \tag{13}$$

in which B is a constant for a given salt and c is the concentration, in

[3] F. Haber and R. Russ, *Z. physik. Chem.*, **47**, 257 (1904).
[4] E. Biilmann, *Bull. soc. chim.*, **41**, 213 (1927).
[5] V. K. LaMer and T. R. Parsons, *J. Biol. Chem.*, **57**, 613 (1923).
[6] F. Hovorka and W. C. Dearing, *J. Am. Chem. Soc.*, **57**, 446 (1935).

equivalents per liter. Some typical values of B are given in Table I. The corrections are added algebraically to the pH values computed with equation (12). The last figure given in the table shows that a "salt error" results from the presence of non-electrolytes in solution as well as from salts.

TABLE I. CONSTANTS, B, FOR COMPUTING THE "SALT ERROR" OF THE QUINHYDRONE ELECTRODE AT 25°

Salt	B	Salt	B
HCl	− 0.0616	$BaCl_2$	− 0.0438
NaCl	− 0.0413	Na_2SO_4	+ 0.0227
LiCl	− 0.0353	$MgSO_4$	+ 0.0206
$MgCl_2$	− 0.0346	Mannitol	+ 0.0237

In making pH measurements with the quinhydrone electrode the solution should be saturated with quinhydrone, *i. e.*, the solid substance should be present. Biilmann and Jensen [7] have found that errors may arise if an unsaturated solution is used. Such errors will be about 0.01 pH unit if the quinhydrone concentration is one-tenth of that of the saturated solution, and the error increases if still less of the substance is present.

It is an interesting fact however that if a solution is saturated with quinhydrone it is not, thereby, saturated with the decomposition products, quinone and hydroquinone. The equilibria involving these substances are:

$$\underset{\text{solid}}{\text{Quinhydrone}} \rightleftarrows \underset{\text{in solution}}{\text{Quinhydrone}} \rightleftarrows \underset{\text{in solution}}{\text{quinone} + \text{hydroquinone}}$$

and in addition there is the mass action relation

$$\mathbf{K} = \frac{(\text{quinone})\,(\text{hydroquinone})}{(\text{quinhydrone})} \tag{14}$$

for the constant, **K**, of which Sørensen, Sørensen and Linderstrom-Lang [8] have found the value 0.263 at 18°. As usual in this book the parentheses represent the activities of the substances enclosed. If the quinhydrone is present as solid its activity is maintained constant. Under these conditions, however, the activities of the decomposition products may vary provided their product remains constant. That such variation occurs when the composition of the solution in which the substances are dissolved is changed, is indicated by the necessity of using a correction for the "salt error," discussed in the previous paragraph.

[7] E. Biilmann and A. L. Jensen, *Bull. Soc. chim.*, **41**, 151 (1927).

[8] S. P. L. Sørensen, M. Sørensen and K. Linderstrom-Lang, *Compt. Rend. Lab. Carlsberg*, **14**, No. 14 (1921).

If the solution is saturated with two of the substances involved in equation (14) their activities will remain constant, and the equation requires that the activity of the third will also be constant. For this reason electrodes without a salt error should be formed by surrounding the electrode with solid quinhydrone and solid hydroquinone. This has been shown, experimentally, to be true by the workers just mentioned.

(c) *The Antimony-antimony Trioxide Electrode.* A means for determining pH values which has had considerable use, especially in cases in which accuracy is not necessary, depends upon measurements of the potentials of cells of the form

$$Sb; Sb_2O_3, \text{Solution X} : \text{saturated KCl} : \text{reference electrode} \qquad (15)$$

As indicated, the electrode consists of metallic antimony in contact with solid antimony trioxide. The electrode reaction at such an electrode may be written

$$2Sb + 3H_2O = Sb_2O_3 + 6H^+ + 6e^-$$

The potential will therefore depend not only upon the activity of the hydrogen ion constituent but also on that of the water. This is important only if concentrated solutions are measured, because the activity of water in dilute aqueous solutions changes very little. The electrode has the advantage of simplicity and of not requiring a current of hydrogen. Like the quinhydrone electrode it cannot be used with strong oxidizing or reducing substances. It also cannot be used with very strong acid or basic solutions because of the solubility of the oxide in such solutions. Rather conflicting results are described in the use of the electrode. Most workers report that the change in potential of cell (15) with the pH, *i. e.*, $\Delta E/\Delta pH$, is lower than the value that would be found with the hydrogen electrode although theoretically it should be very nearly the same. Roberts and Fenwick,[9] in a very careful study, find the electrode can be made to give the theoretical change with pH, if unstable forms of antimony oxide are excluded. They find however that the electrode is affected by oxygen. The form of the electrode consisting of a stick of metal covered with a layer of oxide, resulting from the reaction of the metal with air, is much used in practical work. Parks and Beard [10] report that the potential resulting from the use of such an electrode is far from steady, but that the results may be improved by the use of a vacuum-tube electrometer. This would indicate that the electrode is readily polarized. These authors find that the

[9] E. J. Roberts and F. Fenwick, *J. Am. Chem. Soc.*, **50**, 2125 (1928).
[10] L. R. Parks and N. C. Beard, *J. Am. Chem. Soc.*, **54**, 856 (1932).

theoretical values of $\Delta E/\Delta pH$ are obtained throughout the pH range 2 to 7, with deviations if the measurements are extended to higher or lower pH values. In its present state the use of the antimony-antimony trioxide electrode appears to be justified only if more adequate methods for obtaining pH are, for one reason or another, unavailable.

(d) *The Glass "Electrode."* Another method for determining pH values, which has come rapidly into use the last few years, is the glass electrode. As will be seen it has several decided advantages over the other methods available, with, however, definite limitations.

Haber and Klemensiewicz,[11] in 1909 carried out acid-base titrations potentiometrically with the aid of the simple apparatus indicated diagrammatically in Fig. 3, using a thin bulb of glass, *A,* in the figure,

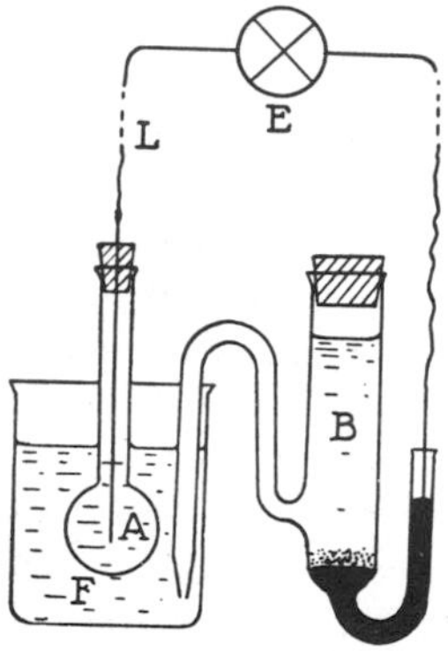

Fig. 3. An Arrangement for Measuring pH using a Glass Electrode.

instead of the usual hydrogen electrode. The bulb, *A,* contained hydrochloric acid into which dipped a platinum wire, *L*. In the diagram *B* represents a reference half cell, and *E* a quadrant electrometer, base being added to acid in the beaker, *F*. The electrometer is an electrostatic instrument which requires inappreciable current for its operation. Readings are thus uninfluenced by the high resistance of the glass membrane in the galvanic cell. Although the trend of the potentials observed by these workers during a titration was the same as that observed with a hydrogen electrode, quantitatively they were somewhat different. Later Hughes[12] showed, using an arrangement similar to that shown in Fig. 3, that the potential follows a change of pH quan-

[11] F. Haber and Z. Klemensiewicz, *Z. physik. Chem.*, **67**, 385 (1909).
[12] W. S. Hughes, *J. Am. Chem. Soc.*, **44**, 2860 (1922); *J. Chem. Soc.*, **1928**, 491.

titatively through a wide range of pH values in the same manner as a cell including the usual hydrogen electrode. He also demonstrated clearly the fact that the glass electrode is uninfluenced by strong oxidizing or reducing agents which seriously affect the results with any other method for obtaining pH values. Mrs. Kerridge [13] demonstrated the usefulness of the glass electrode in connection with solutions of biological origin. Such solutions contain substances which influence the accuracy of the hydrogen electrode, but have no effect on the glass electrode. However the high resistance of the available glass made measurements difficult and inaccurate. Much attention to insulation was necessary because very slight electrical leakage produced decided changes in the potential measurements.

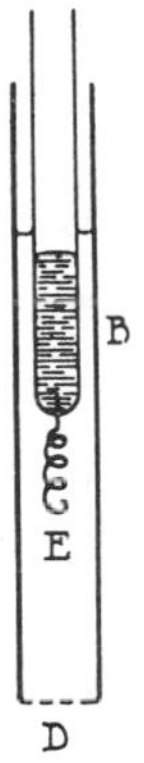

Fig. 4. A Membrane Type Glass Electrode.

MacInnes and Dole [14] designed and tested the type of glass electrode shown in Fig. 4. In this design a thin membrane, *D*, of soft glass is fused over the end of the tube, *B*. In the tube is placed a solution of hydrochloric acid and a silver-silver chloride electrode, *E*. The membranes are readily made by blowing a bubble of the glass and then placing the tube, *B*, the end of which has been heated to redness, in contact with the bubble. The design of electrode has the advantage of compactness, and since the membranes may be made as thin as 0.001 mm., of relatively low resistance. It was found, however, that modern commercial glasses do not act quantitatively in the measurement of pH, older and relatively soft glass being better in this respect. The new

[13] P. T. Kerridge, *Biochem. J.*, **19**, 611 (1925).

[14] D. A. MacInnes and M. Dole, *Ind. Eng. Chem., Anal. Ed.*, **1**, 57 (1929); *J. Am. Chem. Soc.*, **52**, 29 (1930).

form of electrode had the advantage that it made possible a fairly accurate comparison of various samples of glass. The most suitable of the glasses tested contained the three components, Na_2O, CaO and SiO_2 in the proportions of 22, 6 and 72 per cent, respectively. It is an interesting fact that this corresponds to the composition of a mixture of these components having the lowest melting point on Morey and Bowen's[15] phase diagram. This glass is now commercially available as "015 glass." Using this material, MacInnes and Dole[16] and later MacInnes and Belcher[17] made studies of the range of its effectiveness for the determination of pH values. To this end they made direct comparisons of the glass and hydrogen electrodes in solutions the pH

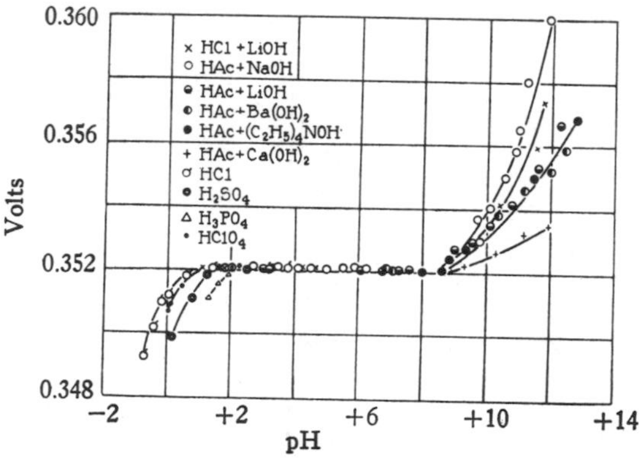

Fig. 5. The Potential Difference Between the Glass and Hydrogen Electrodes as a Function of the pH.

of which could be varied at will. The results of a series of such measurements are shown in Fig. 5. If a glass electrode is effective in the measurement of pH, the difference observed between the potential determinations with it and with a hydrogen electrode will be a constant independent of the pH of the solution. This corresponds to a horizontal line in the figure. Deviations from such a horizontal line represent, in these measurements, errors of the glass electrode since in pH measurements the hydrogen electrode is standard. Errors are apparent in the figure above pH = 9, and in very acid solutions. It will also be observed that the errors in alkaline solution depend upon the nature of the positive ions present in the solution, and in the acid solutions upon the nega-

[15] G. W. Morey and A. R. Bowen, *J. Soc. Glass Tech.*, **9**, 226 (1925).
[16] D. A. MacInnes and M. Dole, *J. Am. Chem. Soc.*, **52**, 29 (1930).
[17] D. A. MacInnes and D. Belcher, *ibid.*, **53**, 3315 (1931).

tive ions. The greatest deviation observed at pH = 11 is 6 mv., which corresponds to 0.1 pH unit. This is about the limit of accuracy in pH determinations that can be obtained using indicators. Between pH = 1 and pH = 9 the electrode functions accurately within the limit of sensitivity of the electrometer, ± 0.1 mv., or ± 0.002 pH unit. The measurements represented in Fig. 5 were made at a total electrolyte concentration of about tenth normal. If the concentration is larger than this the deviation of the glass electrode from the hydrogen electrode increases, and the range of pH in which the glass electrode is accurate shrinks.

Electrodes of the type shown in Fig. 4 are comparatively fragile, and attempts have been made to construct less easily breakable forms. Kahler and DeEds [18] tested relatively thick bulbs and discovered an effect that must be considered in all glass electrode work. If the resistance of the supporting tube is not of a higher order of magnitude than that of the thinner glass bulb or membrane the measurements may be affected by the presence of a film of moisture on the supporting tube and may also be dependent upon the depth of immersion of the bulb. This effect can be overcome by coating the tube with insulating material or by use of a harder glass for the supporting tube.

A form of glass electrode which is not fragile, which does not suffer from the difficulty described by Kahler and DeEds, and has advantages, particularly for routine pH measurements, is the "durable" or "condenser" type described by MacInnes and Belcher,[19] Fig. 6. The thin tube, *G*, of 015 glass is connected at *a* and *b* to harder glass of which the rest of the apparatus is made. A silver-silver chloride electrode, *P*, dips into the hydrochloric acid solution that bathes the outside of the tube of active glass, *G*. The solution, the pH of which is desired, is placed inside the tube. The waviness of that tube is to allow for the difference between the coefficients of expansion of the two types of glass used in the construction of the apparatus. The figure also shows a convenient stopcock for forming a liquid junction between the solution the pH of which is to be measured and the saturated potassium chloride solution, which also fills the reference calomel electrode, *C*.

The results shown in Fig. 5 and other investigations, have demonstrated that the glass electrode will yield accurate values of pH within the range of about 1 to 9. Comparison of the hydrogen and glass electrodes cannot be made with solutions which contain "poisons," oxidizing or reducing substances, or if the solution is highly viscous, because in

[18] H. Kahler and F. DeEds, *J. Am. Chem. Soc.*, **53**, 2998 (1931).
[19] D. A. MacInnes and D. Belcher, *Ind. Eng. Chem., Anal. Ed.*, **5**, 199 (1933).

such cases the hydrogen electrode does not function accurately. Since in the cases in which it can be tested the glass electrode responds only to pH, confidence in the electrode has been acquired for results in which direct comparisons are not possible.. A certain amount of caution must be exercised in the kinds of materials which are brought into contact with the glass surface. At least temporary errors can arise from the use of dehydrating agents, such as chromic acid solutions or alcohol. Accurate pH values of strongly dehydrating solutions can therefore not be obtained. No such difficulty is, however, encountered with complex mixtures, such as whole blood, or highly viscous solutions, if the water in such solutions or suspensions has substantially the activity of the pure liquid, and water is the continuous phase.

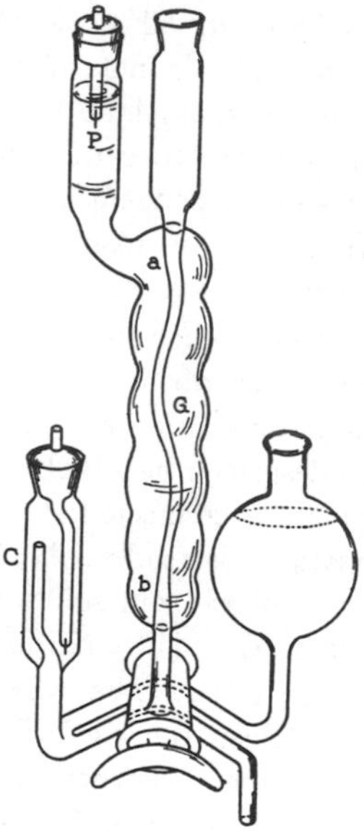

Fig. 6. A "Durable" Type of Glass Electrode.

As will be evident from the foregoing discussion the glass "electrode" is not strictly speaking an electrode at all, if following general usage, an electrode is understood as a metallic conductor of electricity in contact with an electrolytic conductor. The term "glass electrode" has however come into general use. The reason why a membrane of certain types of glass responds to changes of pH in the same way as a hydrogen electrode is not fully understood, though certain facts are quite clear. The operation of the glass electrode is related to the presence of dissolved water in the glass. The resistance of glass membranes can be very greatly increased by drying either with heat or desiccating agents. Furthermore a large portion at least of the conductance through the surface of soft glass at ordinary temperatures is carried out by the hydrogen ion or proton, as has been shown by Quittner[20] and by Burgess.[21] The latter found that with very thin electrodes acid formed, in fair accordance with Faraday's law, at the glass surface when current was passed through the glass. Theories to account for the deviations have been proposed by Horovitz,[22] Dole,[23] Gross and Halpern[24] and others. None of these theories is very satisfactory.

Since the resistance of a galvanic cell including a glass electrode varies in practice from 5 to 200 megohms it is necessary to use an electrometer, or an amplifying apparatus involving vacuum tubes. A serious difficulty with the earlier forms of vacuum tube circuits, and one not completely eliminated from all but the best of the more modern ones, is that during measurements appreciable grid currents pass through the glass electrode, which due to its high resistance, is easily polarized.

The Standardization of the pH Scale. As has been mentioned, Sørensen in defining pH in terms of the equation

$$\mathrm{pH} = -\log C_{\mathrm{H}^+}$$

used the current "classical" theory of electrolytes for computing the hydrogen ion concentrations of acids for obtaining the constant E_0, in the equation

$$E = \mathrm{E}_0 - \frac{RT}{\mathbf{F}} \ln C_{\mathrm{H}^+} = \mathrm{E}_0 + \frac{R'T}{\mathbf{F}}\mathrm{pH} \tag{15}$$

Since potentiometric methods usually yield activities rather than concentrations it has, more recently, been usual to define pH by

$$\mathrm{pH} = -\log a_{\mathrm{H}^+} \tag{2}$$

in which a_{H^+} is the activity of the hydrogen ion component. The difficulties with the concept of single ion activities such as a_{H^+} have been outlined in Chapter 13. It is, however, desirable to bring the pH scale in as close accord as possible with thermodynamic quantities because pH methods are utilized in studying equilibria involving the hydrogen

[20] F. Quittner, *Ann. Physik.* (IV) **85**, 745 (1928).
[21] L. L. Burgess (unpublished work in the author's laboratory).
[22] K. Horovitz, *Z. Physik.*, **15**, 369 (1923).
[23] M. Dole, *J. Am. Chem. Soc.*, **53**, 4260 (1930); **54**, 2120, 3095 (1932).
[24] P. Gross and O. Halpern, *J. Chem. Phys.*, **2**, 136 (1934).

ion component. For this reason Bjerrum and Unmack[25] and Guggenheim and Schindler[26] have recently proposed replacing equation (4) by

$$\mathrm{pH} = \frac{E - \mathrm{E}_0 - E_L}{R'T/\mathbf{F}} \tag{16}$$

in which E_L is the liquid junction potential and is, in their discussion, computed by the Henderson equation (equation 21, Chapter 13). The serious limitations of that equation, especially when applied to concentrated solutions, have been outlined in its derivation. Also the use of equation (16) would involve the computation of a liquid junction potential in connection with every pH measurement. This would be time-consuming even if the compositions of the solutions whose pH values are to be measured were completely known, and would otherwise be impossible. It seems unlikely therefore that the suggestion will be generally adopted.

The following method for obtaining E_0[27] uses the unmodified equation

$$\mathrm{pH} = \frac{E - \mathrm{E}_0}{R'T/\mathbf{F}} \tag{4}$$

and our knowledge of the ionization equilibria of weak acids. For acetic acid, for instance, there are available accurate determinations of the ionization constant, from the work of Harned and Ehlers as outlined in Chapter 11, and from the determinations of Shedlovsky and MacInnes to be described in Chapter 18. Measurements were made on dilute "buffer solutions" (to be discussed in the next section) consisting of a weak acid, HA, and its sodium salt, as "solution X" in cells of type (3), *i.e.*

$$\text{(Pt) } H_2;\ \text{HA, NaA : saturated KCl : KCl } (0.1\,N),\ \text{HgCl; Hg}$$

From these measurements, and assuming a value of E_0, a series of values of the quantity

$$\mathrm{p}\mathbf{K}'' = \mathrm{pH} - \log \frac{[A^-]}{[HA]} \tag{17}$$

was computed. In this expression $[A^-]$ and $[HA]$ are the stoichiometric concentrations of salt and acid corrected (as will be shown below) for the equilibrium

$$HA \rightleftarrows H^+ + A^-$$

[25] N. Bjerrum and A. Unmack, *Kgl. Danske Vidensk. Selsk. Math. Phys. Med.*, **IX**, 1 (1929).

[26] E. A. Guggenheim and T. D. Schindler, *J. Phys. Chem.*, **38**, 533 (1934).

[27] D. A. MacInnes, D. Belcher and T. Shedlovsky, *J. Am. Chem. Soc.*, **60**, 1094 (1938).

Empirically, it is found that for the relatively dilute solutions used

$$p\mathbf{K}'' = p\mathbf{K}_0 - B\sqrt{\omega} \qquad (18)$$

in which $p\mathbf{K}_0$ and B are constants. In addition we have the equation, analogous to equation (7) of Chapter 11,

$$\mathbf{K} = \frac{[H^+][A^-]f_\pm^2}{[HA]f_u} \qquad (19)$$

in which $f_\pm$ is the mean ion activity coefficient, f_u the activity coefficient of the undissociated acid and $\mathbf{K}$ is the thermodynamic ionization constant. This can be rearranged to give

$$-\log [H^+]f_\pm - \log \frac{[A^-]}{[HA]} = p\mathbf{K} + \log (f_\pm/f_u) \qquad (20)$$

in which $p\mathbf{K} = -\log \mathbf{K}$. It will be observed (a) that the quantity on the left-hand side of equation (20) corresponds to $p\mathbf{K}''$ in equation (17), and (b) that the term $-B\sqrt{\omega}$ in equation (18) has the form of the Debye-Hückel limiting law for the term $-\log (f_\pm/f_u)$, which appears in equation (20), since f_u, at least for dilute solutions, is nearly unity. It will be shown that the constant B is not far from the Debye-Hückel constant A. The method for selecting E_0 is therefore to adjust it until the pH values computed with equation (4) and used in equations (17) and (18) will yield a value of $p\mathbf{K}_0$ equal to the thermodynamic constant $p\mathbf{K}$. It is important to realize that $p\mathbf{K}_0$ is an empirical constant valid only for the ionic strengths investigated, and is not necessarily the limit in $p\mathbf{K}''$ that would be obtained if the ionic strength were given very low values. The method just outlined for obtaining E_0 yields pH values which are not equal to $-\log[H^+]f_\pm$ but are probably as close as they can be adjusted to such equality using the unmodified equation (4). This method for obtaining the constant E_0 is essentially that suggested by Cohn, Heyroth and Menkin.[28]

Table II gives the experimental results of measurements on mixtures of acetic acid and sodium acetate together with the results of computations such as are outlined above. The ionic strength, ω, was found from the relation

$$\omega = [NaAc] + [H^+] = [NaAc] + (\text{antilog pH})/f$$

the activity coefficient, f, being obtained from the Debye-Hückel limiting law. The $p\mathbf{K}''$ (observed) values in column 6 of the table were

[28] E. J. Cohn, F. F. Heyroth and M. F. Menkin, *J. Am. Chem. Soc.*, **50**, 696 (1928).

TABLE II. THE DETERMINATION OF E_0 FOR THE pH SCALE AT 25°, WITH SODIUM ACETATE-ACETIC ACID BUFFERS

Concentrations (mols per liter)		emf	pH	Ionic strength	pK″	
CH_3COOH	CH_3COONa	1 atm. H_2		ω	Observed	Computed
0.009000	0.001000	0.5636	3.851	0.001147	4.739	4.737
.008000	.002000	.5811	4.147	.002075	4.729	4.730
.01800	.002000	.5614	3.814	.002163	4.730	4.729
.01600	.004000	.5799	4.127	.004081	4.718	4.719
.01160	.004717	.5921	4.333	.004767	4.717	4.716
.01115	.006181	.5994	4.456	.006219	4.708	4.710
.01055	.007594	.6056	4.561	.007625	4.701	4.706
.01008	.009134	.6113	4.658	.009159	4.699	4.710
.01000	.01000	.6140	4.703	.01002	4.701	4.699
.01015	.01278	.6195	4.796	.01280	4.694	4.691

obtained from equation (17) and the concentrations [Ac^-] and [HA] were computed from the stoichiometric concentrations by addition and subtraction respectively of [H^+]. The resulting p**K**″ values may be represented by the equation

$$p\mathbf{K}'' = 4.756 - 0.5774\sqrt{\omega} \tag{21}$$

as can be seen by comparing the observed and computed values in the last two columns of Table II. The constant 4.756 is the negative logarithm of the thermodynamic ionization constant obtained by MacInnes and Shedlovsky [29] and by Harned and Ehlers.[30] The E_0 value resulting from these computations is 0.3358 volt at 25°, when a tenth normal calomel electrode is used as reference electrode. This may be compared with the Sørensen value of 0.3378 for the same constant. The newer constant yields values of 0.03 of a pH unit higher than the older one. A similar series of measurements on chloroacetic acid and its sodium salt gave the closely agreeing value of E_0 of 0.3357. Measurements on acetic acid buffers were also made at 12° and 38°. The resulting E_0 values follow the empirical equation

$$E_0 = 0.3358 - 4.6 \times 10^{-5}(t - 25) \tag{22}$$

in which t is the temperature. The E_0 values are also given in Table III,

TABLE III. VALUES OF THE CONSTANT, E_0, FOR THE HYDROGEN ELECTRODE AND OF E_0^Q FOR THE QUINHYDRONE ELECTRODE FOR USE IN pH DETERMINATIONS AT VARIOUS TEMPERATURES WITH THE 0.1 NORMAL CALOMEL ELECTRODE

Temperature (°C.)	E_0	E_0^Q
12	0.3364	− 0.3725
15	(0.3362)	− 0.3705
20	(0.3360)	− 0.3671
25	0.3358	− 0.3636
30	(0.3356)	− 0.3601
35	(0.3355)	− 0.3565
38	0.3354	− 0.3543

[29] D. A. MacInnes and T. Shedlovsky, *J. Am. Chem. Soc.*, **54**, 1429 (1932).
[30] H. S. Harned and R. W. Ehlers, *ibid.*, **54**, 1350 (1932).

together with interpolated values, in brackets, at 5° intervals. In addition the table contains values of E_0^q for use with equation (12) involving the quinhydrone electrode and the 0.1 *N* calomel electrode as reference. These constants were obtained by a combination of E_0 values with the standard potentials, E_0, of the quinhydrone electrode, from the work of Hovorka and Dearing [31] at 25° and the temperature coefficient from the work of Harned and Wright.[32]

Buffer Solutions. There are many occasions in which it is necessary or desirable to keep the pH of a solution substantially constant in spite of the addition or loss of acidic or basic materials. This may be most readily accomplished by means of "buffer solutions." A buffer solution usually consists of a mixture of a weak acid or a weak base and its salt. The properties of such solutions were first discussed by Henderson [33] and by Washburn.[34] Buffer solutions are used practically in such diverse fields as electroplating, photography, detergent action and bacterial growth. Many organic and inorganic substances can be prepared only in narrow ranges of pH, and the yields of many others may be increased by the choice of an appropriate pH value for the reacting substances during their formation.

The pH value which a given mixture of weak acid, HA, and salt NaA (in which Na may be regarded as representative of any univalent positive ion) will produce, may be readily estimated if the ionization constant of the acid, **K**, is known. In terms of concentrations the **K** is

$$\mathbf{K} = \frac{(H^+)(A^-)}{(HA)} = \frac{[H^+][A^-]f_\pm^2}{[HA]f_u} \tag{23}$$

If

$$\mathbf{K}' = \frac{\mathbf{K} \cdot f_u}{f_\pm} \tag{24}$$

then

$$\mathbf{K}' = \frac{[H^+][A]f_\pm}{[HA]} \tag{25}$$

which may be put in the form

$$\log \mathbf{K}' = \log [H^+]f_\pm + \log \frac{[A^-]}{[HA]} \tag{26}$$

[31] F. Hovorka and W. C. Dearing, *J. Am. Chem. Soc.*, **57**, 446 (1935).
[32] H. S. Harned and D. D. Wright, *ibid.*, **55**, 4849 (1933).
[33] L. J. Henderson, *Am. J. Physiol.*, **15**, 257 (1906); **21**, 173 (1908); *J. Am. Chem. Soc.*, **30**, 954 (1908).
[34] E. W. Washburn, *ibid.*, **30**, 31 (1908).

We have seen in the preceding section that the pH scale may be adjusted so that, to a fair approximation

$$pH = -\log [H^+] f_\pm$$

and, if in addition $p\mathbf{K}' = -\log \mathbf{K}'$ then

$$pH = p\mathbf{K}' + \log \frac{[A^-]}{[HA]} \tag{27}$$

From this development it can be seen that the pH produced by a mixture of a weak acid and its salt will depend primarily upon the ionization constant, **K**, of the weak acid. In addition, since $\mathbf{K}'$ differs from **K** by the activity coefficient $f_\pm$ (f_u can be assumed to be nearly unity) the pH values will depend upon the ionic strength, ω, of which $f_\pm$ is a function. Finally, pH is determined by the concentration ratio $[A^-]/[HA]$, which, for weak acids, and for not too dilute solutions, can be replaced by the stoichiometric ratio, $[Na]/[HA]$, of salt to acid.

As a measure of the relative effectiveness of buffer solutions in maintaining the pH constant when small amounts of acid or base are added, Van Slyke [35] has suggested the use of a quantity called the "buffer value" defined by

$$\frac{dB}{dpH} \tag{28}$$

in which dB represents the change of concentration in equivalents of a strong base, such as potassium hydroxide, and dpH the corresponding change in pH. Thus a solution has a buffer value of unity if it requires a gram equivalent of strong base per liter to produce an increase of one unit in pH. Since the effect on the pH of adding a strong acid is quantitatively just the reverse of that of adding a strong base we also have the relation

$$\frac{dB}{dpH} = -\frac{dC_A}{dpH} \tag{29}$$

in which C_A represents the concentration in equivalents of the strong acid.

A formula for computing the buffer value may be obtained as follows. If to a liter of a solution of a weak acid of concentration C, B equivalents of a strong base are added, the stoichiometric concentrations of acid and its salt will be respectively, $(C - B)$ and B.

[35] D. D. Van Slyke, *J. Biol. Chem.*, **52**, 525 (1922).

In the equilibrium

$$H^+ + A^- \rightleftarrows HA$$

the concentration $[A^-]$ will be $(B + [H^+])$ and that of the acid, $[HA]$, will be equal to $(C - B - [H^+])$, so that equation (25) gives

$$\mathbf{K}' = \frac{[H^+](B + [H^+]) f_\pm}{(C - B - [H^+])} \tag{30}$$

Since for weak acids $[H^+]$ is usually negligible in comparison with B and $(C - B)$ we have

$$\mathbf{K}' = \frac{[H^+] B f_\pm}{(C - B)} \tag{31}$$

which may be put in the form

$$pH = p\mathbf{K}' + \log \frac{B}{(C - B)} \tag{32}$$

On differentiation, equation (32) yields for the buffer value, dB/dpH, the relation

$$\frac{dB}{dpH} = 2.3 \frac{(C - B)B}{C} \tag{33}$$

Using the ratio, $\nu = B/C$, equation (33) becomes

$$\frac{dB}{dpH} = 2.3\, C\,(1 - \nu)\nu \tag{34}$$

indicating that the buffer value is proportional to the sum, C, of the concentrations of salt and acid and is a function of the ratio, ν. A further differentiation of equation (33) indicates that dB/dpH has a maximum value when $B = 0.5\, C$ or $\nu = 0.5$. Equations (33) and (34) should not be used for very dilute solutions or for values of ν near zero or unity since in these cases the approximation represented by equation (31) is not sufficient.

In measurements of pH, instead of using a reference electrode, such as the tenth normal calomel electrode the E_0 value of which is known, it is frequently convenient to establish the value of E_0 in the equation

$$pH = \frac{E' - E_0}{R'T/\mathbf{F}}$$

by the use of buffers pH values of which have been accurately determined. Some useful buffers for this purpose are listed in Table IV, and their pH values, on the basis given in the preceding section of this book, are given at 12, 25 and 38°. Of these buffers the one made

TABLE IV. THE pH VALUES, TO ± 0.005 UNIT, OF SOME BUFFERS AT DIFFERENT TEMPERATURES

Buffer	12°	25°	38°
$CH_3COOH(0.01N)$, $CH_3COONa(0.01N)$	4.710	4.700	4.710
$CH_3COOH(0.1N)$, $CH_3COONa(0.1N)$	4.650	4.640	4.635
Potassium acid phthalate (0.05 molar)	4.000	4.000	4.015

from potassium acid phthalate is particularly convenient for use with glass electrodes as the solution can be readily prepared. For hydrogen electrodes it is not so suitable since, particularly at higher temperatures, the phthalate may be reduced.

A list of buffer mixtures which have been studied by various workers, with their approximate range of pH values, is given in Table V.

TABLE V. COMPOSITION AND pH RANGE OF SOME BUFFER SOLUTIONS

Composition	Approximate pH Range	Reference
Glycine, NaOH, HCl	1.0 – 3.7	*1*
Citric acid, NaOH, HCl	1.0 – 5.0	*1*
HCl, KCl	1.1 – 3.1	*2*
Potassium acid phthalate, HCl	2.2 – 3.8	*2*
Na_2HPO_4, citric acid	2.2 – 8.0	*3*
Sodium diethylbarbiturate, sodium-acetate, HCl, NaCl	2.6 – 9.6	*4*
Sodium phenylacetate, phenylacetic acid	3.2 – 4.7	*5*
Acetic acid, sodium acetate	3.8 – 5.6	*6*
Potassium acid phthalate, NaOH	4.0 – 6.2	*2*
Citric acid, NaOH	5.0 – 6.7	*1*
$Na_2HPO_4 \cdot 2H_2O$, KH_2PO_4	5.3 – 8.0	*1*
NaOH, KH_2PO_4	5.8 – 8.0	*2*
$Na_2HPO_4 \cdot 2H_2O$, KH_2PO_4	6.8 – 8.0	*7*
$Na_2B_4O_7 \cdot 10H_2O$, H_3BO_3	6.8 – 8.4	*8*
Sodium diethylbarbiturate, HCl	6.8 – 9.9	*4*
Boric acid, NaOH, HCl	7.6 – 9.2	*2*
Boric acid, KCl, NaOH	7.8 – 10.0	*9*
Glycine, NaCl, NaOH	8.5 – 12.9	*1*
Na_2CO_3, $Na_2B_4O_7 \cdot 10H_2O$	9.2 – 11.0	*10*
Boric acid, NaOH	9.2 – 12.3	*1*
$Na_2HPO_4 \cdot 2H_2O$, NaOH	11.0 – 12.0	*10*

1. S. P. L. Sørensen, *Ergeb. Physiol.*, **12**, 393 (1912); A. E. Walbum, *Biochem. Z.*, **107**, 219 (1920).
2. W. M. Clark and H. A. Lubs, *J. Biol. Chem.*, **25**, 479 (1916).
3. T. C. McIlvaine, *ibid.*, **49**, 183 (1921).
4. L. Michaelis, *Biochem. Z.*, **234**, 139 (1931).
5. W. L. German and A. I. Vogel, *J. Chem. Soc.*, **1935**, 912.
6. A. A. Green, *J. Am. Chem. Soc.*, **55**, 2331 (1933).
7. A. B. Hastings and J. Sendroy, Jr., *J. Biol. Chem.*, **61**, 695 (1924).
8. L. Michaelis, *ibid.*, **87**, 33 (1930).
9. S. Palitzsch, *Bull. Inst. Océanographique* No. 409 *monaco.*
10. I. M. Koltoff and J. J. Vleeschhouwer, *Biochem. Z.*, **189**, 191 (1927).

[*Comment to 1961 Edition*] The recent work in the fields covered by this chapter is treated in *Electrometric* pH *Determinations*: *Theory and Practice* by R. G. Bates, published by John Wiley & Sons, Inc. in 1954.

Chapter 16

The Standard Potentials of Electrode Reactions "Oxidation-Reduction" Potentials

In Chapters 10 and 14 the determination of standard potentials from galvanic cells without and with liquid junctions, respectively, has been discussed. With a few exceptions the standard potentials so far considered have been of the elements against hypothetical molar activities of their ions. Other, and more complicated, types of electrode reactions than the ionization of an element are of course possible. There are, for example, the "oxidation-reduction" processes which appear, in current usage, to be restricted to two soluble components in contact with the electrode, although, strictly speaking, every electrode reaction is either an electrochemical oxidation or an electrochemical reduction.

A typical "oxidation-reduction" electrode is

$$(\mathrm{Pt});\ \mathrm{TlCl},\ \mathrm{TlCl_3} \tag{1}$$

at which there is the electrochemical equilibrium

$$\mathrm{Tl^+} = \mathrm{Tl^{+++}} + 2e^- \tag{2}$$

A convenient method for obtaining the standard potential of this electrode reaction is to utilize the galvanic cell:

$$(\mathrm{Pt});\ \mathrm{TlCl},\ \mathrm{TlCl_3} : \mathrm{KCl},\ \mathrm{AgCl};\ \mathrm{Ag} \tag{3}$$

Omitting the process at the liquid junction, the cell reaction occurring during the passage of two faradays, is

$$\mathrm{Tl^+} + 2\mathrm{AgCl} = 2\mathrm{Ag} + \mathrm{Tl^{+++}} + 2\mathrm{Cl^-} \tag{4}$$

reaction (2) occurring at one electrode and

$$\mathrm{AgCl} + e^- = \mathrm{Ag} + \mathrm{Cl^-} \tag{5}$$

at the other. For reaction (4) equation (10), Chapter 10, gives

$$E = E_0 - \frac{RT}{2\mathbf{F}} \ln \frac{(\mathrm{Tl^{+++}})(\mathrm{Cl^-})^2}{(\mathrm{Tl^+})} \tag{6}$$

which may be rearranged to

$$E = E_{01} - \frac{RT}{2\mathbf{F}} \ln \frac{(\mathrm{Tl}^{+++})}{(\mathrm{Tl}^{+})} - E_{02} - \frac{RT}{\mathbf{F}} \ln (\mathrm{Cl}^{-}) \tag{7}$$

in which E_{01} represents the standard potential of the thallous-thallic electrode, and E_{02} that of the silver-silver chloride electrode. In such cases as we are considering it is frequently the practice to keep one half-cell constant for reference. For instance if the reference electrode is the combination Ag; AgCl, KCl (*0.1 N*) the term $E_{02} + \frac{RT}{\mathbf{F}} \ln (\mathrm{Cl}^{-})$ will remain unchanged. There will, however, in general be a liquid junction potential. For the cases we are considering equation (10), Chapter 10, may be modified as follows. Let

$$a\mathrm{A} + b\mathrm{B} + \cdots\cdot = k\mathrm{K} + l\mathrm{L} + \cdots\cdot + n\,\mathrm{e}^{-} \tag{8}$$

be the reaction at one electrode and

$$n\,\mathrm{e}^{-} + p\mathrm{P} + q\mathrm{Q} = r\mathrm{R} + s\mathrm{S} \tag{9}$$

be that at the reference electrode. The sum of these two will be the cell reaction. Applying the equation just mentioned, and recalling that there is a liquid junction potential E_L we have

$$E = E_0 + E_L - \frac{RT}{n\mathbf{F}} \ln \frac{a_K^k \cdot a_L^l \cdots\cdot}{a_A^a \cdot a_B^b \cdots\cdot} + \frac{RT}{n\mathbf{F}} \ln \frac{a_P^p \cdot a_Q^q \cdots}{a_R^r \cdot a_S^s \cdots} \tag{10}$$

Since, however, the activities of the substances entering into the electrochemical reaction of the reference electrode are fixed, the last term of equation (10) will be constant. If now, we define

$$E_R = E_{02} + \frac{RT}{n\mathbf{F}} \ln \frac{a_R^r \cdot a_S^s \cdots\cdot}{a_P^p \cdot a_Q^q \cdots\cdot}$$

in which E_R is the potential of the reference electrode, equation (10) becomes, since $E_0 = E_{01} - E_{02}$

$$E = E_{01} + E_L - E_R - \frac{RT}{n\mathbf{F}} \ln \frac{a_K^k \cdot a_L^l \cdots}{a_A^a \cdot a_B^b \cdots} \tag{11}$$

This equation is not thermodynamic, for reasons made clear in foregoing chapters. In the first place the liquid junction potential E_L and the potential of the reference electrode E_R cannot be computed without non-thermodynamic assumptions. In general also, some of the activities, a_A, a_B, etc. are those of individual ions, which cannot be obtained without assumptions of the same type. However, in spite of these limitations many valuable electrochemical investigations are based on equation (11), as will be seen in this and following chapters.

A great number of researches have been made on "oxidation-reduction" and other electrode processes and a complete discussion of them would be out of place in this book. It will be possible to consider only a few typical cases of theoretical interest or of unusual accuracy. Due partly to experimental difficulties, but also, it must be admitted, to poorly conceived and hastily carried out researches, the results in this field compare unfavorably with those in many other branches of electrochemistry. One difficulty is the liquid junction that is present in nearly every cell used. The various means employed to overcome this difficulty will be discussed below.

The Standard Potential of the Mercurous-Mercuric Electrode. A method for obtaining standard potentials of oxidation-reduction electrodes which utilizes the best procedure so far developed in this field is the one that was used by Popoff and associates. The method may be illustrated by the determination of the standard potential of the mercurous-mercuric electrode. The type of cell used by Popoff, Riddick, Worth and Ough[1] was

$$(Pt); Hg(ClO_4)_2(m_2),\ Hg_2(ClO_4)_2(m_3),\ HClO_4(m_1) \underset{a}{:} HClO_4(m_1);\ H_2(Pt) \quad (12)$$

The molality, m_1, of the perchloric acid was kept the same throughout the cell. There is a liquid junction at a which, however will become smaller and smaller if the molality ratio, $(m_2 + m_3)/m_1$ is progressively decreased, and in the limit will be zero. The procedure adopted by these workers was, therefore, to make a series of measurements on cells in which the ratio m_2/m_3 was kept constant, and the ratio $(m_2 + m_3)/m_1$ decreased. The potential corresponding to values of m_2 and m_3 as they approach zero was then obtained by extrapolation. A few typical data are given in Table I.

TABLE I. POTENTIALS OF THE CELL

$(Pt)\ H_2; Hg\ (ClO_4)_2\ (m_2),\ Hg_2\ (ClO_4)_2\ (m_3),\ HClO_4\ (0.08m): HClO_4\ (0.08m); H_2\ (Pt)$

$m_2 = m_3$	Observed emf, volts E	E''_0 (equation 17)
0.006	−0.90598	−0.9019
0.004	−0.90087	−0.9020
0.002	−0.89167	−0.9019
0.0005	−0.87379	−0.9018

[1] S. Popoff, J. A. Riddick, V. I. Worth and L. D. Ough, *J. Am. Chem. Soc.*, **53**, 1195 (1931).

The reaction accompanying the passage of two faradays through cell (12) will be

$$Hg_2^{++} + 2H^+ = H_2 + 2Hg^{++} \tag{13}$$

for which equation (10), Chapter 10, yields, after separating the electrode reactions,

$$E = E_0 - \frac{RT}{2\mathbf{F}} \ln \frac{(Hg^{++})^2}{(Hg)_2^{++}} + \frac{RT}{\mathbf{F}} \ln \frac{(H^+)}{(H_2)^{1/2}} \tag{14}$$

which may be rewritten, for measurements corrected to one atmosphere pressure of hydrogen

$$E' = E_0 - \frac{RT}{2\mathbf{F}} \ln \frac{m_2^2}{m_3} - \frac{RT}{2\mathbf{F}} \ln \frac{\gamma_2^2}{\gamma_3} + \frac{RT}{\mathbf{F}} \ln m_1 \gamma_1 \tag{15}$$

in which γ_2, γ_3 and γ_1 are the activity coefficients of the mercuric, mercurous and hydrogen ion constituents, respectively. These authors also found that the potential of the cell:

$$\text{(Pt) } H_2;\ HClO_4\ (m) : HCl\ (m);\ H_2 \text{ (Pt)} \tag{16}$$

in which the molality of the two acids was the same, was very nearly equal to the potential computed from equation (26a), Chapter 13, for the liquid junction, in a series of measurements in which m varied from 0.02 to 1 molar. Since the potentials of the two electrodes would be expected to cancel, this tends to indicate, in agreement with the earlier work by Schuhmann,[2] that, in this range, the activities of hydrogen ion from the two acids are equal. The values of γ_1 may be therefore taken equal to those for hydrochloric acid in Table V, Chapter 8. For convenience we may define

$$E_0'' = E' + \frac{RT}{2\mathbf{F}} \ln \frac{m_2^2}{m_3} - \frac{RT}{\mathbf{F}} \ln m_1 \gamma_1 \tag{17}$$

from which equation (15) yields

$$E_0'' = E_0 - \frac{RT}{2\mathbf{F}} \ln \frac{\gamma_2^2}{\gamma_3} \tag{18}$$

and thus $E_0'' = E_0$ at infinite dilution. Values of E_0'' for $m_1 = 0.08$ are given in Table I, and are seen to be constant, within the experimental error, through a wide range of molalities of the mercury salts. The liquid junction potentials in cell (12) are probably, therefore, very

[2] R. Schuhmann, *J. Am. Chem. Soc.*, **46**, 58 (1924).

small. Values of E_0'' for other molalities, m_1, of perchloric acid are given in Table II.

TABLE II. VARIATION OF E_0'' FOR THE MERCUROUS-MERCURIC ELECTRODE WITH THE MOLALITY, m_1, OF PERCHLORIC ACID

Molality of perchloric acid m_1	E_0'' volts
1.0	− 0.9072
0.4	− 0.9012
0.2	− 0.9008
0.12	− 0.9011
0.08	− 0.9019
0.04	− 0.9033
0.02	− 0.9038

Since the Debye-Hückel limiting law for both γ_2 and γ_3 of equation (18) is

$$-\log \gamma = 4A\sqrt{\omega}$$

a fairly satisfactory evaluation of E_0 may therefore be obtained by plotting E_0'' as a function of the ionic strength and extending the resulting line, which is straight for the lower points, to the zero ordinate. This yields, at 25°

$$\text{(Pt)}; \ Hg_2^{++}, \ Hg^{++} \qquad E_0 = -0.906 \qquad (19)$$

The Standard Potential of the Thallous-Thallic Electrode. A procedure quite similar to that used by Popoff and associates as just described has been employed by Sherrill and Haas [3] in the determination of the standard potential of the thallous-thallic electrode. These authors used the cells

$$\text{(Pt)}; \ TlClO_4(m_2), \ Tl(ClO_4)_3(m_3), \ HClO_4(m_1) \underset{a}{:} HClO_4(m_1); \ H_2\,\text{(Pt)} \qquad (20)$$

with, however, relatively low values of the total molality $(m_2 + m_3)$ of the thallium salts compared with that of the acid m_1. The liquid junction at the point a was therefore small. A wide range of ratios m_2/m_3 of the molalities of the two salts was employed. A consideration of the reaction occurring in the cell, and application of equation (10), Chapter 10, yields the equation

$$E = E_0 - \frac{RT}{2\mathbf{F}} \ln \frac{(Tl^{+++})}{(Tl^{+})} + \frac{RT}{\mathbf{F}} \ln \frac{(H^{+})}{(H_2)^{1/2}} \qquad (21)$$

which may be written, for one atmosphere pressure of hydrogen gas

$$E' = E_0 - \frac{RT}{2\mathbf{F}} \ln \frac{m_2}{m_3} - \frac{RT}{2\mathbf{F}} \ln \frac{\gamma_2}{\gamma_3} + \frac{RT}{\mathbf{F}} \ln m_1 \gamma_1 \qquad (22)$$

[3] M. S. Sherrill and A. J. Haas, *J. Am. Chem. Soc.*, **58**, 952 (1936).

If, once more, γ_1, the hydrogen ion activity in perchloric acid, is taken equal to the mean ion activity of hydrochloric acid, values of E_0'' defined by

$$E_0'' = E_0 - \frac{RT}{2\mathbf{F}} \ln \frac{\gamma_2}{\gamma_3} \tag{23}$$

may be computed from the data. E_0'' values were found to depend but slightly upon the relatively small molalities, m_2 and m_3, and on their ratios. Average values are given for different molalities, m_1, of perchloric acid in Table III.

TABLE III. E_0'' VALUES FOR THALLOUS-THALLIC ELECTRODE AT 25°

Molality of Perchloric Acid m_1	E_0'' volts
1.2205	− 1.2634
0.9465	− 1.2598
0.7160	− 1.2569
0.5000	− 1.2540

This is a case in which accurate data can be only roughly interpreted because of a lack of sufficient knowledge, in this instance of the activity coefficients of mixed electrolytes of different ionic types, in relatively concentrated solutions. Furthermore, due to the fact that the thallic compounds hydrolyze, and that thallic hydroxide precipitates if the acid concentration is lowered, the limiting E_0 value has little importance, especially as no adequate methods for making the necessary extrapolation are available.

The Standard Potential of the Ferrous-Ferric Electrode. Many researches have been carried out for the purpose of determining the standard potential of the ferrous-ferric electrode, one of the most recent being that of Popoff and Kunz,[4] utilizing the procedure already described for mercurous-mercuric and thallous-thallic electrodes. Bray and Hershey[5] have however shown that these authors failed to consider the effect of hydrolysis reactions, of which they consider

$$Fe^{+++} + H_2O = FeOH^{++} + H^+$$

to be the most important. Bray and Hershey have also made an estimate of the errors arising from this source and give a value of the standard potential of the ferrous-ferric electrode.

A more accurate value of that constant can however probably be obtained from the equilibrium measurements of Schumb and Sweetzer.[6]

[4] S. Popoff and A. H. Kunz, *J. Am. Chem. Soc.*, **51**, 382 (1929).
[5] W. C. Bray and A. V. Hershey, *ibid.*, **56**, 1889 (1934).
[6] W. C. Schumb and S. B. Sweetzer, *J. Am. Chem. Soc.*, **57**, 871 (1935).

The use of a chemical equilibrium has already been illustrated in the determination of the standard potential of tin, page 255. In the case under discussion the reaction

$$Fe\,(ClO_4)_3 + Ag \rightleftarrows AgClO_4 + Fe\,(ClO_4)_2$$

or

$$Fe^{+++} + Ag \rightleftarrows Fe^{++} + Ag^{+}$$

was allowed to proceed to equilibrium, and the resulting molalities determined, at a series of ionic strengths. Hydrolysis was minimized by the presence of free perchloric acid. From these data values of the function

$$\mathbf{K}' = \frac{[Fe^{++}]\,[Ag^{+}]}{[Fe^{+++}]} \tag{24}$$

were computed. The thermodynamic mass action constant, **K**, was estimated by using a process analogous to the Hitchcock extrapolation. These authors found that the results could be expressed by the empirical formula

$$\log \mathbf{K}' + 2.02\omega^{1/2} = -\,0.275 + 1.645\omega - 0.316\omega^2$$

in which, however, the term $2.02\,\omega^{\frac{1}{2}}$ comes from the Debye-Hückel limiting law and is equal to

$$(2^2 + 1^2 - 3^2) \cdot 0.506\omega^{1/2}$$

This method yields the value 0.531 for the thermodynamic mass action constant **K**. A hypothetical cell of the form

$$(Pt);\ Fe^{++},\ Fe^{+++} :: Ag^{+};\ Ag$$

will have a potential of zero if the ion constituents are present at their equilibrium values. The standard potential of the ferrous-ferric electrode may therefore be obtained from the relation

$$E = 0 = E_{01} - \frac{RT}{\mathbf{F}} \ln \frac{(Fe^{+++})}{(Fe^{++})} - E_{02} + \frac{RT}{\mathbf{F}} \ln (Ag^{+})$$

in which E_{01} is the standard potential of the ferrous-ferric electrode, and E_{02} is that of the silver electrode. Rearranging this equation gives

$$E_{02} = E_{01} - \frac{RT}{\mathbf{F}} \ln \mathbf{K} \tag{25}$$

and, with the value -0.799 volt for E_{02}, the standard potential of the silver electrode, yields, at 25°

$$(Pt);\ Fe^{++},\ Fe^{+++} \qquad E_0 = -\,0.783 \tag{26}$$

This is higher than the directly determined value, the measurement of which is, as has already been stated, much influenced by hydrolysis.

The Standard Potential of the Copper-Cuprous Ion Electrode. Other examples of standard potentials which cannot be readily obtained from direct measurements but which may be computed from the results of equilibrium determinations are the cuprous-cupric and copper-cuprous electrodes. The equilibrium

$$2Cu^+ \rightleftarrows Cu^{++} + Cu$$

has been studied by various workers including Fenwick [7] and Heinerth.[8] The last-mentioned worker directly determined the concentrations of cuprous and cupric salts in equilibrium with copper, using the sulphates and perchlorates. The results with the sulphate he considered most reliable since there was evidence that the perchlorates were somewhat reduced. From his measurements at a series of temperatures Heinerth computed values of

$$\mathbf{K} = \frac{m_{CuSO_4}}{m^2_{Cu_2SO_4}} \qquad (27)$$

which were found to be constant at each temperature, for varying values of the molality of cupric sulphate, within the experimental error, indicating that the differences between activities and molalities tend to cancel. From these data the result

$$\mathbf{K} = 1.190 \times 10^6$$

at 25°, was obtained. The hypothetical galvanic cell

$$Cu;\ Cu^{++}\ (m_1)\ ::\ Cu^+\ (m_2);\ Cu$$

will therefore have a potential of zero if the two molalities are in the proportions given by equation (27). For this case

$$E_{01} - \frac{RT}{2\mathbf{F}} \ln (Cu^{++}) = E_{02} - \frac{RT}{\mathbf{F}} \ln (Cu^+) \qquad (28)$$

$$E_{02} = E_{01} - \frac{RT}{2\mathbf{F}} \ln \frac{(Cu^{++})}{(Cu^+)^2} = E_{01} - \frac{RT}{2\mathbf{F}} \ln \mathbf{K} \qquad (29)$$

in which E_{01} and E_{02} are the standard potentials of copper-cupric ion and copper-cuprous ion electrodes respectively. For the former the value of $E_0 = -0.339$ volt at 25° is given in Chapter 10. We have therefore, at 25° from equation (27)

$$Cu;\ Cu^+ \qquad E_0 = -0.519 \qquad (30)$$

[7] F. Fenwick, *J. Am. Chem. Soc.*, **48**, 860 (1926).
[8] E. Heinerth, *Z. Elektrochem.*, **37**, 61 (1931).

To obtain the standard potential of the electrode (Pt); Cu^+, Cu^{++} which corresponds to the electrode reaction

$$Cu^+ = Cu^{++} + e^-$$

use may be made of a principle which is due to Luther.[9] If the two reactions

$$Cu + H^+ = Cu^+ + \tfrac{1}{2} H_2$$
$$Cu^+ + H^+ = Cu^{++} + \tfrac{1}{2} H_2$$

take place under standard conditions the corresponding changes in the Gibbs free energies are $\Delta Z_2 = -E_{02}\mathbf{F}$ and $\Delta Z_3 = -E_{03}\mathbf{F}$ in which E_{03} is, by definition, the standard potential of the cuprous-cupric electrode. If, on the other hand the reaction occurs in one stage

$$Cu + 2H^+ = Cu^{++} + H_2$$

the change in the Gibbs free energy will be $\Delta Z_1 = -2E_{01}\mathbf{F}$.

Since $\Delta Z_1 = \Delta Z_2 + \Delta Z_3$

then $2E_{01} = E_{02} + E_{03}$

from which $E_{03} = 2E_{01} - E_{02}$

and, substituting the values just given, we have, at 25°

$$\text{(Pt); } Cu^+, Cu^{++} \qquad E_0 = -0.159 \qquad (31)$$

A Table of Standard Potentials of Electrode Reactions. A large number of electrode processes are possible, and many researches dealing with them have appeared in the literature. However, a discussion of only a few are desirable in this book since a few typical examples illustrate the principles involved in dealing with them. A table of standard potentials is nevertheless of interest since it is possible with their aid to obtain Gibbs free energies of many electrochemical reactions, to compute equilibrium constants, etc. Table IV contains standard potentials mostly from results of recent investigations. The earlier work has been summarized by Abegg, Auerbach and Luther,[10] by Drucker,[11] and by Gerke.[12]

9 R. Luther and D. R. Wilson, *Z. physik. Chem.*, **34**, 488 (1900), **36**, 385 (1901).

10 R. Abegg, Fr. Auerbach and R. Luther, "Messungen elektromotorischer Kräfte galvanischer Ketten mit wässrigen Elektrolyten," Wilhelm Knapp, Halle, 1911.

11 C. Drucker, "Messungen elektromoterischerten, Kräfte galvanischer Ketten mit wässrigen Elektrolyten" W10 Verlag Chemie, G. M. B. H., Berlin, 1929.

12 R. H. Gerke, "Potentials of Electrochemical Reactions," International Critical Tables, **6**, 332, McGraw-Hill Book Co., New York, 1929.

TABLE IV. STANDARD POTENTIALS OF SOME TYPICAL ELECTRODE REACTIONS AT 25°

Electrode	Electrode Reaction	Standard Potential at 25°	Reference
(Pt) H_2 (g); H_2O (l), OH^-	$H_2 + 2OH^- \rightarrow H_2O + 2e^-$	+ 0.8279	*1*
Pb; PbO (s), OH^-	$Pb + 2OH^- \rightarrow PbO + H_2O + 2e^-$	+ 0.5785	*2*
(Pt) H_2 (g); H^+	$H_2 \rightarrow 2H^+ + 2e^-$	± 0.0000	–
Hg; HgO (s), OH^-	$Hg + 2OH^- \rightarrow HgO + H_2O + 2e^-$	– 0.0976	*3*
Sb; Sb_2O_3 (s), H^+	$2Sb + 3H_2O \rightarrow Sb_2O_3 + 6H^+ + 6e^-$	– 0.1445	*4*
(Pt); Cu^+, Cu^{++}	$Cu^+ \rightarrow Cu^{++} + e^-$	– 0.159	*5*
(Pt); MnO_2 (s), MnO_4^{--}, OH^-	$MnO_2 + 4OH^- \rightarrow MnO_4^- + 2H_2O + 3e^-$	– 0.587	*6*
(Pt); Fe^{++}, Fe^{+++}	$Fe^{++} \rightarrow Fe^{+++} + e^-$	– 0.783	*7*
(Pt); Hg_2^{++}, Hg^{++}	$Hg_2^{++} \rightarrow 2Hg^{++} + 2e^-$	– 0.906	*8*
(Pt); MnO_2 (s), Mn^{++}	$Mn^{++} + 2H_2O \rightarrow MnO_2 + 4H^+ + 2e^-$	– 1.236	*9*
Au; Au_2O_3 (s), H^+	$2Au + 3H_2O \rightarrow Au_2O_3 + 6H^+ + 6e^-$	– 1.36_0	*10*
Pb; PbO_2 (s), Pb^{++}	$Pb^{++} + 2H_2O \rightarrow PbO_2 + 4H^+ + 2e^-$	– 1.467	*11*
(Pt); MnO_2 (s), MnO_4^-	$MnO_2 + 2H_2O \rightarrow MnO_4^- + 4H^+ + 3e^-$	– 1.586	*12*
(Pt); Ce^{+++}, Ce^{++++}	$Ce^{+++} \rightarrow Ce^{++++} + 2e^-$	– 1.609	*13*
(Pt); $PbSO_4$ (s), PbO_2 (s), SO_4^{--}	$PbSO_4 + H_2O \rightarrow PbO_2 + 4H^+ + SO_4^{--} + 2e^-$	– 1.685	*14*

1. Chapter 14.
2. D. F. Smith and H. K. Woods, *J. Am. Chem. Soc.*, **45**, 2632 (1923).
3. Chapter 14.
4. E. J. Roberts and F. Fenwick, *J. Am. Chem. Soc.*, **50**, 2125 (1928).
5. Chapter 16.
6. L. V. Andrews and D. J. Brown, *J. Am. Chem. Soc.*, **57**, 254 (1935).
7. W. C. Schumb and S. B. Sweetzer, *ibid.*, **57**, 871 (1935).
8. S. Popoff, J. A. Riddick, V. I. Worth and L. D. Ough, *ibid.*, **53**, 1195 (1931).
9. D. J. Brown and H. A. Liebhafsky, *ibid.*, **52**, 2595 (1930).
10. R. H. Gerke and M. D. Rourke, *ibid.*, **49**, 1855 (1927).
11. D. J. Brown and J. C. Zimmer, *ibid.*, **52**, 1 (1930).
12. D. J. Brown and R. F. Tefft, *ibid.*, **48**, 1128 (1926).
13. A. A. Noyes and C. S. Garner, *ibid.*, **58**, 1265 (1936).
14. W. J. Hamer, *ibid.*, **57**, 9 (1935).

Oxidation-Reduction Potentials of Organic Substances. So far the discussion of oxidation-reduction potentials has been mainly concerned with inorganic substances. Most oxidations and reductions of organic compounds are irreversible and such reactions cannot therefore be studied by thermodynamic methods. A limited class of organic substances, however, undergo reversible oxidations and reductions, and their reactions may be followed by potentiometric methods. The greater part of these substances possess a quinoid structure in the oxidized form and the corresponding benzenoid structure in the reduced form. The simplest of these compounds are quinone and hydroquinone for which the structural formulas usually given are:

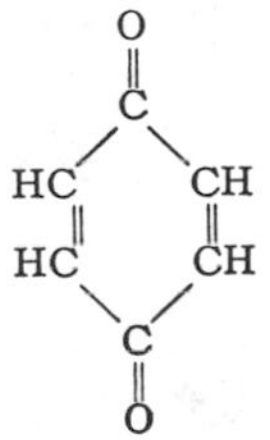

quinone

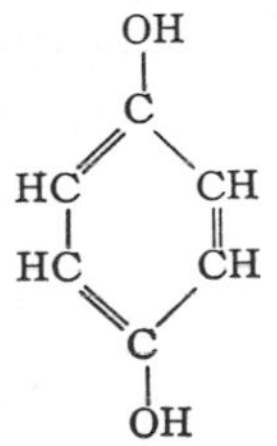

hydroquinone

The determination of the standard potential of the electrode reaction

$$\text{hydroquinone} = \text{quinone} + 2H^+ + 2e^- \tag{32}$$

has been discussed in Chapter 10, and the utility of the reaction for the determination of pH values was considered in Chapter 15. These quinoid and benzenoid substances include many dyestuffs, and also substances of biological interest. Under ordinary conditions the electrochemical reaction for such systems may be represented by the formula:

$$\text{Red} = \text{Ox} + 2e^-$$

the abbreviations Ox and Red denoting the oxidized and reduced forms of the substance. It will be observed that the electrochemical oxidation represented involves *two* electrons. The stepwise reaction, taking place with one electron at a time which occurs in certain electrochemical reactions of organic substances will be discussed later.

The studies on organic oxidation-reduction reactions are usually made with the aid of a galvanic cell of the type:

(Au); Red, Ox, buffer : KCl saturated : KCl, HgCl; Hg

Most of the researches have been carried out using a form of apparatus which makes possible a change of the proportion of oxidized to reduced material, or *vice versa,* by the addition of a suitable reagent, *i.e.,* by a titration. Also, as many substances react with oxygen, it is usually necessary to exclude air. A suitable apparatus is shown diagrammatically in Fig. 1. The solution being studied is placed in

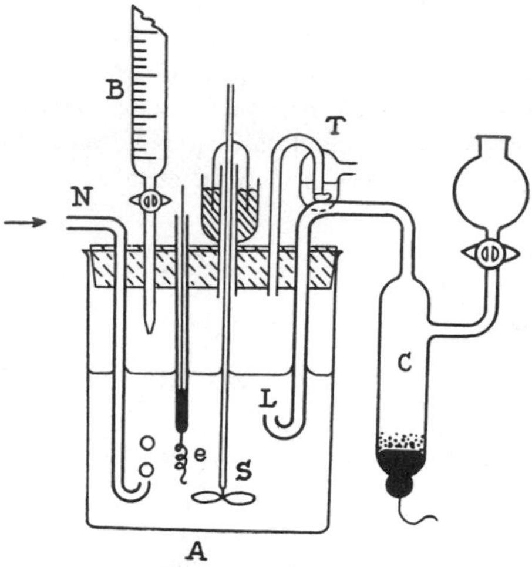

Fig. 1. Arrangement for Oxidation-Reduction Titration.

the vessel *A*. Gas, usually pure nitrogen, is led in through the tube *N* and passes out by the trap *T*. Reagent is added from the burette *B*, and the solution may be agitated with the stirrer, *S*. The potential is measured between the electrode, *e* (which may be made of platinum, gold, or mercury) and the reference calomel electrode, *C*, a liquid junction being formed at the surface, *L*. From measurements of the type just described, along with determinations and control of the pH values of the solutions, much valuable information about the properties of the substances capable of undergoing reversible oxidations and reductions may be obtained. As will be shown in the following paragraphs, it is, for instance, possible to study the ionization relations of the substances, and to estimate their ionization constants.

Equation (32) is an example of a type of electrode reaction which may be represented by

$$H_2R = Ox + 2H^+ + 2e^- \tag{33}$$

Here the reduced substance H_2R is a weak dibasic acid, for which there are two ionization equilibria

$$H_2R \rightleftarrows H^+ + HR^- \quad \text{and} \quad HR^- \rightleftarrows H^+ + R^{--}$$

the mass action expressions of which are

$$\mathbf{K}_1 = \frac{(H^+)\,(HR^-)}{(H_2R)} \qquad \mathbf{K}_2 = \frac{(H^+)\,(R^=)}{(HR^-)} \tag{34}$$

$\mathbf{K}_1$ and $\mathbf{K}_2$ being ionization constants. Using equation (11) we have

$$E = E_0 - \frac{RT}{n\mathbf{F}} \ln \frac{(Ox)\,(H^+)^2}{(H_2R)} \tag{35}$$

in which $E_0 = E_{01} + E_L - E_R$, and possible variations in the liquid junction potential E_L will be ignored. According to our postulate the oxidized form, Ox, exists in only one state but the reduced substance can be present in three, *i.e.*, H_2R, HR^- and R^{--}. The stoichiometric concentration $[H_2R]_s$ will however be

$$[H_2R]_s = [H_2R] + [HR^-] + [R^=] \tag{36}$$

Now assuming, for the purpose of simplification, that activities and concentrations of these substances are the same, a value of the concentration $[H_2R]$ in equation (35) may be obtained with the aid of equations (34) and (36). Such an equation is

$$[H_2R] = \frac{[H_2R]_s\,(H^+)^2}{(H^+)^2 + \mathbf{K}_1(H^+) + \mathbf{K}_1\mathbf{K}_2} \tag{37}$$

which when substituted in equation (35) yields [13]

$$E = E_0 - \frac{RT}{2\mathbf{F}} \ln \frac{[Ox]}{[H_2R]_s} - \frac{RT}{2\mathbf{F}} \ln \left[(H^+)^2 + \mathbf{K}_1(H^+) + \mathbf{K}_1\mathbf{K}_2\right] \tag{38}$$

It can be readily seen that if $\mathbf{K}_1$ and $\mathbf{K}_2$ are very small, as they are for hydroquinone, equation (38) reduces to equation (10), Chapter 15,

$$E = E_0 - \frac{RT}{2\mathbf{F}} \ln \frac{[Q]}{[Hy]} - \frac{RT}{\mathbf{F}} \ln (H^+)$$

for the quinhydrone electrode in which the oxidized substance, Q, is quinone and the reduced substance, Hy, is hydroquinone. Sheppard [14] as a matter of fact has estimated values of 1.8×10^{-10} and 4×10^{-12} for $\mathbf{K}_1$ and $\mathbf{K}_2$ for hydroquinone. Equation (38) also helps to explain why the quinhydrone electrode is accurate only for the lower pH values.

[13] The same equation is obtained if, following W. M. Clark, "The Determination of Hydrogen Ions," 3rd ed., The Williams and Wilkins Co., Baltimore, 1928, the electrical reaction is assumed to be

$$R^{--} = Ox + 2e^-$$

with the exception that E_0 will then contain the constant term

$$RT/2\mathbf{F} \cdot \ln \mathbf{K}_1\mathbf{K}_2$$

[14] S. E. Sheppard, *Trans. Amer. Electrochem. Soc.*, **39**, 429 (1921).

If the pH is increased and (H^+) approaches $\mathbf{K}_1$ the term $\mathbf{K}_1(H^+)$ in equation (38) will begin to be of appreciable magnitude compared to $(H^+)^2$. However, with quinhydrone, other disturbing effects influence the results in alkaline solutions. The fact that quinone and hydroquinone are in equilibrium with quinhydrone has been ignored in the development of equation (38). It has been discussed on page 199.

Equation (38) may be put in the form

$$E = E_0'' - \frac{RT}{2\mathbf{F}} \ln \frac{[\text{Ox}]}{[H_2R]_s} \tag{39}$$

in which E_0'' is a constant at constant pH.[15] Under this condition the measured potential, E, will be a function only of the relative proportions of the oxidized and reduced material. Quite a number of experimental tests of equation (39) have been published using various substances as the oxidized and reduced materials. One of the more accurate is that of LaMer and Baker [16] on the hydroquinone-quinone reaction, equation (32). Using an apparatus similar to that shown in Fig. 1, these workers measured the potentials of cells which may be represented by

(Pt); Hy, Q, HCl (0.2 *N*) : KCl (saturated) : HCl (0.2 *N*); H_2 (Pt).

The proportion of quinone, Q, to hydroquinone, Hy, was progressively changed by the addition of potassium dichromate in 0.2 normal hydrochloric acid. The reaction involved some of the acid present and an appropriate correction was made. If equation (39) is valid for the interpretation of these experiments a plot of the values of the measured potential E against values of log [Q]/[Hy] should be a straight line with a slope equal to 2.303 $RT/2\mathbf{F}$. That this is true is shown in Fig. 2. Within the limits of the experimental error all of the points fall on the line which has the slope required by the theory. This establishes without question the number of electrons, *i.e.*, two, which take part in the reaction represented by equation (32).

As has already been stated E_0'' in equation (39) is a function of the hydrogen ion concentration, represented sufficiently well for this purpose as a function of the pH. If the oxidized substance does not ionize and the reduction product is a dibasic acid equations (38) and (39) give

[15] In the general case this equation may be given the form

$$E = E_0'' - \frac{RT}{2\mathbf{F}} \ln \frac{[\text{Ox}]_s}{[\text{Red}]_s} \tag{39a}$$

in which $[\text{Ox}]_s$ and $[\text{Red}]_s$ are the stoichiometric concentrations of the oxidized and reduced substances respectively. E_0'' will be constant at a constant pH value, but will depend upon the ionization relations of the oxidized and reduced substances.

[16] V. K. LaMer and L. E. Baker, *J. Am. Chem. Soc.*, **44**, 1954 (1922).

$$E_0'' = E_0 - \frac{RT}{2\mathbf{F}} \ln \left[(\mathrm{H}^+)^2 + \mathbf{K}_1(\mathrm{H}^+) + \mathbf{K}_1\mathbf{K}_2 \right] \qquad (40)$$

This affords in certain cases a method for estimating the ionization constants $\mathbf{K}_1$ and $\mathbf{K}_2$. The rapid decomposition of quinone in alkaline

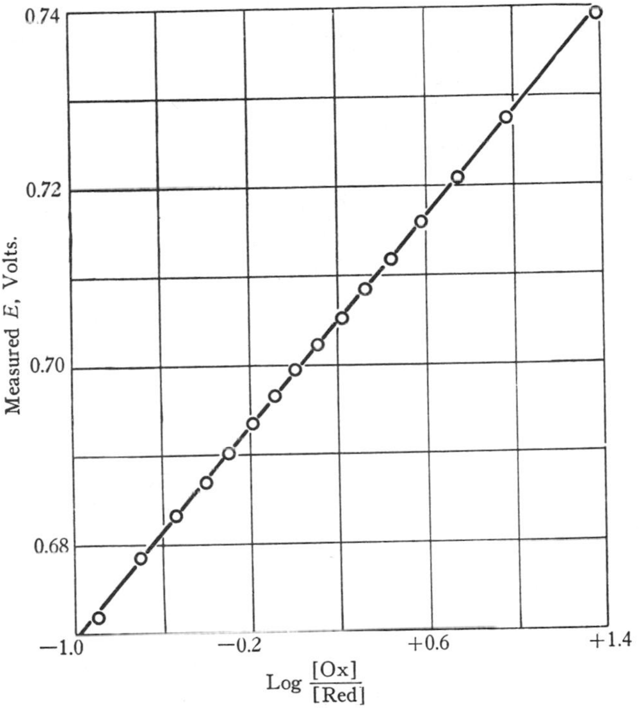

Fig. 2. Titration of Hydroquinone with Potassium Dichromate in a 0.2 Normal Hydrochloric Acid Solution.

solutions makes the hydroquinone-quinone system unsuitable for such a test. A system which fulfills the conditions of being un-ionized in the oxidized form and a dibasic acid in the reduced state is indigo and its reduction product, indigo white, studied by Sullivan, Cohen and Clark.[17] After establishing that equation (39) applies to that system when the proportion of oxidized to reduced substances was changed, measurements were made with the ratio $[\mathrm{Ox}]/[\mathrm{H_2R}]_s$ equal to unity and the pH of the solution altered by the addition of acid or alkali. The four

[17] M. X. Sullivan, B. Cohen and W. M. Clark, *Pub. Health Reports*, **38**, 666 (1923), reprinted as *Hygienic Laboratory Bull.*, **151**, 57 (1928).

possible sulphonates of indigo were studied. The results for the tetrasulphonate are shown in Fig. 3, where the value of E_0'' is plotted against

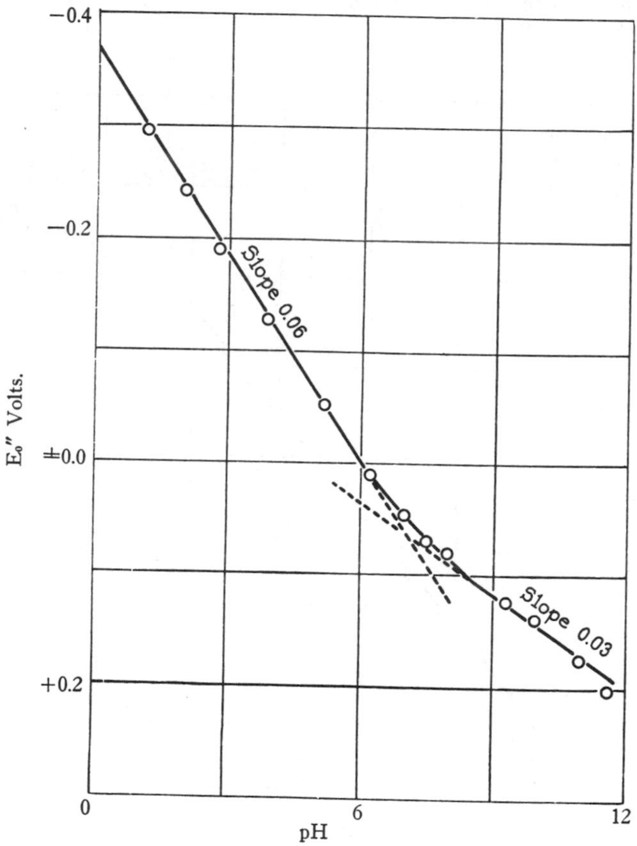

Fig. 3. The pH Dependence of E_0'' for the Indigotetrasulphonate Oxidation-Reduction System.

the pH of the solution. It will be seen that the E_0'' values fall mainly upon two straight lines. This can be readily interpreted with the aid of equation (40) as follows. At low pH values (*i.e.*, at high hydrogen ion concentrations) the term $(H^+)^2$ is large in comparison with $\mathbf{K}_1(H^+)$ and the term $\mathbf{K}_1\mathbf{K}_2$ is of still lower order. Therefore as long as $\mathbf{K}_1$ is negligible in comparison with (H^+) the plot of E_0'' against pH will be a straight line with a slope of 2.3026 $RT/\mathbf{F}$, which, since the experiments were carried out at 30°, has a value of 0.0602. There is another range of pH values in which $(H^+)^2$ is negligible in comparison with $\mathbf{K}_1(H^+)$. Here equation (40) shows that the variation of E_0'' with pH should be a straight line with a slope of 2.3026 $RT/2\mathbf{F}$ or 0.0301, a prediction which was verified by experiment. Differentiation of equation (40) with respect to pH yields

$$\frac{dE_0''}{d\text{pH}} = 2.3026 \frac{RT}{2\mathbf{F}} \frac{2(H^+)^2 + \mathbf{K}_1(H^+)}{[(H^+)^2 + \mathbf{K}_1(H^+) + \mathbf{K}_1\mathbf{K}_2]}$$

an equation from which the two results given above are immediately evident. In addition if $\mathbf{K}_1 = (H^+)$ and $\mathbf{K}_1$ is much greater than $\mathbf{K}_2$ then

$$\frac{dE_0''}{d\text{pH}} = 2.3026 \frac{RT}{\mathbf{F}} \times \frac{3}{4}$$

The value of the pH will therefore correspond to p$\mathbf{K}_1$ when the slope is, at 30°, 0.0602 × ¾ or 0.45, which is intermediate between the values 0.602 and 0.301. Substantially the same result may be obtained by extending the two straight lines as shown, until they intersect. The resulting p$\mathbf{K}_1$ obtained for the tetrasulphonate of indigo is 6.96, corresponding to an ionization constant of 11.2×10^{-8}.

Another interesting example is that of the β-sulphonate of anthraquinone and its reduction product, investigated by Conant, Kahn, Fieser and Kurtz.[18] Their E_0'' values for that compound are plotted against the corresponding pH values in Fig. 4. It will be seen that the plot consists of three straight lines, one with a slope of 0.06, another with a slope of 0.03 and the third horizontal, *i.e.*, with a slope of zero. These lines correspond, from equation (40), to the conditions that $(H^+) \gg \mathbf{K}_1$; $(H^+) \gg \mathbf{K}_2$; and $\mathbf{K}_1\mathbf{K}_2 \gg (H^+)$, respectively the symbol $\gg$ having its usual significance that the first term of the inequality is very much greater than the second. A rough estimate of p$\mathbf{K}_1$ and p$\mathbf{K}_2$ may be made graphically, as shown by extending the straight lines until they cross and reading the corresponding pH values. There is no independent evidence that ionization constants obtained in this way are correct, but they are probably of the right order of magnitude.

The examples discussed have all been of compounds which are un-ionized in the oxidized form and dibasic acids in the reduced form. There are other possibilities. Both the oxidized and reduced substances may ionize, and the ionization may be either acidic or basic. However, the principles used in dealing with the experimental data are sufficiently outlined above. The formulas used for interpreting oxidation-reduction

[18] J. B. Conant, H. M. Kahn, L. F. Fieser and S. S. Kurtz, *J. Am. Chem. Soc.*, **44**, 1382 (1922).

processes involving different types of ionization are discussed by W. M. Clark[19] and by Michaelis.[20]

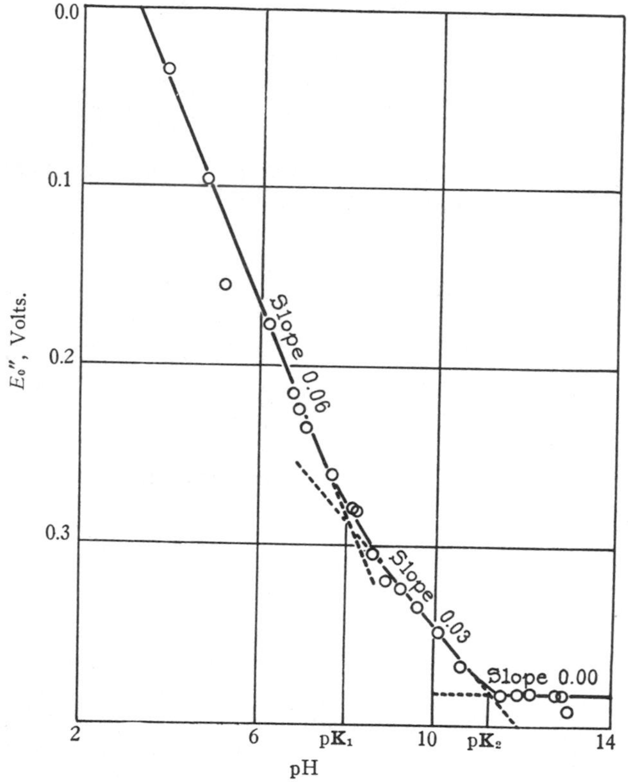

Fig. 4. The pH Dependence of E_0'' for the Oxidation-Reduction System: β-Sulphonate of Anthraquinone.

Although until recently all the reversible organic oxidation-reduction reactions were found to be of the type

$$\text{Red} = \text{Ox} + 2e^-$$

i.e., two electrons or valence steps being involved, Michaelis[21] and Elema[22] have shown that in certain cases, involving substituted

[19] W. M. Clark, *Hygienic Laboratory Bull.*, **151**, 11 (1928).

[20] L. Michaelis, "Oxidation-Reduction Potentials," translated from the German by L. B. Flexner, J. B. Lippincott Co., Philadelphia, 1930.

[21] L. Michaelis, *J. Biol. Chem.*, **92**, 211 (1931); **96**, 703 (1932).

[22] B. Elema, *Rec. trav. chim. Pays-Bas.*, **50**, 807 (1931).

quinones and the corresponding benzenoid compounds, the electrochemical reaction may occur in two steps, involving one electron each. A substance, known as a "semiquinone" is formed as an intermediate step. In such cases there is also an intermediate change of color between the usually highly colored oxidized substance and the nearly colorless reduced form. A single stage of reduction of a quinoid substance results, however, in a substance containing an odd number of electrons, with the properties of an organic radical. The evidence for semiquinones is clearest for substances which form cations in strong acid solutions or anions in strong alkalis. With a substituted *p*-phenylene diamine, for instance, the successive steps of oxidation may be represented by

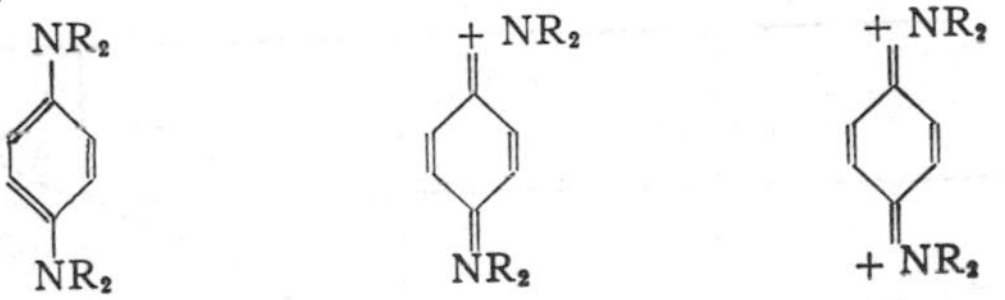

the second stage denoting the organic radical mentioned above. The odd proton may be vibrating between the nitrogen atoms, due to the fact that aside from the proton the molecule is symmetrical. This vibration may, in turn, be responsible for the additional color change observed if a semiquinone is formed.

An equation for the potential E, observed during an oxidation-reduction titration, which includes a consideration of the formation of semiquinones may be obtained as follows.[23]

Representing the oxidized substance by o, the reduced by r and the semiquinone by s we have the reaction

$$r + o = 2s$$

for which the law of mass action gives

$$\mathbf{K} = \frac{[s]^2}{[r]\,[o]} \tag{41}$$

in which **K** is a constant. If $[a]$ is the total concentration of substance then

$$[r] + [s] + [o] = [a] \tag{42}$$

Let x represent the number of mols of oxidant added in a titration starting with a solution of r alone. We thus have, at any point of the titration

$$[s] + [2o] = x \tag{43}$$

[23] L. Michaelis, *Chem. Rev.*, **16**, 243 (1935).

since the formation of o represents two steps of oxidation and s only one step. Equations (41), (42) and (43) are sufficient for obtaining equation (44) which follows, in which $X = [x]/[a]$ [24]

$$E = E_0'' - \frac{RT}{2\mathbf{F}} \ln \frac{X}{2 - X} - \frac{RT}{2\mathbf{F}} \ln \frac{X - 1 \pm \sqrt{(X-1)^2 + 4X(2-X)/\mathbf{K}}}{1 - X \pm \sqrt{(X-1)^2 + 4X(2-X)/\mathbf{K}}} \quad (44)$$

This interesting equation has the following properties. If in equation (41) **K** is very small, *i.e.*, if very little semiquinone is formed, then the equation approaches the following form

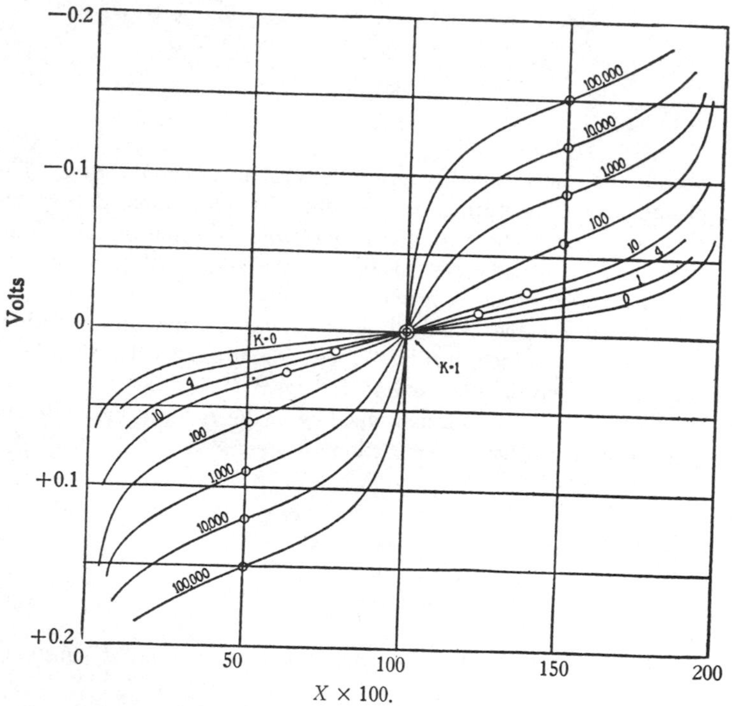

Fig. 5. Theoretical Electrometric Titration Curves of Organic Substances, Showing Two-stage Oxidation Resulting from Presence of Semiquinones.

[24] To obtain equation (44) divide equation (42) by $[a]$ and put $[r] = \rho[s]$ and $[o] = \tau[s]$. Modify equation (43) in the same manner and eliminate $[s]/[a]$ between the two equations. With equation (41), which may be put in the form, $\mathbf{K}\rho\tau = 1$, solve for values of ρ and τ, and substitute in equation (39a) which may be given the form

$$E = E_0'' - \frac{RT}{2\mathbf{F}} \ln \frac{\tau}{\rho}$$

Rearrangement then gives equation (44)

$$E = E_0'' - \frac{RT}{2\mathbf{F}} \ln \frac{X}{2 - X}$$

which is equivalent to equation (39a) for oxidation-reduction equilibria involving two valence steps. This equation corresponds to the line labelled $\mathbf{K} = 0$ in Fig. 5. In this plot values of X are the abscissae and E the ordinate, assuming E_0'' equal to zero. If, however, the value of $\mathbf{K}$ is appreciable, *i.e.*, larger amounts of semiquinone are formed, the curves approach the form

$$E = E_0'' - \frac{RT}{\mathbf{F}} \ln \frac{[\text{Ox}]}{[\text{Red}]}$$

in two separate steps involving one electron each. The results of such a titration are shown in Fig. 6. Here the results of potential measure-

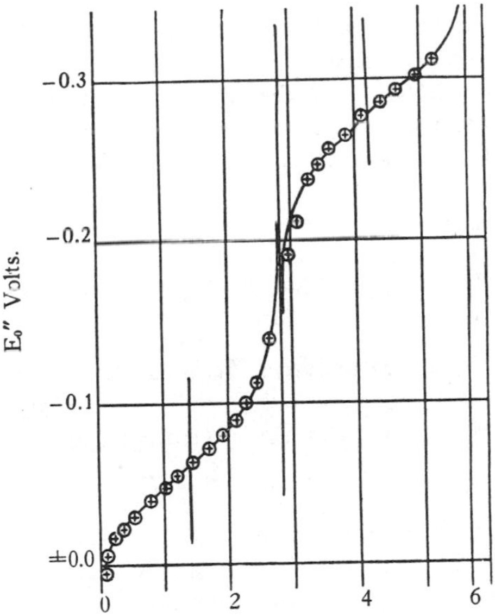

Fig. 6. Potentiometric Titration of α-Oxyphenazine with Quinone.

ments obtained by Michaelis, Hill and Schubert [25] by titrating reduced α-oxyphenazine with a solution of quinone are plotted. The two separate steps in the oxidation are clearly shown.

[*Comment to 1961 Edition*] An important recent addition to the field covered by this chapter is *Oxidation-Reduction Potentials of Organic Systems* by William Mansfield Clark, published by Williams and Wilkins Company (Baltimore) in 1960.

[25] L. Michaelis, E. S. Hill and M. P. Schubert, *Biochem. Z.*, **255**, 66 (1932). L. Michaelis, *loc. cit.*, 261.

Chapter 17

Potentiometric Titrations

Potentiometric titrations may frequently be used as convenient and accurate methods of analysis, some typical examples of which will be considered in this chapter. In addition the experimental procedures adopted for such methods yield results of decided theoretical interest. Examples have been given in Chapter 16 of the use of the potentiometric titration procedure in the investigation of the mechanisms of oxidation-reduction processes. It will be shown that the methods may also be of service in determining ionization constants, and solubilities of slightly soluble substances. A decided advantage of potentiometric titrations over the usual indicator methods is that the progress of a titration may be followed from beginning to end, whereas, in general, indicators are of service in deciding upon the end point only.

The discussion of potentiometric titrations will deal first with acidimetry, by the direct and differential procedures, after which methods depending upon precipitations and oxidation-reduction reactions will be considered.

Acidimetry by Potentiometric Methods. (a) *The Direct Method.* The use of the potentiometric procedure for the analysis of acids and bases depends primarily upon the electrochemical reaction

$$2H^+ + 2e^- = H_2$$

Many of the details of the method will be familiar from the discussion of the determination of pH values. A suitable galvanic cell for the titration of an acid, HA, with a base is

$$\text{(Pt) } H_2;\ \text{HA } (C) : \text{KCl saturated} : \text{KCl, HgCl; Hg} \qquad (1)$$

During the titration the concentration, C, of the acid is progressively changed by the addition of an alkali, and the potential of the cell is followed.

An apparatus designed by J. H. Hildebrand,[1] for the potentiometric titration of acids and bases, illustrates the principles involved in the "direct" method. The apparatus is shown in Fig. 1. The solution to

[1] J. H. Hildebrand, *J. Am. Chem. Soc.*, 35, 847 (1913).

be titrated is placed in vessel *A*. Connection is made by means of a liquid junction at *a* with the reference calomel electrode *C*. The hydrogen electrode is formed, as shown, from a piece of platinized platinum, *P*. A current of hydrogen gas entering by the tube *T*, passes over the electrode and bubbles out into the solution. Titrating reagent is added from the burette *B*. Although oxygen from the air is not excluded the end points of the titrations and the general forms of the titration curves are fairly close to those obtained by more elaborate and precise apparatus.

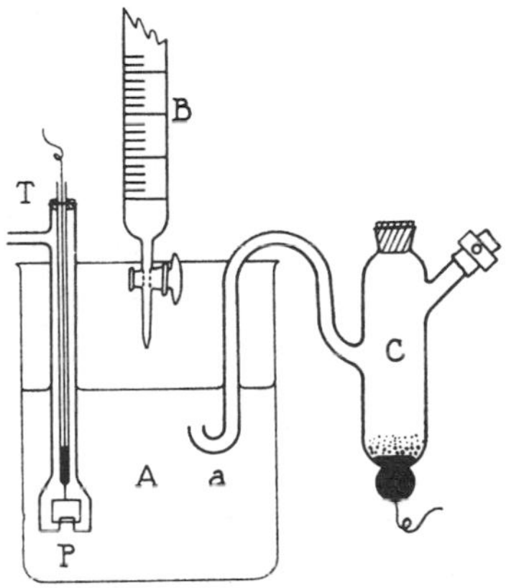

Fig. 1. Arrangement for Potentiometric Titration.

If a solution of a strong alkali is added to that of a strong acid, such as hydrochloric acid, in vessel *A* of Fig. 1 the potential of the cell will increase. A plot of the potential as ordinates against the volume of added alkali will have the form shown in curve (a) of Fig. 2, from the work of Böttger.[2] The scale of pH values is also represented on the right-hand side of the figure. It will be seen that the potential at first changes very little as base is added. The change of potential however becomes progressively greater and at the end of the titration there is a sudden jump in the emf. The potentials obtained during the titration of acetic acid, which are typical of those obtained with weaker acids, are shown by curve (b) of the same figure. These results were obtained by LaMer and Parsons.[3] The shape of the curve is somewhat different and the jump in potential at the end of the titration is less pronounced.

[2] W. Böttger, *Z. physik. Chem.*, **24**, 253 (1897).
[3] V. K. LaMer and T. R. Parsons, *J. Biol. Chem.*, **57**, 613 (1923).

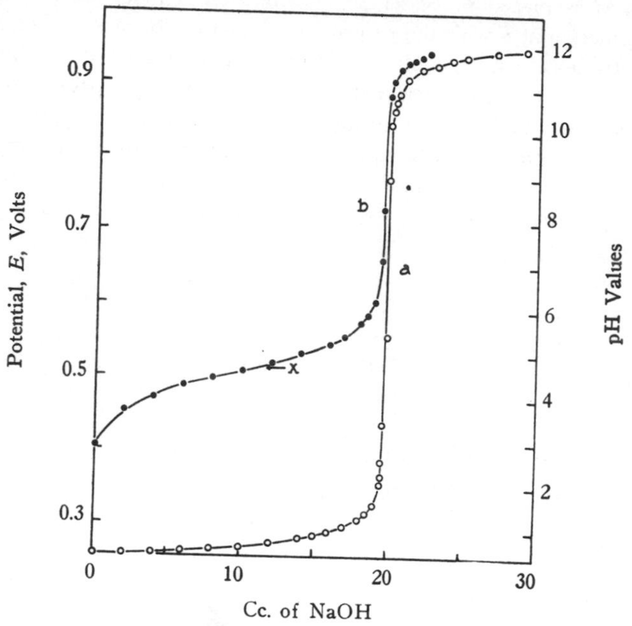

Fig. 2. Electrometric Titration Curves for Mono-basic Acids, E Referred to the Saturated Calomel Electrode.

a. Titration of 20 cc. of 0.5 Normal Hydrochloric Acid with Approximately 0.5 Normal Sodium Hydroxide.

b. Titration of 20 cc. of 0.2 Normal Acetic Acid with Approximately 0.2 Normal Sodium Hydroxide.

The fundamental reason for the utility of the potentiometric method in acid-base titrations may be seen with the aid of the equation

$$E = \text{E}_0 + \frac{R'T}{\mathbf{F}}\text{pH} \tag{2}$$

For the purpose of the following rough computation the pH of a strong acid, HA, may be considered to be

$$\text{pH} = -\log C_{\text{HA}} \tag{3}$$

in which C_{HA} is the concentration of the free acid. Equation (2) thus becomes,

$$E = \text{E}_0 - \frac{R'T}{\mathbf{F}}\log C_{\text{HA}} \tag{4}$$

Thus if vessel, A, of Fig. 1 contains 100 cc. of 0.01 normal hydrochloric acid and the solution is titrated with 0.1 normal sodium hydroxide, the addition of 9 cc. of the alkali will reduce the concentration of the free acid from 0.01 to 0.001 normal with a change of potential of the galvanic cell of about 59 millivolts at 25°. A further addition of 0.9 cc. of the alkali will bring the free acid concentration to 0.0001 normal, with another change of potential of 59 millivolts. Another rise of emf of about 59 millivolts is produced by only 0.09 cc. of the alkali. This approximate computation must not be pushed too far since near the neutral point the ionization of the water begins to have its effect on the potentials measured. In any case the most rapid change of pH with the addition of alkali takes place at or very near the equivalence point of the titration, with a resulting rapid change of the potential of the galvanic cell. Somewhat beyond the end point of the titration the pH is given, approximately, by

$$\text{pH} = -\log \frac{\mathbf{K}_w}{C_{\text{BOH}}} \tag{5}$$

in which C_{BOH} is the concentration of free strong base, so that equation (4) becomes

$$E = \text{E}_0 - \frac{R'T}{\mathbf{F}} \log \mathbf{K} + \frac{R'T}{\mathbf{F}} \log C_{\text{BOH}} \tag{6}$$

Thus the change of the concentration of free base of one order of magnitude will produce an increase of potential of 59 millivolts at 25°, *i.e.*, the increase of base concentration from 10^{-4} to 10^{-3} mol per liter will produce the same effect on the potential as the change from 10^{-3} to 10^{-2} mol per liter. A plot of E against the volume of added base will be very closely of the form given in Fig. 2 (*a*). A more adequate discussion of the theory of the acid-base titration, including the ionization equilibrium of water, is given later in this chapter.

It is possible to estimate, roughly at least, the values of the ionization constants of weak acids and bases by means of potentiometric titration methods. Since the ionization constant **K** of a weak acid HA is given by

$$\mathbf{K} = \frac{(\text{H}^+)(\text{A}^-)}{(\text{HA})}$$

this may, within the limitations of pH measurements as discussed in Chapter 15, be put into the form

$$\text{p}\mathbf{K} = \text{pH} - \log \frac{(\text{A}^-)}{(\text{HA})} = \text{pH} - \log \frac{[\text{A}^-]}{[\text{HA}]} - \log f \tag{7}$$

Here $\text{p}\mathbf{K} = -\log \mathbf{K}$, and f is an activity coefficient. If we define

$$\text{p}\mathbf{K}' = \text{p}\mathbf{K} + \log f \tag{8}$$

then equation (6) becomes [4]

$$\mathrm{p}\mathbf{K}' = \mathrm{pH} - \log \frac{[\mathrm{A}^-]}{[\mathrm{HA}]} \tag{9}$$

Thus the p**K**′ value is equal to the pH when the ratio of free acid to salt is unity, *i.e.*, the titration is half completed. The p**K**′ value for acetic acid in the titration represented in Fig. 2 is shown by the point *x* on the curve (b) and corresponds to a value of about p**K**′ = 4.6 whereas the limiting thermodynamic value, p**K**, is 4.75. The estimate of p**K**′ may obviously be made at other points in the titration curve by computing the appropriate values of the ratios $[\mathrm{A}^-]/[\mathrm{HA}]$. Or this process may be reversed, and the curve (b) obtained from the observed value of p**K**′ and these ratios. The same procedure may also be extended to polybasic acids. A titration curve of malonic acid from the work of Gane and Ingold [5] is shown in Curve I of Fig. 3. Here

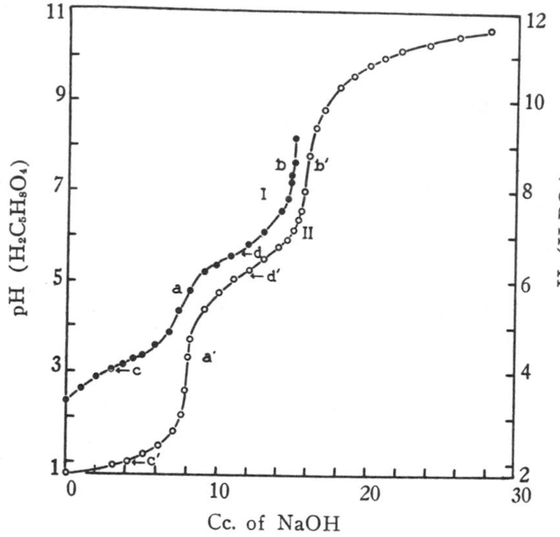

Fig. 3. Electrometric Titration Curves for Polybasic Acids.

Curve. I. Titration of 150 cc. of 0.050 molar malonic acid with 0.9868 normal sodium hydroxide.

Curve II. Titration of 22.72 cc. of phosphoric acid with 0.5 normal sodium hydroxide.

[4] The corresponding equation for a weak base may be readily seen to be

$$\mathrm{p}\mathbf{K}_w - \mathrm{p}\mathbf{K}'_\mathrm{B} = \mathrm{pH} + \log \frac{[\mathrm{B}^+]}{[\mathrm{B}]} \tag{9a}$$

[5] R. Gane and C. K. Ingold, *J. Chem. Soc.*, **1931,** 2153.

there are obviously two points of inflection, at *a* and *b*, corresponding to the end points of the titration of the first and second acid hydrogen. The p**K**′ values corresponding to the half titration points for the two stages of ionization are also indicated by *c* and *d* and are roughly 3.1 and 5.6. By carrying out similar titrations at a series of concentrations of acid these workers obtain limiting values of these constants of p**K**′ = 2.8 and p**K**′ = 5.7 which are at least of the order of magnitude of values of these constants derived from conductance measurements. Fig. 3 also contains a plot of an electrometric titration of phosphoric acid from measurements by Böttger.[6] Here again two points of inflection, at *a*′ and *b*′ are observed.

It must be remarked, however, that the curves shown in Fig. 3 are hardly typical. Two definite points of inflection are observed for a dibasic acid only if the two ionization constants are of quite different orders of magnitude. Auerbach and Smolczyk[7] have shown, theoretically, that the two points of inflection will not appear unless the ratio $\mathbf{K}_1/\mathbf{K}_2$ is greater than 16. However, it is safe to say that the ratio must be considerably greater than that if the inflection points are to be obtained experimentally with any accuracy. Also, if the two ionization constants are not quite different the values of the constants determined by the potentiometric titration method described above may be somewhat in error, as is shown by the authors just mentioned and by Simms.[8]

According to the definition given in equation (8)

$$p\mathbf{K}' = -\log(\mathbf{K}/f) \tag{10}$$

in which **K** is the thermodynamic ionization constant and *f* is an activity coefficient which should, presumably, approach unity as the ionic strength is decreased. However, it must be recalled that the galvanic cells involved in these titrations contain complex liquid junctions, with all their attendant uncertainties. Experiment demonstrates nevertheless that ionization constants of the correct order of magnitude, at least, are obtained by the potentiometric titration mehod.

For the purpose of locating the end points of titrations it is frequently desirable to plot values of $\Delta E/\Delta V$, instead of E, against the volume of titrating reagent V. Here ΔE is the change of potential produced by the increment, ΔV, of reagent. In Fig. 4 the curves (d) and (c) represent the data of curves (a) and (b) of Fig. 2 plotted in this way. The end point of the titration is at, or very near, the maximum ordinate in curves of the form of (c) and (d).

[6] W. Böttger, *Z. physik. Chem.*, **24**, 253 (1897).
[7] Fr. Auerbach and E. Smolczyk, *Z. physik. Chem.*, 110, 65 (1924).
[8] H. S. Simms, *J. Am. Chem. Soc.*, **48**, 1239 (1926).

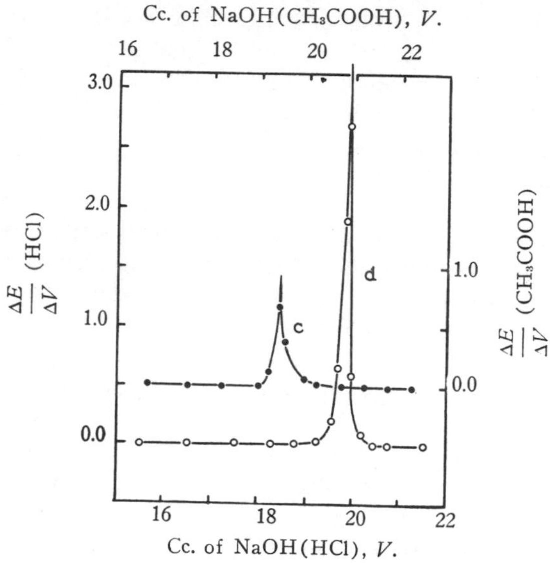

Fig. 4. Change in Slope of the EMF-Volume Curves for the Electrometric Titrations of Fig. 2.

(b) *The Differential Method.* It is possible, by means of the "differential" method of potentiometric titration to obtain values of $\Delta E/\Delta V$, corresponding to the curves (c) and (d) of Fig. 4 directly. The differential method was suggested in principle by Cox,[9] and has been developed by MacInnes and associates.[10] The procedure involved is to use two electrodes of the same type, but at each point in the titration to hold a small portion of the solution around one electrode from reacting with the titrating reagent until a potential measurement can be made. The potential difference ΔE produced by the addition of ΔV of titrating fluid may thus be found directly. A convenient apparatus for carrying out such titrations is shown in Fig. 5. One of the electrodes, *e*, is surrounded by tube, *A*, and the other, *e′*, is in the bulk of the solution to be titrated. A stream of gas from the tube, *G*, operates the simple lift pump which circulates the liquid and brings the solution in the tube, *A*, to the same concentration as the bulk of the solution. Operation of the stopcock shuts off the gas flow with the result that

[9] D. C. Cox, *J. Am. Chem. Soc.*, **47**, 2138 (1925).
[10] D. A. MacInnes and P. T. Jones, *ibid.*, **48**, 2831 (1926).
D. A. MacInnes, *Z. physik. Chem.*, **130A**, 217 (1927).
D. A. MacInnes and M. Dole *J. Am. Chem. Soc.*, **51**, 1119 (1929).
D. A. MacInnes and I. A. Cowperthwaite, *ibid.*, **53**, 555 (1931).

the titrating reagent from the burette, *B*, is added only to the bulk of the solution. There will now be a difference in composition of the solutions surrounding the electrodes, *e* and *e'*, which will result in a difference in potential, ΔE. A reverse turn of the stopcock once more starts the gas flow causing an equalization of the composition of the solution around the two electrodes, when the procedure just described

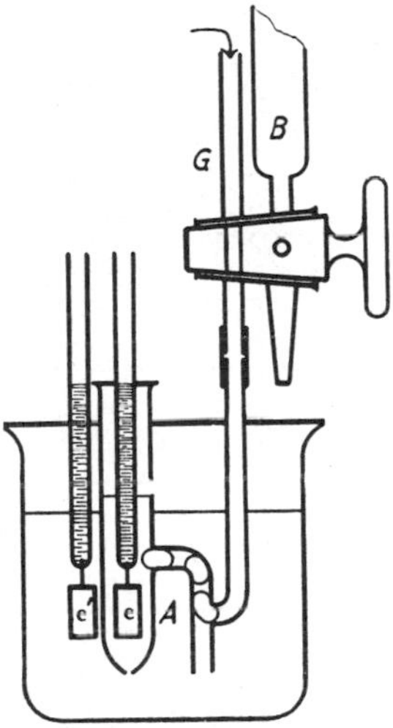

Fig. 5. The Arrangement for a Differential Potentiometric Titration.

may be repeated. Thus at each stage of the titration one electrode (the "retarded" electrode) is kept behind the other when an increment of reagent is added but is allowed to catch up before the next increment is added. Typical differential titration curves for hydrochloric and acetic acids using quinhydrone electrodes are shown in Fig. 6 (a) and (b) where the volume of titration fluid is plotted as abscissa against the value of ΔE as ordinate.[11] In curve (d)[12] of Fig. 6 is shown the result of titrating phosphoric acid by the differential potentiometric method, the two maxima corresponding to the first and second dissociation of the acid being clearly evident. Curve (c)[13] gives a plot of the results of a similar titration of aspartic acid for which the first and second dissociation constants are 1.5×10^{-4} and 2.5×10^{-10}. It will be seen that in this case the end point corresponding to the first dissociation is very definite, but that although there is evidence of an

[11] From unpublished results by D. A. MacInnes and P. T. Jones.
[12, 13] From results obtained by L. Valik under the direction of the author.

end point for the second dissociation it cannot be located with any accuracy, due to the flatness of the curve through the experimental points. The differential method has advantages, especially for work

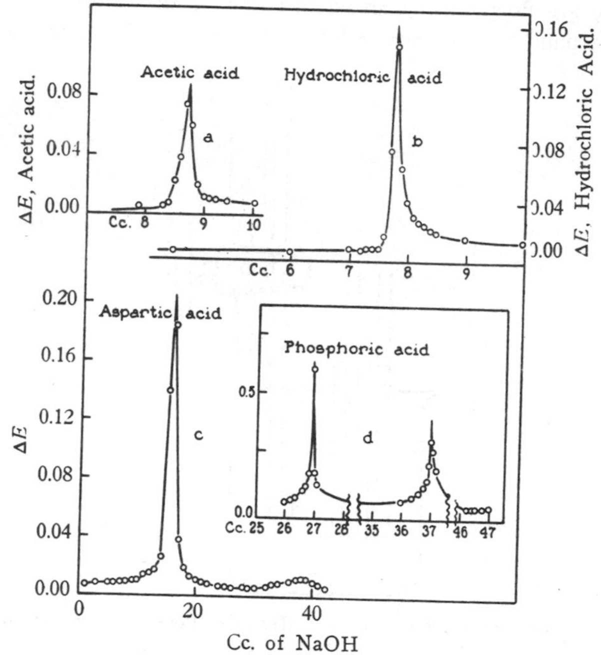

Fig. 6. Electrometric Titration Curves of Some Typical Acids by the Differential Method.

requiring precision. In the first place the significant potentials measured are not the relatively small differences between larger emf values as in the direct method. In addition the solutions titrated are not contaminated by material from the liquid junction connecting to the reference electrode and no material is lost by diffusion into the liquid junction. Furthermore although uncertainty due to liquid junction potentials is not entirely eliminated it is reduced to a minimum since in the differential method the two solutions giving rise to a difference of potential ΔE are in all cases very nearly the same.

Analyses of acid solutions, with accuracy of about 0.003 per cent by differential electrometric titration, have been carried out by MacInnes and Cowperthwaite [14] using hydrogen electrodes. The apparatus

[14] D. A. MacInnes and I. A. Cowperthwaite, *J. Am. Chem. Soc.*, 53, 555 (1931).

used was essentially that shown in Fig. 5 except that access of oxygen from the air was prevented, and the solution titrated and the titrating fluid were both saturated with hydrogen. To attain this precision it was however necessary to utilize weight burettes to bring the titration near its end point, the final adjustment being carried out volumetrically with titrating reagent diluted to one-tenth or one-hundredth of its original concentration.

A natural question arises as to whether holding out a portion of the solution while each increment of titrating fluid is added can have an influence on the end point observed. It can be readily shown that the error involved is negligible. Let ν be the volume of titrating reagent at the end point, $\Delta\nu$ the volume of the last increment, and v and V, respectively, the volume of the solution retained around the "retarded" electrode and the total volume. The error δ in per cent, will be given by the relation

$$\delta < \frac{\Delta\nu}{\nu} \times \frac{v}{V} \times 100$$

since before adding the last increment the titration is within the fraction $\Delta\nu/\nu$ of completion, and the increment is added to all but the small proportion, v/V, of the total volume. Thus if 50 cc. of titrating fluid are used, with 0.1 cc. increments, and the volume of solution around the retarded electrode is 4 cc. of a total of 200 cc. the error will be about 0.004 per cent, which is of course negligible for ordinary volumetric work. The error is still smaller when weight burettes are used, as described in the preceding paragraph.

The Theory of the Potentiometric Method for Acidimetry. In the preceding discussion it has been assumed that in a direct electrometric titration the equivalence point of the titration is the same as the point of inflection on a curve, such as Fig. 2, connecting the emf with the volume of titrating fluid. In the differential method the equivalence point has been assumed to correspond to a maximum in a curve of the form of Fig. 6. As a matter of fact it is not necessarily true that the equivalence point corresponds to the end point indicated by the potentiometric titration, though as will be seen the difference is never large. A discussion of this matter incidentally leads to a more complete consideration of the theory of potentiometric titration of acids and bases. The subject has been investigated by Eastman,[15] Roller [16] and Kilpi.[17] In the following paragraphs the discussion is limited to

[15] E. D. Eastman, *J. Am. Chem. Soc.*, **47**, 332 (1925); **56**, 2646 (1934).
[16] P. S. Roller, *ibid.*, **50**, 1 (1928); **54**, 3485 (1932); **57**, 98 (1935).
[17] S. Kilpi, *Z. physik. Chem.*, **172A**, 277; **173A**, 427; **174A**, 441 (1935); *Z. anal. Chem.*, **104**, 390 (1936).

acid-base titrations that are used in actual practice and is based mainly on the treatment by Kilpi. Although the equations have been developed for the titration of a weak acid with a strong base they are also applicable to the titration of a weak base with a strong acid by obvious substitution of constants. For the purpose of this derivation activities and concentrations will be considered to be identical. In the titration of a weak acid, HA, with a strong base, $B^+ + OH^-$, let C represent the total concentration of acid and (B^+) be the concentration of the positive ion of the base. The necessity for electrical neutrality of the solution gives the expression

$$(B^+) + (H^+) = (A^-) + (OH^-) \tag{11}$$

The negative ion constituent of the acid will, throughout the titration, have the concentration C (it being assumed that the addition of the base produces an inappreciable change in the concentration) so that

$$C = (HA) + (A^-) \tag{12}$$

In addition we have the relations

$$\frac{(H^+)(A^-)}{(HA)} = \mathbf{K}_A \quad (13) \qquad \text{and} \qquad (H^+)(OH^-) = \mathbf{K}_w \quad (14)$$

in which $\mathbf{K}_A$ is the ionization constant of the weak acid and $\mathbf{K}_w$ is the corresponding constant for water. By eliminating (HA) between equations (12) and (13) and introducing the resulting expressions and equation (14) into equation (11) the result is

$$B = (B^+) = \frac{C\mathbf{K}_A}{(H^+) + \mathbf{K}_A} + \frac{\mathbf{K}_w}{(H^+)} - (H^+) \tag{15}$$

Differentiating with respect to pH yields

$$\frac{dB}{2.3d(\text{pH})} = \frac{C(H^+)\mathbf{K}_A}{(H^+ + \mathbf{K}_A)^2} + \frac{\mathbf{K}_w}{(H^+)} + (H^+) \tag{16}$$

which is an alternative expression to that of Chapter 15 for the "buffer capacity" and is the form given it by Van Slyke.[18] Since near the end point of the titrations we are considering that the solutions are alkaline as a result of hydrolysis, the term (H^+) in equation (16) is small compared with the other terms and may be neglected in the following.

[18] D. D. Van Slyke, *J. Biol. Chem.*, **52**, 525 (1922).

Differentiating once more with respect to pH and equating to zero gives the expression

$$(H^+)^3 - (H^+)^2 \frac{\mathbf{K}_A^2 C - 3\mathbf{K}_A\mathbf{K}_w}{\mathbf{K}_A C + \mathbf{K}_w} + (H^+)\frac{3\mathbf{K}_A^2\mathbf{K}_w}{\mathbf{K}_A C + \mathbf{K}_w} + \frac{\mathbf{K}_A^3\mathbf{K}_w}{\mathbf{K}_A C + \mathbf{K}_w} = 0 \qquad (17)$$

The real roots of this cubic equation correspond to maxima or minima of buffer value, $dB/2.3d(\text{pH})$. The potentiometric end point of the titration will, of course, correspond to a minimum in the buffer value. It is possible with the aid of this equation to predict the relation between $\mathbf{K}_A C$ and $\mathbf{K}_w$ at which such minima appear in a titration curve.

For the purpose of investigating the relation between the potentiometric and stoichiometric end points of a titration the assumption will be made that $\mathbf{K}_w$ is negligible compared with $\mathbf{K}_A C$ and equation (17) can then be put in the form

$$(H^+)_p^2 = \frac{\mathbf{K}_A\mathbf{K}_w}{C}\left[1 + (H^+)_p^3 \frac{C}{\mathbf{K}_w\mathbf{K}_A^2} + \frac{3}{\mathbf{K}_A}(H^+)_p\right] \qquad (18)$$

in which $(H^+)_p$ represents the hydrogen ion concentration at the *potentiometric end point*. Since the values of the coefficients of $(H^+)_p^3$ and $(H^+)_p$ are such that the factor in brackets differs little from unity, values of $(H^+)_p$ may be obtained by a short series of approximations. The hydrogen ion concentration at the *stoichiometric end point* may be obtained as follows. Consider the hydrolysis of the salt BA

$$B^+ + A^- + H_2O \rightleftarrows B^+ + OH^- + HA \qquad (19)$$

for which the mass action relation is

$$\mathbf{K}_h = \frac{\mathbf{K}_w}{\mathbf{K}_A} = \frac{(HA)\,(OH^-)}{(A^-)} \qquad (20)$$

That $\mathbf{K}_h = \mathbf{K}_w/\mathbf{K}_A$ may be seen by multiplying the numerator and denominator by (H^+). From equation (19) equivalent quantities of hydroxyl ion and undissociated acid are formed so that $(HA) = (OH^-)$. Also from equation (12)

$$(A^-) = C - (HA) = C - (OH^-) \qquad (21)$$

Equation (20) can therefore be put into the form

$$(OH^-)_s = \frac{\mathbf{K}_w}{(H^+)_s} = \sqrt{\frac{\mathbf{K}_w^2}{4\mathbf{K}_A^2} + \frac{C\mathbf{K}_w}{\mathbf{K}_A}} - \frac{\mathbf{K}_w}{2\mathbf{K}_A} \qquad (22)$$

$(OH^-)_s$ and $(H^+)_s$ referring to the stoichiometric end points.

TABLE I. COMPARISON OF THE HYDROGEN ION CONCENTRATIONS AT THE POTENTIOMETRIC AND STOICHIOMETRIC END POINTS AT 0.1 NORMAL IN THE TITRATION OF ACIDS OF DIFFERENT STRENGTHS

Ionization Constant of Acid K_A	Hydrogen Ion Concentrations, (mols per liter) At Potentiometric end Point $(H^+)_p$	At Stoichiometric end Point $(H^+)_s$
10^{-5}	1.000×10^{-9}	1.000×10^{-9}
10^{-6}	3.163×10^{-10}	3.161×10^{-10}
10^{-7}	1.002×10^{-10}	1.000×10^{-10}
10^{-8}	3.182×10^{-11}	3.167×10^{-11}
10^{-9}	1.020×10^{-11}	1.005×10^{-11}

Values of $(H^+)_p$ and $(H^+)_s$ are listed in Table I for various values of the ionization constants $\mathbf{K}_A$, all at the concentration C of 0.1 normal. The ionization constant of water $\mathbf{K}_w$ is assumed to have the value of 10^{-14}. It will be seen that the hydrogen ion concentrations at the stoichiometric and potentiometric end points are not very different even for a value of $\mathbf{K}_A C$ as low as 10^{-10}.

Valik[19] made differential potentiometric titrations of aspartic acid, one series of results being given in curve (c) of Fig. 6. The volume of solution of alkali necessary to titrate the second acid dissociation for which the ionization constant is 2.5×10^{-10} should be exactly equal to that for the first unless there is a difference between the potentiometric and stoichiometric end points. Within the rather large limit of error, this was found to be true, but the end point could not be located with accuracy due to the flatness of the curve, as shown above. Differences between the stoichiometric and potentiometric end points are predicted for titrations of weak acids or weak bases. Such a difference increases the weaker the acid or base, but the difficulty of locating the end point also increases. It may be safely concluded that within the accuracy to which the potentiometric end point of a titration can be established it is identical with the stoichiometric end point.

Potentiometric Titrations Depending Upon Precipitation Reactions. There are a number of convenient and accurate titrations that depend upon precipitation reactions. For instance the progress of the reaction

$$KCl + AgNO_3 = AgCl + KNO_3$$

may be followed by measuring the potential of a galvanic cell one electrode of which is reversible to the silver or chloride ion constituent. In this particular case the accuracy of the method may be made to approach that achieved in work on atomic weights. Lange and

[19] L. Valik, unpublished work under the direction of the author.

Schwartz [20] made direct potentiometric titrations making use of galvanic cells of the type

$$Ag; AgNO_3\ (C) : KNO_3 : KCl,\ HgCl;\ Hg \qquad (23)$$

They carried out the titration by adding a solution of a halide, such as potassium chloride or bromide to the silver nitrate solution. The cell involved the liquid junction

$$AgNO_3 : KNO_3$$

so that a certain amount of silver solution tended to be lost by diffusion. To overcome this difficulty these workers provided a counter current of halide-free potassium nitrate solution from the liquid junction into the solution being titrated. As described for the precision titrations in acidimetry, the reaction was carried near to the end point by using weight burettes, and the final adjustment was carried out using a dilute titrating fluid. A plot was made of values of ΔE against V in which ΔE is the change of potential produced by the addition of the increment ΔV of reagent, and the end point was taken to be that corresponding to the maximum. Precision of about 0.003 per cent was attained.

Potentiometric titrations of chloride solutions of this precision may more readily and directly be obtained using the differential method. For this purpose MacInnes and Dole [21] used the apparatus shown in Fig. 5. The electrodes e and e' were, however, made of silver-plated platinum. The certainty with which the end point may be determined is shown in Fig. 7 in which the change of potential ΔE produced by a drop of the diluted titrating reagent is plotted against the number of drops. The use of the differential method dispenses with the necessity of a counter flow of liquid from the salt bridge and decreases the uncertainty due to liquid junction potentials. This method of analysis has been of service in the determination of transference numbers by the Hittorf [22] method, which requires an analytical method capable of yielding results of high precision.

The direct method of potentiometric titration has been applied to various precipitation reactions, among which may be mentioned the determination of magnesium in dolomite,[23] the precipitation of zinc with ferrocyanide [24] and the titration of chloride, bromide and thiocyanate [25] with mercurous perchlorate.[26]

20 E. Lange and E. Schwartz, *Z. physik. Chem.*, **129**, 111 (1927).
21 D. A. MacInnes and M. Dole, *J. Am. Chem. Soc.*, **51**, 1119 (1929).
22 D. A. MacInnes and M. Dole, *J. Am. Chem. Soc.*, **53**, 1357 (1931); G. Jones and L. T. Prendergast, *ibid.*, **58**, 1476 (1936).
23 J. H. Hildebrand and H. S. Harned, *ibid.*, **35**, 867 (1913).
24 E. Müller, *Z. angew. Chem.*, **32**, 351 (1919).
25 E. J. A. H. Verzijl and I. M. Koltoff, *Rec. trav. chim.*, **43**, 380, 389, 394 (1924).
26 E. Müller and H. Aarflot, *Rec. trav. chim.*, **43**, 874 (1924).

The Theory of Potentiometric Titrations Involving Precipitation Reactions. The theory of potentiometric titrations involving a precipitation reaction may be indicated by dealing with a typical case, that of the reaction of silver nitrate with an alkali halide, for instance potassium chloride. The following discussion is substantially that of Lange and Schwartz.[27] It is convenient for the purpose of the discussion to

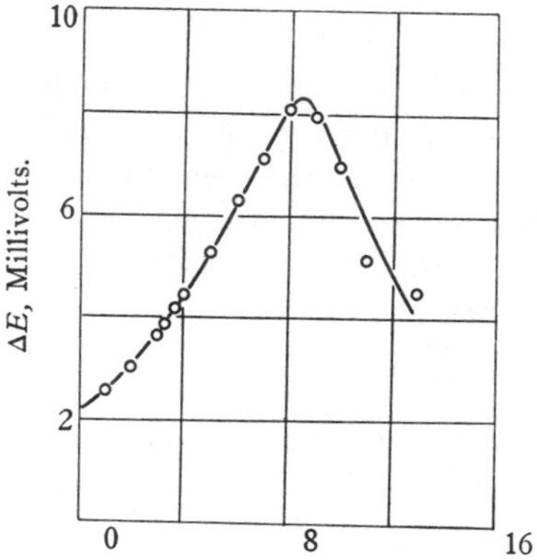

Fig. 7. Change of Potential in the Differential Titration of Potassium Chloride with Silver Nitrate.

reverse the usual procedure by starting at the end point of the titration, and studying the effect of adding reagent. A solution saturated with silver chloride has a solubility product, s, equal to

$$s = (Ag^+)(Cl^-) \tag{24}$$

in which (Ag^+) and (Cl^-) are activities. In terms of concentrations, $[Ag^+]$ and $[Cl^-]$, this is

$$[Ag^+][Cl^-] = s/f_{\pm}^2 = s' \tag{24a}$$

in which $f_{\pm}$ is the mean ion activity coefficient and s′ is substantially constant at a constant ionic strength. At the end point $[Ag^+] = [Cl^-]$.

If a small amount, P equivalents, of silver nitrate is added a portion of the silver chloride will precipitate, the new condition being

$$(\sqrt{s'} + \Delta[Ag^+])(\sqrt{s'} - \Delta[Cl^-]) = s' \tag{25}$$

in which $\Delta[Ag^+]$ and $\Delta[Cl^-]$ are changes of concentration produced by the addition of P equivalents of the silver nitrate. This amount of

[27] E. Lange and E. Schwartz, *Z. physik. Chem.*, **129A**, 111 (1927).

silver ion constituent (*a*) increases the concentration $\Delta[Ag^+]$ and (*b*) leaves the solution as precipitate in amount equal to $\Delta[Cl^-]$ so that

$$P = V(\Delta[Ag^+] + \Delta[Cl^-]) \tag{26}$$

in which V is the volume of the solution. Eliminating $\Delta[Cl^-]$ between equations (25) and (26) we have

$$(\sqrt{s'} + \Delta[Ag^+])(\sqrt{s'} - P/V + \Delta[Ag^+]) = s' \tag{27}$$

from which

$$\Delta[Ag^+] = -\sqrt{s'} + \frac{P}{2V} \pm \sqrt{s' + \frac{P^2}{4V^2}} \tag{28}$$

Now the potential of a cell:

$$\text{Ag; AgCl (at end point) : AgCl (in excess } AgNO_3\text{); Ag} \tag{29}$$

will be, neglecting the liquid junction, and at constant ionic strength

$$E = \frac{RT}{\mathbf{F}} \ln \frac{\sqrt{s'} + \Delta[Ag^+]}{\sqrt{s'}} \tag{30}$$

which with equation (28) gives

$$E = \frac{RT}{\mathbf{F}} \ln \left[\frac{P}{2V\sqrt{s'}} \pm \sqrt{1 + \frac{P^2}{4V^2 s'}}\right] \tag{31}$$

However, both in the direct and differential methods for potentiometric titration we are primarily interested in the changes of electromotive force ΔE with small additions of the titrating agent, P. Differentiation of equation (31) yields

$$dE = \frac{RT}{\mathbf{F}} \cdot \frac{dP}{\sqrt{4V^2 s' + P^2}} \tag{32}$$

so that for *small* increments, p, of the titrating reagent, and if $P = np$ in which n is an integer, it follows that

$$\frac{\Delta E}{p} = \frac{RT}{\mathbf{F}} \frac{1}{\sqrt{4V^2 s' + (np)^2}} \tag{33}$$

A plot of values of ΔE as ordinates against values of n as abscissae is of the form shown in Fig. 7. It will be seen that ΔE will have its maximum value when $n = 0$, *i. e.*, at the end point of the titration.

Normally, of course, the titration does not start at the end point.

Let n_0 be the value of n at the beginning of the titration and n_e be the value of n at the end point, then

$$n = n_e - n_0$$

Equation (33) will thus take the form

$$\frac{\Delta E}{p} = \frac{RT}{\mathbf{F}} \frac{1}{\sqrt{4V^2 s' + (n_e - n_0)^2 p^2}} \tag{33a}$$

which will have a maximum when $n_0 = n_e$ and the curve of values of ΔE against values of n will be symmetrical around this maximum. The curve is of the form shown in Fig. 7. It will be seen that the equation contains the solubility product s′ which suggests that potentiometric titrations may be used for obtaining the solubilities of slightly soluble substances. However a more accurate method for obtaining solubility products from potentiometric data is given in the following.

A potentiometric titration method for obtaining the solubility of silver chloride in potassium nitrate solutions has been used by Brown and MacInnes.[28] These workers used the cell

$$\text{Ag; AgCl, } \underbrace{\underset{C_1}{AgNO_3},\ \underset{C_2}{KNO_3}}_{a} : \underbrace{\underset{C_3}{AgNO_3},\ \underset{C_4}{KNO_3}}_{b}\text{, AgCl: Ag} \tag{34}$$

Initially the concentrations C_1 and C_2 in solution a were equal to C_3 and C_4 in solution b. However, small increments of potassium chloride solution were added to solution a, and solution b was kept constant and, with the silver-silver chloride electrode immersed in it, as a reference half cell. In addition to equation (24a) we have the condition of electrical neutrality,

$$[Ag^+] + [K^+] = [Cl^-] + [NO_3^-] \tag{35}$$

At the equivalence point $[K^+] = [NO_3^-]$ and $[NO_3^-] - [K^+]$ is a measure of the distance from that point. If n is the number of increments of potassium chloride solution added to V liters (assumed to be constant) of solution, and n_e is the value of n at the equivalence point, then

$$[NO_3^-] - [K^+] = (n_e - n)\frac{p}{V} \tag{36}$$

which with equation (35) becomes

$$[Cl^-] - [Ag^+] = (n - n_e)\frac{p}{V} \tag{37}$$

[28] A. S. Brown and D. A. MacInnes, *J. Am. Chem. Soc.*, **57**, 459 (1935).

Since the ionic strength remains practically constant the potential of cell (34) is given by

$$E = \frac{RT}{\mathbf{F}} \ln \frac{[Ag^+]}{[Ag^+]_i} = \frac{RT}{\mathbf{F}} \ln y \tag{38}$$

in which the subscript i refers to the reference electrode, and

$$y = [Ag^+]/[Ag^+]_i$$

Since n equals zero at the beginning of the titration, from equation (37)

$$[Cl^-]_i - [Ag^+]_i = -\frac{n_e p}{V} \tag{39}$$

Substituting $[Cl^-] = s'/[Ag^+]$ and $[Ag^+] = [Ag^+]_i y$, converts equation (37) into

$$\frac{s'}{[Ag^+]_i y} - [Ag^+]_i y = (n - n_e)\frac{p}{V} \tag{40}$$

and equation (39) into

$$[Ag^+]_i = \frac{n_e p}{V} + \frac{s'}{[Ag^+]_i} \tag{41}$$

Combining these two equations and solving for $n/(1 - y)$ gives

$$\frac{n}{1 - y} = n_e + \frac{Vs'}{p[Ag^+]_i} \cdot \frac{1 + y}{y} \tag{42}$$

or

$$\frac{n}{1 - y} = n_e + X\frac{1 + y}{y} \tag{42a}$$

where

$$X = \frac{Vs'}{p[Ag^+]_i} \tag{43}$$

and y and n are measured quantities. Thus a plot of $n/(1 - y)$ against $(1 + y)/y$ should be a straight line with intercept n_e and slope X. Such a plot, for the precipitation of AgCl from KCl, is shown in Fig. 8.

The preceding analysis was based on the assumption of a constant volume of solution throughout the titration, a condition unrealizable experimentally. A slight correction must be thus applied to the measured E values to take account of this volume change. For the details of this correction the reader is referred to the original paper. From the slope and intercept, X and n_e, of the curve in Fig. 8, the solubility of AgCl is readily calculable since, substituting equation (43) in equation (41) gives

$$\sqrt{s'} = \frac{p}{V}\sqrt{X(X + n_e)} \tag{44}$$

For the precipitation under consideration the numerical values were

$$X = 0.2986, \quad n_e = 3.738$$
$$p = 4.8440 \times 10^{-6} \text{ and } V = 0.33859 \text{ liter}$$

so that $\sqrt{s'} = 1.571 \times 10^{-5}$ mol per liter. This is the solubility of silver chloride at 25° in a solution of 0.02859 normal potassium nitrate. The limiting solubility $\sqrt{s}$ of silver chloride may be obtained from the equation $\sqrt{s} = \sqrt{s'} \cdot f_{\pm}$ if an estimate is made of the activity coefficient $f_{\pm}$. Assuming that it is sufficiently closely represented by the activity coefficients of silver nitrate a value of $f = 0.833$ at $C = 0.02859$ mol

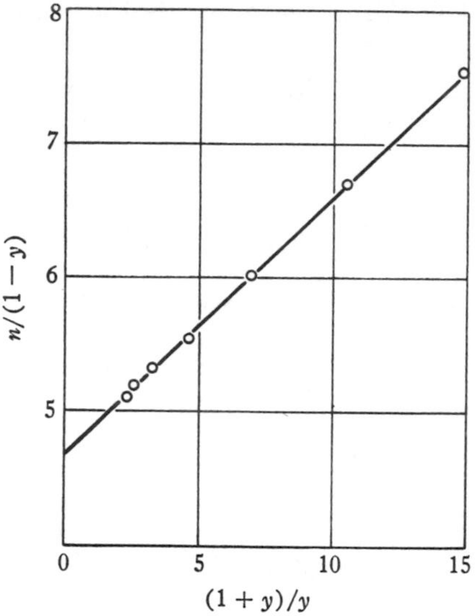

Fig. 8. Plot used in the Potentiometric Determination of the Solubility of Silver Chloride.

per liter has been interpolated from the data in Table III of Chapter 8. This yields a value of $\sqrt{s} = 1.309 \times 10^{-5}$ mol per liter. This is slightly higher than the value obtained by Neuman[29] by a method which consisted in increasing the concentrations of reacting components until a trace of solid appeared. It agrees excellently with the figure 1.306×10^{-5} mol per liter obtained from the results of conductance measurements of Kohlrausch[30] by a short interpolation from his results at other temperatures than 25°.

[29] E. W. Neuman, *J. Am. Chem. Soc.*, **54**, 2195 (1932).
[30] F. Kohlrausch, *Z. physik. Chem.*, **64**, 129 (1908).

Oxidation-Reduction Titrations. The theory and an outline of the practice of oxidation-reduction titrations for organic substances have already been described in Chapter 16. A full discussion of the possible analytical methods depending upon potentiometric titration is not possible here. For detailed accounts of specific methods the reader is referred to more extended treatises on the subject.

Such potentiometric titrations may be of course either reductions or oxidations, and the methods are useful whether or not there is a color change at the end point, and no internal or external indicator, other than the change of potential, is necessary.

Potentiometric Titrations Using Irreversible Electrodes. If a high order of accuracy is not necessary it is occasionally possible to utilize in potentiometric titrations a simple arrangement of two electrodes made of different metals, both electrodes being immersed directly in the solution to be titrated. Since in most titrations it is only necessary to locate the "break," *i. e.*, the sudden change of potential, at the end point, it is not important to know whether the measured potentials correspond to the reversible values or not. The principle is illustrated by the three curves in Fig. 9, from the work of Furman and Wilson.[31]

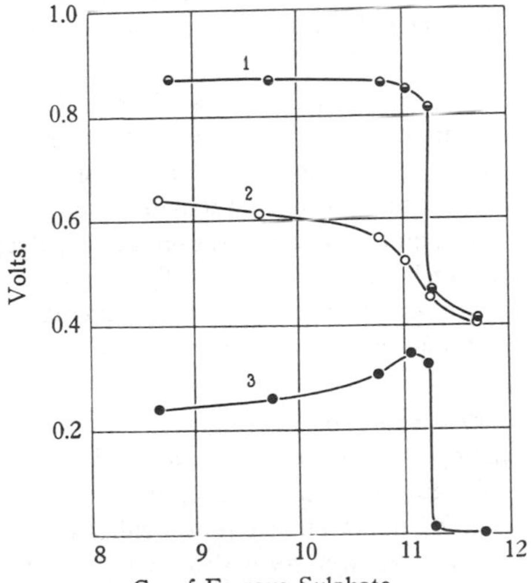

Fig. 9. Titration Curves: 1, Using Reversible Electrodes; 2 and 3, Using an Irreversible Electrode.

[31] N. H. Furman and E. B. Wilson, Jr., *J. Am. Chem. Soc.*, **50**, 277 (1928).

Curve 1 illustrates the changes of potential observed when ferrous sulphate was added to the potassium bichromate in a galvanic cell, initially of the form

$$(Pt);\ K_2Cr_2O_7 : KCl,\ HgCl;\ Hg \qquad (45)$$

The two electrodes, platinum and calomel, used in this titration are reversible, or very nearly so. Curve 1 may be contrasted with curve 2 in which the platinum electrode of cell (45) has been replaced by one made of metallic tungsten. It will be seen that the trend of the potentials is quite different. The mechanism of the reaction at the tungsten electrode is not understood, but it is undoubtedly far from equilibrium. Curve 3 represents the potentials obtained during a titration in which the platinum and tungsten electrodes were placed directly in the bichromate solution, *i. e.,* potentials of the system, initially

$$(Pt);\ K_2Cr_2O_7;\ (W) \qquad (46)$$

as ferrous sulphate was added. This last curve is, roughly, the difference between the other two, and shows a pronounced drop in potential at the end of the titration. It will be evident that although the potential of the tungsten electrode is changing near the end point of the titration it does not vary as rapidly as does that of the platinum electrode. Thus the tungsten electrode may be used as a somewhat shifting reference.

The idea of using two dissimilar metals in a potentiometric titration originated in an observation by Hostetter and Roberts [32] that whereas a platinum electrode assumed nearly the reversible potential during a titration of a ferrous salt with dichromate, the potential changing abruptly at the end point, the substitution of a palladium electrode resulted in a nearly unchanging potential. Quite a number of combinations of metal electrodes have been used in oxidation-reduction titrations including platinum and platinum-rhodium [33] and platinum and amalgamated gold.[34]

Use has also been made of two different metals as electrodes in acid-base titrations by a number of workers. For instance Fuoss [35] found that the combinations antimony-lead, antimony-amalgamated copper, bismuth-silver and copper-copper oxide will all serve more or less effectively. In this case, as well as with oxidation-reduction titrations, most observers report that, although the end-point of a titration can

[32] J. C. Hostetter and H. S. Roberts, *J. Am. Chem. Soc.,* **41,** 1337 (1919).
[33] H. H. Willard and F. Fenwick, *J. Am. Chem. Soc.,* **44,** 2504 (1922).
[34] N. H. Furman, *ibid.,* **50,** 273 (1928).
[35] R. M. Fuoss, *Ind. Eng. Chem. Anal. Edition* **1,** 125 (1929).

be determined with some accuracy the actual potentials are far from reproducible.

The oxygen electrode, although definitely proved to be irreversible,[36] has had limited application in connection with electrometric titrations.[37, 38, 39] Although by no means a precise instrument, the oxygen electrode has been resorted to for use in acid-base titrations in systems containing oxidizing agents such as permanganates and chromates or metals below hydrogen in the electrochemical series. However, such titrations may now be much more conveniently and accurately carried out with the aid of the glass electrode, described in Chapter 15.

[36] See for example:
W. T. Richards, *J. Phys. Chem.*, **32**, 990 (1928).
A. K. Goard and E. K. Rideal, *Trans. Farad. Soc.*, **19**, 740 (1924).
H. V. Tartar and V. E. Wellman, *J. Phys. Chem.*, **32**, 1171 (1928).
The first two of these include summaries of previous work and references to earlier literature.

[37] H. T. S. Britton, *J. Chem. Soc.*, **127**, 1896, 2148 (1925); **130**, 147 (1928).

[38] N. H. Furman, *J. Am. Chem. Soc.*, **44**, 2685 (1922).

[39] J. A. V. Butler and G. Armstrong, *Trans. Farad. Soc.*, **29**, 862 (1933).

Chapter 18

The Interionic Attraction Theory of Conductance of Aqueous Solutions of Electrolytes

The "classical" theory of conductance as proposed and supported by Arrhenius and his followers is outlined briefly in Chapter 3. It will be recalled that according to that theory the decrease of the equivalent conductance, Λ, of an electrolyte with increasing concentration was considered to be due to a decrease in the relative number of ions, the proportion being given by the ratio

$$\alpha = \frac{\Lambda}{\Lambda_0} \quad (1)$$

in which α, according to this theory, is the "degree of dissociation," and Λ_0 is the equivalent conductance at infinite dilution. It was also shown that the tacit assumption involved in equation (1) is that the mobilities of the ions do not change with the concentration. According to the Arrhenius theory a 0.001 normal solution of potassium chloride, for instance, is about 98 per cent dissociated, and a 0.1 normal solution 86 per cent, in the form of ions. With the Debye-Hückel theory, however, it has been found that the thermodynamic properties of aqueous solutions of strong electrolytes may be more readily accounted for if, as is shown in Chapters 7 and 8 of this book, it is assumed that such electrolytes are substantially completely dissociated in solution. Since strong electrolytes can obviously not be assumed to be completely dissociated for thermodynamic properties and incompletely dissociated to explain conductance, it is of interest to see whether interionic attractions and repulsions, which are the basis for the Debye-Hückel thermodynamic theory, will not also be of service in explaining the conductance phenomena. As a matter of fact the assumption made by Arrhenius and his followers, that highly charged ions, in close proximity, have no influence on one anothers' properties, was the object of early attack by opponents of the ionic theory and was the source of the real weakness of the ionic theory in the form given it by Arrhenius.

Since, then, it is not possible to account for the decrease of the equivalent conductance of strong electrolytes with increasing concentration by postulating a change of the number of ions carrying current, that decrease must be looked for in a diminution of the mobilities of the ions. According to Debye and Hückel[1] and Onsager[2] interionic attractions and repulsions lead to two "effects" both of which result in the lowering of ionic mobilities with increasing ion concentrations. These are the *electrophoretic effect* and the *time of relaxation effect* which will be discussed in the order given.

The Electrophoretic Effect. According to the Debye-Hückel theory an ion is surrounded by an "ionic atmosphere" distributed with radial symmetry around the ion as center. This ion atmosphere, it will be recalled, is due to the fact that interionic attractions and repulsions tend to produce a slight preponderance of negative ions in the vicinity of a positive ion, and *vice versa*. Although the ion atmosphere is treated as a reality in mathematical discussions it actually is the result of a time average of a distribution of the ions. Each ion serves as a center of an ion atmosphere, and the relative position of each ion with respect to the other charged bodies in the solution influences the atmospheres of all the other ions.

One result of the presence of an ionic atmosphere is that there is a potential field distributed with radial symmetry around the ion, the value of the potential being given by the formula, (14) of Chapter 7, for a singly charged ion, and equation (33a) of the same chapter, *i. e.*,

$$\psi = \frac{z_i\epsilon}{D}\,\frac{e^{\kappa a_i}}{1 + \kappa a_i}\cdot\frac{e^{-\kappa r}}{r} \tag{2}$$

for a multiply charged ion. From this equation and the Poisson relation

$$\frac{1}{r^2}\,\frac{d}{dr}\left(r^2\,\frac{d\psi}{dr}\right) = -\,\frac{4\pi}{D}\,\rho \tag{3}$$

the corresponding distribution of the space charge, ρ, in the ion atmosphere as a function of the distance, r, may be obtained by eliminating ψ between equations (2) and (3). The resulting expression is

$$\rho = -\,\frac{z_i\epsilon\, e^{\kappa a_i}}{4\pi(1 + \kappa a_i)}\,\kappa^2\,\frac{e^{-\kappa r}}{r} \tag{4}$$

If a potential gradient of intensity E is impressed on the solution a selected ion will tend to move with a velocity that will be denoted by v_i. This velocity is independent of the presence of other ions, and is determined by the limiting mobility U_0 of the ion constituent. The ion atmosphere, however, being of opposite charge to the ion itself will

[1] P. Debye and E. Hückel, *Physik. Z.*, **24**, 185, 305 (1923).
[2] L. Onsager, *ibid.*, **27**, 388 (1926); **28**, 277 (1928).

tend to move in a reverse direction. Each element of the ion atmosphere will be acted on by a force F per unit volume given by the relation

$$\mathrm{F} = \rho\,\mathrm{E} \tag{5}$$

which with equation (4) gives

$$\mathrm{F} = -\,\mathrm{E}\,\frac{z_i\epsilon\, e^{\kappa a_i}}{4\pi(1 + \kappa a_i)}\,\kappa^2\,\frac{e^{-\kappa r}}{r} = -\,A\,\frac{e^{-\kappa r}}{r} \tag{6}$$

in which A is a constant for a given solution and potential gradient. Around the selected ion the atmosphere may be considered to be arranged, as is shown in Fig. 1, in spherical shells of thickness, dr,

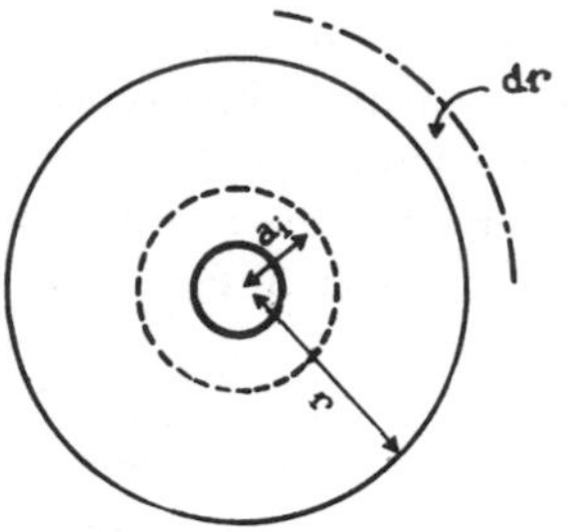

Fig. 1. Electrophoretic Effect.

in each of which the force per unit volume F is constant. The total force dF acting on such a shell will be

$$dF = 4\pi r^2\,\mathrm{F}\,dr \tag{7}$$

As a result of this force the shell and contents, including the ion, will tend to move with a velocity dv_2 in a direction contrary to the motion of the ion.

The velocity of a sphere of radius r in a fluid of viscosity, η, moving under a force F, is given by Stokes' Law [3]

$$v = \frac{F}{6\pi\eta r} \tag{8}$$

for a derivation of which, treatises on hydrodynamics may be consulted. It is strictly applicable only to a sphere moving in a medium of infinite extent, made up of particles of a smaller order of magnitude than the sphere. From equations (7) and (8) therefore the contribution, dv_2, of a spherical shell of thickness, dr, to the velocity v_2 will be

$$dv_2 = -\,\frac{2A}{3\eta}\,e^{-\kappa r}\,dr \tag{9}$$

[3] G. G. Stokes, *Cambridge Phil. Soc. Trans.*, 9, 5 (1856); see also Weyssenhoft, *Ann. Physik.*, 62, 1 (1920); H. Lamb, "Hydrodynamics," 6th ed., Cambridge University Press, 1932.

Integrating this expression between the distance of closest approach a_i and infinity and substituting the value of A from equation (6) yields

$$v_2 = -\frac{E\epsilon z_i}{6\pi\eta} \cdot \frac{\kappa}{1 + \kappa a_i} \tag{10}$$

The resultant velocity of the ion will be, for solutions dilute enough so that κa_i is small compared with unity,

$$v = v_1 + v_2 = v_1 - \frac{E\epsilon z_i \kappa}{6\pi\eta} \tag{11}$$

The factors determining the velocity v_1 are, so far, unknown. The velocity, v_1, however, determines the limiting conductance, λ_0, of an ion. The velocity v_2, which has an opposite sign to that of v_1 is the retardation due to the electrophoretic effect.

As already mentioned it has been customary to deal with the ionic atmosphere as if it were a reality, and the derivation just given assumes that an electric force acting on the ion atmosphere will produce a motion of the solvent. However, the effect of a potential gradient cannot be directly on the solvent, but must have its influence indirectly through the ions. The fundamental explanation of the electrophoretic effect must therefore be sought in a modification of inter-reactions between ions and solvent produced by the ion atmosphere, and the latter is, as we have seen, due in turn to a time average of the distribution of the ions.

It is of interest to estimate the influence of the electrophoretic effect on the equivalent conductance of typical electrolytes. From equation (11) for a singly charged ion

$$v_2 = -\frac{E\epsilon\kappa}{6\pi\eta} \tag{12}$$

Therefore for a potential gradient E of one volt per centimeter (10^8 absolute units) utilizing[4] $\kappa = \beta\sqrt{C} = 0.3286 \times 10^8 \cdot \sqrt{C}$ at 25°, the value of the viscosity of water, $\eta = 0.008949$ poise,[5] also at 25°, and converting to electromagnetic units with the constant 2.9979×10^{10} we obtain

$$v_2 = 0.00030995\sqrt{C}$$

For an equivalent of ion, consisting of **N** ions, each carrying a charge ϵ, the decrease, $\Delta\lambda$, of the equivalent conductance of an ion constituent,[6]

[4] Values of β are given in Table I, Chapter 7.
[5] *International Critical Tables,* 5, 10 (1929).
[6] For a definition of this term see equation (6a), Chapter 4.

λ, due to the electrophoretic effect at 25° will be

$$\Delta\lambda = -\mathbf{N}\,0.00030995\,\epsilon\sqrt{C} = -\mathbf{F}\,0.00030995\sqrt{C} \qquad (13)$$
$$= 29.92\sqrt{C}$$

The equivalent conductance, Λ, of potassium chloride at 25° decreases 2.9 units from infinite dilution to a concentration of 0.001 normal. The contribution of the electrophoretic effect to this decrease, for the two ions involved is by equation (13) 1.83 units. Thus the greater part of the decrease of the equivalent conductance of this salt is, according to these deductions, due to the electrophoretic effect.

The Time of Relaxation Effect. The other mechanism tending to decrease the equivalent conductance of an electrolyte when the ion concentration is increased is, as already mentioned, the "time of relaxation effect." The underlying assumptions made in deriving an expression for the effect are as follows. Around a selected ion the atmosphere has spherical symmetry, regions having the same values of space charge or potential being arranged in spheres with the ion as center. If the ion is suddenly moved the ion atmosphere will tend to move with it. The adjustment of the ion atmosphere to the new condition will take place rapidly, but, however, not instantaneously. An attempt to provide a picture of the effect is given in Fig. 2. If an ion, initially at the point *a*

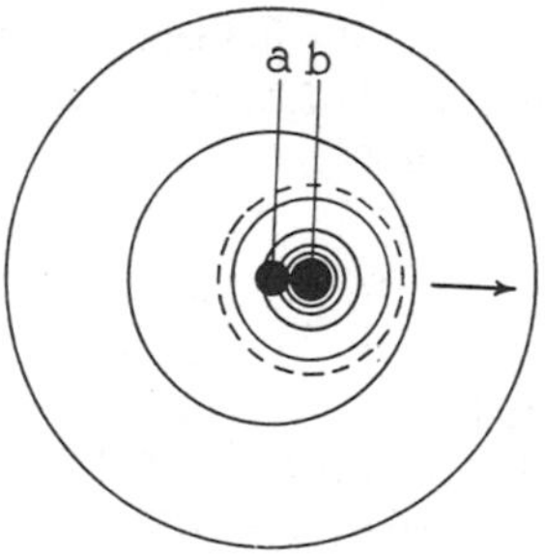

Fig. 2. Time of Relaxation Effect.

is suddenly moved to the point *b* the ion atmosphere, originally symmetrical about position *a,* tends to rearrange itself to be symmetrical about *b.* In the figure the part inside the dotted circle has adjusted to the new position of the ion, but the part outside that circle is asymmetrically arranged with respect to the ion. The result is that the adjusted portion of the atmosphere tends to exert an electrostatic attraction in a contrary direction to the motion of the ion, which is represented as being in the direction of the arrow. If an ion is moving steadily the ion atmosphere will be under the influence of a permanent distortion the effect of which is to decrease the velocity of the ion under a given external force. The effect, being purely electrostatic, is independent of the viscosity of the medium. It does, however, depend upon the limiting conductance λ_0, of the ion constituent. A derivation of the relations for the time of relaxation effect will not be included in this book as it is a quite elaborate problem in statistical mechanics. The reader is referred to the original articles.[7] Onsager's equation for the equivalent conductance, λ, of an ion constituent which has a valence z_i for a solution containing c equivalents of solute per liter is

$$\lambda = \lambda_0 - \left[\frac{0.9834 \times 10^6}{(DT)^{3/2}} w\lambda_0 + \frac{28.94\, z_i}{(DT)^{1/2}\eta}\right] \sqrt{(z^+ + z^-)c} \tag{14}$$

in which z^+ and z^- represent respectively the valence of the positive and negative ion constituents. D and η are the dielectric constant and the viscosity of the solvent and T is the absolute temperature. Also

$$w = z^+ z^- \frac{2q}{1 + \sqrt{q}}\ ; \quad q = \frac{z^+ z^-(\lambda_0^+ + \lambda_0^-)}{(z^+ + z^-)(z^+ \lambda_0^- + z^- \lambda_0^+)} \tag{15}$$

in which λ_0^+ and λ_0^- are the limiting equivalent conductances of the positive and negative ions. The first term in the brackets in equation (14) accounts for the time of relaxation effect and the second for the electrophoretic effect.

The Validity of Onsager's Equation for Aqueous Solutions. It is, of course, important to determine with what accuracy and within what range of concentration Onsager's equation, represented by expression (14), is confirmed by the results of experiments. The remainder of this chapter will be concerned with the evidence of conductance and transference measurements on aqueous solutions as to the validity of Onsager's equation. It will be well to recall that the equation refers to the *ionic part only* of a dissolved electrolyte. For a univalent ion equation (14) takes the form

$$\lambda = \lambda_0 - \left[\frac{8.147 \times 10^5}{(DT)^{3/2}} \lambda_0 + \frac{40.93}{(DT)^{1/2}\eta}\right] \sqrt{c} \tag{16}$$

which for a given solvent medium and temperature becomes

$$\lambda = \lambda_0 - [\theta\lambda_0 + \tfrac{1}{2}\sigma]\sqrt{c} \tag{17}$$

[7] L. Onsager, *Physik, Z.*, **28**, 277 (1927), **27**, 388 (1926).

in which θ and σ are constants. Thus the expression connecting the concentration c with the equivalent conductance, Λ, of a uni-univalent electrolyte, for which

$$\Lambda = \lambda^+ + \lambda^-$$

is

$$\Lambda = \Lambda_0 - [\theta\Lambda_0 + \sigma]\sqrt{c} \qquad (18)$$

For aqueous solutions at 25° the constants of equation (18) have the values $\theta = 0.2273$ and $\sigma = 59.78$. These are based on a value of 0.008949 poise [8] for the viscosity of water at that temperature and on a value of D of 78.55 for the dielectric constant of water from the equations given on page 144.

In the derivation of equation (14) a number of simplifying assumptions of a physical nature, and mathematical approximations, such as taking only the first terms of a series, were made. The expressions are therefore strictly valid only as limiting equations and may be expected to hold only for very dilute solutions. It is of interest that Kohlrausch observed, empirically, that the conductances of dilute strong electrolytes follow the relation

$$\Lambda = \Lambda_0 - K\sqrt{c} \qquad (19)$$

in which K is a constant. Onsager's equation (18) leads to an expression of this form but also predicts, for each electrolyte, the value of the constant K. Onsager showed, mainly from Kohlrausch's data, that equation (18) is the limiting relation between Λ and the concentration c for strong electrolytes.

The most recent and accurate test of the equation is, however, due to Shedlovsky.[9, 10] Using the methods for determining conductances outlined in Chapter 3, he was able to make measurements, with precision, at salt concentrations as low as 0.00003 normal. Some of his results for aqueous solutions of potassium chloride at 25° are given in Table I. A sensitive method for testing the validity of the Onsager equation is to compute values of the limiting conductance, Λ_0, with its aid, for which purpose the equation may be conveniently rearranged in the form

$$\Lambda_0 = \frac{\Lambda + \sigma\sqrt{c}}{1 - \theta\sqrt{c}} \qquad (20)$$

[8] *International Critical Tables*, **5**, 10 (1929).

[9] T. Shedlovsky, *J. Am. Chem. Soc.*, **54**, 1411 (1932).

[10] T. Shedlovsky, A. S. Brown and D. A. MacInnes, *Trans. Electrochem. Soc.*, **66**, 165 (1934).

TABLE I. THE EQUIVALENT CONDUCTANCE OF POTASSIUM CHLORIDE AT 25°

Concentration $c \times 10^4$	Equivalent conductance observed Λ	Limiting conductance calculated by equation (20) Λ_0
0.32576	149.37	149.91
1.0445	148.95	149.92
2.6570	148.42	149.95
3.3277	148.23	149.94
3.5217	148.16	149.92
4.6948	147.93	149.97
6.0895	149.56	149.88
8.4200	147.27	150.00
9.2856	147.11	149.97
11.321	146.80	149.96
14.080	146.50	150.02
15.959	146.30	150.05
20.291	145.76	149.99
20.568	145.75	150.01
23.379	145.52	150.06
27.848	145.04	150.00
28.777	145.03	150.07
32.827	144.68	150.06

The values of Λ_0 computed in this way from Shedlovsky's measurements are given in the last column of Table I. It will be seen that the values are constant within a small limit of error up to a concentration, c, of about 0.0014 mol per liter, above which there is a definite deviation. The range and precision of the agreement between the theory and the measurements is also shown in Fig. 3, based on Shedlovsky's investiga-

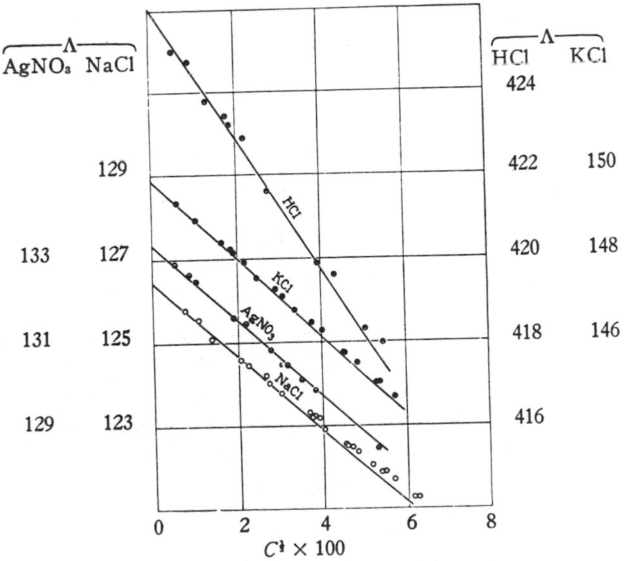

Fig. 3. The Equivalent Conductance of Some Uni-univalent Electrolytes in Very Dilute Solutions at 25°.

tions, in which the equivalent conductance, Λ, of a number of typical uni-univalent strong electrolytes is plotted with the square root of the concentration as abscissae. A plot of Onsager's equation is also shown for each electrolyte which for the examples under consideration is a straight line with a slope equal to $(\theta\Lambda_0 + \sigma)$. It is evident that in each case this line clearly represents the observations, within the small experimental error, up to a concentration of about 0.001 mol per liter, above which concentration there is, in most cases, a slight but definite upward deviation.

A further important test of Onsager's expression, equation (14), is of its validity when used with conductance data on salts of the higher valence types. For aqueous solutions of a bi-univalent electrolyte, such as calcium chloride at 25°, this equation may be readily seen, by using the values for the dielectric constant and viscosity already given, to reduce to

$$\Lambda = \Lambda_0 - \left(\frac{1.2673\Lambda_0}{1 + t_0^- + 0.8165\sqrt{1 + t_0^-}} + 109.819\right)\sqrt{c} \qquad (21)$$

in which the limiting transference number, t_0^-, is equal to the ratio λ_0^-/Λ_0 and c is the concentration in equivalents per liter. For a salt of tri-univalent type such as lanthanum chloride the corresponding equation is

$$\Lambda = \Lambda_0 - \left(\frac{9}{1 + 2t_0^- + 0.8660\sqrt{1 + 2t_0^-}}\Lambda_0 + 169.08\right)\sqrt{c} \qquad (22)$$

The method for obtaining the limiting transference numbers, t_0^-, will be discussed later in this chapter. For calcium and lanthanum chlorides at 25° equations (21) and (22) become respectively

$$\Lambda = \Lambda_0 - (0.49073\Lambda_0 + 109.82)\sqrt{c} \qquad (23)$$

$$\Lambda = \Lambda_0 - (0.75131\Lambda_0 + 169.08)\sqrt{c} \qquad (24)$$

In Fig. 4 the results of recent determinations of the equivalent conductance, Λ,[11, 12, 13, 14] of the three different valence types are plotted as functions of the square root of the equivalent concentration and, for comparison, the straight lines representing the corresponding Onsager equations. It will be observed that the equation in each of these cases accurately represents the data for the more dilute solutions.

[11] T. Shedlovsky, A. S. Brown and D. A. MacInnes, *Trans. Electrochem. Soc.*, **66**, 165 (1934).

[12] T. Shedlovsky and A. S. Brown, *J. Am. Chem. Soc.*, **56**, 1066 (1934).

[13] G. Jones and C. F. Bickford, *ibid.*, **56**, 602 (1934).

[14] T. Shedlovsky, *private communication.*

The evidence just given, which is typical of that obtained from all recent measurements, shows that the Onsager equation is valid for very dilute aqueous solutions of strong electrolytes. This fact is important as it lends additional and strong support to the correctness and utility of the interionic attraction theory. As has already been emphasized Onsager's equation is a limiting equation and deviations from it, even for completely dissociated electrolytes, are to be expected as the concentration is increased.

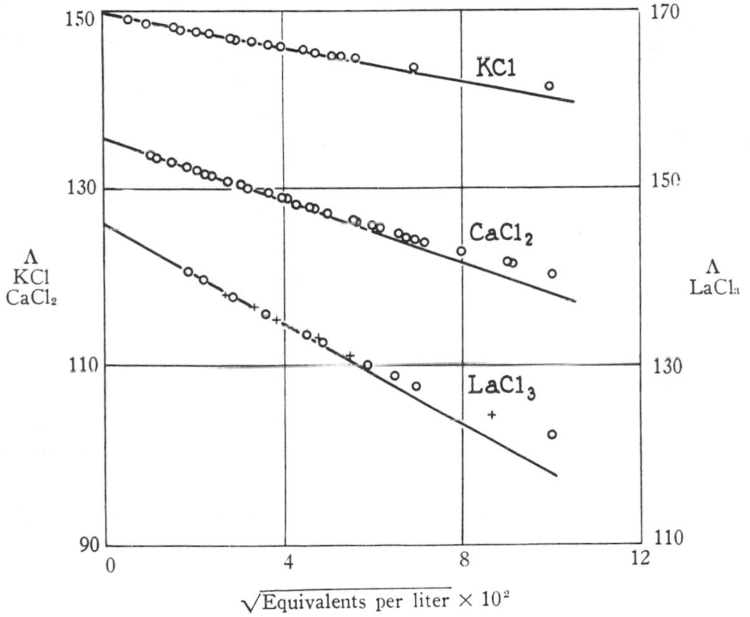

Fig. 4. The Equivalent Conductance of Different Valence Type Chlorides at 25°. +Jones and Bickford. o Shedlovsky.

Evidence from Transference Numbers for the Correctness of the Onsager Equation. An important and quite sensitive test of the validity of the Onsager equation, and of the ideas underlying it, is afforded by transference number measurements. It will be recalled from the discussion in Chapter 3, that the Arrhenius theory of ionization assumes that ions have the same mobilities at all concentrations. The transference number, t^+, of the cation constituent of a binary electrolyte is equal to

$$t^+ = \frac{U^+}{U^+ + U^-} \tag{25}$$

in which U^+ and U^- are the ion constituent mobilities. Thus if mobilities are independent of the concentration the transference number must be constant also. As may be seen by reference to Table IV of Chapter 4, transference numbers are not constant but change with concentration, a fact which, in itself, is sufficient to throw doubt on the Arrhenius theory. The interionic attraction theory, on the other hand, requires decreases of ion mobilities with increasing ion concentrations. This leads, as the following discussion will show, to changes in the transference number with concentration, even for very dilute salt solutions. With equation (6a) of Chapter 4, equation (25) becomes

$$t^+ = \frac{\lambda^+}{\lambda^+ + \lambda^-} \tag{26}$$

which for uni-univalent electrolytes may with equation (17) be put in the form

$$t^+ = \frac{\lambda_0^+ - (\theta\lambda_0^+ + \frac{1}{2}\sigma)\sqrt{c}}{\lambda_0^+ + \lambda_0^- - [\theta(\lambda_0^+ + \lambda_0^-) + \sigma]\sqrt{c}} \tag{27}$$

This equation shows that even for solutions dilute enough for the Onsager equation to hold, the transference numbers should in general change with the concentration, if the interionic attraction theory is valid.

By differentiating equation (27) with respect to $\sqrt{c}$, and including the condition that c approaches the value zero, the expression

$$\left(\frac{dt}{d\sqrt{c}}\right)_{c\to 0} = \frac{t_0^+ - 0.5}{\Lambda_0}\,\sigma \tag{28}$$

is obtained, in which $t_0^+ = \lambda_0^+/(\lambda_0^+ + \lambda_0^-)$. It follows therefore that a plot of transference numbers of uni-univalent electrolytes against $\sqrt{c}$ should enter the axis of zero concentration with a slope given by equation (28). The limiting slope should thus be proportional to the deviation of t_0^+ from 0.5 and inversely proportional to the limiting equivalent conductance, Λ_0, of the electrolyte.

A plot of transference numbers of uni-univalent electrolytes is shown in Fig. 5 and is based on Longsworth's data given in Table IV, Chapter 4. The scale of ordinates is the same for all the electrolytes but the ordinates themselves have been shifted to make the plot more compact. The circles represent the experimental results and the smooth curves have been drawn through them to the limiting value t_0^+ of the transference number, obtained as will be shortly described. The straight line starting from t_0^+ has in each case the limiting slope given by equation (28). It will be seen that for the four chlorides the curve through the experimental points evidently merges into the line having the limiting slope. Thus for these substances the interionic attraction theory predicts the sign and, for very dilute solutions, the magnitude of the change of the transference number with the concentration. Of the

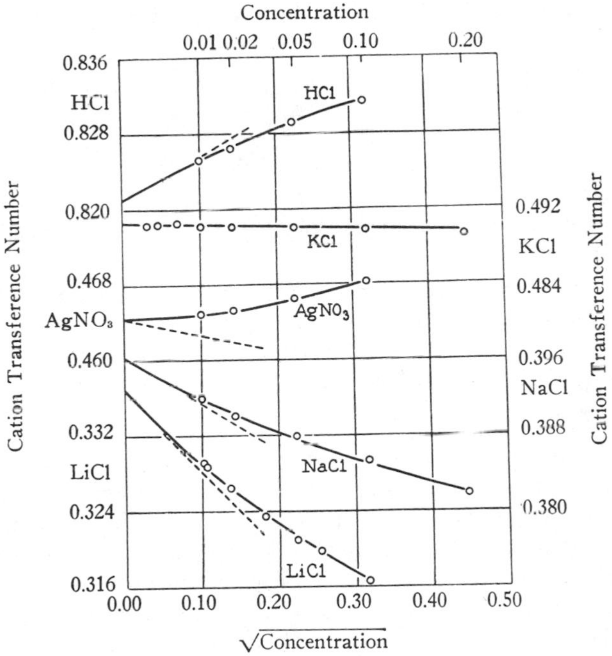

Fig. 5. The Transference Numbers of Uni-univalent Electrolytes as a Function of the Square Root of the Concentration.

other electrolytes, for which data are available, sodium acetate shows much the same agreement with the theory as that shown by hydrochloric acid. The cation transference numbers of ammonium chloride and potassium bromide and iodide change little with concentration, and have a minimum value at a concentration of about c equals 0.05 equivalent per liter. However the variation in dilute solutions is closely that predicted by equation (28).

The transference numbers of certain electrolytes, however, show variations from the predictions of the theory. The data for silver nitrate are plotted in Fig. 5 and it is seen that the variation of the transference number of that substance with concentration is in the opposite direction from that required by the theory. A somewhat similar behavior is observed with the data on potassium nitrate, the observed values changing much more rapidly with increasing concentration than would be expected from equation (28). However, as will be seen from a study of the conductance measurements, nitrates are somewhat abnormal.

The variation of the cation transference number with the concentration for electrolytes of higher valence type is given by the equation[15]

$$\lim_{c \to 0}\left(\frac{dt_0^+}{d\sqrt{c}}\right) = \frac{(z^+ + z^-)\,t_0^+ - z^+}{\Lambda_0}\,\sigma\,\sqrt{z^+ + z^-} \qquad (28a)$$

in which z^+ and z^- are the valences of the positive and negative ions. Data suitable for testing this equation are available for calcium and lanthanum chlorides and sodium sulphate. Rough qualitative agreement, only, with the relation is observed.

It must be emphasized that the transference number measurements are at concentrations at which the Onsager equation, on which expressions (28) and (28a) are based, is only approximately valid. In general the transference data lend strong support to the interionic attraction theory of electrolytic conductance.

Empirical Extensions of the Onsager Equation to Higher Concentrations. As has been explained, the Onsager equation is strictly valid only as a limiting expression. In the derivation of the equation higher terms in mathematical series were neglected, and such complications as interactions between the electrophoretic and time of relaxation effects were not considered. Onsager[7] found that agreement with the experimental data on conductance measurements can be extended to somewhat higher concentrations than with the limiting equation by the use of the modified formula

$$\Lambda = \Lambda_0 - (\theta\Lambda_0 + \sigma)\sqrt{c} + bc \qquad (29)$$

in which b is an empirical constant. A more useful equation has, however, been suggested by Shedlovsky[16] who observed that for most strong uni-univalent electrolytes values of Λ_0', computed from the Onsager equation in the form

[15] L. G. Longsworth, *J. Am. Chem. Soc.*, **57**, 1185 (1935).
[16] T. Shedlovsky, *J. Am. Chem. Soc.*, **54**, 1405 (1932).

$$\Lambda'_0 = \frac{\Lambda + \sigma\sqrt{c}}{1 - \theta\sqrt{c}} \tag{30}$$

are linear with respect to the concentration c. In the concentration region in which Onsager's equation holds Λ'_0 is obviously equal to Λ_0. Under other conditions Λ'_0 varies with c. A plot of Λ'_0 values for solutions of sodium and potassium chlorides and hydrochloric acid as ordinates against the concentration c is shown in Fig. 6 and is based

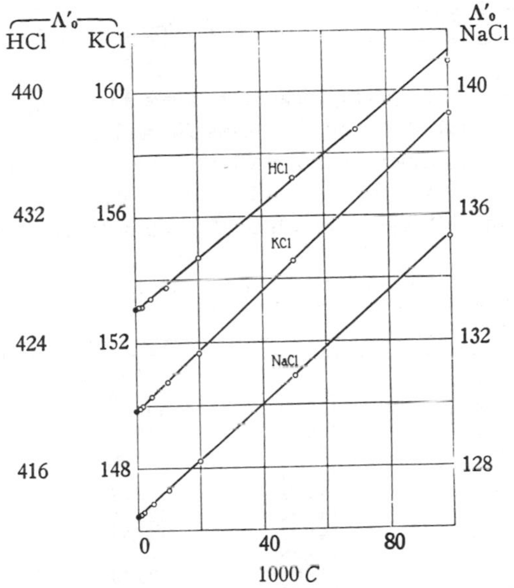

Fig. 6. Plot of Λ_0' for Three Uni-univalent Electrolytes.

on Shedlovsky's data, given in Table III. It is evident that the plots for these substances are, very nearly at least, straight lines. Shedlovsky's observation may be put in the form of the equation

$$\frac{\Lambda + \sigma\sqrt{c}}{1 - \theta\sqrt{c}} = \Lambda_0 + Bc \tag{31}$$

in which B is an empirical constant, which, incidentally, has a value not far from the factor $(\theta\Lambda_0 + \sigma)$ in equation (18), a fact which is frequently of service in preliminary computations and in interpolations.

For the conductance of solutions of the greater number of the strong uni-univalent electrolytes equation (31) expresses the data very nearly within the experimental error. It does not, however, serve for the data on nitrates and chlorates. We have already seen that nitrates are abnormal with respect to the relation of transference numbers to the concentration. Furthermore the conductance data on solutions of higher valence type, such as calcium chloride, lanthanum chloride, or sodium sulphate, show pronounced deviations from linearity when plotted as is shown in Fig. 6. In certain cases the data can be accurately expressed by the addition of terms to equation (31) giving

$$\frac{\Lambda + \sigma\sqrt{c}}{1 - \theta\sqrt{c}} = \Lambda_0 + Bc + Dc \log c - Ec^2 \tag{32}$$

Although this equation is empirical it reduces to Onsager's equation for very small values of c, and the terms Bc and $Dc \log c$ are of the form required by consideration of factors neglected in the derivation of the simple equation.[7, 17] The utility of an equation of the form of (32) is shown in Table II in which the conductance data of Shed-

TABLE II. OBSERVED AND COMPUTED VALUES OF THE EQUIVALENT CONDUCTANCE OF POTASSIUM AND SODIUM CHLORIDES IN WATER AT 25° C.

Concentration, c	KCl Equivalent Conductance, Λ		NaCl Equivalent Conductance, Λ	
	Observed	Calculated	Observed	Calculated
0	149.86	149.86	126.45	126.45
0.0001	148.94	148.92	125.59	125.57
0.0002	148.57	148.55	125.24	125.22
0.0005	147.81	147.79	124.50	124.50
0.001	146.95	146.94	123.74	123.74
0.002	145.78	145.77	122.67	122.68
0.003	144.91	144.92	121.88	121.89
0.004	144.19	144.21	121.21	121.22
0.005	143.55	143.58	120.65	120.66
0.006	143.02	143.04	120.15	120.17
0.007	142.52	142.54	119.69	119.71
0.008	142.06	142.08	119.26	119.29
0.009	141.65	141.67	118.86	118.88
0.01	141.27	141.28	118.51	118.53
0.02	138.34	138.32	115.76	115.77
0.03	136.27	136.25	113.83	113.81
0.04	134.67	134.64	112.34	112.30
0.05	133.37	133.35	111.06	111.03
0.06	132.30	132.28	110.01	109.97
0.07	131.33	131.34	109.06	109.02
0.08	130.44	130.45	108.23	108.18
0.09	129.65	129.65	107.45	107.42
0.10	128.96	128.96	106.74	106.74
0.11	128.29	128.29	......	......
0.12	127.69	127.65	105.48	105.50
0.14	......	......	104.42	104.42
0.16	......	......	103.43	103.44
0.18	......	......	102.52	102.54
0.20	......	......	101.70	101.68
0.22	......	......	100.96	100.92

[17] L. Onsager and R. Fuoss, *J. Phys. Chem.*, **36**, 2689 (1932).

lovsky[10] on sodium and potassium chloride at 25° at a series of concentrations are compared with the results of computations using the semi-empirical equation

$$\frac{\Lambda + 59.78\sqrt{c}}{1 - 0.2273\sqrt{c}} = 149.86 + 141.9c + 29.24c \log c - 180.6c^2 \quad (33)$$

for the data on potassium chloride and

$$\frac{\Lambda + 59.78\sqrt{c}}{1 - 0.2273\sqrt{c}} = 126.45 + 95.79c - 65.29c^2 \quad (34)$$

for sodium chloride. It will be seen that the agreement is excellent throughout the concentration range studied. A similar equation for the equivalent conductance of hydrochloric acid at 25° is

$$\frac{\Lambda + 59.78\sqrt{c}}{1 - 0.2273\sqrt{c}} = 426.16 + 169c \quad (35)$$

This will be found useful in computations given later in this chapter.

The Limiting Conductance, Λ_0, of Aqueous Solutions of Electrolytes. Since the value of the limiting equivalent conductance, Λ_0, was used in the computation of the degree of dissociation, α, according to the theory of Arrhenius, methods for obtaining that constant were considered from the early days of the ionic theory. Many such methods for making the necessary extrapolation, based on empirical or semi-empirical equations, have been proposed. The earlier methods are of historic interest only and have been critically discussed by Bates.[18] In spite of the fact that the simple Arrhenius theory can no longer be considered valid the evaluation of Λ_0 retains interest, as it is, as we have just seen; an important constant in Onsager's equation. The more recent methods for obtaining Λ_0 values are based on the assumption that the Onsager equation holds for very dilute solutions of strong electrolytes.

When accurate data on equivalent conductances are available at very low concentrations a simple graphical method for obtaining Λ_0 is, following Kohlrausch's early suggestion, to plot values of Λ as a function of the square root of the concentration and extend to the zero axis. The resulting line for strong electrolytes at concentrations below about 0.002 equivalent per liter has, in general, very nearly the slope required by the Onsager equation. Such plots are shown in Figs. 3 and 4. The extrapolation may of course be carried out

[18] S. J. Bates, *J. Am. Chem. Soc.*, 35, 519 (1913).

analytically using the method of least squares. A difficulty with the method just outlined is, however, that the greatest weight is given to the data obtained on very dilute solutions, where the experimental errors are greatest.

Another method, also based on the assumption that Onsager's equation is the true limiting relation, depends upon Shedlovsky's equation, which for uni-univalent electrolytes takes the form

$$\frac{\Lambda + \sigma\sqrt{c}}{1 - \theta\sqrt{c}} = \Lambda_0 + Bc \tag{31}$$

The method of extrapolation consists in plotting values of Λ'_0 defined by equation (30) against the concentration, c, and extending the line to the zero axis as shown in Fig. 6. The extrapolation may, obviously, also be carried out analytically. The values of the limiting conductance, Λ_0, given in Table III were obtained by the methods just described.

A Table of Equivalent Conductances of Electrolytes at 25°. Table III gives values of the equivalent conductances of typical electrolytes at round concentrations between 0.0005 and 0.1 normal at 25°. When not measured at the precise concentration given the values have been interpolated by means of equations of the form of (31). With the exceptions indicated by asterisks the values have been obtained with apparatus substantially as described in Chapter 3. Of these substances the alkalis present great experimental difficulties, but the values are probably correct to the number of places given. The data on nickel sulphate are included as an example of a bi-bivalent salt. All the values given are based on the conductance of 0.1 demal potassium chloride, at 25° as obtained by Jones and Bradshaw.[19]

Limiting Equivalent Conductances of the Ion Constituents in Aqueous Solution. The equivalent conductance, λ^-, of the negative ion constituent of a binary electrolyte is defined by equation (6b) of Chapter 4, as follows

$$\lambda^- = t^-\Lambda \tag{36}$$

in which t^- is the anion transference number and Λ the equivalent conductance of the salt. If salts such as sodium and potassium chloride were completely dissociated and no interionic attraction existed, the equivalent conductance of, for instance, the chloride ion constituent would be independent both of the concentration and of the nature of the positive ion with which it is associated. However, there is abundant evidence of the effects of interionic attractions so that λ values will

[19] See page 47.

TABLE III. THE EQUIVALENT CONDUCTANCES OF SOME ELECTROLYTES AT 25°

Electrolyte	Concentration, Equivalents per Liter								Reference
	0.0000	0.0005	0.001	0.005	0.01	0.02	0.05	0.10	
NaCl	126.45	124.50	123.74	120.65	118.51	115.76	111.06	106.74	*1*
KCl	149.86	147.81	146.95	143.55	141.27	138.34	133.37	128.96	*1*
LiCl	115.03	113.15	112.40	109.40	107.32	104.65	100.11	95.86	*2,3*
HCl	426.16	422.74	421.36	415.80	412.00	407.24	399.09	391.32	*2*
NH_4Cl	149.7	.....	.....	.....	141.28	138.33	133.29	128.75	*4,5*
KBr	151.9	.....	.....	146.09	143.43	140.48	135.68	131.39	*6*
KI	150.3_8	.....	.....	144.37	142.18	139.45	134.97	131.11	*4*
NaI	126.94	125.36	124.25	121.25	119.24	116.70	112.79	108.78	*7*
NaO_2CCH_3	91.0	89.2	88.5	85.72	83.76	81.24	76.92	72.80	*8,4*
$NaO_2CCH_2CH_3$	85.92	84.24	83.54	80.90	79.05	76.63	.....	.....	*9*
$NaO_2C(CH_2)_2CH_3$	82.70	81.04	80.31	77.58	75.76	73.39	69.32	65.27	*9*
KNO_3	144.96	142.77	141.84	138.48	132.82	132.41	126.31	120.40	*2*
$KHCO_3$	118.00	116.10	115.34	112.24	110.08	107.22	.....	.....	*10*
$AgNO_3$	133.36	131.36	130.51	127.20	124.76	121.41	115.24	109.14	*2*
NaOH*	248	246	245	240	237	233	227	221	*11,12,13*
$\frac{1}{2}CaCl_2$	135.84	131.93	130.36	124.25	120.36	115.65	108.47	102.46	*14*
$\frac{1}{2}BaCl_2$	139.98	135.96	134.34	128.02	123.94	119.09	111.48	105.19	*14*
$\frac{1}{2}SrCl_2$	135.80	131.90	130.33	124.24	120.29	115.54	108.25	102.19	*14*
$\frac{1}{2}MgCl_2$	129.40	125.61	124.11	118.31	114.55	110.04	103.08	97.10	*14*
$\frac{1}{2}Ca(OH)_2$*	258[17]	.....	.....	233	226	214	.....	.....	*15,16*
$\frac{1}{2}Na_2SO_4$	129.9	125.74	124.15	117.15	112.44	106.78	97.75	89.98	*4,18*
$\frac{1}{2}NiSO_4$*	....	118.7	113.1	93.2	82.7	72.3	59.2	50.8	*19*
$\frac{1}{3}LaCl_3$	145.9	139.6	137.0	127.5	121.8	115.3	106.2	99.1	*18,20*
$\frac{1}{4}K_4Fe(CN)_6$	184	.....	167.24	146.09	134.83	122.82	107.70	97.87	*21*

1. Shedlovsky, Brown and MacInnes, *Trans. Electrochem. Soc.*, **66**, 165 (1934).
2. T. Shedlovsky, *J. Am. Chem. Soc.*, **54**, 1411 (1932).
3. MacInnes, Shedlovsky and Longsworth, *ibid.*, **54**, 2758 (1932).
4. L. G. Longsworth, *private communication.*
5. L. G. Longsworth, *J. Am. Chem. Soc.*, **57**, 1185 (1935).
6. L. G. Longsworth, *private communication*, see also (20).
7. P. A. Lasselle and J. G. Aston, *J. Am. Chem. Soc.*, **55**, 3067 (1933).
8. D. A. MacInnes and T. Shedlovsky, *ibid.*, **54**, 1429 (1932)
9. D. Belcher and T. Shedlovsky, *private communication.*
10. Shedlovsky and MacInnes, *J. Am. Chem. Soc.*, **57**, 1705 (1935).
11. G. H. Jeffery and A. I. Vogel, *Phil. Mag.*, **17**, 582 (1934).
12. M. Randall and C. C. Scalione, *J. Am. Chem. Soc.*, **49**, 1486 (1927)
13. J. Goworecka and M. Hlasko, *Roczniki. Chem.*, **12**, 403 (1932).
14. T. Shedlovsky and A. S. Brown, *J. Am. Chem. Soc.*, **56**, 1066 (1934).
15. T. Noda and A. Miyoshi, *J. Soc. Chem. Ind. Japan*, **35**, *Suppl. Bndg.*, 317 (1932).
16. F. M. Lea and G. E. Bessey, *J. Chem Soc*, **1937**, 1612.
17. From sum of ion mobilities.
18. T. Shedlovsky, *private communication.*
19. K. Murata, *Bull. Chem. Soc. (Japan)*, **3**, 47 (1928).
20. G. Jones and C. F. Bickford, *J. Am. Chem. Soc.*, **56**, 602 (1934).
21. G. Jones and F. C. Jelen, *J. Am. Chem. Soc.*, **58**, 2561 (1936).

be expected to change with concentration even in dilute solutions. The data given in Table IV, Chapter 4, and Table III of this chapter, on solutions of chlorides are available for the study of the effect of the nature of the ion of opposite charge on the equivalent conductance of the chloride ion. From these data values of the product $\lambda_{Cl} = t_{Cl}\Lambda$ are given in Table IV. It will be seen that the λ_{Cl} values change

TABLE IV. EQUIVALENT CONDUCTANCES OF THE CHLORIDE ION CONSTITUENT FROM DIFFERENT ELECTROLYTES AT 25°

Electrolyte	Concentration, equivalents per liter 0.001	0.002	0.005	0.01	0.02	0.05	0.10
KCl	74.88	74.28	73.22	72.07	70.56	68.03	65.79
NaCl	.....	.....	.....	72.05	70.54	67.92	65.58
HCl	.....	.....	.....	72.06	70.62	68.16	65.98
LiCl	.....	.....	.....	72.02	70.52	67.96	65.49

markedly with the concentration. With c at 0.01 equivalent per liter, which is the lowest concentration for which there are data for more than one electrolyte, the values of λ_{Cl} agree, almost within experimental error. At higher concentrations, however, these values show decided deviations, increasing as the concentration increases. The data tend to show that as the salt concentration is lowered the differences between the λ_{Cl} values from different salts tend to decrease and disappear completely in the limit. A similar table of values of the equivalent conductances of the potassium ion may be prepared from the data in Table IV, Chapter 4, and Table III, of this chapter. Such a table would show that although λ_K values are different at higher concentrations they also approach the same value in dilute solutions. From these results and others it is possible to come to the conclusion that at infinite dilution the equivalent conductance, λ_0, values are a property of the ion constituent alone and are independent of the associated ions. This is known as *Kohlrausch's law of the independent mobility of ions.*

A sensitive method for obtaining the limiting value of the conductance of chloride ion constituent, $\lambda_{0_{Cl}}$, is shown in Fig. 7. Here values of $\lambda'_{0_{Cl}}$ defined by

$$\lambda'_{0_{Cl}} = \frac{t_{Cl}\Lambda + \frac{1}{2}\sigma\sqrt{c}}{1 - \theta\sqrt{c}} = \frac{\lambda_{Cl} + \frac{1}{2}\sigma\sqrt{c}}{1 - \theta\sqrt{c}} \tag{37}$$

are plotted as functions of the concentration c. It will be seen that the curves through the points for each electrolyte are very nearly straight lines and that they converge to the same point on the zero axis. A graphical extrapolation leads to a value of

$$\lambda_{0_{Cl}} = 76.34 \tag{38}$$

the greatest weight being given to the values of $\lambda'_{0_{Cl}}$ from potassium chloride, since the measurements extend to very dilute solutions.

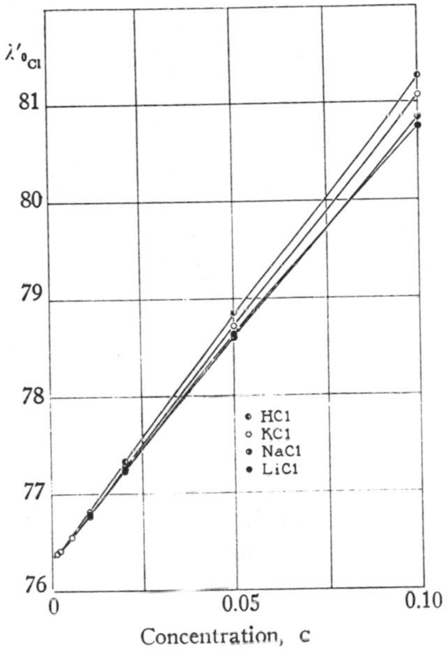

Fig. 7. Values of λ'_0 for the Chloride Ion Constituent as Functions of the Concentration.

According to Kohlrausch's law, for which experimental evidence has just been given, the limiting equivalent conductances, Λ_0, of the salts are additive functions of the limiting ion conductances, λ_0. The values of $\lambda_{0_{Cl}}$ may be, therefore, used to obtain the limiting conductances of the cations of all chlorides for which values of the limiting conductance, Λ_0, are available, and the values for the cations may be, in turn, used to obtain the λ_0 values for other anions. From the Λ_0 values in Table III, the limiting ion conductances at 25° have thus been collected in Table V. With this table it is possible to obtain by addition the limiting equivalent conductances of salts on which conductance data are not available, but whose positive and negative ions appear in the table. This table can also be used to obtain Λ_0 values of certain weak electrolytes such as acetic acid and carbonic acid. The limiting conductances for such electrolytes cannot be determined by direct measurements of conductance, since the extrapolation from such data is inaccurate or impossible.

TABLE V. LIMITING ION CONDUCTANCES[1] AT 25°

Cation	λ^+_0	Anion	λ^-_0
K^+	73.52	Cl^-	76.34
Na^+	50.11	Br^-	78.4
H^+	349.82	I^-	76.8_5
Ag^+	61.92	NO_3^-	71.44
Li^+	38.69	HCO_3^-	44.48
NH_4^+	73.4	OH^-	198
Tl^+ [2]	74.7	$CH_3CO_2^-$	40.9
$\frac{1}{2}Ca^{++}$	59.50	$CH_2ClCO_2^-$ [4]	39.7
$\frac{1}{2}Ba^{++}$	63.64	$CH_3CH_2CO_2^-$ [4]	35.81
$\frac{1}{2}Sr^{++}$	59.46	$CH_3(CH_2)_2CO_2^-$	32.59
$\frac{1}{2}Mg^{++}$	53.06	ClO_4^- [2]	68.0
$\frac{1}{3}La^{+++}$	69.6	$C_6H_5CO_2^-$ [5,6]	32.3
$\frac{1}{3}Co(NH_3)_6^{+++}$ [3]	102.3	$\frac{1}{2}SO_4^{=}$	79.8
....	......	$\frac{1}{3}Fe(CN)_6^{---}$ [2]	101.0
....	......	$\frac{1}{4}Fe(CN)_6^{----}$	110.5

[1] Values, except where otherwise indicated, were computed from data in Table III.
[2] R. A. Robinson and C. W. Davies, *J. Chem. Soc.*, **1937**, 574.
[3] G. S. Hartley and G. W. Donaldson, *Trans. Farad. Soc.*, **33**, 457 (1937).
[4] T. Shedlovsky, A. S. Brown, D. A. MacInnes, *Trans. Electrochem. Soc.*, **66**, 165 (1935).
[5] B. Saxton and H. F. Meier, *J. Am. Chem. Soc.*, **56**, 1918 (1934).
[6] F. G. Brockman and M. Kilpatrick, *J. Am. Chem. Soc.*, **56**, 1483 (1934).

Another use of the constants given in Table V is to obtain limiting transference numbers, the values for the positive and negative ion constituents, $t_0^{\pm}$, being given by the relations

$$t_0^+ = \frac{\lambda_0^+}{\lambda_0^+ + \lambda_0^-} \tag{39}$$

$$t_0^- = \frac{\lambda_0^-}{\lambda_0^+ + \lambda_0^-} \tag{39a}$$

The Conductance of Aqueous Solutions of Weak Electrolytes. In the foregoing portion of this chapter it has been shown that the interionic attraction theory on which Onsager's equation is based is of great utility in the interpretation of data on strong electrolytes. As has been repeatedly mentioned, it has not been found necessary, in dealing with such electrolytes, to introduce the assumption that they are incompletely dissociated. There is, however, a large group of electrolytes, of which acetic acid is typical, for which it is not possible to assume complete ionization of the aqueous solutions, and it is thus necessary to complicate the treatment by introducing a degree of dissociation. In dealing with data on such solutions the original Arrhenius theory was, as has been indicated in Chapter 3, moderately successful. It will be shown, however, that this apparent agreement with the early theory is due, from our present point of view, to a partial compensation of two errors.

Aqueous solutions of salts are nearly all strong electrolytes. The weak electrolytes, with few exceptions, are acids and bases. The weak acids include the acids of the less strongly electronegative elements and practically all organic acids. The weak bases include a few inorganic compounds such as ammonia, and a large number of organic compounds derivable from ammonia or ammonia analogues.

In a solution of a weak acid which may be represented by HA (A being the negative radical) the equilibrium

$$\mathrm{HA} = \mathrm{H}^+ + \mathrm{A}^-$$

exists between the ions and the undissociated acid. For such an equilibrium, equation (9), Chapter 6, yields the following relation for the chemical potentials μ_{HA}, etc.

$$\mu_{\mathrm{HA}} = \mu_{\mathrm{H}} + \mu_{\mathrm{A}} \tag{40}$$

which with equation (26b) of the same chapter gives

$$RT \ln \frac{C_{\mathrm{H}}C_{\mathrm{A}}}{C_{\mathrm{HA}}} \frac{f_{\mathrm{H}}f_{\mathrm{A}}}{f_{\mathrm{HA}}} = \mu^{\circ}_{\mathrm{HA}} - \mu^{\circ}_{\mathrm{H}} - \mu^{\circ}_{\mathrm{A}} \tag{41}$$

Since the μ° terms are constant at a given temperature we may define the ionization constant **K** by the relation

$$\mu^{\circ}_{\mathrm{HA}} - \mu^{\circ}_{\mathrm{H}} - \mu^{\circ}_{\mathrm{Cl}} = RT \ln \mathbf{K} \tag{42}$$

which with equation (41) is

$$\frac{C_{\mathrm{H}}C_{\mathrm{A}}}{C_{\mathrm{HA}}} \frac{f_{\mathrm{H}}f_{\mathrm{A}}}{f_{\mathrm{HA}}} = \mathbf{K} \tag{43}$$

This expression differs from equation (7), Chapter 11, in the substitution of concentrations, C, for molalities, m. Both expressions are in common use due to the fact that emf measurements are usually made on solutions prepared and reported on a molality, m, basis, whereas conductance measurements are more frequently stated in terms of concentrations, C, *i.e.* mols per liter of solution.

If the solution of a monobasic acid, HA, at the concentration C, is ionized to the extent α, *i.e.* the degree of dissociation, the concentrations, C_{H}, C_{A} and C_{HA} will be respectively $C\alpha$, $C\alpha$ and $C(1 - \alpha)$ so that equation (43) becomes

$$\frac{C\alpha^2}{(1 - \alpha)} \cdot \frac{f_{\mathrm{H}}f_{\mathrm{A}}}{f_{\mathrm{HA}}} = \mathbf{K} \tag{44}$$

which is equivalent to the Ostwald dilution law, equation (17), Chapter

3, without the assumption that the constituents are all normal solutes. However, as we have seen, Arrhenius, and following him, Ostwald computed α from the ratio Λ/Λ_0, a procedure which makes no allowance for the change of ion mobilities with the changing ion concentration. As has been shown in this chapter, such changes are to be expected even in very low concentrations. A more nearly correct value of the degree of dissociation, α, than those obtained by means of the Arrhenius relation, Λ/Λ_0, may be obtained by the use of the expression

$$\frac{\Lambda}{\Lambda_e} = \alpha \tag{45}$$

in which Λ is the equivalent conductance at the concentration C and Λ_e is the equivalent conductance of the electrolyte in a completely dissociated condition, and at the ion concentration $C\alpha$.

The following method for computing Λ_e will make this conception clear. As an example the value of Λ_e for acetic acid as a function of the ion concentration will be obtained. The computation depends upon two assumptions the evidence for which has been considered in this chapter. The assumptions are (*a*) aqueous solutions of sodium chloride, sodium acetate and hydrochloric acid are completely dissociated, and (*b*) at low ion concentrations the equivalent conductance, λ, of the ion constituents of strong electrolytes are independent of the nature of the associated ions, *i.e.*, they follow Kohlrausch's law of independent ion migration. Thus if completely dissociated acetic acid were capable of existence the value of its equivalent conductance $\Lambda_{e\,\mathrm{HAc}}$ would be in accord with the relation [20, 21, 22]

$$\Lambda_{e\,\mathrm{HAc}} = \Lambda_{\mathrm{HCl}} + \Lambda_{\mathrm{NaAc}} - \Lambda_{\mathrm{NaCl}} \tag{46}$$

since, in terms of λ values

$$\Lambda_{e\,\mathrm{HAc}} = \lambda_{\mathrm{H}} + \lambda_{\mathrm{Ac}} = \lambda_{\mathrm{H}} + \lambda_{\mathrm{Cl}} + \lambda_{\mathrm{Na}} + \lambda_{\mathrm{Ac}} - \lambda_{\mathrm{Na}} - \lambda_{\mathrm{Cl}} \tag{47}$$

all the terms except λ_{H} and λ_{Ac} cancel. The equivalent conductances, Λ, in equation (46) are all known as functions of the concentration, C, so that values of Λ_e for completely dissociated acetic acid may be obtained as a function of the ion concentration, $C\alpha$. With the aid of the relation (46) and equations (34) and (35) for sodium chloride and hydrochloric acid and

$$\Lambda_{\mathrm{NaAc}} = 91.00 - 80.46\sqrt{C} + 90.\,C(1 - 0.2273\sqrt{C}) \tag{48}$$

for sodium acetate,[23] and recalling that the concentration C of a strong

[20] C. W. Davies, *J. Phys. Chem.*, **29**, 977 (1925).
[21] M. S. Sherrill and A. A. Noyes, *J. Am. Chem. Soc.*, **48**, 1861 (1926).
[22] D. A. MacInnes, *ibid.*, **48**, 2068 (1926).
[23] D. A. MacInnes and T. Shedlovsky, *J. Am. Chem. Soc.*, **54**, 1429 (1932).

electrolyte must be replaced by the ion concentration, $C\alpha$, for a weak electrolyte

$$\Lambda_{e\,\mathrm{HAc}} = 390.71 - 148.57\sqrt{C\alpha} + 163.2\,C\alpha(1 - 0.2273\sqrt{C\alpha}) \quad (49)$$

The computation of the degree of dissociation with the aid of equations (45) and (49) requires a short series of approximations, since the value of Λ_e chosen must be that corresponding to the ion concentration $C\alpha$. A first approximation may be made using $\Lambda_e = \Lambda_0$ in equation (45), and with this preliminary value of α an estimate of Λ_e from equation (49) may be obtained which in turn may be substituted in equation (45). This procedure may be continued until repetition fails to change the result. The conductance data on acetic acid of Shedlovsky and MacInnes [24] and the results of computations are given in Table VI.

TABLE VI. EQUIVALENT CONDUCTANCES AND IONIZATION CONSTANT VALUES OF ACETIC ACID AT 25°

Total concn. $C \times 10^3$	Λ	Ion concn. $C\alpha \times 10^3$	Degree of Dissociation α	Λ_e	$\mathbf{K}' \times 10^5$	$\mathbf{K} \times 10^5$
0.028014	210.38	0.015107	0.53926	390.13	1.768_2	1.752_2
.11135	127.75	.036491	.32771	389.79	1.778_7	1.753_8
.15321	112.05	.044049	.28751	389.72	1.777_5	1.750_2
.21844	96.493	.054101	.24767	389.60	1.781_0	1.750_7
1.02831	48.146	.12727	.12377	389.05	1.797_4	1.750_8
1.36340	42.227	.14803	.10857	388.92	1.803_4	1.752_6
2.41400	32.217	.20012	.082900	388.63	1.809_0	1.750_3
3.44065	27.199	.24092	.070022	388.43	1.814_0	1.749_6
5.91153	20.962	.31929	.054011	388.10	1.823_0	1.748_7
9.8421	16.371	.41557	.042224	387.72	1.832_0	1.746_9
12.829	14.375	.47591	.037096	387.52	1.834	1.743
20.000	11.566	.5975	.029875	387.16	1.840	1.738
50.000	7.358	.9524	.019048	386.30	1.849	1.721
52.303	7.202	.9754	.018649	386.18	1.854	1.723
100.000	5.201	1.3496	.013496	385.40	1.846	1.695
119.447	4.760	1.4763	.012359	385.18	1.847	1.689
200.000	3.651	1.8992	.0094960	384.52	1.821	1.645
230.785	3.392	2.0371	.0088268	384.26	1.814	1.633

The values of degrees of dissociation, α, computed as has just been described, are given in column five of the table. With these degrees of dissociation values of the "dissociation function," $\mathbf{K}'$, defined by the relation

$$\mathbf{K}' = \frac{\alpha^2 C}{(1 - \alpha)} \quad (50)$$

are given in the sixth column of the table. It will be seen that these $\mathbf{K}'$ values are not constant but increase with increasing concentration. The dissociation function $\mathbf{K}'$ is, from equation (44), connected with

[24] Shedlovsky and MacInnes, *loc. cit.*

the thermodynamic ionization $\mathbf{K}$ by

$$\mathbf{K}' \frac{f_H f_A}{f_{HA}} = \mathbf{K} \tag{51}$$

Since undissociated acetic acid is a non-electrolyte its activity coefficient, f_{HA}, may with little error, be assumed to be unity. At very low concentrations the ion activity coefficients may be expected to follow the limiting Debye-Hückel relation, which at 25° is

$$-\log f_H = -\log f_A = 0.5056\sqrt{C\alpha} \tag{52}$$

in which the product $C\alpha$ is the ion concentration. With this relation equation (51) may be put into the form

$$\log \mathbf{K} = \log \mathbf{K}' - 1.011\sqrt{C\alpha} \tag{53}$$

Thus, if our reasoning is correct, a plot of the logarithm of the dissociation function $\mathbf{K}'$ against $\sqrt{C\alpha}$ should be a straight line with a slope of 1.011 and an intercept equal to the logarithm of the thermodynamic dissociation constant $\mathbf{K}$. Such a plot is shown in Fig. 8. The straight

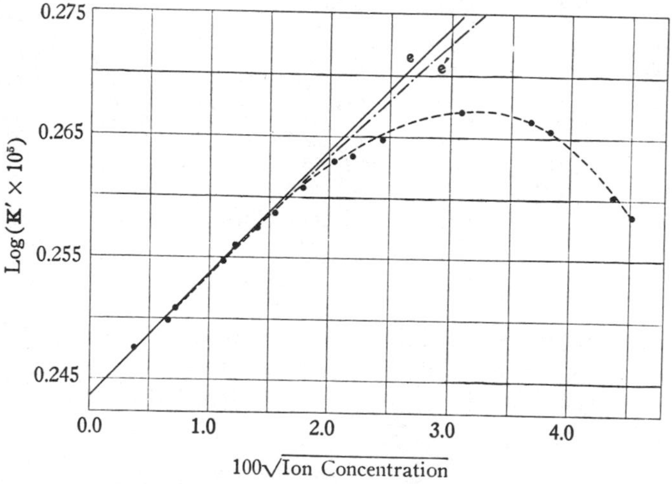

Fig. 8. The Variation of the Dissociation Function, $\mathbf{K}' = \alpha^2 C/(1 - \alpha)$, of Acetic Acid with the Square Root of the Ion Concentration.

line, e, has the theoretical slope required by equation (53) and passes accurately through the six points corresponding to the measurements at the lower concentrations. The range of ion concentration involving these six points is 0.000015 to 0.00015 mol per liter. These results and others of the same nature afford an accurate test of the *limiting* Debye-Hückel relation

$$-\log f = A\sqrt{C\alpha}$$

The agreement of the experimental results with the theory is also shown in the constancy of the values of the thermodynamic ionization constant, **K**, in the first six or eight figures in the last column of Table VI. However, at somewhat higher concentrations there is a rapid divergence from the requirements of the simple theory, as is shown by the decided bending away of the curve through the experimental points from the straight line *e*. The use of the Debye-Hückel expression (13), Chapter 7, in the form

$$-\log f = \frac{A\sqrt{C\alpha}}{(1 - \beta a_i\sqrt{C\alpha})}$$

produces agreement to but slightly higher ion concentrations, as is shown by the curve *e'*, which corresponds to the reasonable value of the "distance of closest approach," a_i of 4 Å for the hydrogen and acetate ions.

It must, however, be recalled that as the concentration of acetic acid is increased the solvent progressively changes from pure water to a mixture of water and undissociated acetic acid. The properties of undissociated acetic acid are quite different from those of water. For instance the acid has a dielectric constant less than ten per cent that of water, and its presence in aqueous solutions would be expected to influence the mobilities as well as the activities of the ions.

As already stated the limiting value of **K'** is the thermodynamic ionization constant, **K**, which in this case is 1.753×10^{-5}. Another method for obtaining thermodynamic ionization constants is given in Chapter 11, depending on measurements of the electromotive force of concentration cells without liquid junction. Using that method Harned and Ehlers found 1.754×10^{-5} for the ionization constant of acetic acid at 25°. However, that constant is based on molalities, *m*, rather than concentrations, *C*. The relation between the ionization constants may be readily shown to be

$$\rho_0 \mathbf{K}_m = \mathbf{K}_C \qquad (54)$$

in which ρ_0 is the density of the solvent, in this case water. The correction lowers the value of the constant based on the emf measurements to 1.749×10^{-5}. However, the two results are in substantial agreement, although the determinations are based on quite different principles and experimental procedures.

A comparison of the results of recent measurements, by the conductance and electromotive force methods, of the ionization constants of some typical weak acids is given in Table VII. The values from the electromotive force measurement have been taken from Table II of Chapter 11, but have been put on the concentration, C, basis by means of equation (54). The figure for carbonic acid is, however, from the work of MacInnes and Belcher.[25] The ionization constants from conductance measurements have been taken from Table VIII of this chapter, which follows. The agreement of the determinations by the two methods, considering the very different techniques used, appears to be very satisfactory and may be regarded as evidence for the great utility of the theories on which the computations are based.

TABLE VII. A COMPARISON OF IONIZATION CONSTANTS, K_C, AT 25°, AS DETERMINED FROM CONDUCTANCE MEASUREMENTS AND FROM THE POTENTIALS OF GALVANIC CELLS WITHOUT LIQUID JUNCTIONS

Acid	$K_C \times 10^5$ Conductance	$K_C \times 10^5$ Electromotive force
Carbonic	0.0431	0.0452
Acetic	1.753	1.749
Chloroacetic	139.6	137.4
Propionic	1.343	1.332
n-Butyric	1.506	1.510
Lactic	13.87	13.70

The reason why Ostwald's dilution law, equation (17), Chapter 3, is moderately successful in accounting for the conductances of weak electrolytes is now evident. Arrhenius' equation, $\alpha = \Lambda/\Lambda_0$, yields degrees of dissociation which are too low. This error, from our present point of view, was more or less offset by the tacit assumption made by Arrhenius and Ostwald, that activity coefficients are unity, whereas, for dilute solutions at least, they are less than unity.

The results of recent determinations of the thermodynamic ionization constants of weak acids determined from conductance measurements which have been interpreted substantially as described above, are given in Table VIII. The relations of some of these values to the molecular structures of the corresponding compounds are discussed in Chapter 21.

Electrolytic Conductances at High Potentials and High Frequencies. All the conclusions concerning the conductances of solutions of electrolytes outlined in this and the preceding chapters have been based on measurements of conductances made using alternating current

[25] D. A. MacInnes and D. Belcher, *J. Am. Chem. Soc.*, 55, 2630 (1933).

TABLE VIII. THERMODYNAMIC IONIZATION CONSTANTS OF ACIDS AT 25° FROM CONDUCTANCE MEASUREMENTS

Acid	$K \times 10^5$	Reference
Acetic	1.753	*12*
Monochloroacetic	139.6	*11,10*
Propionic	1.343	*16*
n-Butyric	1.506	*16*
Benzoic	6.30	*1,2,3*
o-Chlorobenzoic	119.7	*2**
m-Chorobenzoic	15.06	*2**
p-Chlorobenzoic	10.4	*2**
o-Bromobenzoic	140	*8*
p-Bromobenzoic	10.7	*8*
p-Fluorobenzoic	7.22	*8*
Phenylacetic	4.88	*4*
o-Chlorophenylacetic	8.60	*3*
m-Chlorophenylacetic	7.24	*3*
p-Chlorophenylacetic	6.45	*4*
o-Bromophenylacetic	8.84	*3*
p-Bromophenylacetic	6.49	*4*
p-Iodophenylacetic	6.64	*4*
p-Methoxyphenylacetic	4.36	*3*
Acrylic	5.50	*13*
Lactic	13.87	*15*

Acid	$K_1 \times 10^5$	Reference
Carbonic	0.0431	*14*
Malonic	139.$_7$	*5*
Succinic	6.63	*5*
Glutaric	4.54	*5*
Adipic	3.72	*5*
Pimelic	3.10	*5*
Suberic	2.99	*5*
Methylmalonic	8.47	*6*
Ethylmalonic	10.9	*6*
n-Propylmalonic	10.3	*6*
Dimethylmalonic	7.06	*6*
Methylethylmalonic	15.4	*6*
Diethylmalonic	70.8	*6*
Ethyl-*n*-propylmalonic	78.4	*6*
Di-*n*-propylmalonic	92.0	*6*
Phenylmalonic	277	*9*
Cyclopropane-1,1-dicarboxylic	150	7
Cyclobutane-1,1-dicarboxylic	7.55	7
Cyclopentane-1,1-dicarboxylic	5.96	7
Cyclohexane-1,1-dicarboxylic	3.54	7

* See also 8.

1. F. G. Brockman and M. Kilpatrick, *J. Am. Chem. Soc.*, **56**, 1483 (1934).
2. B. Saxton and H. F. Meier, *ibid.*, **56**, 1918 (1934).
3. J. F. J. Dippy and F. R. Williams, *J. Chem. Soc.*, **1934**, 1888.
4. J. F. J. Dippy and F. R. Williams, *ibid.*, **1934**, 161.
5. G. H. Jeffery and A. I. Vogel. *ibid.*, **1935**, 21.
6. G. H. Jeffery and A. I. Vogel, *ibid.*, **1936**, 1756.
7. W. L. German, G. H. Jeffery and A. I. Vogel, *ibid.*, **1935**. 1624.
8. J. F. J. Dippy, F. R. Williams and R. H. Lewis, *ibid.*, **1935**, 343.
9. S. Basterfield and J. W. Tomecko, *Can. J. of Research*, **8**, 447 (1933).
10. B. Saxton and T. W. Langer, *J. Am. Chem. Soc.*, **55**, 3638 (1933).
11. T. Shedlovsky, A. S. Brown and D. A. MacInnes, *Trans. Electrochem. Soc.*, **66**, 165 (1934).
12. D. A. MacInnes and T. Shedlovsky, *J. Am. Chem. Soc.*, **54**, 1429 (1932).
13. W. L. German, G. H. Jeffery and A. I. Vogel, *J. Chem. Soc.*, **1937**, 1604.
14. T. Shedlovsky and D. A. MacInnes, *J. Am. Chem. Soc.*, **57**, 1705 (1935).
15. A. W. Martin and H. V. Tartar, *J. Am. Chem. Soc.*, **59**, 2672 (1937).
16. D. Belcher, *J. Am. Chem. Soc.*, **60**, 2744 (1938).

at low potentials and at relatively low frequencies, *i.e.*, up to 8000 per second. Interesting and important phenomena appear, however, if either the potential or the frequency is greatly increased. The first of these is called the *Wien effect* and the second the *Debye-Falkenhagen effect.* Both of these effects may, as will be seen, be explained, not quite unambiguously, by the interionic attraction theory of conductance.

The Wien Effect. If, instead of using potentials of the order of one volt per centimeter in the measurement of electrolytic conductance, voltages of several hundred thousand times this are employed, the conductances of solutions of electrolytes are no longer constant but tend to increase with the potential. Under these conditions Ohm's law is evidently no longer valid. This increase of conductance at high potentials is called the *Wien effect.* The passage of high potentials for any appreciable length of time would obviously produce great heating effects. It is, therefore, necessary to employ a method in which the current passes for a minute fraction of a second. Surges lasting only 10^{-6} second are actually used. The principle of the method employed by Wien and associates is to measure the time integral of the heating effect produced by a single condenser discharge. The circuit used by Malsch and Wien[26] is shown in Fig. 9 and consists of an

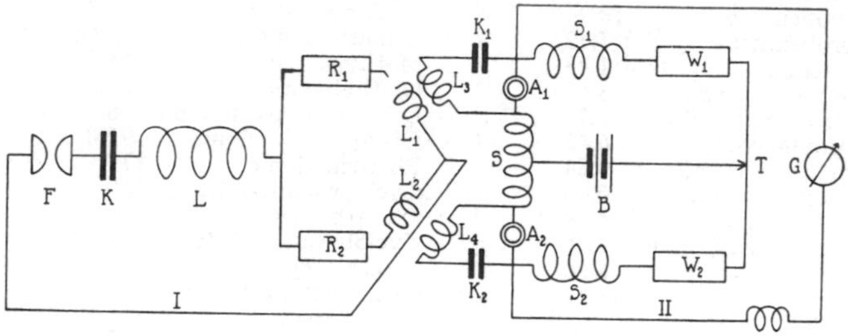

Fig. 9. Barretter and Oscillation Circuits for Measuring the Wien Effect.

oscillation circuit, I, and a "barretter" bridge circuit, II. The oscillation circuit involves a condenser, K, a spark gap, F, the inductances, L, L_1 and L_2, the resistance R_1 and the adjustable resistance R_2. The barretter bridge is coupled to the oscillation circuit by means of the inductances, L_3 and L_4. In the barretter bridge K_1 and K_2 are large condensers, S, S_1, S_2 and W_1 and W_2 are respectively inductances and resistances. B is a battery and T is a slide wire. Important parts of this bridge are the filament electric lamps, A_1 and A_2. These are carefully matched with regard to their resistances, their increase of resistance with increasing voltage, and their rates of cooling.

In operation the potential from the battery, B, is first adjusted to give the maximum sensitivity of the lamp bulbs to slight currents passing through them. The condenser, which has been charged from an outside source, is then discharged in the oscillation circuit, I. If the resistances, R_1 and R_2 are equal the whole circuit is symmetrical and there will be no deflection of the galvanometer, G. However, if that condition is not satisfied there will be unequal heating of the lamp bulbs from the high frequency current induced in the arms of the barretter bridge. The variable resistance, R_1, may then be adjusted until the galvanometer shows no deflection. If the resistance, R_2, is a

[26] J. Malsch and M. Wien, *Ann. Physik.* [4], 83, 305 (1927).

conductance cell containing an electrolyte, this resistance is found to decrease with the potential induced in the oscillating circuit, I.

Some of the results obtained by Wien [27] are shown in Figs. 10 and

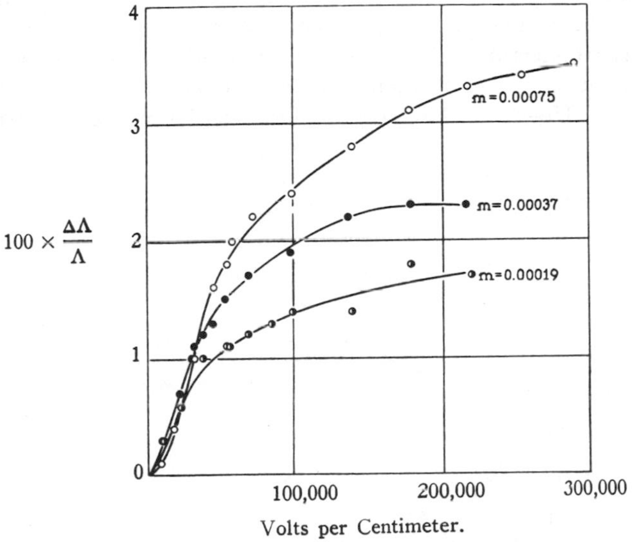

Fig. 10. The Wien Effect for Lithium Ferricyanide at Three Concentrations.

11. In Fig. 10 the percentage increase of the equivalent conductances of three solutions of lithium ferricyanide are plotted as ordinates against the applied potential, *E*. It will be seen that the effect increases markedly with the concentration, but that all the values tend toward maxima at very high potentials. The increase of the effect when salts of increasing high valence type are used is shown in Fig. 11. Here the solutions used all had the same specific conductance ($L = 4.6 \times 10^{-5}$). However, as can be observed from this plot, the increases in the equivalent conductance when high potentials were used in the measurements were found to be much greater if the solutions contained polyvalent ions.

It is of great interest that the curves shown in Figs. 10 and 11 indicate that the equivalent conductances are approaching maxima which would be observed if the electrophoretic and time of relaxation effects were overcome by the intense electric fields. At one volt per centimeter an ion with an equivalent conductance, Λ, of 100 moves at a rate of about two centimeters per hour. At, for instance, 300

[27] M. Wien, *Ann. Physik.*, [5] 1, 400 (1929).

kilovolts per centimeter the rate is several meters *per second*. It is therefore possible that in the latter case the ion moves so rapidly that the ion atmosphere cannot form, or if it does partially form, the ion is removed from its restraining effects. In support of the inter-ionic attraction theory of conductance as outlined earlier in this chapter the Wien effect is found to be greatest under conditions in which the electrophoretic and time of relaxation effects are expected to be large, *i.e.*, at high concentrations and for solutions containing highly charged ions. However, the force of these conclusions is somewhat

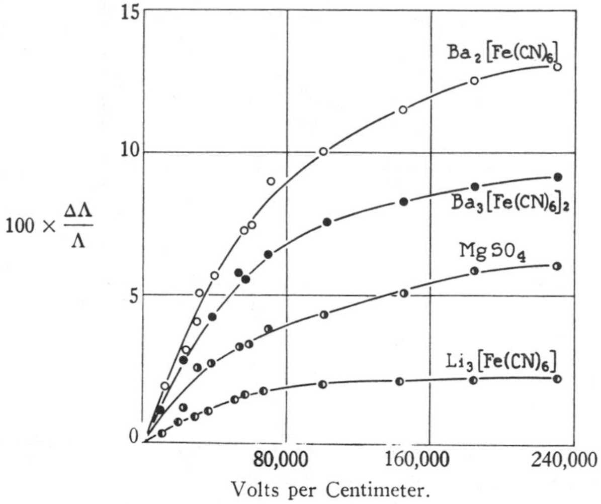

Fig. 11. The Wien Effect for Different Valence Type Salts at Concentrations having the Same Low Field Conductance.

lessened by the fact that increases of conductance in large fields are also observed with weak acids, such as acetic acid. Due to the low ion concentrations existing in solutions of such acids both the electrophoretic and time of relaxation effects should be small, so that but small increases of conductance would be expected if the effect of large potentials is that of overcoming the restraints due to the ion atmosphere. Nevertheless, large increases of conductance have been found by Schiele [28] when high potentials have been used with solutions of weak acids. Some typical results obtained by this worker, who used a slight modification of the method of Malsch and Wien, are shown in Fig. 12. In this figure

[28] J. Schiele, *Ann. Physik.*, [5] 13, 811 (1932).

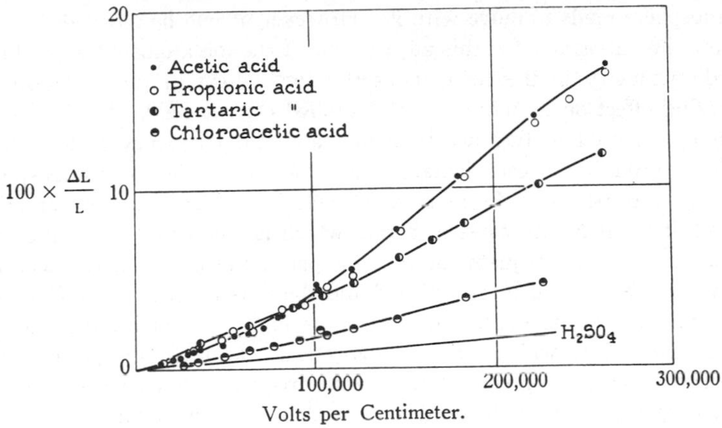

Fig. 12. The Wien Effect for Weak Acids.

the percentage increases of conductance with increasing high frequencies for three weak acids are plotted as functions of the field strength. The plot also includes, for comparison, measurements on sulphuric acid. It will be observed by comparing the scale of ordinates of the figure with those of Figures 10 and 11, that the effect with sulphuric acid is about what would be expected for a strong electrolyte of its valence type. The Wien effect for the weaker acids is very much greater. From these data the conclusion seems inescapable that these high potentials have the effect of producing a temporary ionization of the weak electrolytes, an explanation which has been proposed and discussed theoretically by Onsager.[29]

The Debye-Falkenhagen Effect, The Dispersion of Conductivity at High Frequencies. From theoretical considerations Debye and Falkenhagen [30] predicted that, if electrolytic conductances were measured with alternating current with very high frequencies, higher values of the conductances would be found than if relatively low frequencies, say 1000 to 8000 cycles per minute, were used. The predicted increase of conductivity was demonstrated experimentally by Sack [31] and is known as the *Debye-Falkenhagen effect.* As has already been explained in this chapter, the decrease of equivalent conductance of a strong electrolyte with concentration, due to the time of relaxation effect, arises from the fact that, when an ion moves, its ion

[29] L. Onsager, *J. Chem. Phys.*, **2**, 599 (1934).
[30] P. Debye and H. Falkenhagen, *Physik. Z.*, **29**, 121, 401 (1928).
[31] H. Sack, *Physik. Z.*, **29**, 627 (1928).

atmosphere tends to move with it. However, it will be recalled that the finite time necessary for this adjustment of the ion atmosphere produces a dissymmetry in the field about the ion, which in time produces a braking effect on its motion in the applied electric field. If alternating current is used in the measurements and the frequency is low there will be produced at each instant a dissymmetry in the ionic atmosphere which corresponds to the momentary velocity of the ion. On the other hand, if the field alternates at a rate which is comparable with the time necessary for the adjustment to take place, the dissymmetry will not have time enough to be established and the braking effect on the motion of the ion will be decreased. At very high frequencies the time of relaxation effect will therefore, theoretically at least, disappear. The observed changes of conductance with frequency have been found to agree, qualitatively, with the theory of Debye and Falkenhagen.

The principles of the methods used in demonstrating the effect cannot be described without too much of a digression at this point. Resonance circuits similar to those to be described in Chapter 22 have been mostly employed. So far it has been possible only to compare the effect for a given solution with that for a reference solution, for which aqueous potassium chloride has been used. Since from Onsager's theory the time of relaxation effect increases with the valence of the ions, a solution containing polyvalent ions should show an increase of conductance with increasing frequency of the measuring current that is greater than that of potassium chloride at the same concentration. This has been found to be the case, in good agreement with the predictions of Debye and Falkenhagen, by Brendel, Mittelstaedt and Sack,[32] Goldammer and Sack,[33] Brendel [34] and others, and recently by Arnold and Williams.[35]

[32] B. Brendel, O. Mittelstaedt and H. Sack, *Physik. Z.*, **30**, 576 (1929).
[33] R. Goldammer and H. Sack, *ibid.*, **31**, 345 (1930).
[34] B. Brendel, *ibid.*, **32**, 327 (1931).
[35] O. M. Arnold and J. W. Williams, *J. Am. Chem. Soc.*, **58**, 2613, 2616 (1936).

Chapter 19

The Conductance of Electrolytes in Non-Aqueous and Mixed Solvents

As was shown in Chapter 3, the ionic theory in the form given it by Arrhenius was fairly satisfactory for aqueous solutions of weak electrolytes, and though admittedly inadequate for similar solutions of strong electrolytes it served as a guide for the investigations of a generation of research workers. However, the utility of the theory was not found to extend to solutions of electrolytes in non-aqueous solvents. Many measurements[1] of conductance and other properties of such solutions of electrolytes were carried out, but the attempts to interpret the results led to few generalizations of definite value.

The reasons for the comparative lack of success of the Arrhenius theory with data on non-aqueous solvents are now clear. As has already been stated, that theory rested on two tacit assumptions, both now known to be incorrect. These assumptions may be stated as: (*a*) activity coefficients all have the value of unity and (*b*) for a given solvent and temperature, ion mobilities are constants independent of concentration. As we have seen in the preceding chapter the errors produced by the incorrectness of these two assumptions tend, in the case of the interpretation of the data on aqueous solutions of weak electrolytes, to compensate, producing apparent agreement with the Arrhenius theory. On leaving water as a solvent one must go, with few but important exceptions, to solvents having considerably lower dielectric constants. Thus water has a dielectric constant of 78.6 and methyl and ethyl alcohol have constants of 31.5 and 24.3 respectively. In water, therefore, interionic attractions and repulsions are relatively small. Changes of activity coefficients and of ion mobilities with ion concentrations will therefore be low when compared with most other solvents. The importance of the dielectric constant in the value of the activity coefficient is seen, for instance, in equation (22) of Chapter 7, and on the equivalent conductance, λ, of an ion constituent in equation (14), Chapter 18. To anticipate the main conclusion of this chapter, the interionic attraction theory is of even greater utility in

[1] For the early work see, P. Walden, "Elektrochemie nichtwässriger Lösungen," Johann Ambrosius Barth, Leipzig, 1924.

dealing with the results of measurements on salt solutions in non-aqueous solvents than it has been for the corresponding water solutions.

Partly because of the great practical importance of aqueous solutions, and also because of the greater experimental difficulties in working with other solvents, non-aqueous solutions have received relatively little attention. The data on concentration cells without liquid junctions, involving mixed and non-aqueous solvents, are discussed in Chapter 12. Considering the possible extensions of the field such data are very few. Many more investigations have been made on the conductances of salts in non-aqueous and mixed solvents. However, not many of these researches were carried out with modern accuracy. In the following only the results of measurements that are of unusual precision or of theoretical interest will be discussed. The order adopted will be that of decreasing dielectric constant of the solvent.

The Conductance of Strong Electrolytes in Methyl and Ethyl Alcohol. Careful studies of the conductances of electrolytes in methyl and ethyl alcohol have been carried out by Hartley and associates.[2] A plot of the equivalent conductance, Λ, values for a series of sulphocyanates in methyl alcohol as functions of the square root of the concentration are given in Fig. 1.[3] It will be seen that the plots are all straight lines as required by Onsager's equation for uni-univalent electrolytes, equation (18), Chapter 18. Since methyl alcohol at 25° has a dielectric constant of 31.5 [4] and a viscosity of 0.00545 [5] poise, that equation takes the form:

$$\Lambda = \Lambda_0 - (0.892\,\Lambda_0 + 155.0)\sqrt{c} \tag{1}$$

However, although the results of the conductance measurements follow an equation of the form $\Lambda = \Lambda_0 - K\sqrt{c}$ the value of the constant K does not in every case correspond to that predicted by Onsager's equation. A comparison of the observed values of K, (the slopes of the lines in plots similar to Fig. 1) and the computed values from equation (1) are given in Table I. Although the data for a number of the salts agree with the predictions of the theory almost within the experimental error, pronounced deviations are observed particularly for cesium salts and some of the nitrates. However, as mentioned in Chapter 18, nitrates

[2] A. Unmack, E. Bullock, D. A. Murray-Rust and H. Hartley, *Proc. Roy. Soc.*, **A132**, 427 (1931).
C. P. Wright, D. A. Murray-Rust and H. Hartley, *J. Chem. Soc.*, **1931**, 199.
M. Barak and H. Hartley, *Z. physik. Chem.*, **A165**, 272 (1933).
A. Unmack, D. M. Murray-Rust and H. Hartley, *Proc. Roy. Soc.*, **A127**, 228 (1928).
E. D. Copley, D. M. Murray-Rust and H. Hartley, *J. Chem. Soc.*, **1930**, 2492.
T. H. Mead, O. L. Hughes and H. Hartley, *ibid.*, **1933**, 1207.

[3] A. Unmack, D. M. Murray-Rust and H. Hartley, *loc. cit.*

[4] G. Akerlof, *J. Am. Chem. Soc.*, **54**, 4125 (1932).

[5] H. Hartley and H. R. Raikes, *J. Chem. Soc.*, **127**, 524 (1925).

are abnormal, even in aqueous solutions, when their conductance and transference data are compared with, for instance, the halogen salts.

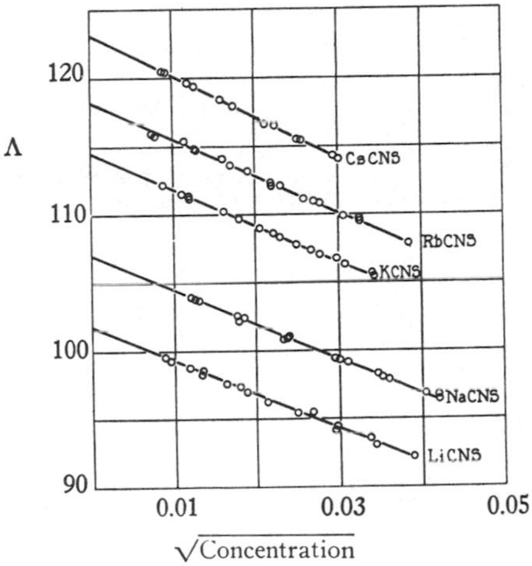

Fig. 1. Change in the Equivalent Conductance of Some Alkali Cyanates with the Square Root of the Concentration in Methyl Alcohol at 25°.

Although there is thus observed to be fair agreement with the Onsager equation for the conductance of uni-univalent salts in methyl alcohol, this accord does not extend to the results of bi-univalent salts, such as magnesium sulphocyanate. For such salts in methyl alcohol,

TABLE I. TEST OF ONSAGER'S EQUATION FOR SOME UNI-UNIVALENT SALTS IN METHYL ALCOHOL AT 25°

Salt	Λ_0	*K*		Deviation (per cent)
		Observed	Onsager	
LiCNS	101.8	253	246	+ 3
NaCNS	107.0	255	250	+ 2
KCNS	114.5	268	257	+ 4
RbCNS	118.2	271	260	+ 4
CsCNS	123.2	304	265	+ 15
NH_4CNS	118.7	279	261	+ 7
LiCl	90.9	224	236	− 5
NaCl	96.9	230	241	− 5
KCl	105.0	261	249	+ 5
RbCl	108.6	281	252	+ 12
CsCl	113.6	293	256	+ 15
$LiNO_3$	100.2	250	244	+ 3
$NaNO_3$	106.4	288	250	+ 2
KNO_3	114.5	345	257	+ 34
$RbNO_3$	118.1	355	260	+ 37
$CsNO_3$	122.9	379	265	+ 43

equation (14), Chapter 18, becomes

$$\Lambda = \Lambda_0 - \left\{ 2.8815 \frac{\Lambda_0/(\lambda_0^- + \Lambda_0)}{1 + \sqrt{\frac{2}{3} \frac{\Lambda_0}{\lambda_0^- + \Lambda_0}}} \Lambda_0 + 164.4 \right\} \sqrt{3c} \qquad (2)$$

In Table II observed values of K and those computed from equation

TABLE II. TEST OF ONSAGER'S EQUATION FOR SOME BI-UNIVALENT SULPHOCYNATES IN METHYL ALCOHOL AT 25°

Salt	Λ_0	λ_0^-	K Observed	K Onsager
$Mg(CNS)_2$	120	59	2000	519
$Ca(CNS)_2$	122	61	1400	524
$Sr(CNS)_2$	122	61	980	524
$Ba(CNS)_2$	125	61.5	850	526

(2) are compared. In no case is there even approximate agreement. In this regard methyl alcohol solutions differ sharply from aqueous solutions, since with the latter typical uni-bivalent and even uni-trivalent salts yield solutions the conductances of which are in good accord with the Onsager theory. Since accurate transference data are not available in methyl alcohol solutions, the value of λ_0^- for each salt in equation (2) may be somewhat in error. However, no adjustment of this parameter could alter the computed values of K to bring them in much closer agreement with the observed values.

Hartley and associates [6, 7] have also studied the conductances of salts in ethyl alcohol. With that solvent the equivalent conductances, at least for dilute solutions, follow the relation

$$\Lambda = \Lambda_0 - K\sqrt{c} \qquad (3)$$

The value of K is, in most cases, considerably greater than that predicted by Onsager's equation which for ethyl alcohol at 25° takes the form [8]

$$\Lambda = \Lambda_0 - (1.321\,\Lambda_0 + 88.2)\sqrt{c} \qquad (4)$$

Table III from the work of Barak and Hartley [9] gives some typical results. It is evident from these figures that there are still greater deviations from Onsager's relation for salts dissolved in ethyl alcohol

[6] M. Barak and H. Hartley, *Z. physik. Chem.*, **A165**, 272 (1933).

[7] E. D. Copley, D. M. Murray-Rust and H. Hartley, *J. Chem. Soc.*, **1930**, 2492.

[8] The constants in this equation are based on a value of 24.3 (G. Akerlof, *loc. cit.*) for the dielectric constant of ethyl alcohol at 25° and on 0.0109 poise for the viscosity. (Averaged from the results of: J. W. Ingham, *J. Chem. Soc.*, **1928**, 1917, F. E. King and J. R. Partington, *Trans. Farad. Soc.*, **23**, 522 (1924), and L. C. Connell, R. T. Hamilton and J. A. V. Butler, *Proc. Roy. Soc.*, **A147**, 418 (1934)).

[9] M. Barak and H. Hartley, *Z. physik. Chem.*, **165A**, 272 (1933).

TABLE III. TEST OF ONSAGER'S EQUATION FOR SOLUTIONS OF UNI-UNIVALENT SALTS IN ETHYL ALCOHOL AT 25°

Salt	Λ_0	*K* Observed	*K* Onsager	Deviation (per cent)
LiCl	39.2	166	140	19
NaCl	42.5	197	144	37
LiI	43.4	156	146	7
NaI	47.3	176	151	17
KI	50.8	209	155	35
RbI	51.8	228	157	45
$LiNO_3$	42.7	171	145	18
$AgNO_3$	46.25	336	149	126
NH_4NO_3	47.5	233	151	54

than in methyl alcohol. It is also seen that the deviations are greater, the higher the atomic weight of the cation.

Although few data have been quoted they appear to be a good sample of results that are found with solvents having dielectric constants above about 25. In general, an equation of the form

$$\Lambda = \Lambda_0 - K\sqrt{c}$$

holds for measurements in dilute solutions (below say, c = 0.002 equivalent per liter). Increasing deviations of *K* from the values predicted by Onsager's equation are, however, to be expected as the dielectric constant is lowered, or as the valence of the ion is increased. Certain ion constituents, such as the silver and nitrate ions, which deviate from normal behavior in water, show still greater deviations in non-aqueous solvents.

It is of interest to see whether Kohlrausch's law of independent ion migration which has been shown (page 340) to hold accurately for aqueous solutions is also valid for methyl alcohol solutions. Since transference data are not available a test similar to that for water solutions is not yet possible. If, however, limiting equivalent conductances are independent of the ions with which they are associated the differences of, for instance, the limiting conductances of the sodium and lithium salts of an acid HX should be independent of the nature of the radical X, since

$$\Lambda_{0NaX} - \Lambda_{0LiX} = \lambda_{0Na} + \lambda_{0X} - \lambda_{0Li} - \lambda_{0X} = \lambda_{0Na} - \lambda_{0Li}$$

The accuracy with which this relation holds is shown in Table IV from

TABLE IV. TEST OF KOHLRAUSCH'S LAW OF INDEPENDENT ION MIGRATION FOR SALT SOLUTIONS IN METHYL ALCOHOL AT 25° VALUES OF $\Delta\lambda_0^+$

	Na − Li	K − Li	Rb − Li	Cs − Li
CNS	5.2	12.7	16.4	21.4
Cl	6.0	14.1	17.7	22.7
NO_3	6.2	14.3	18.1	22.7

the work of Hartley and associates. It will be observed that the differences of the limiting ion conductances for each pair of positive ion constituents are, probably within the errors of experiment and extrapolation, independent of the nature of the anions.

Since values of Λ_0 are available for a number of salts in water, and methyl and ethyl alcohols, a test is possible, for these solvents, of Walden's [10] rule connecting the equivalent limiting conductance of an electrolyte, Λ_0, with the viscosity, η, of the solvent in which it is dissolved. The relation is as follows:

$$\Lambda_0\eta = K \tag{5}$$

K is a constant for a given solute. That, at least as tested by these data, the rule is only a rough approximation, is shown by Table V

TABLE V. TEST OF WALDEN'S RULE, $\Lambda_0\eta = K$, FOR WATER AND METHYL AND ETHYL ALCOHOL AT 25°

Salt	Solvent	Limiting conductance Λ_0	Viscosity (poise) η	$\Lambda_0\eta$
LiCl	H_2O	115.03	0.008949	1.029
LiCl	CH_3OH	90.9	0.00545	0.495
LiCl	C_2H_5OH	39.2	0.0109	0.427
NaCl	H_2O	126.45	0.008949	1.132
NaCl	CH_3OH	69.9	0.00545	0.528
NaCl	C_2H_5OH	42.5	0.0109	0.463
LiCNS	CH_3OH	101.8	0.00545	0.555
LiCNS	C_2H_5OH	44.5	0.0109	0.485
Tetraethyl ammonium chloride	CH_3OH	121.7	0.00545	0.663
	C_2H_5OH	51.9	0.0109	0.566
Tetraethyl ammonium picrate	CH_3OH	116.7	0.00545	0.636
	C_2H_5OH	54.95	0.0109	0.599

which gives typical examples. If the rule held, the product, $\Lambda_0\eta$, given in the last column of the table should be a constant for each salt. If the ions can be assumed to be spherical in shape and to have the same radius, r, in each of the solvents, Walden's rule would follow from Stokes' Law, equation (8), Chapter 18;

$$v = \frac{F}{6\pi\eta r} \tag{6}$$

in which, it will be recalled, F is the force acting on a sphere of radius, r, moving in a medium made up of particles with radii of a lower order of magnitude than that of the sphere. The velocity, v, of an ion moving in an electric field of strength, E, is, at infinite dilution

$$v = U_0 E$$

[10] P. Walden, *Z. physik. Chem.*, **55**, 207 (1906); see also P. Walden and G. Busch, *ibid.*, **140A**, 89 (1929).

in which u_0 is the ion mobility. With equation (6a) of Chapter 4 this becomes

$$v = \frac{\lambda_0}{\mathbf{F}} E \tag{7}$$

The force, F, in equation (6), on an ion of charge ϵ is $E\epsilon$, so that equating the expression for the velocity we have

$$\lambda_0 \eta = \frac{\epsilon \mathbf{F}}{6\pi r} \tag{8}$$

Thus for a uni-univalent electrolyte

$$\Lambda_0 \eta = (\lambda_0^+ + \lambda_0^-)\eta = \frac{\epsilon \mathbf{F}}{6\pi}\left(\frac{1}{r^+} + \frac{1}{r^-}\right) \tag{9}$$

The failure of Walden's rule to hold except as an approximation indicates that the ions are not sufficiently large when compared with those of the solvent medium for Stokes' Law to be valid, or that the radii of the ions (r^+ and r^-) vary from solvent to solvent, or that, possibly, the deviations are due to both these causes. It is known, from the determinations of "true" transference numbers described in Chapter 4 and from other evidence, that ions are hydrated in aqueous solution, and it is probable that they are solvated in other media. It is unlikely, therefore, that the radii of the ions would remain constant in different solvents so that the failure of Walden's rule to be more than an approximation is not surprising.

The Conductance of Weak Electrolytes in Methyl and Ethyl Alcohol. The foregoing discussion has concerned solutions of electrolytes which are strong electrolytes in water solution, and remain at least highly ionized in methyl and ethyl alcohols. However, as with solutions in water, there are also weak electrolytes in alcohols and similar solvents. The electrolytes of this type that have been carefully studied are all acids, though bases in these solvents are known. The reason why some electrolytes are completely dissociated and others are not remains to be explained. To anticipate the conclusions of a later section of this chapter all uni-univalent electrolytes with radii 4 Å or greater, should be completely dissociated in solvents of dielectric constant greater than about 30, *if electrostatic forces alone are operative,* and no electron rearrangement takes place when the ions combine or dissociate. If this is true then dissociation of say, acetic acid, must involve, even in water, more than a simple separation of charged particles. An adequate explanation of the dissociation of weak electrolytes will probably involve shifts of electron levels, *i.e.,* quantum jumps. It may be significant that, in the great majority of cases, a weak electrolyte has an ion in common with the solvent.

Careful studies of the conductances of weak acids, and salts of weak acids in methyl and ethyl alcohol have been made by Goldschmidt and associates.[11-15] They were, in general, interpreted by the authors in terms of the Ostwald dilution law, equation (17), Chapter 3,

$$\frac{c\alpha^2}{(1 - \alpha)} = \mathbf{K} \tag{10}$$

the dissociation being computed from the Arrhenius relation Λ/Λ_0. As pointed out in the previous chapter this procedure involves a partial compensation of two errors. Fortunately the results of these workers may be, to a large extent, reinterpreted in terms of the interionic attraction theory. The computation of the thermodynamic ionization constant of picric acid in methyl alcohol according to the newer conceptions is outlined in Table VI, the procedure being the same as that

TABLE VI. THE COMPUTATION OF THE THERMODYNAMIC DISSOCIATION CONSTANT OF PICRIC ACID AT 25° IN METHYL ALCOHOL

Total concn. c	Λ	Ion Concn. $c\alpha$	α	Λ_e	$K' \times 10^4$
0.1	9.323	0.005871	0.05871	158.80	3.662
0.05	12.81	0.003912	0.07823	163.75	3.320
0.025	17.48	0.002591	0.1037	168.65	2.996
0.0125	23.85	0.001724	0.1379	172.95	2.758
0.00625	32.12	0.001138	0.1820	176.45	2.532
0.003125	43.25	0.0007523	0.2408	179.65	2.384
0.001563	57.17	0.0004892	0.3131	182.60	1.946

$\mathbf{K} = 1.84_5 \times 10^{-4}$

given in Chapter 18 for the determination of the ionization constant of acetic acid in aqueous solution. The degrees of dissociation given in the fourth column are computed from the ratio, $\alpha = \Lambda/\Lambda_e$, in which Λ_e is given by

$$\Lambda_{eHP} = \Lambda_{HCl} + \Lambda_{NaP} - \Lambda_{NaCl} \tag{11}$$

the Λ values all being at the ion concentration ($c\alpha$), in which c is the concentration of the acid. From the relation

$$\mathbf{K} = \frac{c\alpha^2 f^2}{(1 - \alpha)} = \mathbf{K}' f^2 \tag{12}$$

[11] H. Goldschmidt, *Z. physik. Chem.*, **91**, 46 (1916).
[12] H. Goldschmidt, E. Marum and L. Thomas, *ibid.*, **132A**, 257 (1928).
[13] H. Goldschmidt, E. Marum and L. Thomas, *ibid.*, **129A**, 223 (1927).
[14] H. Goldschmidt and H. Aarflot, *ibid.*, **117A**, 312 (1925).
[15] H. Goldschmidt and F. Aas, *ibid.*, **112A**, 423 (1924).

in which **K**′ is the dissociation function and **K** the thermodynamic ionization constant, the expression

$$\log \mathbf{K} = \log \mathbf{K}' - 2A\sqrt{c\alpha} \tag{13}$$

follows if the limiting Debye-Hückel relation (26), Chapter 7, is valid. Thus, as was found for acetic acid in water, a plot of log **K**′ against $\sqrt{c\alpha}$ should be a straight line with a slope of $2A$, in which A is the Debye-Hückel constant which, in this case for methyl alcohol at 25°, equals 1.99. That this relation holds for picric acid is shown in Fig. 2 in which the **K**′ and α values have been taken from Table VI.

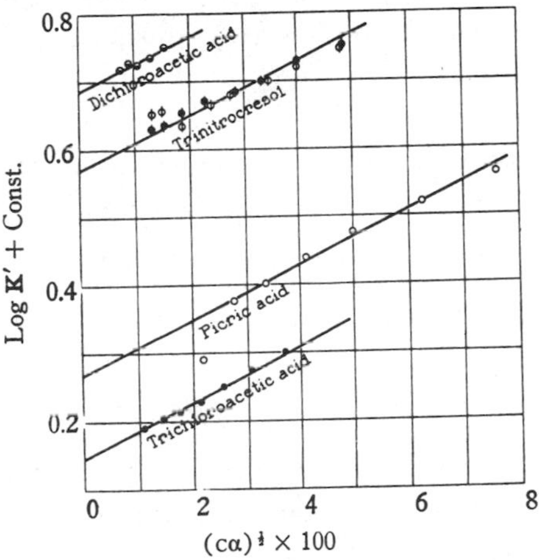

Fig. 2. Extrapolation for the Dissociation Constants of Some Acids in Methyl Alcohol at 25°. The Straight Lines Represent the Limiting Debye-Hückel Relation.

The figure also shows the corresponding lines for trichloroacetic acid and trinitrocresol. It will be seen that the lines are all parallel and have the slope required by the theory. This calculation indicates, indirectly, that the electrolytes, the conductances of which are used in computing Λ_e as described above, are substantially completely dissociated.

However, computations made for the purpose of determining the thermodynamic ionization constants of weak electrolytes in *ethyl alcohol* solutions, carried out as just described, gave curves instead of straight lines for a plot of values of $-\log \mathbf{K}'$ against $\sqrt{c\alpha}$, indicating as at least one possibility, that the electrolytes such as HCl, NaCl, and the sodium salt of picric acid are not completely dissociated, and their equivalent conductances cannot, therefore, be used in obtaining Λ_e values. The divergence of the conductances of these electrolytes from the predictions of the Onsager equation, shown in Table III, is also evidence of incomplete dissociation. The method for determining thermodynamic ionization constants from conductance measurements described above is therefore of no use for solutions of electrolytes in ethyl alcohol, and in other solvents with similar or lower dielectric constants. Under these conditions, however, a method of computation described and used by Fuoss and Kraus [16, 17, *] may be advantageously used. The method is as follows. The degree of dissociation, α, is, as usual, obtained from the relation Λ/Λ_e. For low ion concentrations, Λ_e may be computed from Onsager's equation, (18), Chapter 18,

$$\Lambda_e = \Lambda_0 - (\theta\Lambda_0 + \sigma)\sqrt{c\alpha}$$

The degree of dissociation, α, is therefore given by the expression:

$$\alpha = \frac{\Lambda}{\Lambda_0 - (\theta\Lambda_0 + \sigma)\sqrt{c\alpha}} \tag{14}$$

To deal with this implicit equation these authors define the variable, z, equal to

$$z = (\theta\Lambda_0 + \sigma)\Lambda_0^{-3/2}\sqrt{c\Lambda} \tag{15}$$

by means of which the denominator of equation (14) may be readily shown to be equal to Λ_0 times the continued fraction

$$F(z) = 1 - z(1 - z(1 - z(1 - \cdots\cdots)^{-1/2})^{-1/2})^{-1/2} \tag{16}$$

Thus equation (14) becomes

$$\alpha = \frac{\Lambda}{\Lambda_0 F(z)} \tag{17}$$

Fuoss [18] has computed a convenient interpolation table giving $F(z)$ for values of z up to 0.209. In order to obtain values of the ionization constant, **K**, and of the limiting conductance, Λ_0, from the conductance data a rough estimate of the latter is made by plotting Λ values against the $\sqrt{c}$ and extrapolating. With this tentative value, Λ_0, a preliminary estimate of the degree of dissociation, α, is made by computing z,

[16] R. M. Fuoss, *J. Am. Chem. Soc.*, **57**, 488 (1935).
[17] R. M. Fuoss and C. Kraus, *ibid.*, **55**, 476, 2390 (1933).
* Recently T. Shedlovsky, *J. Franklin Inst.*, **225**, 739 (1938), has proposed a more convenient method of computation.
[18] R. M. Fuoss, *J. Am. Chem. Soc.*, **57**, 488 (1935).

(equation (15)) obtaining $F(z)$, and using equation (17). This value of α may be used in obtaining the activity coefficient, f, from the Debye-Hückel limiting equation

$$-\log f = A\sqrt{c\alpha}$$

With equation (17) equation (12), becomes

$$\mathbf{K} = \frac{cf^2\left(\dfrac{\Lambda}{\Lambda_0 F(z)}\right)^2}{1 - \dfrac{\Lambda}{\Lambda_0 F(z)}} \tag{18}$$

which may be rearranged to give

$$\frac{F(z)}{\Lambda} = \frac{1}{\mathbf{K}\Lambda_0^2} \cdot \frac{c\Lambda f^2}{F(z)} + \frac{1}{\Lambda_0} \tag{19}$$

so that if values of $F(z)/\Lambda$ are plotted against $c\Lambda f^2/F(z)$ and the result is a straight line, the slope of this line is $1/(\mathbf{K}\Lambda_0^2)$ and the intercept is $1/\Lambda_0$. This computation may then be repeated with this more precise value of Λ_0. Fuoss and Kraus have shown that conductance data on a considerable variety of electrolytes and in a number of solvents show a linear relationship when plotted in this manner. A plot of this type is shown in Fig. 3 for Goldschmidt's data [19] on the halogen acids in

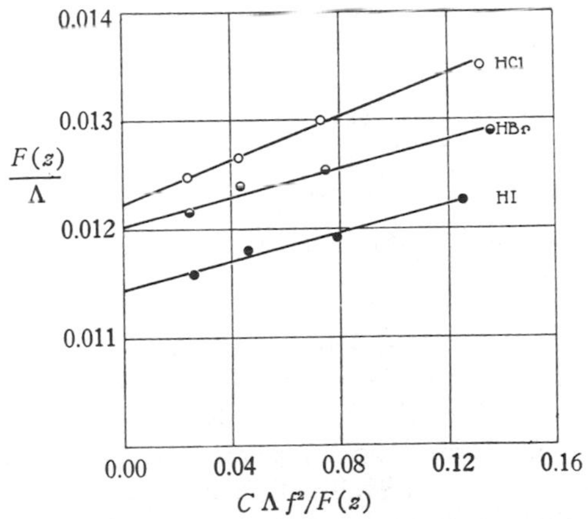

Fig. 3. Plots of Equation (19) for the Hydrogen Halides in Ethyl Alcohol at 25°.

ethyl alcohol at 25°. These yield the values of **K** and Λ_0 shown in Table VII.

[19] H. Goldschmidt and P. Dahll, *Z. physik. Chem.*, **114A**, 1 (1925).

TABLE VII. LIMITING CONDUCTANCES AND DISSOCIATION CONSTANTS OF THE HALOGEN ACIDS IN ETHYL ALCOHOL AT 25° AS CALCULATED BY THE METHOD OF FUOSS AND KRAUS

Acid	Limiting Conductance Λ_0	Dissociation Constant **K**
HCl	81.8	0.015
HBr	83.1	0.022
HI	87.4	0.020

A weakness of this type of computation becomes evident when it is applied to electrolytes which are only slightly ionized. In such cases straight lines are obtained from which the slopes, $1/(\mathbf{K}\Lambda_0^2)$, can be accurately estimated. However, the scale of the intercept, $1/\Lambda_0$, is, for weak electrolytes, so small that Λ_0 cannot be found with accuracy. An example is shown in Fig. 4 which is based on Goldschmidt's data

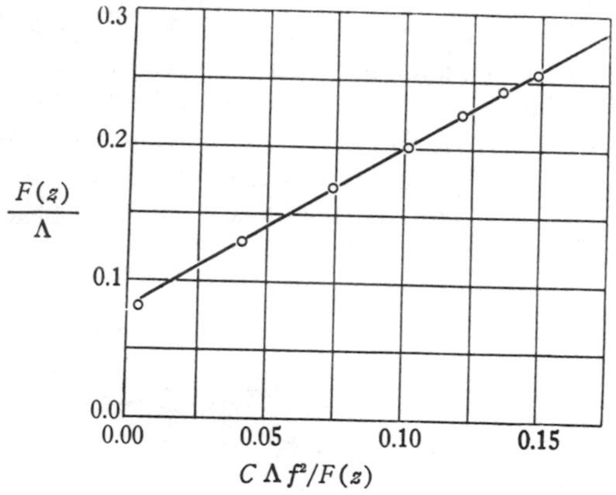

Fig. 4. Plot of Equation (19) for Picric Acid in Ethyl Alcohol at 25°.

for picric acid. In this case, however, it is possible to estimate Λ_0 independently and thus arrive at a value of **K**. This is done by obtaining, by the method of Fuoss and Kraus as just described, Λ_0 values for sodium picrate, sodium chloride and, as already shown, hydrochloric acid. These are all sufficiently ionized in ethyl alcohol so that the limiting equivalent conductances may be found with some accuracy. With these data it is evident that the limiting equivalent conductance of picric acid may be obtained using Kohlrausch's law of independent ion migration, *i.e.*,

$$\Lambda_{0HP} = \Lambda_{0HCl} - \Lambda_{0NaCl} + \Lambda_{0NaP}$$

in which HP and NaP represent picric acid and sodium picrate respectively. This computation yields 84.7 for the Λ_0 value of picric acid. With the slope, $1/(\mathbf{K}\Lambda_0^2)$, of the plot shown in Fig. 4, this gives 1.17×10^{-4} for the ionization constant, **K**, at 25° and in ethyl alcohol.

The Conductance of Salts in Solvents of Low Dielectric Constant. In order to approach a discussion of phenomena that are encountered in solvents of low dielectric constants, *i.e.*, less than 25, it will be of service to consider as examples the experimental results of Kraus and Fuoss [20] who have determined the conductance of a single salt (tetraisoamylammonium nitrate) in mixtures of widely varying composition of dioxane and water. The dielectric constants of these mixtures covered the range of values from the dielectric constant of 2.2 for pure dioxane to 78.6 for water. The experimental results are plotted in Fig. 5, in

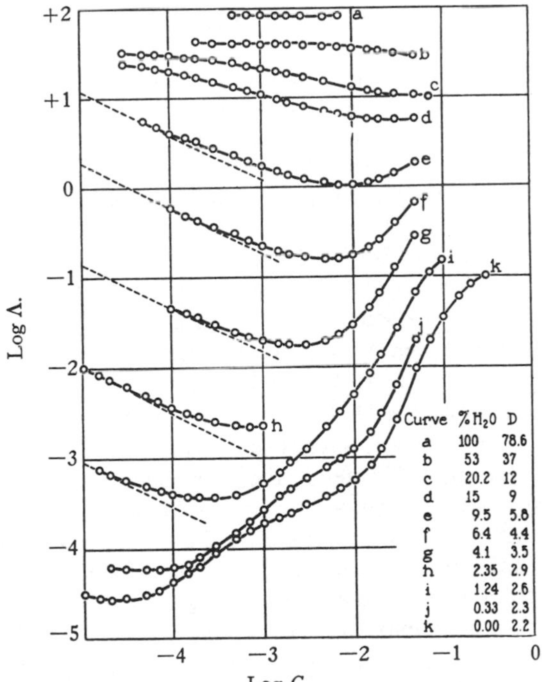

Fig. 5. The Conductance of Tetraisoamylammonium Nitrate in Dioxane-Water Mixtures at 25°.

[20] C. A. Kraus and R. M. Fuoss, *J. Am. Chem. Soc.*, **55**, 21 (1933).

which values of the logarithm of the equivalent conductance, log Λ, are plotted as ordinates against the logarithm of the concentration. The curves show a number of peculiarities that are of interest in connection with the discussion which follows. The curve corresponding to 100 per cent water is nearly flat, and is close to that which would be found if the equivalent conductance data of any strong electrolyte were plotted in this manner. The curve for a solvent containing 53 per cent water, which mixture has a dielectric constant of 37, indicates lower equivalent conductances, but has nearly the same shape as that for pure water. When, however, the solvent contains 15 per cent water and has a dielectric constant of 9, a distinct minimum is observed in the curve, and as this percentage is still further lowered the minimum persists, but moves progressively to the region of smaller concentrations. These results are, roughly speaking, typical logarithmic conductance-concentration curves for uni-univalent electrolytes in both pure solvents and mixtures of solvents having equivalent dielectric constants.

The explanation of these results proposed by Kraus and Fuoss [21] in a series of papers, is based on two assumptions. The first of these is that electrolytes that are completely dissociated in water or any other solvents of high dielectric constant will be more or less associated into ion pairs in solvents of low dielectric constants. Ion pairs, AB, are considered to form *entirely by electrostatic forces* from the charged ions A^+ and B^-, and the complexes are assumed to take no part in the conduction. Though no sharp division has been made experimentally these ion pairs are considered to differ from the undissociated portion of a weak electrolyte in that no electron shift has occurred in their formation. The second assumption is that, as the concentration of the ion pairs increases, a proportion of them will combine with ions by electrostatic forces, to form "triple ions."

A modification of the Debye-Hückel theory to include the possible formation of ion pairs was suggested as early as 1926 by Bjerrum.[22] As has been shown in Chapter 7, the unmodified form of the Debye-Hückel theory, leads, even with water solutions, to absurdities, for electrolytes with small ions, or salts of the higher valence types; and the "extended theory" was necessary to account for these cases.

Strong attractions of oppositely charged ions exist, when such ions are in close proximity, if (a) the ions are small, (b) they are of high valence, and (c) the dielectric constant of the solvent is small. These conditions all tend to favor the formation of ion pairs. Bjerrum's

[21] R. M. Fuoss and C. A. Kraus, *J. Am. Chem. Soc.*, **55**, 476, 1019, 2387 (1933); **57**, 1 (1935).
C. A. Kraus, *Trans. Electrochem. Soc.*, **66**, 179 (1934).

[22] N. Bjerrum, *Det. Kgl. Danske viden.*, [VII] **9**, 2 (1926).

theoretical discussion deals essentially with the same problem as that of the "extended theory," and may be considered as an alternative solution of the problem. The following treatment is based on Bjerrum's original paper. More rigorous, but more elaborate, discussions of the same subject have been published by Fuoss[23] and by Kirkwood[24] both of which agree, essentially, with Bjerrum's treatment, which follows.

We are concerned with the effects close to an ion, which will be assumed to hold a charge $z_1\epsilon$. According to the Boltzmann equation, already used in equations (1) and (28), Chapter 7, the time average number, dn_2, of ions in an elementary volume, dV, near such an ion will be given by the relation

$$dn_2 = n_2 e^{-\frac{W}{kT}} dV \tag{20}$$

in which W is the work necessary to separate a pair of the ions. If the ions are spherical and in close proximity the work, W, will be $+ z_1z_2\epsilon^2/Dr$,[25] in which D is the dielectric constant and r is the distance separating the centers of the two charged bodies. Equation (20) thus becomes

$$dn_2 = n_2 e^{\frac{-z_1z_2\epsilon^2}{DrkT}} dV \tag{21}$$

Now if the volume, dV, is a spherical shell, thickness dr, with the positive ion in the center, then

$$\frac{dn_2}{dr} = 4\pi n_2 r^2 e^{\frac{-z_1z_2\epsilon^2}{DrkT}} \tag{22}$$

The relations of the value of the differential coefficient dn_2/dr to the distance r and to the charges on the ions are of considerable interest. If the charges, $z_1\epsilon$ and $z_2\epsilon$, are zero the curve of dn_2/dr values will be of the form of I in Fig. 6. When the ions have charges of the same sign the relation will be of the form of Curve II of the figure. Curve III, the most important for the present discussion, is for the case of oppositely charged ions, and has a minimum. The value of r at this minimum may be found by differentiating equation (22) with respect to r and equating to zero with the result

$$r_{\min} = q = \frac{z_1z_2\epsilon^2}{2DkT} \tag{23}$$

[23] R. M. Fuoss, *Trans. Farad. Soc.*, **30**, 967 (1934).

[24] J. G. Kirkwood, *J. Chem. Phys.*, **2**, 767 (1934).

[25] This equation involves two assumptions: (a) that the effect of all ions other than the pair under consideration may be neglected and (b) that the effective dielectric constant for small values of r is the macroscopic value, D. Neither of these assumptions can be more than an approximation.

This minimum value of r is frequently referred to as the "Bjerrum distance" and given, as shown, the symbol q. Bjerrum makes the arbitrary assumption that ions inside the sphere with radius q are associated, and that those outside this sphere are free. The value of q evidently depends upon the ratio of the electrostatic potential energy of an ion pair, $z_1z_2\epsilon^2/Dr$, to the thermal kinetic energy of an ion, which is $\frac{3}{2}kT$, *i.e.*, to the ratio of the energies tending to hold an ion pair together and tending to knock it apart. According to equation (23) the distance

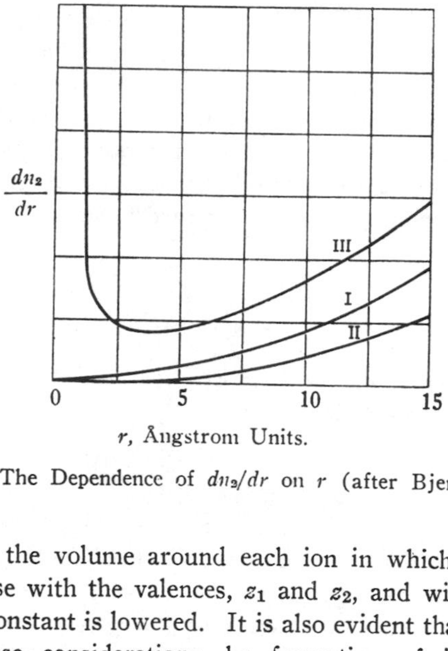

Fig. 6. The Dependence of dn_2/dr on r (after Bjerrum).

q, and therefore the volume around each ion in which ion pairs can exist, will increase with the valences, z_1 and z_2, and will also increase as the dielectric constant is lowered. It is also evident that there cannot, according to these considerations, be formation of ion pairs if q is equal to or smaller than the distance of closest approach, a_i, of the ions, since, unless $q > a_i$, there is no available volume in which they can form.

It is evidently of interest to obtain an expression for computing the extent of association of free ions into ion pairs as a function of the concentration, the distance of closest approach, the dielectric constant, etc. The following discussion will be restricted to uni-univalent electrolytes for which $z_1 = -z_2 = 1$. To obtain the time average number of ion pairs for each ion of opposite charge it is necessary to integrate through the volume between spheres with radii, a_i, the distance of closest approach, and the Bjerrum distance, q. For this purpose it is convenient to define

$$y = \frac{\epsilon^2}{DrkT} \text{ and } b = \frac{\epsilon^2}{Da_ikT}$$

with which equation (22) may be given the form

$$dn_2 = -4\pi n_2\left(\frac{\epsilon^2}{DkT}\right)^3 \frac{e^y}{y^4}\,dy \tag{24}$$

The number of ion pairs, per ion of given charge, is obtained by integrating between $y = b$ and $y = 2$, the latter corresponding to the condition that $r = q$. This integral is the degree of association which will be represented by θ, thus

$$\theta = \int dn_2 = \frac{4\pi c\mathbf{N}}{1000}\left(\frac{\epsilon^2}{DkT}\right)^3 \int_2^b e^y y^{-4} dy \tag{25}$$

Bjerrum[26] and Fuoss and Kraus[27] have evaluated, for various values of b, the integral

$$\int_2^b e^y y^{-4} dy = Q(b) \tag{26}$$

some values of which are given in Table VIII. This means, of course,

TABLE VIII. VALUES OF $Q(b) = \int_2^b e^y y^{-4} dy$

b	$Q(b)$	b	$Q(b)$
1	− 1.090	12	13.41
1.5	− 0.316	15	101.8
2	± 0.000	17	390
2.5	+ 0.188	20	3,900
3	0.325	25	2.24×10^5
4	0.550	30	1.55×10^7
5	0.755	40	1.03×10^{11}
6	1.041	50	9.6×10^{14}
7	1.417	60	9.55×10^{18}
8	1.996	70	1.113×10^{23}
9	2.950	80	1.42×10^{27}
10	4.547		

that the degree of association of an electrolyte may be computed if the ion size, the dielectric constant, and the concentration are known.

Bjerrum, further, applies the law of mass action to the equilibrium

$$A^+ + B^- = AB$$

[26] N. Bjerrum, *Det. Kgl. Danske viden.*, [VII], **9**, 2 (1926).
[27] R. M. Fuoss and C. A. Kraus, *J. Am. Chem. Soc.*, **55**, 1019 (1933).

Since the concentration of the ions is $c(1 - \theta)$ and of the ion pairs is $c\theta$ the mass law gives

$$\frac{c(1 - \theta)^2 f^2}{\theta} = \mathbf{K} \tag{27}$$

in which f is the activity coefficient of the ions, that of the ion pairs being assumed to be unity. For low values of θ and c this reduces to $\theta = c\mathbf{K}^{-1}$ so that equation (25) may be given the limiting form:

$$\mathbf{K}^{-1} = \frac{4\pi \mathbf{N}}{1000}\left(\frac{\epsilon^2}{DkT}\right)^3 Q(b) \tag{28}$$

from which the mass law constant, **K**, may evidently be obtained for any value of the distance of closest approach, a_i, and dielectric constant, D. The relation of log **K** to log D for a distance of closest approach, a_i, of 6.4 Å is shown in Fig. 7. As can be seen from the

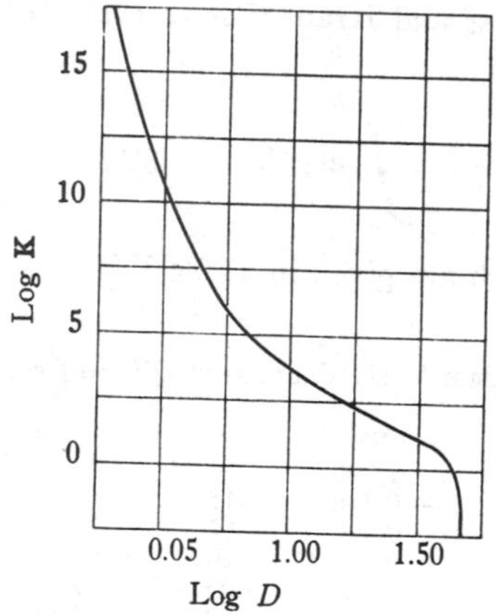

Fig. 7. The Variation of Log **K** with Log D when a_i=6.4Å. (after Fuoss and Kraus)

figure, **K** becomes smaller and smaller as the dielectric constant is decreased, indicating increased association into ion pairs. As D is increased the curve drops rapidly and becomes asymptotic at an ordinate corresponding to a value of about 44. This means that according to this equation, ions with a distance of closest approach of 6.4 Å or greater will be completely dissociated in solvents with a value of D greater than 44. For ions of higher valence types, or for ions of smaller sizes than those assumed above, the dielectric constant at which complete dissociation occurs would be higher. If we assume that the degree of dissociation, $\alpha = (1 - \theta)$, is sufficiently closely given, for low conducting solutions by the Arrhenius ratio, Λ/Λ_0, then equation (10) becomes

$$\frac{c\alpha^2}{(1 - \alpha)} = \frac{c(\Lambda/\Lambda_0)^2}{(1 - \Lambda/\Lambda_0)} = \mathbf{K} \tag{29}$$

and still more roughly

$$c\alpha^2 = c(\Lambda/\Lambda_0)^2 = \mathbf{K} \tag{30}$$

from which $d \log \Lambda / d \log c = -\frac{1}{2}$. Thus if the mass law holds between ions and ion pairs a plot of log Λ against log c would be expected, for dilute solutions and low dielectric constants to have the slope $-\frac{1}{2}$. That this is true is shown in Fig. 5 where the dotted lines have that slope and are parallel to at least a portion of the curve for the solvents of low dielectric constant.

As already mentioned, in addition to ion pairs, Kraus and Fuoss assume that combinations between ion pairs and ions may take place as follows

$$\begin{aligned} AB + A^+ &\leftrightarrows ABA^+ \\ AB + B^- &\leftrightarrows BAB^- \end{aligned} \tag{31}$$

The complexes ABA^+ and BAB^- are called "triple ions." The mass action constants for the two equilibria are assumed to have the same value, thus

$$\frac{(ABA^+)}{(AB)\,(A^+)} = \frac{(BAB^-)}{(AB)\,(B^-)} = \frac{1}{k} \tag{32}$$

The reactions (31) introduce new ions into the solution without greatly changing the value of (A^+), with the result that the equivalent conductance decreases less rapidly with increasing concentration, and at higher concentrations it actually increases. Assuming ions at concentrations low enough so that ion atmosphere effects may be neglected and a low degree of ionization, a relation between the equivalent conductance and the concentration may be obtained as follows. The total equivalent conductance, Λ, will be due to the equivalent conductance, Λ_1, due to the ions A^+ and B^- and to Λ_2 arising from the ions ABA^+ and BAB^-. The value of Λ_1 may be found from equation (30) giving

$$\Lambda_1 = \Lambda_{01}\sqrt{\frac{\mathbf{K}}{c}} \tag{33}$$

in which Λ_{01} is the limiting equivalent conductance of the ions A^+ and B^-. Equation (32) gives for the concentration $c' = (ABA^+) = (BAB^-)$

$$\frac{c'}{c^2(1 - \alpha)\alpha} = \frac{1}{k} \tag{34}$$

or approximately

$$\frac{c'}{c^2\alpha} = \frac{1}{k} \tag{35}$$

Therefore the equivalent conductance Λ_2 will be given by

$$\Lambda_2 = \Lambda_{02}\frac{c'}{c} = \Lambda_{02}\frac{c\alpha}{k} \tag{36}$$

in which Λ_{02} is the limiting equivalent conductance of the ions ABA^+ and BAB^-. With equation (30) this becomes

$$\Lambda_2 = \frac{\Lambda_{02}}{k}\sqrt{\mathbf{K}c} \tag{37}$$

so that

$$\Lambda = \Lambda_1 + \Lambda_2 = \Lambda_{01}\sqrt{\frac{\mathbf{K}}{c}} + \frac{\Lambda_{02}}{k}\sqrt{\mathbf{K}c} \tag{38}$$

A plot of this equation has the shape shown by all but the lower two curves in Fig. 5. Differentiating indicates that there is a minimum at

$$c_m = \frac{\Lambda_{01}}{\Lambda_{02}}k \tag{39}$$

so that if the constants Λ_{01} and Λ_{02} can be evaluated the mass law constant, k, may be determined from the position of this minimum.

The process of association of ions and ion pairs does not necessarily stop with the formation of ion pairs. In solvents of very low dielectric constants ion pairs may react to form complex molecules according to the equation

$$n\text{AB} = (\text{AB})_n \tag{40}$$

This may account for the shapes of the curves for the conductance data shown in Fig. 5 on solvents with very low dielectric constants, in which, in addition to the minimum just discussed, there are additional points of inflection.

Chapter 20

The Use of Conductance Measurements in Various Physico-Chemical Investigations

Conductance measurements are useful as aids in the solution of many physico-chemical problems. A few of the more important of these applications are (a) determination of the solubilities of certain substances, (b) estimation of the degree of hydrolysis of salts, (c) determination of speeds of reaction, (d) investigation of molecular complexes and (e) conductometric titrations. These will be considered in the order given. The discussions will be brief since the chief purpose of this chapter is to illustrate the use of conductance measurement as an analytical method in other than electrochemical fields of investigation.

Determination, by Conductance Measurements, of the Solubilities of Slightly Soluble Substances. The solubilities of slightly soluble salts can, in many cases, be obtained from determinations of the specific conductance, L, of their saturated solutions. The calculation involves equation (8a) of Chapter 3, which, for this case, may be put in the form

$$S = \frac{1000L}{\Lambda} \tag{1}$$

in which S is the solubility of the substance, in equivalents per liter, and Λ is its equivalent conductance. If the concentration of the saturated substance is sufficiently small, the value of Λ for a binary electrolyte may be assumed to be equal to the limiting equivalent conductance, Λ_0. The latter may be found from the relation

$$\Lambda_0 = \lambda_0^+ + \lambda_0^- \tag{2}$$

the values of λ_0^+ and λ_0^- being the equivalent conductances of the ion constituents obtained from the conductances of the more soluble substances, as explained in Chapter 18. If, however, the substance dissolved gives a solution so strong that Λ in equation (1) cannot, without error, be assumed to be equal to Λ_0, then some independent means must be devised for obtaining the equivalent conductance, Λ, at the concentration S. Due to the fact that most of the λ values, at least for uni-univalent strong electrolytes below a concentration of 0.01 normal, are independent of the oppositely charged ions with which they are associated, a curve, or an equation connecting $\Lambda = \lambda^+ + \lambda^-$ with the concentration can be used. From such a curve, or an appropriate equation, the value of Λ and the corresponding value of S may be found by a short series of approximations from the value of the specific conductance, L. This is essentially the procedure adopted by Kohlrausch.[1] The method is not well adapted to the determination of the solubilities of substances of great solubility.

The conductance method for obtaining the solubility of a very slightly soluble substance may be illustrated by the determination of the solubility of silver chloride in water at 25°. By a short interpolation of Kohlrausch's data at other temperatures the value 1.802×10^{-6} is found for the conductance of a saturated solution of silver chloride at 25°. From the equivalent conductances of the ion constituents given in Table V of Chapter 18, the limiting value of the equivalent conductance, Λ_0, for silver chloride is found to be 138.26 at 25°. Substituting these figures in equation (1) gives for the value of the solubility, S, 1.304×10^{-5} mol per liter at 25°. The estimate may be somewhat improved if instead of employing Λ_0 in the computation, a value of Λ from Onsager's expression,

$$\Lambda = \Lambda_0 - [\theta\Lambda_0 + \sigma]\sqrt{c}$$

is used. For silver chloride at 25° this equation is

$$\Lambda = 138.26 - 91.21\sqrt{c}$$

With the preliminary estimate of c just given, Λ is found to be 137.93, yielding with equation (1) the slightly higher value, 1.306×10^{-5}, for the solubility of silver chloride at 25°. This agrees, within experimental error, with the determination, described in Chapter 17, of the same quantity by Brown and MacInnes using a potentiometric method.

Determination of the Degree of Hydrolysis by Conductance Methods. Since hydrolysis of many salts is accompanied by relatively large changes in the conductance of the solutions, conductivity measurements may frequently be used as the basis for determining the extent to which hydrolysis has taken place. For instance, the chloride of a weak base, BOH, will hydrolyze, in aqueous solution, to form a definite amount of hydrochloric acid and an equivalent quantity of the undissociated base, as follows:

$$B^+ + Cl^- + H_2O \leftrightarrows BOH + H^+ + Cl^- \qquad (3)$$

[1] F. Kohlrausch, *Z. physik. Chem.*, **44**, 197 (1903); **64**, 129 (1908).

The conductance of the resulting solution is, of course, greater than it would be if this reaction did not occur, because the hydrogen ions, H^+, replace some of the less mobile B^+ ions.

The method for obtaining the degree of hydrolysis, h, from conductance measurements is as follows. Let Λ_S represent the equivalent conductance which the salt would exhibit if it were not hydrolyzed, Λ the observed equivalent conductance of the solution, and Λ_{HCl}, that of hydrochloric acid, all at the concentration C. Then, since the undissociated base plays no part in the conductance, its ionization being repressed in the presence of B^+ ion, we have the obvious relation

$$\Lambda = (1 - h)\Lambda_S + h\Lambda_{HCl} \quad (4)$$

from which

$$h = \frac{\Lambda - \Lambda_S}{\Lambda_{HCl} - \Lambda_S} \quad (5)$$

The solutions must, however, be sufficiently dilute so that the conductances of the ion constituents are substantially independent of the associated ions. The value of Λ_S can be obtained by adding enough of the free base to repress the hydrolysis of its salt practically completely, the dissociation of the added weak base being negligible in the presence of its salt. As an example of the use of the method, the degree of hydrolysis of aniline hydrochloride and the hydrolysis constant will be computed using the early conductance determinations of Bredig.[2] The data and the results of the computations are given in Table I. Due to

TABLE I. HYDROLYSIS OF ANILINE HYDROCHLORIDE AT 25°, DETERMINED FROM CONDUCTANCE MEASUREMENTS.

Dilution V	Concentration C	Λ	Λ_{HCl}	Λ_S	$100h$	$K_A \times 10^5$
32	0.03125	104.6	403.2	96.7	3.23	(3.37)
64	0.01563	111.5	408.5	99.9	3.76	2.30
128	0.007810	119.4	413.0	103.0	5.29	2.31
256	0.003906	128.1	416.5	105.1	7.39	2.30
512	0.001952	138.4	419.2	107.2	10.33	2.32
1024	0.000976	151.2	421.3	108.3	13.71	2.13

a change of standards since Bredig's measurements were made it has been necessary to correct his conductance values by a factor of 1.050, obtained by comparing his values of the conductances of potassium chloride solutions with recent values. The equivalent conductances of hydrochloric acid have been interpolated from the data in Table III of Chapter 18. Since the reaction indicated by equation (3) presumably follows the law of mass action we have, if the activity of the solvent is

[2] G. Bredig, *Z. physik. Chem.*, **13**, 289 (1894).

assumed to be constant,

$$\mathbf{K}_h = \frac{(BOH)\,(H^+)}{(B^+)} = \frac{[BOH]\,[H^+] f_H}{[B^+] f_B} \tag{6}$$

Since the activity coefficients tend to cancel and we have the relations,

$$[H^+] = [BOH] = hC \quad \text{and} \quad [B^+] = C(1 - h)$$

equation (6) may be put in the form

$$\mathbf{K}_h = \frac{h^2 C}{(1 - h)} \tag{7}$$

Values of $\mathbf{K}_h$, which will be seen to be quite reasonably constant, are given in the last column of the table.

Determination of Speeds of Reaction with the Aid of Conductance Measurements. Conductance measurements have been found useful in determining the speeds of reactions, particularly those involving organic substances. The large subject of chemical kinetics is not within the range of the topics to be considered in this book. The following few paragraphs will simply indicate, for two simple cases, the utility of the conductance method as a convenient analytical procedure in studying the kinetics of certain types of reactions. Many chemical reactions proceed with the formation of disappearance of ions, or changes in the nature of the ions present. The solutions in which such reactions occur show progressive shifts in their electrolytic conductances which permit one to follow the reactions as they proceed. Some examples of reactions that have been investigated by this means are diazotization,[3] molecular rearrangement,[4] saponification [5] and esterification.[6, 7]

As a typical case we may consider the work of Walker [5] on the saponification of methyl and ethyl acetates by sodium hydroxide, which for the latter may be represented by the equation

$$C_2H_5COOCH_3 + Na^+ + OH^- = C_2H_5OH + Na^+ + CH_3COO^- \tag{8}$$

During this reaction the conductance decreases, since hydroxyl ions are replaced by the slower moving acetate ions. The ester which is decomposed and the alcohol formed are non-electrolytes and thus have, in dilute solution at least, little effect on the conductance of the solution. The reaction involves the ester and the hydroxyl ion and thus would

[3] M. Schümann, *Ber.*, **33**, 527 (1900).
[4] J. A. Muller, *ibid.*, **43**, 2609 (1910).
[5] J. Walker, *Proc. Roy. Soc.*, **78A**, 157 (1906).
[6] J. F. Norris and A. A. Morton, *J. Am. Chem. Soc.*, **50**, 1795 (1928). J. F. Norris, E. V. Fasce and C. J. Staud, *ibid.*, **57**, 1415 (1935). R. F. Nielson, *ibid.*, **58**, 206 (1936).
[7] See also G. Jander and H. Immig, *Ber.*, **69B**, 1282 (1936) for a general discussion.

be expected to be of the second order. The conductances, L_0 and L_∞ at the beginning and end of the reaction are substantially equal to those of sodium hydroxide and sodium acetate at the same concentrations. Since the solutions used are dilute and the ionic strength does not change during the reaction, the conductances of the ion constituents OH^- and CH_3COO^- may be assumed, to very close approximation, to be constant, whether each one is the only negative ion present or is in a mixture of the two. If this is true the specific conductance will be a linear function of the proportion, x, of, say, the acetate ion in the mixture, and x may be obtained from the equation

$$x = \frac{L_0 - L_s}{L_0 - L_\infty} \tag{9}$$

in which L_s is the conductance at the time s.

If C is the concentration of acetate ion constituent at the time s, the differential equation relating the formation of that constituent to the concentration of the reactants is, if the reaction is of the second order,

$$\frac{dC}{ds} = k[C_2H_5COOCH_3][OH^-] \tag{10}$$

in which k is the velocity constant. In Walker's experiments the initial concentrations of the reactants were the same, and may be represented by C_0. Equation (10) can thus be given the form

$$\frac{dC}{ds} = k(C_o - C)^2 \tag{11}$$

and since $C = C_0x$, it becomes

$$\frac{dx}{ds} = kC_o(1 - x)^2 \tag{12}$$

Integrating this expression between the limits zero and s gives

$$k = \frac{1}{C_o s} \cdot \frac{x}{(1 - x)} \tag{13}$$

In Table II are given values of reaction velocity constants computed

TABLE II. DETERMINATION OF THE VELOCITY CONSTANT OF THE SAPONIFICATION OF ESTERS WITH SODIUM HYDROXIDE, AT 25°, FROM CONDUCTANCE MEASUREMENTS. INITIAL CONCENTRATIONS 0.01 NORMAL.

Methyl acetate			Ethyl acetate		
Time (minutes)	x	k	Time (minutes)	x	k
3	0.260	11.7	5	0.245	6.49
5	0.366	11.5	7	0.313	6.51
7	0.450	11.7	9	0.367	6.45
10	0.536	11.5	15	0.496	6.50
15	0.637	11.7	20	0.566	6.52

from equation (13), from Walker's data on the hydrolysis of ethyl and methyl esters. From these data it can be seen that a second order velocity constant is obtained when the saponification reactions of these two typical esters are followed by the conductance method.

Investigation of Molecular Complexes with the Aid of Conductance Measurements. Extensive investigations into the structures of complex salts have been carried out by Werner. From these researches he has been able to deduce important generalizations concerning these compounds.[8, 9] In explaining his experimental results Werner postulates that in addition to the primary valences of an element there are also *auxiliary valences* which are capable of attaching neutral molecules, such as molecules of water or ammonia. The total number of ionized radicals, such as Cl^- and NO_2^-, and of molecules which can be combined with an element is limited to the *coördination* number of the element. This number is apparently determined largely by spatial relations around the central atom. Cobaltic chloride, for instance, can form such complexes as $[Co(NH_3)_6]Cl_3$, $[Co(NH_3)_5H_2O]Cl_3$ and $[Co(NH_3)_3(H_2O)_3]Cl_3$ in all of which the coördination number is six. The part enclosed in brackets is in each of these cases a complex ion with three positive charges. The combined neutral molecules can, however, be successively replaced by negative radicals, producing a change in the charge on the complex ion. This charge is equal to the difference between the *primary valence* of the cobalt, which is three in this case, and the *number of univalent radicals* attached by the auxiliary valences. The following series illustrates this rule.

Name of Complex Salt	Formula
Hexammine cobaltic chloride	$[Co(NH_3)_6]^{+++} + 3Cl^-$
Chloropentammine cobaltic chloride	$[Co(NH_3)_5Cl]^{++} + 2Cl^-$
Dichlorotetrammine cobaltic chloride	$[Co(NH_3)_4Cl_2]^+ + Cl^-$
Trichlorotriammine cobaltic chloride	$[Co(NH_3)_3Cl_3]$
Potassium tetranitrodiammine cobaltiate	$K^+ + [Co(NH_3)_2(NO_2)_4]^-$
Potassium cobalti-nitrite	$3K^+ + [Co(NO_2)_6]^{---}$

A group represented within a bracket is held together by the primary and auxiliary valences and functions as a stable complex ion. In the case of dichlorotetrammine cobaltic chloride, for instance, the complex univalent cation, $[Co(NH_3)_4Cl_2]^+$, behaves as a univalent ion. The two chlorine atoms within the complex are not ions as they are not precipitated by silver nitrate. Similarly, trichlorotriammine cobaltic chloride, $[Co(NH_3)_3Cl_3]$, is a neutral complex, and has but a very

[8] A. Werner, "Neuere Anschauungen auf dem Gebiete der anorganischen Chemie," Vieweg, Braunschweig, ed. 4, 1920.

[9] A useful summary of the researches in this field is given in "The Chemistry of the Inorganic Complex Compounds," by R. Schwarz, translated by L. W. Bass, John Wiley and Sons, New York, 1923.

small conductance. The molar conductances of these complex compounds change with shifts in composition in just the manner that would be expected from the theory as outlined. Data for such a series are shown in Table III and a plot of the data given in the table is

TABLE III. MOLAR CONDUCTANCE VALUES FOR A SERIES OF COMPLEX COBALT SALTS, AT 25°.

$m = 0.001$

	Formula	Λ_m
1.	$[Co(NH_3)_6]^{+++} + 3Cl^-$	461
2.	$[Co(NH_3)_5(NO_2)]^{++} + 2Cl^-$	263
3.	$[Co(NH_3)_4(NO_2)_2]^+ + Cl$	105
4.	$[Co(NH_3)_3(NO_2)_3]$	1.6
5.	$K^+ + [Co(NH_3)_2(NO_2)_4]^-$	106
6.	$2K^+ + [Co(NH_3)(NO_2)_5]^{--}$	Unknown
7.	$3K^+ + [Co(CN)_6]^{---}$	459

shown in Fig. 1. These conductance data are clearly in accord with

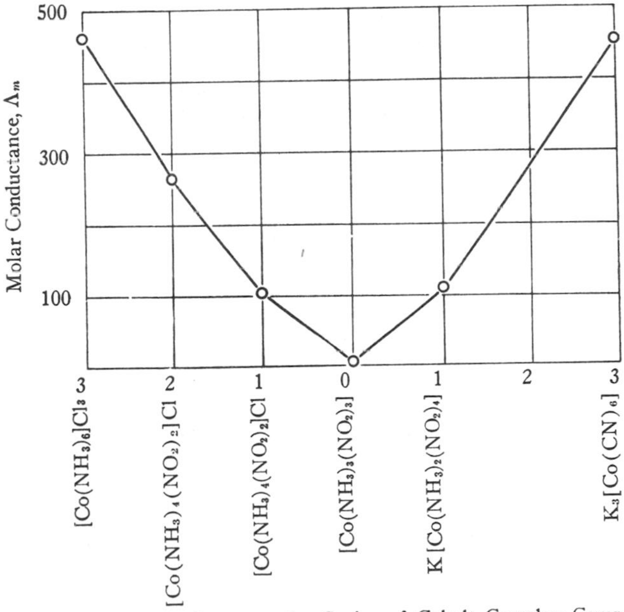

Fig. 1. The Molar Conductance of a Series of Cobalt Complex Compounds.

Werner's description of these compounds. The first and last compounds in the table have molar conductances of the same magnitude as that of, for instance, lanthanum chloride, $LaCl_3$, which yields four ions, and has a molar conductance of 411 at the same concentration and temperature. The second substance, $[Co(NH_3)_5(NO_2)]^{++} + 2Cl^-$, has a molar conductance of 263, closely agreeing with that of calcium chloride (260), whereas the third and fifth compounds have molar conductances corresponding in magnitude to most uni-univalent electrolytes. Many other examples could be given of the utility of conductance measurements in the field of molecular complexes and of related topics. However, the data given are sufficient to illustrate the method and the principles involved.

Conductometric Titrations. Volumetric analyses may be conveniently classified under the headings of neutralization, oxidation-reduction, and precipitation. During the progress of each of these types of reaction there is, in general, some change of electrical conductivity. It is frequently convenient to use this change of conductance to follow the progress of a reaction, and particularly to decide when it has been completed. The conductometric method may sometimes be used in connection with colored solutions, such as dyestuffs, in which ordinary indicators cannot be used. It can also be used to advantage in certain cases in which the potentiometric method of titration, described in Chapter 17, fails or is inconvenient.

A suitable form of cell for conductometric titrations is shown in Fig. 2. The electrodes E and E' are of platinum. A conductance deter-

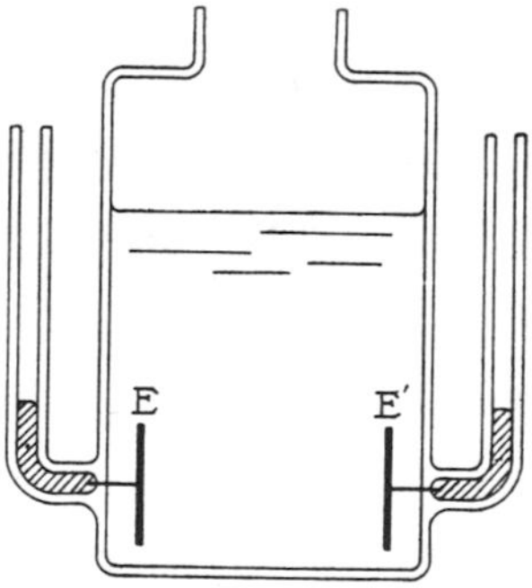

Fig. 2. A Cell for Conductometric Titration.

mination is made after each addition of reagent. It is desirable to maintain the temperature at a constant value, since conductances vary approximately two per cent per degree. The values of the conductances obtained, which may be simply relative, are plotted to give a graph which, as a rule, consists of straight, or nearly straight, lines intersecting at the equivalence point. When a strong acid is titrated with a strong base the conductance of the solution at first decreases, due to the replacement of hydrogen ions, which have a high mobility, by slower moving cations. At the equivalence point this decrease in conductance ceases and further addition of base causes an increase of conductance, since the hydroxyl ions are no longer neutralized but remain free to carry electricity. Fig. 3 in which conductances are plotted as ordinates and

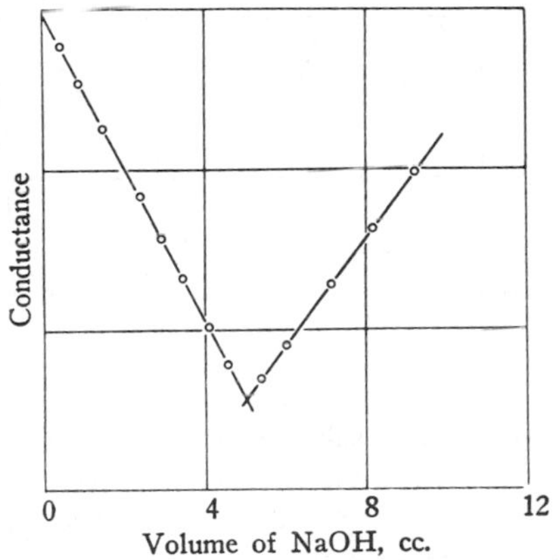

Fig. 3. The Titration of Hydrochloric Acid with 0.1010 Normal Sodium Hydroxide.

quantity of base added as abscissae represents the results of such a titration, from the measurements of Poethke.[10] Since the lines are practically straight and the angle of their intersection is acute, it is evidently easy to obtain the point of intersection from a few measurements on each side. Fig. 4, also from the results of Poethke, shows the results of a titration of sodium hydroxide with acetic acid. In this case there is a rapid drop of conductance due to the replacement of hydroxyl ion by acetate ion. However, after that replacement has been completed, there is little change in the conductance due to the fact that the ionization of the added acetic acid has been repressed by the presence of sodium acetate. The equivalence point may, however, be obtained by extending the two straight lines through the points representing the measurements.

An interesting contrast to the example just considered is given in the titration of a very weak acid, boric for example, with a strong base. A plot of the measurements made in such a titration is shown in Fig. 5.

[10] W. Poethke, *Z. anal. Chem.*, **86**, 45 (1931).

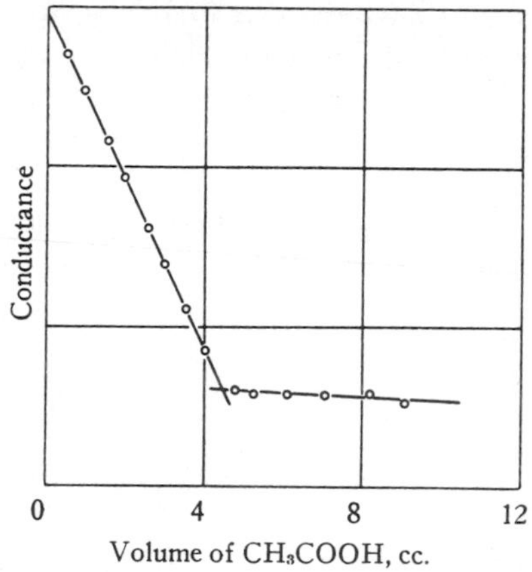

Fig. 4. The Titration of Sodium Hydroxide with 0.11 Normal Acetic Acid.

Here the acid, which yields a nearly non-conducting solution, is transformed progressively into a salt which is a strong electrolyte, after the completion of which process the conductance rises still more rapidly due to the increasing excess of highly conducting base. In contrast to other methods of titration, measurements near the equivalence point have no special significance. They may be, as in this case, worthless in the construction of the intersecting lines, since near the endpoint the products of the reaction may hydrolyze, or otherwise complicate the results, giving, for instance, points lying on the dotted line in the figure.

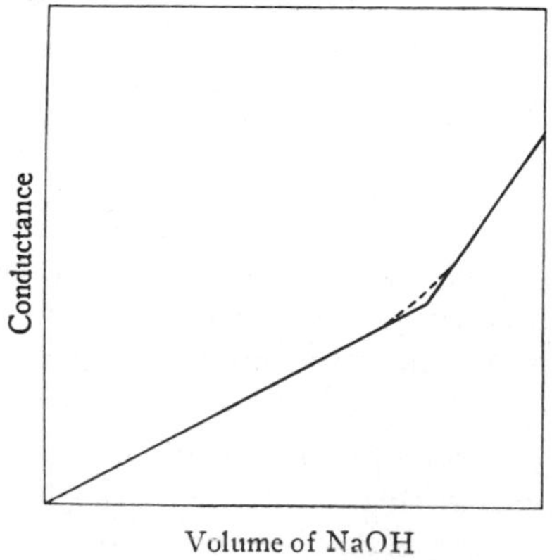

Fig. 5. The Titration of 0.01 Normal Boric Acid with Sodium Hydroxide.

For an acid of intermediate strength, such as acetic, the titration curves with a strong base take the form shown in Fig. 6, from the work of Kolthoff.[11] The neutral salt formed during the first part of the titration represses the ionization of the remaining free acid, and thus decreases the conductance due to that component. The increase of the salt concentration, however, produces a rise in the conductance. The result of these opposing effects is a curve with a minimum, the position of which, as is shown in the figure, depends upon the strength of the acid, and on its concentration.

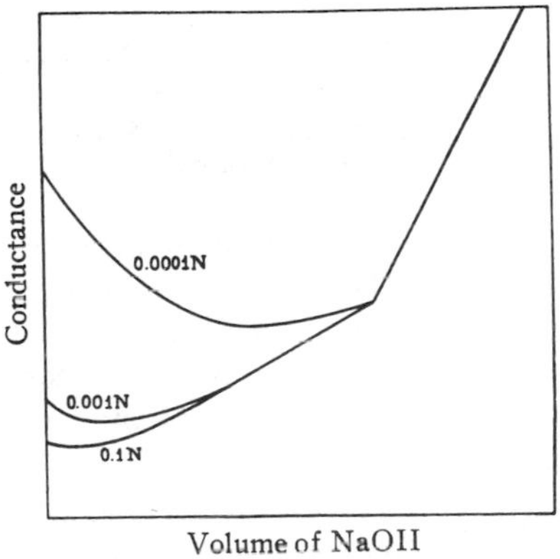

Fig. 6. The Titration of Acetic Acid, at Different Dilutions, with Sodium Hydroxide.

Conductometric methods are useful in the titration of mixtures of strong and weak acids. Thus an analysis of a mixture of acetic and hydrochloric acids can be carried out with fair accuracy. The result of such a titration is shown in Fig. 7, also due to Kolthoff. The added base first reacts with the strong acid producing a break in the curve at the equivalence point, followed by another break in the curve when the weak acid is neutralized. As shown, the first intersection is far from being sharp. This is due to the commencement of the ionization of the acetic acid before all the hydrochloric acid has disappeared. However, the equivalence point can be established by extending the straight portions of the lines as indicated.

The conductometric method of titration is of particular service when the reaction is the replacement of the anion of the salt of a weak acid with the anion of a strong acid. Thus, for example, in the titration of sodium acetate with hydrochloric acid the weak acid is liberated and

[11] I. M. Kolthoff, *Ind. Eng. Chem., Anal Ed.*, **2**, 225 (1930).

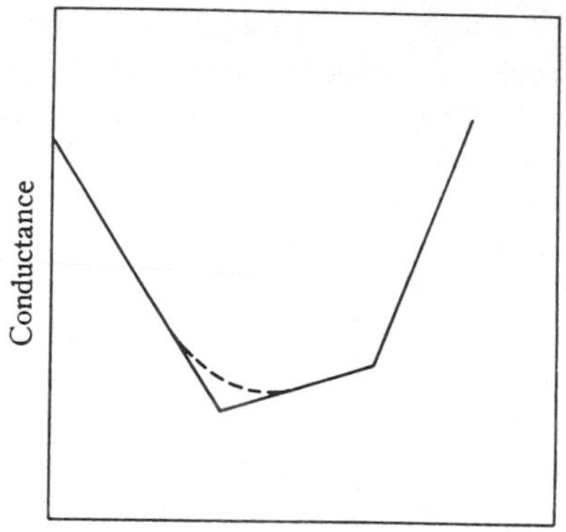

Fig. 7. The Titration of a Mixture of 0.01 Normal Hydrochloric Acid and 0.01 Normal Acetic Acid with Sodium Hydroxide.

sodium chloride is formed. This type of titration is difficult to carry out by any other method. The type of plot obtained is shown in Fig. 8. The first line is nearly horizontal since it results from the change of one strong electrolyte into another. After the equivalence point there is another line of greater slope produced by the increasing concentration of hydrochloric acid.

Oxidation-reduction titration using the conductometric method has been studied by Edgar[12] and others. One example, the oxidation of ferrous iron by a dichromate, may be represented as follows:

$$6Fe^{++} + Cr_2O_7^{--} + 14H^+ = 6Fe^{+++} + 7H_2O + 2Cr^{+++}$$

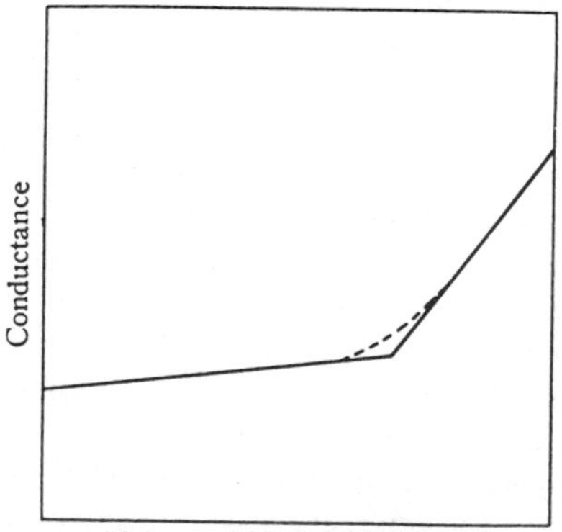

Fig. 8. The Titration of 0.01 Normal Sodium Acetate with Hydrochloric Acid.

[12] G. Edgar, *J. Am. Chem. Soc.*, **39**, 914 (1917). See also G. Jander and J. Harms, *Angew Chem.*, **48**, 267 (1935).

During this titration the highly conducting hydrogen ions disappear as the reaction takes place. As a result, the conductance progressively decreases to the endpoint and then rises very slowly. According to Edgar, the endpoint can be determined more accurately by the conductance method than with potassium ferricyanide as an outside indicator.

The conductometric method may also be used in many titrations involving precipitations. For instance Harned [13] has followed the conductance during the titrations of magnesium, nickel and cobalt sulphates with barium hydroxide. The resulting plots of the conductance against volumes of titrating fluid are all of the form shown in Fig. 9, two

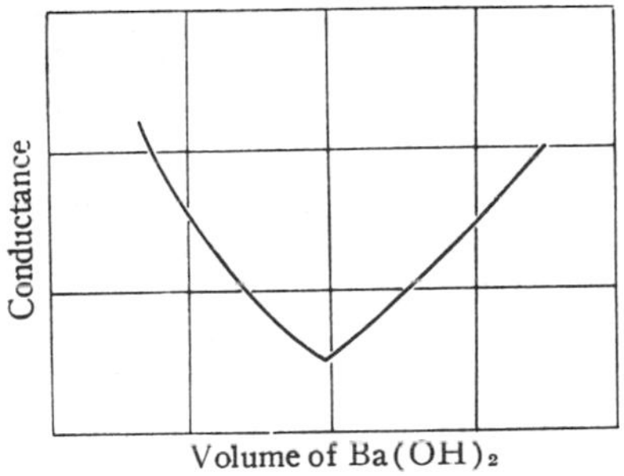

Fig. 9. The Titration of Magnesium Sulphate with 0.1709 Normal Barium Hydroxide.

curves rather than straight lines, meeting at the equivalence point. In a similar titration with cadmium sulphate, however, the results were complicated by the formation of a basic sulphate. In addition, difficulties have been encountered with precipitation titrations due to the slowness of formation of the solid and also due to surface adsorption, both of which would affect the conductance. Kolthoff and Kameda [14] have found that the titration of alkali sulphate with barium chloride is subject to considerable error, which increases with the concentration of the solution used. They attribute part of the error to conductance by the suspended solid. Dilute solutions may, however, be accurately titrated in the presence of alcohol. In brief, conductometric titrations involving precipitations may be carried out in certain cases, but they appear to be subject to sources of error not found with titrations in which no solid separates.

[13] H. S. Harned, *J. Am. Chem. Soc.*, **39**, 252 (1917).
[14] I. M. Kolthoff and T. Kameda, *Ind. Eng. Chem., Anal. Ed.*, **3**, 129 (1931).

Chapter 21

The Effect of Structure and Substitution on the Ionization Constants of Organic Acids and Bases

A large number of ionization constants of organic acids and bases have been determined by the methods already described and efforts have been made by a number of investigators to connect the values of the constants with the molecular structures of the compounds. The effect of the introduction of inorganic atoms and radicals, such as the halogens, into acids and bases has also been extensively studied. Much of the early data was obtained by Ostwald,[1] who made important deductions from the results.

Effect of Substituents on the Ionization Constants of Fatty Acids. Ostwald defined as a *negative* element or group one which, when substituted for a hydrogen in the $-CH_3$ group of acetic acid, raises the ionization constant of that acid, and a *positive* group as one that has the contrary effect. Negative groups are for instance the halogens, $-CN$, $-OH$, $-OCH_3$, and $-COOH$. Positive and negative groups and elements are classed together as *polar*.

Weakly polar groups are exemplified by methyl, $-CH_3$, and phenyl, $-C_6H_5$. Much effort has been expended in the attempt to find relations between the position of substitution of such weakly polar groups and the corresponding ionization constants, but without great success. For instance the increase in the length of a carbon chain of fatty acid has comparatively little effect, the change from acetic to propionic resulting in a shift of the ionization constant from 1.75×10^{-5} to 1.34×10^{-5}, and further lengthening of the chain producing a still smaller effect. The discussion given below will be confined to the effects produced by strongly polar groups, *i.e.,* substituents such as $-Cl$ and $-NH_2$.

Ostwald found that the nearer to the carboxyl group in an acid a substituent is placed the greater is the resultant positive or negative effect. Thus with a fatty acid α-chlor substitution, in for instance propionic acid, giving the compound, $CH_3 \cdot CHCl \cdot COOH$, has a much

[1] W. Ostwald, *Z. physik. Chem.,* **3,** 170, 241, 369 (1889).

larger effect on the strength of the acid than β-substitution, which would result in the compound, $CH_2Cl \cdot CH_2 \cdot COOH$.

Wegscheider [2] has computed a table of empirical factors based on the older data by which the constant of an unsubstituted acid may be multiplied to obtain the corresponding constant of a substituted acid. With chloride substitution these factors are 90, 6.2, 2.0 and 1.3 for the α, β, γ, and δ positions respectively. Derick [3] has proposed a "rule of thirds" which may be stated as follows. If $\mathbf{K}_u$ and $\mathbf{K}_{\alpha}$, $\mathbf{K}_{\beta}$, $\mathbf{K}_{\gamma}$, are the ionization constants of the unsubstituted acid and the α, β, and γ substituted acids, respectively, he finds that

$$\frac{\log \mathbf{K}_u}{\log \mathbf{K}_\alpha} - 1 : \frac{\log \mathbf{K}_u}{\log \mathbf{K}_\beta} - 1 : \frac{\log \mathbf{K}_u}{\log \mathbf{K}_\gamma} - 1 = 1 : \frac{1}{3} : \frac{1}{9}$$

approximately, for several series of strongly negative substituents. As has been mentioned no generally applicable rules seem to apply for weakly polar groups such as the methyl, $-CH_3$, and phenyl, $-C_6H_5$, groups. The same groups may even have effects of opposite polarity in different compounds.

Ives, Linstead and Riley [4] have made extensive studies on the effect of a double bond in the carbon chain of aliphatic acids. In general, the unsaturated acids are stronger than the saturated ones. However, the effect of the double bond decreases as it is removed from the carboxyl. These effects are shown in the results obtained by these authors quoted in Table I. The ionization constants of corresponding unsaturated and saturated acids may be compared in the third and fourth columns of the table. It will be seen that the unsaturated compounds are in all cases stronger acids than the saturated substances but that the ionization constants get nearer the same value as the double bond is displaced further and further along the chain.

TABLE I. EFFECT OF POSITION OF DOUBLE BOND ON THE IONIZATION CONSTANT OF UNSATURATED ALIPHATIC ACIDS

		Ionization Constants, $\mathbf{K} \times 10^5$	
Acid	Formula	Unsaturated acid	Corresponding saturated acid
Acrylic	$CH_2 : CHCOOH$	5.6	1.75
Vinylacetic	$CH_2 : CHCH_2COOH$	4.48	1.51
$\Delta\gamma n$-pentenoic	$CH_2 : CH(CH_2)_2COOH$	2.10	1.56
$\Delta\delta n$-hexenoic	$CH_2 : CH(CH_2)_3COOH$	1.90	1.40

[2] R. Wegscheider, *Monatsh*, **23**, 287 (1902).
[3] C. G. Derick, *J. Am. Chem. Soc.*, **33**, 1152, 1167, 1181 (1911).
[4] D. J. G. Ives, R. P. Linstead and H. L. Riley, *J. Chem. Soc.*, **1933**, 561.

MacInnes[5] has suggested, as a relation between the ionization constant and the position of substitution, the equation:

$$\mathbf{pK} = \mathbf{pK}_0 - s\frac{1}{d} \tag{1}$$

in which $\mathbf{pK}$ is the negative logarithm of the dissociation constant of the substituted acid, d is unity for the α-substituted acid, two for β-substitution, etc. The terms s and $\mathbf{pK}_0$ are constants. In Fig. 1

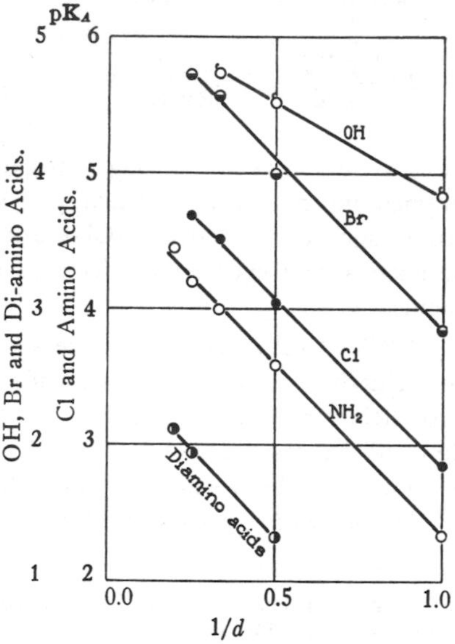

Fig. 1. The Relation Between the Logarithm of the Acid Dissociation Constant, $\mathbf{pK}_A$, in Series of Substituted Fatty Acids, and the Reciprocal of the Number of Carbon Atoms, d, between the Substituent and the Carboxyl Group.

the values of $\mathbf{pK}$ for several series of aliphatic acids are plotted as ordinates against values of $1/d$. It will be seen that the results are straight lines, indicating agreement with equation (1), the intercept on the zero axis being the constant $\mathbf{pK}_0$ and s the slope of the line. The term $\mathbf{pK}_0$ may be interpreted as the $\mathbf{pK}$ value of a substituted acid in which d has a very large value. It corresponds to a value of the ionization, $\mathbf{K}$, of the same order of magnitude but somewhat smaller than that of an unsubstituted long chain acid. It will be seen that the relation holds with fair accuracy for the $-Cl$, $-OH$, and $-NH_2$ sub-

[5] D. A. MacInnes, *J. Am. Chem. Soc.*, **50**, 2587 (1928).

stituted acids, and with less accuracy for the Br substitution. Decided deviations are observed for the iodine compounds, which are not included in the plot. The plot contains points representing, for amino-substitution, values of p**K** for acids in which the substituent has been placed on the terminal $-CH_3$ group of the chain, and also for substitution in a $-CH_2-$ group of a longer chain. It is true however that there is little difference between these two sets of values, as would be expected from what has already been said concerning the small influence of changing the length of a carbon-hydrogen chain on the ionization of a carboxyl group on one end of the chain. The values of the ionization constant of the amino acids given in Fig. 1 have been computed on the "zwitterion" or dipole ion basis, which will be explained in the next section. The data plotted in Fig. 1 are given in Table II. As will be

TABLE II. EFFECT OF THE POSITION OF THE SUBSTITUENT ON THE NEGATIVE LOGARITHM OF THE DISSOCIATION CONSTANT, p**K** IN SERIES OF SUBSTITUTED STRAIGHT CHAIN FATTY ACIDS

Substituent	α	β	γ	δ	ε
Cl	2.86 *(1,2)*	4.10 *(3)*	4.52 *(4)*	4.70 *(4)*	....
Br	2.89 *(3)*	3.99 *(3)*	4.58 *(4)*	4.72 *(4)*	....
I	3.16 *(3)*	4.09 *(3)*	4.64 *(4)*	4.77 *(4)*	....
OH	3.83 *(5)*	4.53 *(6)*	4.72 *(6)*	....	....
NH_2[a]	2.35 *(8)*	3.60 *(9)*	....	4.21 *(9)*	4.43 *(10)*
NH_2[b]	2.36 *(9)*	....	4.03 *(9)*	4.21 *(9)*	....

The substituent is on the terminal carbon except in the (*b*) series of the amino acids where the compounds are of constant chain length, *i.e.*, amino-*n*-valerianic acids. The methods used in the determination of the constants are given after each of the following references.

1. D. D. Wright, *J. Am. Chem. Soc.*, **56**, 314 (1934), concentration cells without liquid junctions (see Chapter 11).
2. T. Shedlovsky, A. S. Brown and D. A. MacInnes, *Trans. Electrochem. Soc.*, **66**, 165 (1934); B. Saxton and T. W. Langer, *J. Am. Chem. Soc.*, **55**, 3638 (1933), conductance.
3. E. Larson, *Z. physik. Chem.*, **A165**, 53 (1933), emf measurements in salt solutions, cells with liquid junctions (at 18°).
4. D. M. Lichty, *Liebig's Annalen*, **319**, 369 (1901), conductance values from International Critical Tables.
5. L. F. Nims, *J. Am. Chem. Soc.*, **58**, 987 (1936), concentration cells without liquid junctions.
6. G. Magnanini, *Gazz. chim. ital.*, **33**, [I] 197 (1893); W. Ostwald, *Z. physik. Chem.*, **3**, 369 (1889), conductance, values as given in International Critical Tables.
7. R. Anschütz and O. Motschmann, *Liebig's Annalen*, **392**, 100 (1912); W. Ostwald, *J. prakt. Chem.*, **32**, 300 (1885), conductance, values as given in International Critical Tables.
8. B. B. Owen, *J. Am. Chem. Soc.*, **56**, 24 (1934), concentration cells without liquid junctions.
9. C. L. A. Schmidt, W. K. Appleman and P. L. Kirk, *J. Biol. Chem.*, **81**, 723 (1929), p**K**′ values, electrometric titration.
10. J. I. Edsall and M. H. Blanchard, *J. Am. Chem. Soc.*, **55**, 2337 (1933), p**K**′ values, electrometric titration.

seen from the references the p**K** values have been determined by different methods. The constants based on the older conductance measurements are "classical" values. Where comparisons are possible these agree with the thermodynamic ionization constants at least to the order of precision used in the table. Fig. 1 also contains p**K**′ values for the acid dissociation constants of a series of diamino acids having the formula $NH_2(CH_2)_nCHNH_2COOH$. These values were obtained by Greenstein [6] and by Schmidt and associates [7, 8] and are given in Table III. (For a definition of p**K**′ see page 303.)

TABLE III. EFFECT OF THE POSITION OF A SECOND AMINO GROUP ON THE NEGATIVE LOGARITHM OF THE DISSOCIATION CONSTANT, pK'_A, OF A SERIES OF AMINO ACIDS

Acid	Position of second NH_2	pK'_A	Reference
α,β-diaminopropionic	β	1.33	*6*
α,δ-diaminovalerianic	δ	1.94	*7*
α,ϵ-diaminocaproic	ϵ	2.17	*8*

The Ionization of Amino Acids, The "Zwitterion" or "Dipole Ion." As is well known amino acids are of importance in that they are the structural units from which proteins are formed, and into which food proteins break down during digestion. The simplest compound of the series is glycine, NH_2CH_2COOH, which like all other amino acids contains an amino and a carboxyl group. The type formula for the series may thus be represented by NH_2RCOOH. Amino acids are able to combine with both acids and bases, *i. e.,* they are amphoteric.

From the theoretical point of view the greatest interest of amino acids is the possibility of the formation of "dipole ions" (a recent and more descriptive term than the German "zwitterion") by an internal rearrangement, or internal salt formation, according to the equilibrium

$$NH_2RCOOH \rightleftarrows {}^+NH_3RCOO^- \quad (2)$$

Although dipole ions were postulated as early as 1916 by Adams [9] and later by Bjerrum,[10] to explain phenomena occurring in solutions of amino acids, it is only recently that good evidence of their existence has been obtained. An important difference between the uncharged molecule and the dipole ion is that the latter has a much larger dipole moment. The evidence favorable to the presence of a large proportion of dipole ions in solutions of amino acids will be given in Chapter 22.

Another difference between the uncharged molecule and dipole ion, the two forms of an amino acid, lies in the reactions with strong acids

6 J. P. Greenstein, *J. Biol. Chem.*, **96**, 499 (1932).
7 W. Schmidt, P. L. Kirk and C. L. A. Schmidt, *ibid.*, **81**, 249 (1929).
8 C. L. A. Schmidt, P. L. Kirk and W. K. Appleman, *ibid.*, **88**, 285 (1930).
9 E. Q. Adams, *J. Am. Chem. Soc.*, **38**, 1503 (1916).
10 N. Bjerrum, *Z. physik. Chem.*, **104**, 147 (1923).

and bases. For the titration with sodium hydroxide, for instance, the two reactions are

$$NH_2RCOOH + Na^+ + OH^- = Na^+ + NH_2RCOO^- + H_2O \quad (3)$$

$$^+NH_3RCOO^- + Na^+ + OH^- = Na^+ + NH_2RCOO^- + H_2O \quad (4)$$

the reaction being with the carboxyl if the uncharged molecule is present, and with the NH_3^+ group if the substance consists of dipole ion. Equation (3) corresponds to the interaction of a weak acid and a strong base, and equation (4) to a strong base with a salt of a weak acid and a weak base. The two assumptions (3) and (4) make a formal difference in the interpretation of experimental data, reversing the role of the acid and basic constants. According to equation (9), Chapter 17, the apparent ionization constant, $\mathbf{K}'_A$, of a weak acid may be found from electrometric titration data by

$$p\mathbf{K}'_A = pH - \log \frac{[A^-]}{[HA]}$$

(in which $p\mathbf{K}'_A = -\log \mathbf{K}'_A$) so that for relation (3)

$$p\mathbf{K}'_A = pH - \log \frac{[NH_2RCOO^-]}{[NH_2RCOOH]} \quad (5)$$

The corresponding equation, (9a) of Chapter 17, for a weak base gives, with reaction (4)

$$p\mathbf{K}_w - p\mathbf{K}'_B = pH + \log \frac{[B^+]}{[B]} = pH + \log \frac{[^+NH_3RCOO^-]}{[NH_2RCOO^-]} \quad (6)$$

Thus as the concentration NH_2RCOOH according to the assumption of reaction (3) is numerically equal to $^+NH_3RCOO^-$ with the assumption of reaction (4), equations (5) and (6) give

$$p\mathbf{K}'_A = p\mathbf{K}_w - p\mathbf{K}'_B \text{ or } \mathbf{K}'_B = \mathbf{K}_w/\mathbf{K}'_A$$

It follows therefore that instead of obtaining the apparent ionization constant $\mathbf{K}'_A$ from the titration data according to the older ideas represented by equation (3) the same data yield, if the substance is in the form of the dipole ion, the corresponding constant of a weak base with the value $\mathbf{K}'_B = \mathbf{K}_w/\mathbf{K}'_A$.

The reactions of a strong acid with an amino acid may be represented by the two possible equations

$$NH_2RCOOH + H^+ + Cl^- = {}^+NH_3RCOOH + Cl^- \quad (7)$$

$$^+NH_3RCOO^- + H^+ + Cl^- = {}^+NH_3RCOOH + Cl^- \quad (8)$$

Reasoning such as that just given indicates that data obtained during electrometric titration with a strong acid gives the ionization constant of the substance as a weak base, $\mathbf{K}'_B$, if equation (7), which follows the older assumption, is assumed, and $\mathbf{K}'_A$, its ionization constant as an acid, if (8) is considered to be the reaction taking place. The positions of the acid and basic constants are thus seen to be reversed if the substance is present as dipole ions rather than uncharged molecules.

From the considerations just given it is evidently of interest and importance to determine the constant of the mass action expression

$$\mathbf{K}_Z = \frac{(^+NH_3RCOO^-)}{(NH_2RCOOH)} \tag{9}$$

which may be found by a method due to Adams,[11] Ebert,[12] and Edsall and Blanchard[13] which follows. The ion $^+NH_3RCOOH$ may give off a hydrogen ion from either the carboxyl or the NH_3^+ group. The two dissociation constants involved are

$$\mathbf{K}_a = \frac{(H^+)\,(^+NH_3RCOO^-)}{(^+NH_3RCOOH)} \qquad \mathbf{K}_b = \frac{(H^+)\,(NH_2RCOOH)}{(^+NH_3RCOOH)} \tag{10}$$

The two forms of the amino acid, ($^+NH_3RCOO^-$ and NH_2RCOOH), may also yield a hydrogen ion to give the NH_2RCOO^- ion, the constants being

$$\mathbf{K}_c = \frac{(H^+)\,(NH_2RCOO^-)}{(^+NH_3RCOO^-)} \qquad \mathbf{K}_d = \frac{(H^+)\,(NH_2RCOO^-)}{(NH_2RCOOH)} \tag{11}$$

From these expressions and equation (9) we have

$$\mathbf{K}_Z = \frac{\mathbf{K}_a}{\mathbf{K}_b} = \frac{\mathbf{K}_d}{\mathbf{K}_c} \tag{12}$$

The constant $\mathbf{K}_1$ for the first stage of ionization obtained by electrometric titration of the amino acid with a strong acid involves both types of ionization of the ion $^+NH_3RCOOH$ so that

$$\mathbf{K}_1 = \frac{(H^+)\{(^+NH_3RCOO^-) + (NH_2RCOOH)\}}{(^+NH_3RCOOH)} = \mathbf{K}_a + \mathbf{K}_b \tag{13}$$

Also the titration of the amino acid with a strong base includes the reaction with both the uncharged and dipole ion involving the constant

$$\mathbf{K}_2 = \frac{(H^+)\,(NH_2RCOO^-)}{\{(^+NH_3RCOO^-) + (NH_2RCOOH)\}} \tag{14}$$

[11] E. Q. Adams, *J. Am. Chem. Soc.*, **38**, 1503 (1916).
[12] L. Ebert, *Z. physik. Chem.*, **121**, 385 (1926).
[13] J. I. Edsall and M. H. Blanchard, *J. Am. Chem. Soc.*, **55**, 2337 (1933).

from which

$$\frac{1}{\mathbf{K}_2} = \frac{1}{\mathbf{K}_c} + \frac{1}{\mathbf{K}_d} \tag{15}$$

The constants $\mathbf{K}_1$ and $\mathbf{K}_2$ can be determined by direct experiment. With one additional relation the constants $\mathbf{K}_a$, $\mathbf{K}_b$, $\mathbf{K}_c$, $\mathbf{K}_d$ and $\mathbf{K}_Z$ may be obtained. This relation is furnished by the assumption that the value of the ionization constant $\mathbf{K}_b$ is the same as that, $\mathbf{K}_E$, of an ester of the same amino acid. Such an ester may be represented by NH_2RCOOR' in which the R′ may, for instance, be an ethyl, $-C_2H_5$, group. The assumption is contained in the equations

$$\mathbf{K}_b = \frac{(H^+)\,(NH_2RCOOH)}{(^+NH_3RCOOH)} = \mathbf{K}_E = \frac{(H^+)\,(NH_2RCOOR')}{(^+NH_3RCOOR')} \tag{16}$$

An ester of an amino acid obviously cannot form a dipole ion, so that its ionization must follow that just indicated. Furthermore, it has been shown earlier in this chapter that the addition of a carbon-hydrogen chain does not have more than a slight effect on an ionization constant. With the assumption that $\mathbf{K}_b = \mathbf{K}_E$ and equations (12) and (13) we have

$$\mathbf{K}_Z = \frac{\mathbf{K}_1}{\mathbf{K}_E} - 1$$

Since for aliphatic amino acids $\mathbf{K}_1$ is always very much larger than $\mathbf{K}_E$, as will be shown by data to be given below, this reasoning leads to the conclusion that the amino acids are almost entirely in the form of dipole ions. From equations (12), (13) and (15) the relations

$$\mathbf{K}_a = \mathbf{K}_1 - \mathbf{K}_E \quad (17); \qquad \mathbf{K}_c = \mathbf{K}_1\mathbf{K}_2/(\mathbf{K}_1 - \mathbf{K}_E) \tag{18}$$

an

$$\mathbf{K}_d = \mathbf{K}_1\mathbf{K}_2/\mathbf{K}_E \tag{19}$$

may readily be derived. These equations thus give the constants, $\mathbf{K}_a$, $\mathbf{K}_b$, $\mathbf{K}_c$, and $\mathbf{K}_d$ for the different types of ionization, as functions of the experimentally determined constants, $\mathbf{K}_1$, $\mathbf{K}_2$, and $\mathbf{K}_E$. Table IV from Edsall and Blanchard [13] gives the negative logarithm of the dissociation constants, $p\mathbf{K}_1$ and $p\mathbf{K}_2$ for a number of amino acids and their esters, $p\mathbf{K}_E$. A number of interesting conclusions follow from the data in this table. It will be seen that the $p\mathbf{K}_1$, $p\mathbf{K}_2$ and $p\mathbf{K}_E$ values for the α-amino acids are all very nearly the same, illustrating once more the fact that the length of the unsubstituted carbon-hydrogen chain has little effect on the ionization constants. The most important point of interest is the value of log $\mathbf{K}_Z$ which is, roughly speaking, independent of the

TABLE IV. NEGATIVE LOGARITHM OF THE DISSOCIATION CONSTANT OF SOME AMINO ACIDS AND THEIR ESTERS, AT 25°

Substance	Formula	pK_1	pK_2	pK_E	pK_Z	pK_d
Glycine	$NH_2CH_2 \cdot COOH$	2.31	9.72	7.73	5.42	4.30
α-Alanine	$CH_3 \cdot CHNH_2 \cdot COOH$	2.39	9.72	7.80	5.41	4.31
α-Aminobutyric acid	$CH_3CH_2CHNH_2 \cdot COOH$	2.55	9.60	7.71	5.16	4.44
Leucine	$(CH_3)_2CHCH_2CH(NH_2) \cdot COOH$	2.34	9.64	7.63	5.29	4.35
β-Alanine	$CH_2NH_2CH_2 \cdot COOH$	3.60	10.19	9.13	5.53	4.66
ε-Aminocaproic acid	$CH_2NH_2(CH_2) \cdot COOH$	4.43	10.75	10.37	5.94	4.81
Aspartic acid	$NH_2CH(COOH) \cdot CH_2 \cdot COOH$	2.08	3.87	6.5	...	...

distance between the carboxyl and amino groups, and corresponds to a value of $\mathbf{K}_Z$ of over 100,000, meaning, if the deductions given above are valid, that the concentration of dipole ions is 10^5 to 10^6 greater than that of the uncharged molecules.

From equations (17) and (18) it is evident that if $\mathbf{K}_1$ is very much larger than $\mathbf{K}_E$ as is the case for the substances listed in the table, then $\mathbf{K}_a = \mathbf{K}_1$ and $\mathbf{K}_c = \mathbf{K}_2$. Values of $p\mathbf{K}_d$ are given in the last column of the table.

As already mentioned, further evidence as to the presence of dipole ions will be given from measurements on dipole moments, in Chapter 22.

The Structure and Ionization of Dicarboxylic Compounds. If an organic acid contains two carboxyl, COOH, groups it is capable of giving off hydrogen ions in two steps, one to form singly and one to form doubly charged anions. Malonic acid, $CH_2(COOH)_2$, for instance may ionize in two stages according to the equations

$$CH_2(COOH)_2 = HOOCCH_2COO^- + H^+ \tag{20}$$
$$HOOCCH_2COO^- = CH_2(COO^-)_2 + H^+ \tag{21}$$

It is possible, by conductance measurements or by electrometric methods, to obtain dissociation constants, which will be denoted by $\mathbf{K}_1$ and $\mathbf{K}_2$, for these two stages of ionization. It is the purpose of this section to consider the relations of the values of these constants to the molecular structures of the molecules. Adams [11] has demonstrated that, from statistical considerations, the theoretical ratio of the two constants, $\mathbf{K}_1/\mathbf{K}_2$, is four. His reasoning is as follows. Representing a dibasic acid by HRR′H in which R and R′ represent the two halves of the molecule to which the ionizable hydrogens are attached, the acid may have a primary dissociation in two different ways

$$HRR'H = H^+ + {}^-RR'H \tag{22}$$

and

$$HRR'H = H^+ + HRR'^- \tag{23}$$

for which there are the mass action expressions

$$\mathbf{K}^{\mathrm{I}} = \frac{(\mathrm{H}^+)\,(^-\mathrm{RR'H})}{(\mathrm{HRR'H})}; \quad \mathbf{K}^{\mathrm{II}} = \frac{(\mathrm{H}^+)\,(\mathrm{HRR'}^-)}{(\mathrm{HRR'H})} \tag{24}$$

However, the experimental mass action constant is

$$\mathbf{K}_1 = \frac{(\mathrm{H}^+)\,(^-\mathrm{RR'H} + \mathrm{HRR'}^-)}{(\mathrm{HRR'H})} \tag{25}$$

and thus

$$\mathbf{K}_1 = \mathbf{K}^{\mathrm{I}} + \mathbf{K}^{\mathrm{II}} \tag{26}$$

Furthermore the ions $^-\mathrm{RR'H}$ and $\mathrm{HRR'}^-$ can both ionize according to the equations

$$^-\mathrm{RR'H} = {}^-\mathrm{RR'}^- + \mathrm{H}^+ \tag{27}$$

$$\mathrm{HRR'}^- = {}^-\mathrm{RR'}^- + \mathrm{H}^+ \tag{28}$$

and the corresponding constants are

$$\mathbf{K}^{\mathrm{III}} = \frac{(\mathrm{H}^+)\,(^-\mathrm{RR'}^-)}{(^-\mathrm{RR'H})}; \quad \mathbf{K}^{\mathrm{IV}} = \frac{(\mathrm{H}^+)\,(^-\mathrm{RR'}^-)}{(\mathrm{HRR'}^-)} \tag{29}$$

The experimentally determined second dissociation constant is

$$\mathbf{K}_2 = \frac{(\mathrm{H}^+)\,(^-\mathrm{RR'}^-)}{(^-\mathrm{RR'H}) + (\mathrm{HRR'}^-)} \tag{30}$$

From equations (29) and (30) we have

$$\mathbf{K}_2 = \frac{\mathbf{K}^{\mathrm{III}}\mathbf{K}^{\mathrm{IV}}}{\mathbf{K}^{\mathrm{III}} + \mathbf{K}^{\mathrm{IV}}} \tag{31}$$

If it is assumed that the acid is symmetrical, *i. e.,* that R′ is the same as R in the formula HRR′H and that the ionization processes represented by (22), (23), (27) and (28) are quite independent of each other, and have the same ionization constant, *i. e.,*

$$\mathbf{K}^{\mathrm{I}} = \mathbf{K}^{\mathrm{II}} = \mathbf{K}^{\mathrm{III}} = \mathbf{K}^{\mathrm{IV}} = \mathbf{K} \tag{32}$$

then

$$\mathbf{K}_1 = 2\mathbf{K} \quad \text{and} \quad \mathbf{K}_2 = \mathbf{K}/2$$

and

$$\mathbf{K}_1/\mathbf{K}_2 = 4 \tag{33}$$

It is evident that unless equation (32) is fulfilled the measured values of the ionization constants will deviate from relation (33). It is, therefore, important to consider the conditions of molecular structure which would make equation (32) possible. With a symmetrical dibasic acid of the type

$$HOOC(CH_2)_nCOOH$$

the two carboxyl groups will evidently be of the same strength, *i. e.*,

$$\mathbf{K}^{I} = \mathbf{K}^{II}$$

However, after one hydrogen has ionized giving

$$^{-}OOC(CH_2)_nCOOH$$

the fact that one end of the molecule now carries a charge will influence, to decrease (by a mechanism to be discussed below), the ionization of the second COOH group unless it is considerably removed from the charged end of the molecule. Thus if the carbon chain is not very long we have

$$\mathbf{K}^{I} = \mathbf{K}^{II} > \mathbf{K}^{III} = \mathbf{K}^{IV}$$

which may be easily shown to lead to

$$\mathbf{K}_1/\mathbf{K}_2 > 4$$

As a matter of fact in all recent measurements of the ionization constants of dibasic acids the first dissociation constant has been found to be considerably greater than four times the second.[14]

The variation of the ratio $\mathbf{K}_1/\mathbf{K}_2$ from the value of four has been explained on the basis just outlined by Bjerrum[15] who has shown as follows, how the ratio, $\mathbf{K}_1/\mathbf{K}_2$, should vary, quantitatively, with the length of the carbon chain. If there is a negative charge on a molecule resulting from the ionization of an acid the charge will attract hydrogen ions and there will be a distribution of the latter around the charge of the ion as a center according to Boltzmann's law (page 138).

$$dn = n_0\, e^{\frac{-W}{kT}}\, dV \tag{34}$$

in which n_0 is the average number of hydrogen ions per unit volume and dn is the number in the volume dV into which the work of intro-

[14] E. E. Chandler, *J. Am. Chem. Soc.*, **30**, 707 (1908), L. Rosenstein, *ibid.*, **34**, 1117 (1912) and E. Q. Adams and L. Rosenstein, *ibid.*, **36**, 1452 (1914), all report results for the ionization of dibasic organic acids in which the ratio $\mathbf{K}_1/\mathbf{K}_2$ is very nearly four. However, Chandler's results have not been verified by later, and presumably more accurate work, and the other measurements are based on colorimetric determinations on indicator solutions, involving many difficulties of interpretation.

[15] N. Bjerrum, *Z. physik. Chem.*, **106**, 219 (1923).

ducing an ion from an infinite distance is W. For molecular distances between the positive charge of a hydrogen ion and the ionized carboxyl group the work, W, will be given, approximately, by

$$W = \frac{-\epsilon^2}{Dr} \tag{35}$$

so that equation (34) becomes

$$\left(\frac{dn}{dV}\right) = n_0 \, e^{\frac{\epsilon^2}{DrkT}} \tag{36}$$

which may also be written

$$[H^+]_r = [H^+] \, e^{\frac{\epsilon^2}{DrkT}} \tag{37}$$

in which $[H^+]$ is the (average) hydrogen ion concentration and $[H^+]_r$ is the effective concentration at a distance r from the carboxyl. Since $[H^+]_r$ for small values of r will be greater than $[H^+]$ this increased hydrogen ion concentration will tend to oppose the ionization of another carboxyl group near the negative charge. If the two COOH groups are separated by a distance a in the molecule then

$$[H^+]_a = [H^+] \, e^{\frac{\epsilon^2}{DakT}} \tag{38}$$

Representing the dissociation constant when $a = \infty$ by $\mathbf{K}_2^*$, then equation (33) may be written

$$\mathbf{K}_1/\mathbf{K}_2^* = 4 \tag{33a}$$

The effect of bringing up the charged group, COO^-, within a distance of a few Ångstrom units will be to increase the effective hydrogen ion concentration in the immediate vicinity of the second ionizing group, thus hindering the formation of the doubly charged ion $^-RR'^-$. The ionization, omitting activity effects, will follow the relation

$$\mathbf{K}_2^* = \frac{[H]_a[^-RR'^-]}{[HRR'^-]} = \frac{[H^+] \, e^{\frac{\epsilon^2}{DakT}}[^-RR'^-]}{[HRR'^-]} \tag{38}$$

However the measured ionization constant is

$$\mathbf{K}_2 = \frac{[H^+][^-RR'^-]}{[HRR'^-]}$$

so that

$$\mathbf{K}_2^* = \mathbf{K}_2 \, e^{\frac{\epsilon^2}{DakT}} \tag{39}$$

which with equation (33a) gives

$$\frac{\mathbf{K}_1}{\mathbf{K}_2} = 4\,e^{\frac{\epsilon^2}{DakT}} \tag{40}$$

It is evident from equation (40) that the distance a may be computed if values of the ratio $\mathbf{K}_1/\mathbf{K}_2$ are available. The equation may be rearranged to give

$$\log \frac{\mathbf{K}_1}{\mathbf{K}_2} - \log 4 = \frac{\epsilon^2}{2.303\,DakT} \tag{41}$$

and for water as solvent at 25° this becomes

$$\log \frac{\mathbf{K}_1}{\mathbf{K}_2} - \log 4 = \frac{3.08 \times 10^{-8}}{a} \tag{42}$$

Unfortunately, few data have been published on second dissociation constants, and different investigators give, in some cases, different orders of magnitude for the same constants.[16] The most reliable determinations of $\mathbf{K}_1$ and $\mathbf{K}_2$ on the series of dibasic acids of the type $HOOC(CH_2)_nCOOH$ appear to be those of Gane and Ingold,[17] who obtained their figures by the method of electrometric titration. Since the two constants in most cases did not greatly differ, the interpretation of the experimental results is, as explained in Chapter 17, difficult and not very accurate. The results are however given in Table V.

TABLE V. THE IONIZATION CONSTANTS OF ALIPHATIC DIBASIC ACIDS AND THE COMPUTATION OF INTERGROUP DISTANCES

Acid	Formula	Ionization Constants $\mathbf{K}_1 \times 10^5$	Ionization Constants $\mathbf{K}_2 \times 10^6$	$\mathbf{K}_1/\mathbf{K}_2$	Distance a Ångstrom units
Malonic	$HOOC(CH_2)\,COOH$	149.	2.0_3	734	2.4_4
Succinic	$HOOC(CH_2)_2COOH$	6.4_1	3.3_3	19.2_5	4.5_2
Glutaric	$HOOC(CH_2)_3COOH$	4.5_3	3.8_0	11.9_2	6.5_0
Adipic	$HOOC(CH_2)_4COOH$	3.8_2	3.8_7	9.8_7	7.8_5
Pimelic	$HOOC(CH_2)_5COOH$	3.2_8	3.7_7	8.7_0	9.1_3
Suberic	$HOOC(CH_2)_6COOH$	3.0_4	3.9_5	7.7_0	10.8_3
Azelaic	$HOOC(CH_2)_7COOH$	2.8_1	3.8_5	7.3_0	11.7_9

The table also includes values of the ratio $\mathbf{K}_1/\mathbf{K}_2$ and of the distance a computed from equation (42). It is of considerable interest that these distances are of the expected order of magnitude. For the higher members of the series the distance a increases roughly 1.3 Ångstrom units for each $^-CH_2{}^-$ group introduced between the two carboxyl groups. This is at least comparable with the distance 1.53 Å as found by x-ray

[16] The data and their relevance to the matter under discussion are summarized in a paper by J. Greenspan, *Chem. Rev.*, **12**, 339 (1933).

[17] R. Gane and C. K. Ingold, *J. Chem. Soc.*, **1931**, 2153.

measurements, between the carbon atoms in the diamond. The steady increase of the *a* values in Table V would appear to indicate that the carbon atoms are connected together in a long straight, or nearly straight chain.

In this connection the effect of substitution of methyl and ethyl groups for hydrogens in the carbon chain is interesting. The data for a series of substituted acids are given in Table VI and are also from

TABLE VI. THE IONIZATION CONSTANTS OF SUBSTITUTED GLUTARIC ACIDS

Acid	Formula	$K_1 \times 10^5$	$K_2 \times 10^7$	*a* Ångstrom units
Glutaric	$HOOC \cdot (CH_2)_3 \cdot COOH$	4.5_3	$38._0$	6.5_0
β-methyl glutaric	$HOOC \cdot CH_2\overset{CH_3}{\underset{H}{\dot{C}}} \cdot CH_2COOH$	5.6_5	5.9_6	2.2_4
β-*n*-propyl glutaric	$HOOC \cdot H_2\overset{C_3H_7}{\underset{H}{\dot{C}}} \cdot CH_2COOH$	4.8_7	4.1_1	2.0_9
β,β-dimethyl glutaric	$HOOC \cdot CH_2\overset{(CH_3)_2}{\dot{C}} \cdot CH_2COOH$	$19._8$	5.1_0	1.5_5
β,β-diethyl glutaric	$HOOC \cdot CH_2\overset{(C_2H_5)_2}{\dot{C}} \cdot CH_2COOH$	$33._8$	0.75	1.5_0

the work of Gane and Ingold. While the glutaric acid has an *a* value of 6.5 Ångstrom units, β-substitution of a methyl or propyl group reduces this to a little over two Ångstroms, and this is reduced still further by substituting two groups in the β-position. The simplest explanation of this observation is that while the glutaric acid has a straight or zig-zag structure

$$HOOC - \overset{H}{\underset{H}{C}} - \overset{H}{\underset{H}{C}} - \overset{H}{\underset{H}{C}} - COOH$$

substitution on the middle carbon atom causes the molecule to bend, bringing the two carboxyl groups closer together as in the formula

$$\begin{array}{l} \quad\; H_2 \\ \quad\; C - COOH \\ RCH \\ \quad\; C - COOH \\ \quad\; H_2 \end{array}$$

Conclusions in this field are, however, highly speculative.

Chapter 22

The Dielectric Constants of Liquids and the Electric Moments of Molecules

Although the dielectric constant, D, has been frequently mentioned in the foregoing chapters its meaning and method of measurement have not been so far considered. The literature concerning dielectric constants, and the related subject of electric moments, is large and rapidly increasing. In the following pages only an outline of the subject can be given, and it will be restricted to material of present electrochemical interest.

The most familiar use of the dielectric constant is in connection with Coulomb's law:

$$F = \frac{\epsilon_1 \epsilon_2}{Dr^2} \tag{1}$$

in which F is the force of repulsion of two small bodies carrying electric charges, ϵ_1 and ϵ_2. In this formula r is the distance separating the centers of the bodies and D is the dielectric constant of the medium in which the bodies are placed. As is well known the repulsion becomes an attraction if the charges ϵ_1 and ϵ_2 have opposite signs. For a vacuum D equals unity by definition. An equally familiar use of the dielectric constant is in connection with the capacities of electrostatic condensers. If C_0 is the capacity of a condenser when its plates are separated by air, which is little different in its effect from a vacuum, then the capacity of the condenser, C, using any other medium is given by

$$C = DC_o \tag{2}$$

in which D is again the dielectric constant of the medium.

The present chapter will consider briefly several of the methods used for determining the dielectric constant. This discussion will be followed by some interpretations of dielectric constant measurements which are of interest in electrochemistry; and finally the evaluation of the electric moment of substances from dielectric constant measurements will be considered.

Methods for Determining the Dielectric Constant. The method used by Drude[1] for determining the dielectric constant depends upon the fact that, unless dispersion (to be discussed below) takes place, the dielectric constant may be determined from the relationship

$$D = \frac{\lambda_0^2}{\lambda^2} \tag{3}$$

obtained from the electromagnetic theory of light by Maxwell, in which λ_0 is the wave length of an electromagnetic wave in air, assumed to be the same as in a vacuum, and λ is the corresponding wave length in another medium, the dielectric constant of which is desired. Drude's apparatus is indicated in principle in Fig. 1. Electromagnetic waves of

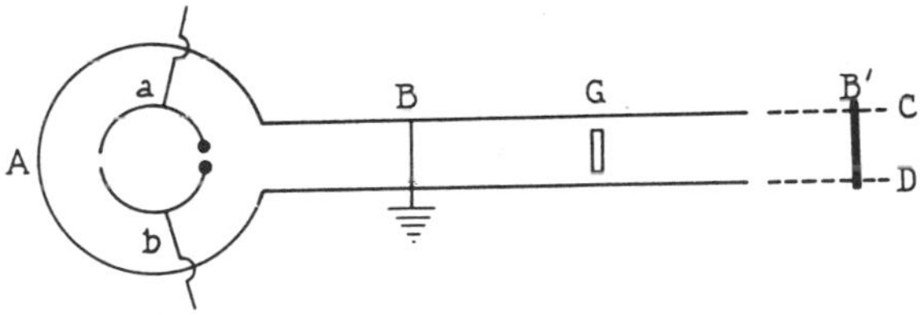

Fig. 1. Diagrammatic Arrangement for Measuring Dielectric Constants by Drude's Method.

a constant length are excited by producing sparks from an induction coil across the gap between the semicircular conductors *a* and *b*. The wire loop, *A*, is connected to the parallel ("Lecher") wires, *CD*, which are connected by the grounded wire, *B*. When the apparatus is operating, standing waves are formed by induction along the Lecher wires. When the wire bridge, *B'*, is at the appropriate point, corresponding to a node in the standing wave, a Geissler bulb will glow with maximum brightness due to the oscillating current induced from the loop *A*. By moving *B'* another point of maximum glow may be found, the difference between the two positions being half the wave length, λ_0. Minima in the glow of the indicator may also be used. Now by replacing the air between the wires by another fluid, the wave length, λ, in the new medium may be determined in the same manner, and the dielectric constant determined with the aid of equation (3). Drude's determinations of dielectric constants agree reasonably well with recent determinations by more accurate methods. The precision is however limited by the unavoidable influences of the surroundings of the wires, and in any

[1] P. Drude, *Wied. Ann.*, **55**, 633 (1895); *Z. physik. Chem.*, **23**, 267 (1897).

case the method requires a large amount of material. A modern variant of Drude's method will be described later.

The Wheatstone bridge, described in Chapter 3, has been adapted by Nernst [2] to the determination of dielectric constants. A more general form of equation (6), Chapter 3, is

$$Z_1 : Z_2 = Z_3 : Z_4 \tag{4}$$

in which the Z values representing the "impedances" of the four arms of the Wheatstone bridge, may be resistances, capacities, or inductances; or they may be combinations of any two of these or of all three.[3] The type of bridge used by Nernst is represented in Fig. 2. The resistances

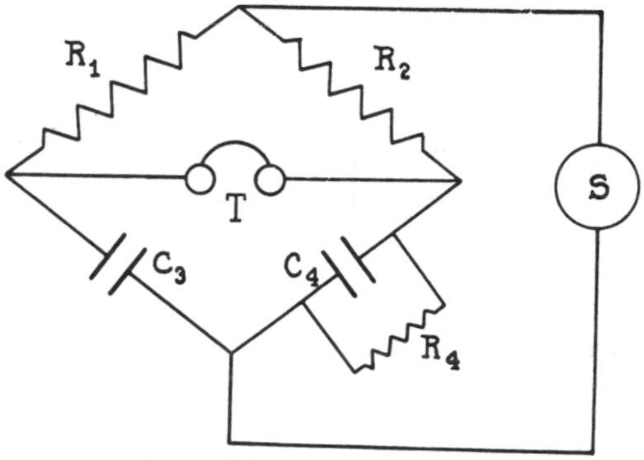

Fig. 2. Nernst's Bridge for the Measurement of Dielectric Constants.

R_1 and R_2 were tubes of an electrolyte consisting of a solution which is molar with respect to both mannite and boric acid. This has been found by Magnanini [4] to have a very low temperature coefficient of conductance. The liquid whose dielectric constant was desired was placed in a vessel of the type shown in Fig. 3. The metal shell, B, and the metal plate, A, held apart by the insulating cap, J, form the condenser C_3, of Fig. 2. The bridge was balanced by adjusting the distance between the plates of an air condenser, C_4, consisting of two sheets of brass, insulated from each other, which could be set at accurately known distances apart. Since most liquids and solutions have at least a slight electrolytic conductance, an adjustable electrolytic resistance, R_4, was

[2] W. Nernst, *Z. physik. Chem.*, **14**, 622 (1894); *Wied. Ann.*, **60**, 600 (1897).
[3] For a full discussion see B. Hague, "Alternating Current Bridge Methods," 2nd. ed. Sir Isaac Pitman and Sons Ltd., London (1930).
[4] G. Magnanini, *Z. Physik. Chem.*, **6**, 58 (1890).

also used. A small inductance coil, S, served as a source of oscillating current and a minimum of sound in a telephone, T, served for adjustment to bridge balance. The condition for a minimum is

$$R_1 : R_2 = C_3 : C_4 \tag{5}$$

so that if R_1, R_2 and C_4 are known C_3 may be obtained. The dielectric constant of a fluid was thus found by measuring the capacity, C, of a vessel, as is shown in Fig. 3, first containing air and then containing the fluid the dielectric constant of which was desired. If, however, the dielectric constant of this fluid was large, so that the ratios R_1 to R_2 and C_3 to C_4 became too great for the required accuracy to be obtained directly, the cell C_3 was filled with a liquid the dielectric constant of which had been determined by the method just described.

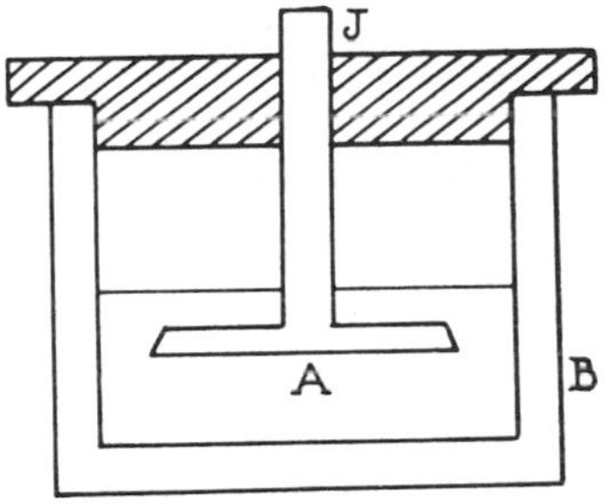

Fig. 3.

Although the Nernst method is subject to many minor sources of error Nernst's results on the dielectric constant of liquids are very close to those obtained by later and more precise measurements. The Wheatstone bridge method has been used by Smyth and associates.[5] In principle their bridge is essentially that shown in Fig. 2. However, a vacuum tube oscillator replaces the induction coil, and precision in locating the balance is attained with the aid of vacuum tube amplification. As in precise conductance measurements care must be exercised to avoid errors due to capacity effects between the different arms of the bridge, and to earth. This may be accomplished by appropriately designed electrostatic screening.[6]

[5] C. P. Smyth, S. O. Morgan and J. C. Boyce, *J. Am. Chem. Soc.* **50,** 1536 (1928). C. P. Smyth, "Dielectric Constant and Molecular Structure," Chemical Catalog Co. (Reinhold Publishing Corp.), New York, 1931.

[6] G. A. Campbell, *Electrical World,* April, 647 (1904). W. S. Shackleton, *Bell System Technical Journal,* **6,** 142 (1927).

A modification of Drude's standing wave method has been used by Drake, Pierce and Dow.[7] Instead of the two Lecher wires, which must be far removed from other objects if accurate results are to be obtained, these authors use one wire surrounded by a concentric metal tube. The apparatus is shown in Fig. 4. The metal tube, *T*, closed at its lower

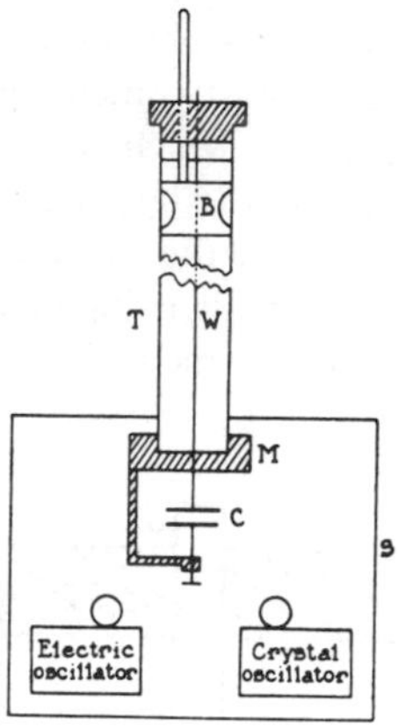

Fig. 4. Arrangement for Measuring Dielectric Constants by Means of Standing Waves.

end by the insulating bottom, *M*, surrounds the wire, *W*. A metallic bridge between the tube and wire, the position of which may be accurately adjusted, is shown at *B*. To the lower end of the wire is connected a small adjustable condenser, *C*. By connecting this condenser to the grid of an electric oscillator, indicated inside the metallic shield, *S*, standing waves can be set up in the tube, *T*. A sensitive galvanometer, recording changes in the plate current of the oscillator, indicates, by sudden variation in its readings, successive half-wave positions, as the bridge, *B*, is moved along the tube. The frequencies of the electric oscillator are accurately determined by comparison with the fundamental frequency or the harmonic frequencies of a piezoelectric quartz crystal oscillator, also shown in position in the shield, *S*. Although the method entirely overcomes errors due to exchange of energy with the surroundings, the theory of the operation of this apparatus is more complicated than that of Drude's apparatus. The reader is referred to the original article for details. The method, however, resembles Drude's in requiring a large volume of material. The

[7] F. H. Drake, G. W. Pierce and M. T. Dow, *Phys. Rev.*, **35**, 613 (1930).

only published results obtained with it are values of the dielectric constant of water as a function of the temperature which are the basis for equation (25b) of Chapter 7.

A more convenient method than that just described, and one which requires smaller amounts of material, has been used by Wyman.[8] The relation between the current, I, and the potential, E, of a circuit through which flows an alternating current of frequency, ν, is given by

$$I = \frac{E}{\sqrt{R^2 + \left(2\pi\nu L - \frac{1}{2\pi\nu C}\right)^2}} \tag{6}$$

in which R, L and C are respectively the "lumped" rather than the "distributed" values of the resistance, the inductance and the capacity. The current will evidently be a maximum when

$$2\pi\nu L = \frac{1}{2\pi\nu C}$$

and the frequency has the value,

$$\nu = \frac{1}{2\pi\sqrt{LC}} \tag{7}$$

This is the natural period of the circuit. In Wyman's method the period of a "resonator" of the general type shown in Fig. 5 is deter-

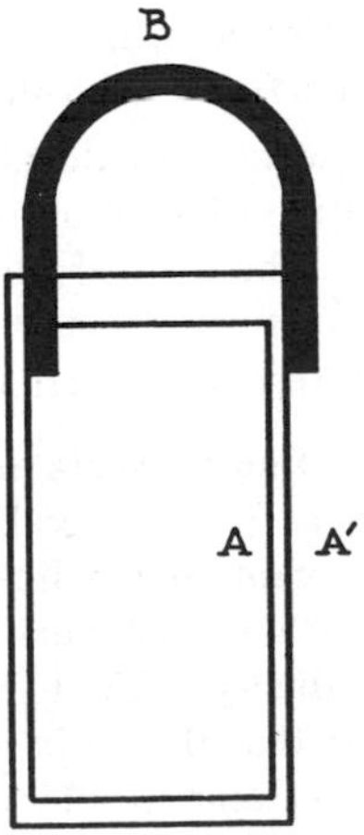

Fig. 5. A Resonator.

[8] J. Wyman, Jr., *Phys. Rev.*, 35, 623 (1930).

mined first in air and then placed in the liquid whose dielectric constant is desired. In this resonator practically all the capacity is located between the concentric cylinders, A and A', and the inductance is mainly confined to the single loop, B. The method for determining the dielectric constant consists first in determining the natural period, ν_0, when the resonator is filled with air and then the frequency, ν_1, when it is placed in the liquid the dielectric constant of which is desired. Since the inductance, L, remains constant the relation between these frequencies is, by equation (7)

$$\frac{C_o}{C_1} = \frac{\nu_1^2}{\nu_o^2} \tag{8}$$

Since the dielectric constant is given by the relation

$$C_1 = DC_o$$

we have

$$D = \frac{\nu_o^2}{\nu_1^2} \tag{9}$$

To determine the resonance frequency of the resonator it is suspended by a fine thread in a vessel containing the medium being studied, and is brought into the field of an oscillator whose frequency is controlled by a variable condenser. When this frequency corresponds to that of the oscillator there is a sudden change of the plate current of the oscillator indicating an absorption of power. In determining high dielectric constants it is frequently desirable to use, as reference, an intermediate liquid whose constant may be found with the aid of equation (9).

It must be emphasized that none of the methods described for determining the dielectric constant yields accurate results for liquids which are appreciably conducting. The methods differ in this regard. Those depending upon standing waves are reputed to be less sensitive to conductance than the bridge or resonance methods.

Dielectric Constants of Some Solvents and Solutions. The dielectric constant of the solvent, as will be recalled from the discussion in Chapters 7 and 18, is important in the interpretation of the thermodynamic and conductance data on solutions of electrolytes, according to the interionic attraction theory. Up to the present time the data which are useful for tests of that theory have mostly been obtained on aqueous and alcoholic solutions, and on solutions in mixtures of dioxane and water. It is to be hoped that in the near future studies will be made on solutions in other solvents the dielectric constants and other relevant properties of which are now, in many cases at least, accurately known. The dielectric constants of a few of the available solvents are listed in Table I.

TABLE I. THE DIELECTRIC CONSTANT OF A SERIES OF LIQUIDS AT 25°, AND THE CORRESPONDING DIPOLE MOMENT, OBTAINED FROM MEASUREMENTS OF THE TEMPERATURE COEFFICIENT OF THE DIELECTRIC CONSTANT IN THE VAPOR PHASE.

Substance	Dielectric Constant D	Dipole Moment $\mu \times 10^{18}$	Reference for D	Reference for μ
Hydrogen cyanide	116.1(20°)	....	*14*	*See 1*
Water	78.55	1.842	*16*	*17*
Deuterium oxide	78.25	....	*20*	....
Nitrobenzene	34.8	4.21	*2, 3*	*4, 5*
Methyl alcohol	31.5*	1.68	*8*	*18, 6*
Benzonitrile	25.20	4.39	*2, 3*	*4*
Ethyl alcohol	24.33	1.69	*9, see also 8*	*18, 19, 6*
n-Propyl alcohol	20.1	1.64	*8*	*6*
i-Propyl alcohol	18.0	1.68	*8*	*15, 6*
Ethylene dichloride	10.36	2.05	*3, 12*	*See 1*
Chlorobenzene	5.63	1.69	*12*	*5, 4*
Chloroform	4.71	1.00	*2, 12*	*See 1*
Diethyl ether	4.24	1.12	*10*	*See 1*
Benzene	2.27	0.00	*3*	*4, see also 1*
Carbon tetrachloride	2.23	0.00	*12, 13*	*17*
Dioxane	2.10*	0.00	*11*	*See 1*

* More recent work tends to indicate a slightly higher value.

1. N. V. Sidgwick, *Trans. Farad. Soc.*, **30**, appendix (1934).
2. A. O. Ball, *J. Chem. Soc.*, **1930**, 570.
3. S. Sugden, *ibid.*, **1933**, 768.
4. L. G. Groves and S. Sugden, *ibid.*, **1934**, 1094.
5. C. P. Smyth and K. B. McAlpine, *J. Chem. Phys.*, **3**, 55 (1935).
6. M. Kubo, see 7.
7. K. Higashi, *Sci. Papers Inst. Phys-Chem. Research (Tokyo)*, **28**, 284 (1936).
8. G. Åkerlof, *J. Am. Chem. Soc.*, **54**, 4125 (1932).
9. R. C. Gore and H. T. Briscoe, *J. Phys. Chem.*, **40**, 619 (1936).
10. J. Wyman, Jr., *J. Am. Chem Soc.*, **55**, 4116 (1933)
11. G. Åkerlof and O. A. Short, *ibid.*, **58**, 1241 (1936).
12. R. M. Davies, *Phil. Mag.*, **21**, 1 (1936); **21**, 1008 (1936).
13. H. O. Jenkins, *J. Chem. Soc.*, **1934**, 480.
14. K. Fredenhagen and J. Dahmlos, *Z. anorg. Chem.*, **179**, 77 (1929).
15. J. D. Stranathan, *J. Chem. Phys.*, **5**, 828 (1937).
16. Chapter 7.
17. R. Sänger, *Physik. Z.*, **31**, 306 (1930).
18. J. B. Miles, *Phys. Rev.*, **34**, 964 (1929).
19. A. L. Knowles, *J. Phys. Chem.*, **36**, 2554 (1932).
20. J. Wyman, Jr., and E. N. Ingalls, *J. Am. Chem. Soc.*, **60**, 1182 (1938).

As already mentioned dielectric constant measurements have given good evidence of the presence of dipole ion in aqueous or alcoholic solutions of amino acids and related compounds. The subject has been studied by Hedestrand[9] and Devoto,[10] but most recent and important investigations have been carried out by Wyman and McMeekin.[11, 12] These authors found for aqueous solutions of amino acids and peptides that the dielectric constant is accurately a linear function of the concen-

[9] G. Hedestrand, *Z. physik. Chem.*, **135**, 36 (1928).
[10] G. Devoto, *Gazz. chim. ital.*, **60**, 520 (1930); **61**, 897 (1932).
[11] J. Wyman, Jr., and T. L. McMeekin, *J. Am. Chem. Soc.*, **55**, 908 (1933); **55**, 915 (1933).
[12] J. Wyman, Jr., *Chem. Rev.*, **19**, 213 (1936).

tration, the equation being

$$D_s = D_0 + \delta C$$

in which D_s is the dielectric constant of the solution, D_0 that of the solvent, C the concentration and δ the "dielectric increment," *i. e.*, the increase of the dielectric constant produced by one mol of the solute. The values of δ for a group of amino acids and peptides are given in Table II.

TABLE II. THE DIELECTRIC INCREMENT, δ, OF AMINO ACIDS AND PEPTIDES, AT 25°

Substance	Formula	Dielectric Increment δ
Glycine	NH_2CH_2COOH	22.58
α-Alanine	$CH_3NH_2CHCOOH$	23.16
α-Aminobutyric acid	$CH_3CH_2NH_2CHCOOH$	23.16
α-Aminovaleric acid	$CH_3(CH_2)_2NH_2CHCOOH$	22.58
β-Alanine	$NH_2(CH_2)_2COOH$	34.56
β-Aminobutyric acid	$CH_3NH_2CHCH_2COOH$	32.36
γ-Aminovaleric acid	$CH_3NH_2CH(CH_2)_2COOH$	54.8
ϵ-Aminocaproic acid	$NH_2(CH_2)_5COOH$	77.5
Glycine dipeptide	$NH_2CH_2CONHCH_2COOH$	70.6
Glycine tripeptide	$NH_2CH_2CONHCH_2CONHCH_2COOH$	113.3
Glycine tetrapeptide	$NH_2CH_2CO(NHCH_2CO)_2COOH$	159.2
Glycine pentapeptide	$NH_2CH_2CO(NHCH_2CO)_2COOH$	214.5
Glycine hexapeptide	$NH_2CH_2CO(NHCH_2CO)_2COOH$	234.2

A number of interesting conclusions may be obtained from these figures. In the first place the values of δ are approximately the same for all the α-amino acids. The two β-amino acids listed in the table have δ values which are not very different, but are larger than those of the α-amino acids. The value of δ increases linearly with the number, n, of carbon atoms separating the amino and hydroxyl groups as is shown in Fig. 6, in which δ is plotted against n. The δ values are thus mostly determined by the spacing in the molecule of the two polar groups and are relatively uninfluenced by the rest of the carbon-hydrogen chain. A similar linear plot is obtained for the glycine polypeptide series if values of δ are plotted against the number of glycine units in the molecule.

It will be observed that the values of δ are all positive. The increases of the dielectric constant produced are larger than those obtained by the addition of any substances that are not amphoteric. As a matter of fact the addition of most substances to water produces a decrease of the dielectric constant. These facts are in favor of the assumption, discussed in Chapter 21, that amino acids are largely present as dipolar ions. If glycine, for instance, has the constitution

$$^{+}NH_3CH_2COO^{-}$$

it has a large "dipole moment," and this would, as we shall see in the next section, tend to produce relatively large increases in the dielectric constant of solutions. As will be shown shortly, electric moments can-

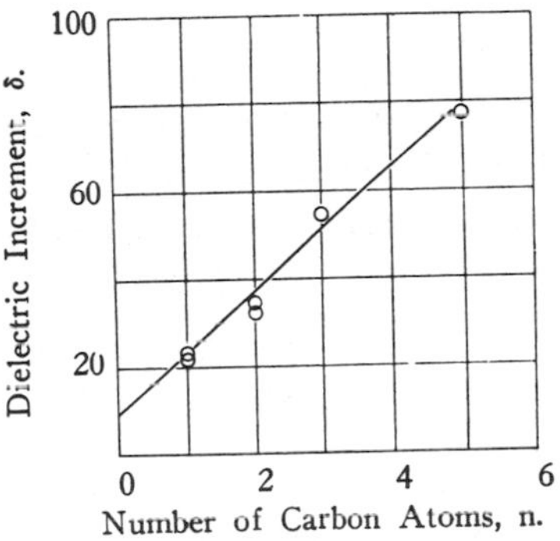

Fig. 6. Plot of the Dielectric Increment for a Series of Amino Acids against the Number of Carbon Atoms separating the Polar Groups.

not be obtained accurately unless the material is in the gaseous condition. Less accurate values may be determined from dielectric constant measurements in non-polar solvents. The actual dipole moments of ampholytes which are soluble only in polar solvents such as water and alcohol cannot therefore be obtained. Further evidence that increase of the dielectric constant is due to the formation of dipole ions and not to the uncharged molecules will be given from estimates on the dipole moments of the amino acid and peptide esters, to be discussed below.

The Interpretation of Dielectric Constant Measurements. Electric Moments. The interpretation of dielectric constant measurements is a matter of considerable difficulty. Although much has been accomplished in this field much more progress is to be expected from investigations which are actively proceeding at the present time. Fundamentally the dielectric constant of a medium is considered to be due to the presence of "electric moments." If two charges $+\epsilon$ and $-\epsilon$, which are of the same magnitude, are separated by a distance, d, they form an electric dipole, m, of moment ϵd. Such moments may be induced due to distortion of the atoms or molecules by an applied electric field, or they may be permanently present in the medium.

Originally, induced moments only were considered to be present in dielectrics, and the early theory was based on that assumption. Let us suppose that a molecule is located in a piece of dielectric through which the uniform electric field of strength E exists. Under these conditions we may assume that an electric moment, m, will be induced parallel to the direction of the field and of the value given by

$$m = \alpha_0 E \tag{10}$$

in which α_0 is the *polarizability,* and is a constant at least for an isotropic body and for not too high electric fields. In crystals α_0 may have different values along different crystal axes.

The polarizability is almost always computed from dielectric constant measurements by means of the Clausius-Mosotti equation

$$\frac{D - 1}{D + 2} = \frac{4\pi n}{3} \alpha_0 \tag{11}$$

in which n is the number of molecules per cubic centimeter. Although it is apparently valid, or very nearly so, for gases at least, there does not appear to be a really satisfactory derivation of the equation. Derivations according to the classical theory and with simplifying assumptions are given by Debye [13] and by Smyth.[14] Jeans [15] obtains a somewhat different form of the equation. Recently Kirkwood,[16] from statistical theory, has shown that the Clausius-Mosotti equation is the limiting form of a more general relation and is valid only at zero density. To avoid a long discussion of a much debated point which has little to do with electrochemistry, the relation between polarizability and dielectric constant will be assumed, with reservations, to be given by equation (11). The equation may be put into a more usable form by multiplying both sides of the equation by the molecular volume, M/ρ, in which M is the molecular weight and ρ the density. This gives

$$\frac{D - 1}{D + 2}\frac{\text{M}}{\rho} = \frac{4\pi}{3}\mathbf{N}\alpha_0 \equiv P \tag{12}$$

in which $\mathbf{N}$ is Avogadro's number and P is the molar polarization. So far, since purely electrical forces are uninfluenced by temperature changes, there is nothing to lead us to suspect that there should be a temperature coefficient of polarization. As a matter of fact the measured polarization of many substances has been found to be independent of the temperature. For such substances the polarization is found to agree

[13] P. Debye, "Polar Molecules," Chemical Catalog Co. (Reinhold Publishing Corp.), New York, 1929.

[14] C. P. Smyth, "Dielectric Constant and Molecular Structure," Chemical Catalog Co. (Reinhold Publishing Corp.), New York, 1931.

[15] J. H. Jeans, "Electricity and Magnetism," 5th ed. Cambridge University Press, 1927.

[16] J. Kirkwood, *J. Chem. Phys.*, 4, 592 (1936).

fairly closely with the molar refraction, P_E, defined by

$$P_E = \frac{n^2 - 1}{n^2 + 2} \frac{M}{\rho} \tag{13}$$

even when P is determined for long electric waves and the refractive index, n, is measured for visible light.

For many substances, however, the molar polarization, P, is decidedly a function of the temperature. Debye [17] in an important paper explained these cases as being due to the presence of *permanent* electric moments in the molecules. Although such molecules are electrically neutral, the center of the positive charges is considered to be at a different point from that of the negative charges. If this is true a molecule can have a polarization arising not only from atom and electron displacements but also by orientation in the electric field. Furthermore, it is to be expected that the amount of orientation in the field will be larger the less disturbed the molecules are by thermal agitation, so that the polarization would thus decrease with rising temperature, as is actually observed. Debye's reasoning is as follows. If a molecule has an electric moment, represented by μ, and it is present in an electrical field of intensity, E, then its potential energy, W, will be $-\mu E$ if it lies in the direction of the field, and will be

$$W = -\mu E \cos \theta \tag{14}$$

if, as is shown in Fig. 7, its moment makes an angle θ with the direction

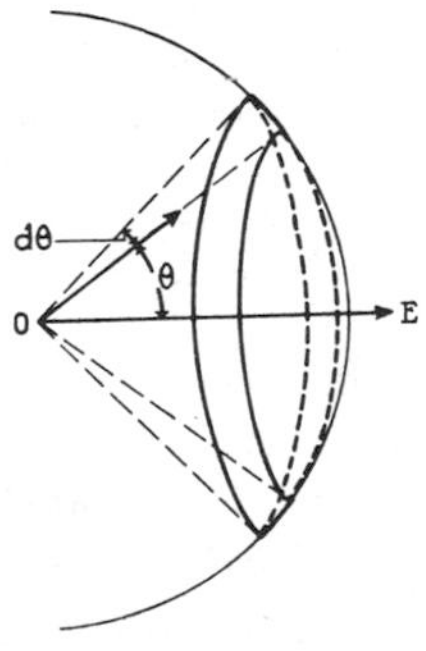

Fig. 7.

of the field. If no force is acting, the orientation of the electric moments will be equally distributed over all directions of space. Thus the num-

[17] P. Debye, *Physik. Z.*, 13, 97 (1912).

ber of components with moments pointing at angles between θ and $d\theta$ to the direction OE of the electric field will be, as is evident from Fig. 7, $A\,2\pi \sin\theta\, d\theta$, in which A is a constant depending upon the number of molecules considered. In a field of intensity E the number will be, by Boltzmann's law

$$A\,e^{-\frac{W}{kT}}\,dV = A\,e^{-\frac{W}{kT}} \cdot 2\pi \sin\theta\, d\theta \tag{15}$$

and with equation (14) it becomes

$$2\pi A\, e^{\frac{\mu E \cos\theta}{kT}} \sin\theta\, d\theta \tag{16}$$

A given molecule pointing in the direction $d\theta$ to the direction of the electric field has a component $\mu \cos\theta$. The average moment, $\bar{m}$, in the direction of the field, of one molecule, will thus be

$$\bar{m} = \frac{2\pi A \int e^{(\mu E/kT)\cos\theta}\, \mu\cos\theta \sin\theta\, d\theta}{2\pi A \int e^{(\mu E/kT)\cos\theta} \sin\theta\, d\theta} \tag{17}$$

the integration being over all possible directions. Substituting $x = \mu E/kT$ and $\xi = \cos\theta$ then equation (17) becomes

$$\frac{\bar{m}}{\mu} = \frac{\int_{-1}^{+1} e^{x\xi}\, \xi d\xi}{\int_{-1}^{+1} e^{x\xi}\, d\xi} \tag{18}$$

By performing the integration and rearranging we obtain

$$\frac{\bar{m}}{\mu} = \frac{e^x + e^{-x}}{e^x - e^{-x}} - \frac{1}{x} \tag{19}$$

which is a form of a function derived by Langevin [18] for computing the mean magnetic moment of gas molecules possessing a permanent magnetic moment. If values of $\bar{m}/\mu$ are plotted as ordinates against corresponding values of x, *i. e.*, $\mu E/kT$, the resulting curve is as shown in Fig. 8. It will be observed that for small values of x, or what amounts to the same thing, of the potential E, the function is approximately linear. For very high values of E, m/μ approaches unity, *i.e.*, the mean moment, $\bar{m}$, of the molecules becomes nearer and nearer the permanent moment μ. If, as is actually the case in the experimental work,

[18] P. Langevin, *J. Phys.* [4], 4, 678 (1905).

$x = \mu E/kT$ is a small number, then equation (19) takes the limiting form

$$\overline{m} = \frac{\mu^2}{3kT}E \tag{20}$$

For sufficiently small field intensities the apparent average moment of one molecule is thus proportional to the field intensity, E. Therefore instead of the simple relation given by equation (10), $m = \alpha_0 E$, the total electric moment is given by

$$m = \left(\alpha_0 + \frac{\mu^2}{3kT}\right)E \tag{21}$$

where α_0 is the polarizability by distortion and $\mu^2/(3kT)$ the polarizability by orientation. With this change equation (12) becomes

$$P \equiv \frac{D-1}{D+2}\frac{M}{\rho} = \frac{4\pi N}{3}\left(\alpha_0 + \frac{\mu^2}{3kT}\right) \tag{22}$$

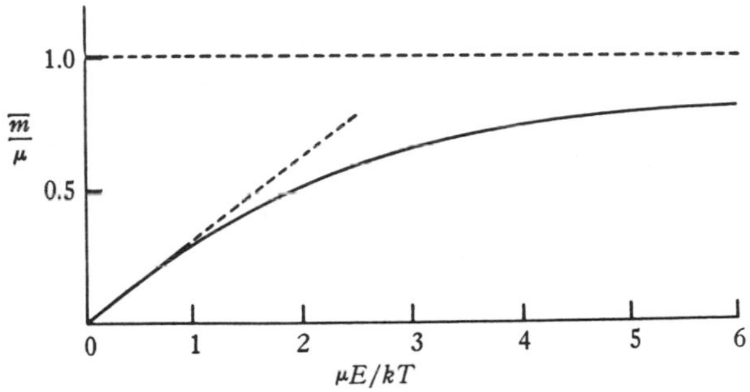

Fig. 8. The Langevin Function.

It is convenient for the purpose of discussion to divide the polarization, P, into three parts, P_E, P_A and P_M, which are respectively the polarizations due to electronic and atomic distortions, and to orientation. Of these P_E is given by equation (13). The polarization, P_A, is ordinarily small. It is determined, in general, by the relation

$$P_A = P - P_E - P_M \tag{23}$$

the last term being computed as will be presently described. Our interest is, however, mainly in the polarization, P_M, which is equal to

$$P_M = \frac{4\pi N}{9} \cdot \frac{\mu^2}{kT} \tag{24}$$

Its value is, as will be observed, inversely proportional to the temperature, which, interpreted physically, means that the tendency of the permanent moments of the molecules to orient with the field is disturbed by thermal vibration.

It is, of course, important to determine to what extent the theory from which equation (22) was obtained represents the results of experiments. The equation may be put into the form

$$P = a + \frac{b}{T} \tag{25}$$

from which, if the theory is valid, measured values of the polarization, P, should be linear functions of the reciprocal of the absolute temperature. Plots of the results of measurements on a number of typical gases are given in Fig. 9. It will be observed that the results may be

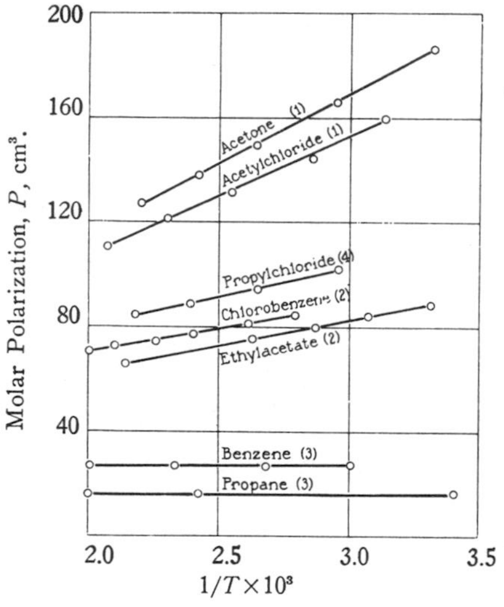

Fig. 9. The Temperature Dependence of the Molar Polarization for a Number of Substances in the Gaseous State.

The numbers on the plot correspond to the following references.
1. C. T. Zahn, *Physik. Z.*, **33**, 686, 730 (1932).
2. L. G. Groves and S. Sugden, *J. Chem. Soc.*, **1934**, 1094.
3. K. R. McAlpine and C. P. Smyth, *J. Am. Chem. Soc.*, **55**, 453 (1933).
4. R. Sänger and O. Steiger, *Helv. Phys. Acta*, **2**, 411 (1929).

accurately represented by straight lines. Of the lines shown those for propane and benzene are horizontal, indicating that the polarizations are independent of the temperature, *i. e.*, that $b = 0$. Since b is equal to $4\pi \mathbf{N}\mu^2/9k$, this means that $\mu = 0$ for these substances. The result is just what would be expected on physical grounds, as propane and benzene both have symmetrical structures. The other lines given in the plot have definite slopes, indicating finite values of the electric moment. It is of great interest that substances, the polarization measurements of which indicate electric moments, are those to which a symmetrical molecular structure cannot readily be assigned.

The values of the electric moment for a number of simple substances from the measurements of the dielectric constant at different temperatures in the gaseous state are given in Table I. The order of magnitude of these moments should be noticed. If as our theory assumes, the moments are due to charges separated by distances comparable to the diameter of a molecule, they should be comparable in magnitude to the product of an elementary charge, 4.770×10^{-10} e.s.u. and the diameter of a molecule which is of the order of several Ångstrom units, *i. e.*, 10^{-8} cm. The figures given in Table I are all small numbers multiplied by 10^{-18}. The factor, 1×10^{-18} e.s.u., cm., is frequently called the "Debye unit."

Many substances cannot be obtained in the vapor state at sufficient pressures to make measurements of dielectric constant possible. To overcome this difficulty a large number of dielectric constant measurements of solutions have been carried out. As solvents substances such as benzene, dioxane or hexane which possess no electric moments, were usually chosen. For binary solutions equation (22) may be given the form

$$\frac{D-1}{D+2} \cdot \frac{\mathrm{M}_1\mathrm{N}_1 + \mathrm{M}_2\mathrm{N}_2}{\rho_s} = P_1\mathrm{N}_1 + P_2\mathrm{N}_2 \tag{26}$$

in which M_1, M_2, N_1, and N_2 are respectively the molecular weights and mol fractions of the two components of the solution and ρ_s is the density of the solution. The quotient $(\mathrm{M}_1\mathrm{N}_1 + \mathrm{M}_2\mathrm{N}_2)/\rho_s$ is evidently the molar volume of the mixture. The terms P_1 and P_2 are the polarizations of the two components. If one of them is known the other may be computed from equation (26). If one of the substances (the solvent) is non-polar it is customary to assume that its polarization, P_1, is independent of the mol fraction, and any deviation from the linear relation indicated by equation (26) is assigned to a variation in P_2. A limiting value of P_2 may be obtained by determining its value for a number of low mol fractions, N_2, and extrapolating to $\mathrm{N}_2 = 0$. In this way the limiting polarization, P_2, should correspond to that of molecules of the solute surrounded only by non-polar molecules of the solvent. The effect of the electric moments on each other should thus be completely eliminated. By obtaining values of P_2 for a series of temperatures the electric moment, μ, may then be computed as has already been described for data on gases. In this way many of the electric moments found in the literature have been obtained. An extensive list of electric moments of substances is given by Sidgwick.[19]

It is unfortunately true, however, that electric moments of substances determined from measurements on the dielectric constants of solutions are not independent of the solvent used, as has been shown by Müller [20] and others. Some typical values of the apparent electric moment, μ, of nitrobenzene in a series of solvents are given in Table III, from the work of Jenkins.[21] For this case, at least, there is a tendency

TABLE III. THE ELECTRIC MOMENT OF NITROBENZENE FROM MEASUREMENTS IN VARIOUS SOLVENTS, AT 25°

Solvent	Dielectric Constant of Solvent	Electric Moment $\times 10^{18}$
n-Hexane	1.887	4.049
Cyclohexane	2.016	3.974
Dekalin	2.162	3.930
Carbon tetrachloride	2.228	3.932
Benzene	2.273	3.936
Carbon disulfide	2.633	3.658
Chloroform	4.722	3.172

for the measured electric moment to decrease as the dielectric constant of the solvent is increased. Although the measured values of the electric moment are usually lower when determined on the substance in solution than when obtained in the gaseous state, this is not always true. For instance, Fairbrother [22] has found that the apparent dipole moments of the three hydrogen halides are slightly greater in solution than in the gaseous state. The data are given in Table IV. It seems probable that

TABLE IV. THE ELECTRIC MOMENT OF THE HYDROGEN HALIDES IN THE GASEOUS STATE AND IN NON-POLAR SOLVENTS

Substance	Solvent	Dipole Moment, $\times 10^{18}$	
		In Solvent	In Gas
HCl	benzene	1.26	1.03
HBr	benzene	1.01	0.79
	chloroform	0.96	
HI	benzene	0.58	0.38
	chloroform	0.50	

there are interactions between the unsymmetrical electric fields of the polar solutes and the symmetrical fields of the non-polar solvents that

[19] N. V. Sidgwick, *Trans. Farad. Soc.*, **30**, appendix (1934).
[20] F. H. Müller, *Physik. Z.*, **33**, 731 (1932); **34**, 689 (1933); **35**, 346 (1934).
[21] H. O. Jenkins, *Nature*, **133**, 106 (1934).
[22] F. Fairbrother, *Trans. Farad. Soc.*, **30**, 862 (1934).

result in real or apparent changes in the electric moments of the polar substances. A number of theoretical investigations [23] have been made to account for the solvent effect on the measured electric moment. The fundamental question as to whether the Clausius-Mosotti equation should be used in the interpretations of dielectric constant measurements on liquids does not, however, appear to have a satisfactory answer.

Measurements of electric moments have been much used in connection with problems of molecular structure. An excellent summary of the conclusions in that field has been published by Sidgwick.[24] Certain questions can be decided from the presence or absence of an electric moment. Thus water, which has the comparatively large moment of 1.842×10^{-18}, must have an unsymmetrical structure. Three possible arrangements are:

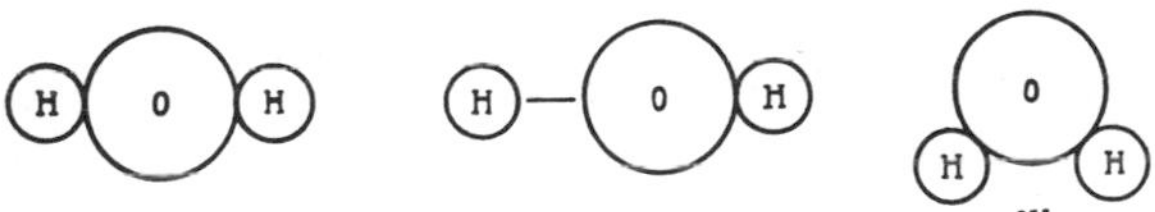

I II III

Fig. 10.

Of these I is symmetrical and would have no electric moment. II is unstable [25] while III is in accord with crystallographic evidence, and would have an electric moment. Similarly Errera [26] showed that of the two possible formulas for the di-halogen substituted products of ethylene:

I
```
H—C—X
  ‖
H—C—X
```

II
```
X—C—H
  ‖
H—C—X
```

in which X is a halogen atom, the structure II must be the correct one as the substances are non-polar.

Anomalous Dispersion at High Frequencies. All the experimental work so far described in this chapter was carried out using in the measurements electric currents of relatively low frequencies, and the theoretical deductions are based on phenomena observed at such

[23] C. V. Raman and K. S. Krishnan, *Proc. Roy. Soc.*, **A117**, 589 (1928).
J. Weigle, *Helv. phys. Act.*, **6**, 68 (1933).
K. Higashi, *Sci. Papers, Inst. Phys-Chem. Research (Tokyo)*, **28**, 284 (1936).
F. C. Frank, *Proc. Roy. Soc.*, **A152**, 171 (1935).
P. Debye, *Physik. Z.*, **36**, 100 (1935).

[24] N. V. Sidgwick, *Chem. Rev.*, **19**, 183 (1936).

[25] P. Debye, "Polar Molecules," p. 67, Chemical Catalog Co. (Reinhold Publishing Corp.), New York, 1929.

[26] J. Errera, *Physik. Z.*, **27**, 764 (1926).

frequencies. Some of these conclusions must, however, be modified if the region of high frequencies is investigated. It will be recalled that when an electric field is applied to a substance polarization will take place from (a) distortion or deformation of the atoms and molecules and from (b) polarization due to orientation if the molecules possess permanent moments. Such moments normally point in all possible directions. Under the influence of an applied electric field, they become more or less oriented in the direction of that field. If the measurements are made with electrical energy of relatively low frequency, the adjustment of the positions of the electric moments must follow the changing field strength, since, until relatively high frequencies are encountered, the values of the measured dielectric constant are independent of the frequency. However, depending, as we shall see, upon the viscosity, the electric moment and the temperature, measurements of polar substances at low wave lengths will show *anomalous dispersion, i.e.*, a decrease of dielectric constant with increasing frequency. The theory of anomalous dispersion has been developed by Debye.[25] In brief his development involves a modification of the Boltzmann equation as used in equation (15), to take account of the special case in which the rotation of the dipoles in an electric field is appreciably opposed by the viscosity of the medium, so that complete adjustment to the field is not possible in the short time available before the electric field changes sign. Debye's modification of equation (22) to take account of phenomena at high frequencies is

$$P_\nu = \frac{D - 1}{D + 2}\frac{M}{\rho} = \frac{4\pi N}{3}\left[\alpha_0 + \frac{\mu^2}{3kT}\frac{1}{1 + i\omega\tau}\right] \tag{27}$$

For a derivation of this equation Debye's monograph may be consulted. It will be observed that equation (27) differs from (22) in that the term $\mu^2/3kT$ is multiplied by the factor $1/(1 + i\omega\tau)$, in which i is the operator $\sqrt{-1}$, ω equals $2\pi\nu$ in which ν is the frequency, P_ν is the corresponding polarization, and τ is the time of relaxation, *i.e.*, the time necessary for the moments to revert practically to random distribution after the removal of the impressed field. By assuming that the molecules are spheres of radius r, and that Stokes' relation for the rotation of a sphere in a medium of viscosity η is valid, Debye obtains the equation

$$\tau = \frac{4\pi\eta r^3}{kT} \tag{28}$$

for the time of relaxation, τ. Several important conclusions may be obtained from equation (27). If ω is very small, that is, if the frequency of the field approaches zero, equation (27) becomes

$$\frac{D_1 - 1}{D_1 + 2}\frac{M}{\rho} = \frac{4\pi N}{3}\left[\alpha_o + \frac{\mu^2}{3kT}\right] \tag{29}$$

in which D_1 is the dielectric constant obtained at low frequencies. On the other hand if the value of ω is very large, equation (27) approaches

$$\frac{D_o - 1}{D_o + 2}\frac{M}{\rho} = \frac{4\pi N}{3}\alpha_o \tag{30}$$

This means of course, that at such frequencies the effect of the permanent moment has vanished, and that only the part of the dielectric constant due to deformation of the atoms and molecules remains. According to this theory non-polar substances, for which μ equals zero, should not show anomalous dispersion, and this is borne out by experiment. The effect will become evident at frequencies at which the product, $\omega\tau$, has an appreciable value. From equation (28) for the time of relaxation, τ, anomalous dispersion should appear at relatively lower frequencies (a) if the molecules have large radii, r, (b) if the viscosity, η, is high, and (c) if the temperature is low, all of which have been observed experimentally. Since equation (27) is complex it follows that the dielectric constant, D, has a "real" and an "imaginary" part. Debye has shown that the "real" part, D', which corresponds to the measured dielectric constant, follows the equation

$$D' = D_o + \frac{D_1 - D_o}{1 + \left(\frac{D_1 + 2}{D_o + 2}\right)^2 \omega^2\tau^2} \tag{31}$$

in which D_1 and D_0 are defined by equations (29) and (30).

Debye has tested his theory as just briefly outlined with the results of Mizushima,[28] and has found that the change of dielectric constant of n-propyl alcohol with frequency observed by that worker leads to a value of the radius, r, for the molecule of 2.2×10^{-8} cm., which, though small, is of the right order of magnitude. Johnstone and Williams [29] have studied the variation of the dielectric constant with frequency of solutions of nitrobenzene in mineral oil. Their results are plotted in Fig. 11 in which the measured dielectric constant of the solution is plotted as a function of the frequency, ν. Curves I, II, and III correspond respectively to 3.2, 9.5 and 20.4 grams of nitrobenzene per

[28] S. Mizushima, *Sci. Papers, Inst. Phys. Chem. Research. (Tokyo)*, **5**, 79, 201 (1927); **9**, 166, 209, (1928); *Physik. Z.*, **28**, 418 (1927).
[29] J. H. L. Johnstone and J. W. Williams, *Phys. Rev.*, **34**, 1483 (1929).

100 grams of oil. It will be seen that as the higher frequencies are approached the measured dielectric constant tends to decrease. These workers find that their results are in accord with a value of the radius, r, of 2.4×10^{-8} cm., in rough accord with X-ray determinations. Bock,[30] however, finds that the Debye theory is not in quantitative accord with measurements on glycerin.

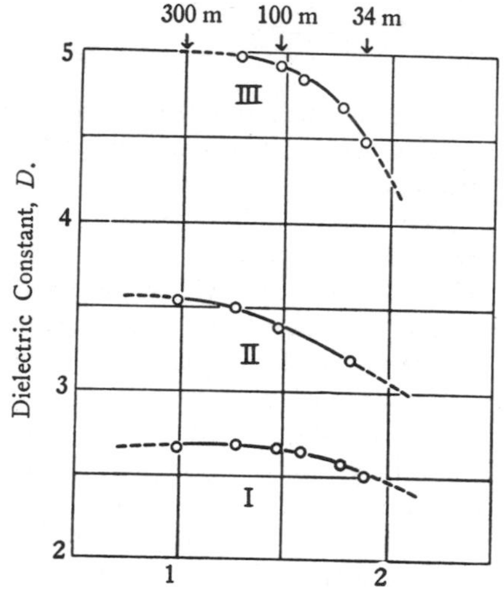

Fig. 11. The Dispersion of Nitrobenzene in Mineral Oil.

Anomalous dispersion in solutions of proteins has been studied by Wyman,[31] by Oncley,[32] and by Ferry and Oncley.[33] Although the phenomenon is particularly evident in such solutions, only polar solvents are available, so that theoretical interpretation is difficult. Interesting semi-empirical relations have been found, for which the reader is referred to the original papers.

[30] Bock, *Z. Physik.*, **31**, 534 (1925).
[31] J. Wyman, Jr., *J. Biol. Chem.*, **90**, 443 (1931).
[32] J. L. Oncley, *J. Am. Chem. Soc.*, **60**, 1115 (1938).
[33] J. D. Ferry and J. L. Oncley, *Ibid.* **60**, 1123 (1938).

[*Comment to 1961 Edition*] The recent work on dielectric constants and related subjects is given in *Dielectric Behavior and Structure* by C. P. Smyth, published by McGraw-Hill Book Company, Inc. in 1955.

Chapter 23

Electrokinetic Phenomena Electro-osmosis, Electrophoresis and Streaming Potentials

As has been repeatedly emphasized in this book, a large portion of electrochemistry deals with processes which occur at phase boundaries, the most important of which are the metal-liquid and liquid-liquid boundaries. In nearly all the cases so far considered the phases in contact have been conductors of electricity. The phenomena observed have been, in many instances, explained by assuming that electric potentials are present at the boundaries, although the magnitude, or even the existence, of potentials at single boundaries can in no case be experimentally demonstrated. Interesting phenomena also occur when one or both of the phases in contact are electrical insulators, and also when one of the phases consists of relatively large particles in suspension, *i. e.,* colloidal solutions. As we shall see it is also useful in this connection to assume the presence of electric potentials at the phase interfaces in order to account theoretically for the experimental results. The observed facts in this field are grouped together as *electrokinetic phenomena.*

The displacement, produced by an applied electromotive force, of a liquid with reference to the surface of a solid, is termed *electro-osmosis. Electrophoresis* (or cataphorèsis) is the movement of suspended solid or liquid particles in an electric field. In both these phenomena mechanical effects are brought about by the external potential. The reverse effects have also been studied. The electromotive force produced by forcing a liquid through capillary tubes or porous plugs is the *streaming potential.* Potentials arising from particles falling through liquids have also been observed. In this chapter an attempt will be made to describe the more important experimental methods used in investigating these phenomena, to outline some typical results, and to develop the none too satisfactory theories of the effects. It should be realized that investigators are active in this field, in some cases with recently improved technique, so that this chapter may be expected to need revision sooner than other parts of this book.

Electro-osmosis. A diagram of an arrangement by means of which *electro-osmosis* may be observed is shown in Fig. 1. In the figure *D*

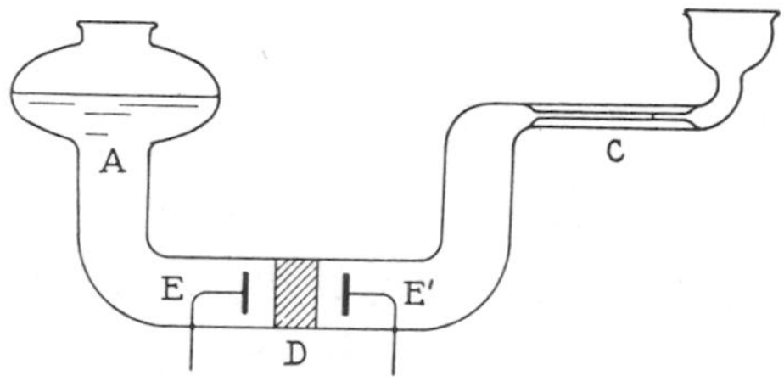

Fig. 1. An Apparatus for Observing Electro-osmosis.

represents a porous plug (which may also be a single capillary tube). The measuring tube of narrow bore, *C*, is horizontal, and the apparatus is filled to such a level that there is no difference in hydrostatic head between the liquid in *C* and the tube *A*. If a potential is applied to the electrodes *E-E′*, the meniscus in tube *C*, will, in general, be found to move in one direction or the other, depending upon the nature of the diaphragm and of the liquid. Electro-osmosis was discovered by Reuss in 1808, only a few years after the demonstration of the electrolytic decomposition of water, and is thus one of the earliest electrochemical effects to be observed. Later experimental studies of electro-osmosis were made by Wiedemann [1] and Quincke.[2] Theoretical investigations of the effect have been made by Helmholtz,[3] Lamb,[4] v. Smoluchowski,[5] J. Perrin,[6] and others. The theoretical discussion of electro-osmosis given below follows the treatment of Perrin.

Let a cylindrical capillary of radius r and of unit area be inserted between the electrodes *E-E*, as is shown in Fig. 2. The capillary surface is assumed to be the seat of a *Helmholtz double layer,* the solid wall having one charge, positive for instance, and the liquid carrying the opposite charge. The charge on the wall is assumed to be fixed and the other movable. An external electromotive force will therefore tend to produce a motion of the charges in the liquid, and to carry the liquid with them. The motion will, however, be opposed by friction within the liquid. Such friction will be proportional to the viscosity, η, of the liquid, the extent of the surface over which the liquid flows, and the

[1] G. Wiedemann, *Pogg. Ann.*, **87**, 321 (1852); **99**, 177 (1856).
[2] G. Quincke, *Pogg. Ann.*, **113**, 513 (1861).
[3] H. Helmholtz, *Wied. Ann.*, **7**, 337 (1879).
[4] H. Lamb, *Phil. Mag.*, (5) (25), 52 (1888).
[5] M. v. Smoluchowski, see particularly his discussion in Graetz, "Handbuch der Elektrizität und des Magnetismus," v. II, p. 366, Barth, Leipzig, 1914.
[6] J. Perrin, *J. Chem. Phys.*, **2**, 601 (1904).

velocity gradients in the liquid. If the volume, V, of liquid flows per second out of the capillary its mean velocity, u, in the capillary will be given by the relation

$$V = \pi r^2 u \tag{1}$$

This velocity is assumed to hold for all the liquid in the tube with the exception of a layer of thickness δ at the surface. The frictional force opposing the movement is due to the viscosity, η, and will be an integral of $\eta(du/dz)$ in which the direction of z is perpendicular to the wall of

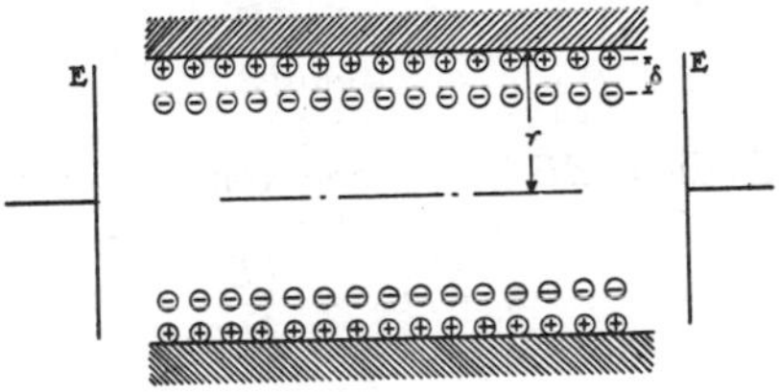

Fig. 2.

the tube. Assuming that the velocity varies linearly in the direction z, and the change in velocity is confined to the layer, δ, then $du/dz = u/\delta$ and the frictional force is, with equation (1)

$$\frac{u\eta}{\delta} = \frac{V\eta}{\pi r^2 \delta} \tag{2}$$

The electric force acting on the movable side of the double layer is equal for unit area to the charge density, σ, times the gradient dE/dx of the applied potential; thus when a steady state is reached

$$\sigma \frac{dE}{dx} = \frac{V\eta}{\pi r^2 \delta} \tag{3}$$

We may, however, regard the Helmholtz double layer as a condenser of potential, ζ, in a medium of dielectric constant D, for which

$$\zeta = \frac{4\pi\delta\sigma}{D} \tag{4}$$

When substituted in equation (3) this gives, for the volume, V, of flow of liquid per second

$$V = \frac{\zeta r^2 D}{4\eta} \frac{dE}{dx} \tag{5}$$

If the capillary is uniform and of length, l, dE/dx may be set equal to E/l in which E is the applied potential, with which equation (5) becomes

$$V = \frac{\zeta r^2 ED}{4\eta l} \tag{6}$$

This equation evidently gives a relation between the flow per second of liquid in a capillary to the more or less hypothetical "zeta potential" ζ, and to experimentally measurable quantities. With the equations (1) and (5) the electro-osmotic velocity is given by the expression

$$u = \frac{\zeta D}{4\pi\eta} \frac{dE}{dx} \tag{7}$$

If the potential gradient dE/dx is one volt per centimeter then

$$u = \text{U}_0 = \frac{\zeta D}{4\pi\eta} \tag{7a}$$

in which U_0 is the *electro-osmotic mobility*. The provisional nature of the assumptions made in obtaining equations (6), (7), and (7a) is obvious. However, Helmholtz [7] and following him, Smoluchowski[8] obtained the same equations, using more elaborate electrical and hydrodynamic theory. A more modern interpretation of the meaning of the zeta potential, and of the Helmholtz double layer, will be given later in this chapter. The experimental data bearing on electro-osmosis can be best discussed together with those on electrophoresis, which immediately follow.

Electrophoresis. Electrophoresis, or cataphoresis as it is sometimes called, is concerned with the movement of colloidal particles in an electric field. A simple apparatus, due to Burton,[9] by which the phenomenon of electrophoresis may be demonstrated, is shown in Fig. 3. The solvent is first poured into the tube, T, followed by the colloidal suspension. If precautions are taken to avoid mixing of the solutions, boundaries between the colloidal suspension and solvent will be present at equal heights, such as *a-a*, and these will be visible if the suspension is colored or turbid. If now a potential is applied at the electrodes E-E' the boundaries will, in general, move. Rough estimates of the mobilities of the colloidal particles may be obtained by measuring the distances passed through by the boundaries. However, there is little or

[7] H. von Helmholtz, *Wied. Ann.*, **7**, 337 (1879).
[8] M. v. Smoluchowski, in Graetz, "Handbuch der Eletrizität und des Magnetismus," v. II, p. 366, Barth, Leipzig, 1914.
[9] E. F. Burton, *Phil. Mag.*, (6) **11**, 425 (1906).

no "adjusting effect" such as is present in the moving boundary determinations of transference numbers. The boundaries are usually more or less disturbed by convection due to density differences arising from the fact that temperature differences will arise from the passage of the current. The position of the boundary may also be difficult to locate due to progressive diffusion. Also, unless the electrodes, *E*-*E'*, are properly chosen, the motion of the boundaries will be disturbed by the evolution of gas at the electrodes and by electrode products, such as the rapidly moving hydrogen and hydroxyl ions.

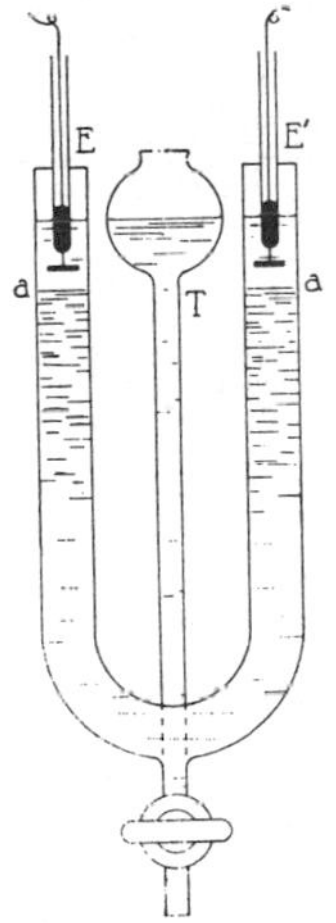

Fig. 3. An Apparatus for Demonstrating Electrophoresis.

The moving boundary method for investigating electrophoresis has, however, been improved by Tiselius to the extent that it has become a powerful tool of investigation, particularly of proteins. Tiselius'[10] early work is outlined in his dissertation which includes a valuable discussion of the conditions governing the movement of a boundary. A particularly important experimental detail is that the composition of the solution in which the colloid is suspended should be, as nearly as possible, of the same composition as the solution with which it is in contact at the boundary, otherwise "boundary anomalies" will be present. These will cause the particles of colloid to move abnormally fast or slow, and may even give rise to additional boundaries.

[10] Arne Tiselius, "The Moving Boundary Method of Studying the Electrophoresis of Proteins," Uppsala, 1930.

Tiselius'[11] most recent apparatus is shown in Figs. 4 and 5. Since the use of this apparatus will certainly throw much light on the phenome-

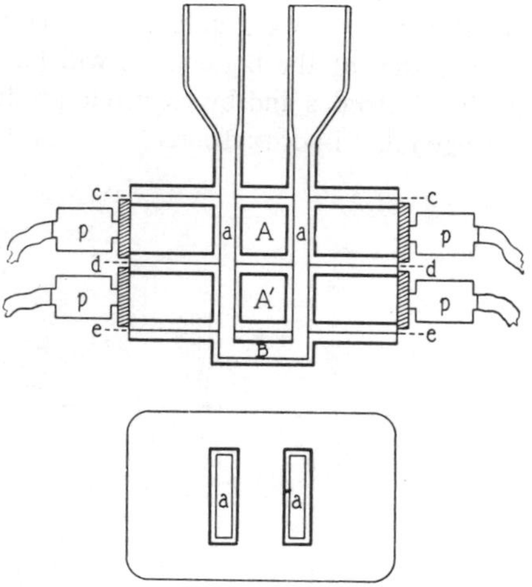

Fig. 4. The U-Tube Section of the Tiselius Electrophoresis Apparatus.

non of electrophoresis in the near future it will be described here in considerable detail. The U-tube of earlier types of apparatus is replaced by the vessel *A-A'* of Fig. 4. Through the center of this vessel runs the channel *a-a* of rectangular cross-section. A plan of one of the sections, *A* or *A'*, is shown in the lower half of the figure. The vessel is divided into sections, *A*, *A'*, and *B* which may be slid over one another, along the planes *c-c*, *d-d*, and *e-e*, by utilizing one or more of the small pneumatic pumps, *p*, *p*... A boundary is formed initially by filling the channel a little above the plane *d-d* with the colloidal suspension. The section *A'* is then pushed to one side, the excess material in section *A* is pipetted out and the section and the rest of the apparatus except the immediate region of the electrodes is filled with solution (usually a buffer) of the same composition as that containing the suspensions. The section *A'* is then pushed back into the position shown. When current is passed through the channel the boundaries move upward or downward depending upon the charges on the colloidal particles and the direction of the current.

[11] A. Tiselius, *Särtryck ur Svensk Kemisk Tidskrift*, **50**, 58 (1938).

A more complete diagram of Tiselius' apparatus (modified in details by Longsworth) is shown in Fig. 5. The vessel *A-A'* is attached, as

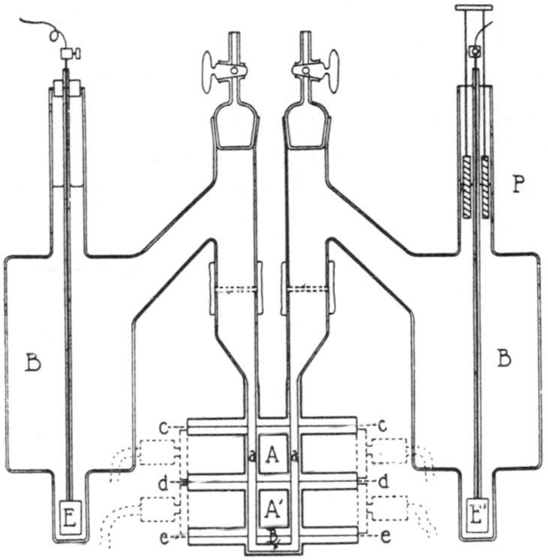

Fig. 5. The Complete Tiselius Electrophoresis Apparatus.

shown, to two large electrode chambers, *B-B'*, at the bottom of which are the reversible silver-silver chloride electrodes *E-E'* which must have a large current carrying capacity. Initially the narrow tubes surrounding these electrodes are filled with a strong chloride solution. The large volumes of the vessels *B-B'* are due to the necessity of keeping the products of the electrical reactions at *E* and *E'* from entering the channel in the vessel *A-A'* containing the moving boundaries.

Since, as has been mentioned, there is little or no "adjusting effect" in moving boundaries of colloidal suspensions, there is no automatic correction for the effects of diffusion and convection as in properly chosen boundaries between salt solutions. Tiselius has, however, utilized a simple scheme by means of which convection currents are almost completely suppressed. Such currents are, as already mentioned, due to differences of density of the liquid produced by the passage of electricity. Because of the greater path for heat conduction the center of the channel through which an electric current is passing is hotter than the edges. This usually means that the liquid becomes less dense as the center is approached. However, by making the measurements near the temperature of maximum density of water, 4°, the solutions in the channel may have very nearly the same density throughout in spite of the gradients of temperature. In this manner the greatest source of error in earlier work in electrophoresis by the moving boundary method has been largely eliminated.

The greatest present utility of the electrophoretic method is in the study of proteins and related compounds. Since the suspensions of most of these substances are colorless the relatively diffuse boundaries are not readily visible. Their positions can, however, be followed by the Tœpler "schlieren"[12] or the Lamm[13] "scale" method. Certain proteins which were until recently thought to be single compounds have been shown by the Tiselius method to be complex. For instance horse serum globulin yields three boundaries indicating three compounds having different electrophoretic mobilities. The measurements of Stenhagen[14] on the electrophoretic mobilities of the constituents of human blood plasma as functions of the pH are shown in Fig. 6, and are

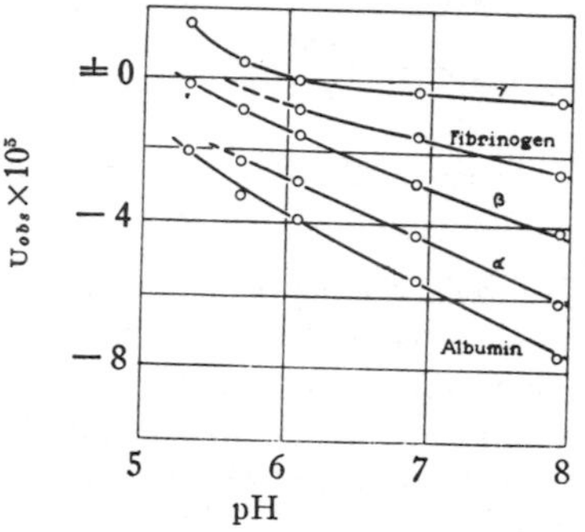

Fig. 6. The Electrophoretic Mobilities of the Constituents of Human Blood.

illustrative of the type of data that may be obtained with the aid of the new technique. It will be seen that the mobilities are functions of the pH value of the solution in which they are in suspension. The mobilities change sign at the "isoelectric point."

The utility of the Tiselius apparatus in revealing the presence of closely related substances is indicated in Fig. 7, which is from a series of photographs of boundaries in the channel *a-a* of Fig. 4 made visible

[12] For an adequate description see M. Toepler, *Handwörterbuch der Naturwissenschaften*, **8**, 924 (1913). H. Schardin, "Das Toeplersche Schlierenverfahren. Grundlagen für seine Anwendung und quantitative Auswertung," V D I-Verlag, G.M.B.H., Berlin N W 7, 1934.

[13] O. Lamm, *Z. phys. Chem.*, **138**, 313 (1928).

[14] E. Stenhagen, *Biochem. J.*, **32**, 714 (1938).

by the "schlieren" technique. The solution studied was a mixture of albumins from the eggs of the duck and the guinea hen, in a buffer at pH = 5.20. The photographs were made at intervals of 30 minutes. It will be seen that two boundaries are visible on the second exposure,

Fig. 7. Schlieren Photographs of Boundaries in a Mixture of Proteins.

and that they continue to separate as they rise farther in the channel. The leading and following boundaries are due to the duck and guinea hen albumins respectively, since the observed mobilities, -4.34×10^{-5} and -3.77×10^{-5} at zero degrees agree excellently with determinations made when only one type of albumin was present.

By another device, also due to Tiselius, the apparatus shown in Fig. 5 may be used to make separations of components having different mobilities, even if the boundaries tend to move in the same direction. If two boundaries start upward at slightly different rates from the plane *d-d* they will, obviously, not be far apart before they are both completely out of the channel in which they are moving. However, by lifting the plunger, *P*, by a clockwork mechanism a mass flow of solution through the apparatus may be produced at such a rate that the slower moving boundary will remain stationary with respect to the walls of the channel *a-a*. Under these conditions the faster moving boundary will get farther and farther away from the slower one. This may be continued until the two boundaries are removed from each other by the distance, say, between the planes *c-c* and *d-d*. The faster moving component may then be isolated by moving section *A* of the vessel to one side.

If the individual particles of a colloidal suspension are visible the microscopic method may be used for studying electrophoresis. For this purpose quite a number of arrangements have been utilized. A typical apparatus is shown in Fig. 8, and represents Abramson's[15] modification of Northrop and Kunitz's[16] design. The suspension under observation is contained in the channel *a-b,* which is usually of rectangular cross-section. This channel is connected to the vessels which contain

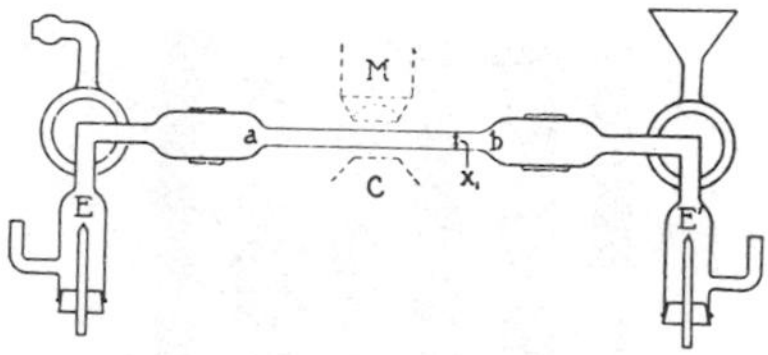

Fig. 8. The Microscopic Method of Studying Electrophoresis.

electrodes E and E', which should not give off gas when the current is passing. A portion of the suspension in the channel *a-b* is illuminated by the condenser, *C,* and may be viewed by the microscope, *M.* The latter will be in sharp focus at only one plane. It is however fitted with a screw adjustment by which it may be raised and lowered through distances that may be accurately measured. One end of the apparatus is closed so that there will be no mass flow of fluid through the channel. When an electromotive force is applied to the electrodes the motion of individual particles may be observed quantitatively by means of a scale in the eye-piece of the microscope. The potential gradient under the influence of which the particles are moving may be computed from measurements of the current passing through the apparatus, the conductance of the fluid in the channel and its cross-section. The observed mobilities, u_{obs}, may then be computed from the relation

$$u_{obs} = v_{obs}/(dE/dz)$$

in which (dE/dz) is the potential gradient and v_{obs} is the observed velocity.

Data obtained with an apparatus of this type on particles of an oil emulsion are plotted in Fig. 9, in which the observed mobilities, in centimeters per second per volt, of the particles are ordinates and the distances from the top surface of the channel are abscissae. The results are from the measurements of Ellis,[17] who first applied the method. It is evident that the measured mobilities of the particles are not constant, but change markedly with the distance of the particles from the walls of the channel. As is shown in the figure, the direction of the

[15] H. A. Abramson, *J. Gen. Physiol.,* **12,** 469 (1929); *J. Phys. Chem.,* **36,** 1454 (1932).
[16] J. H. Northrop and M. Kunitz, *J. Gen. Physiol.,* **7,** 729 (1925).
[17] R. Ellis, *Z. phys. Chem.,* **78,** 321 (1912).

motion may even reverse near the walls of the channel. The explanation of these observations is that, although the electrophoretic mobilities, u_e, of the particles are really constant, the motion is greatly influenced by the electro-osmotic flow of the suspending fluid due to charges

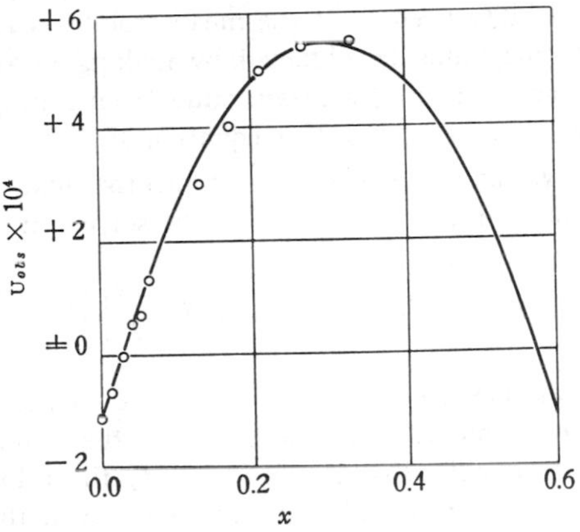

Fig. 9. The Observed Electrophoretic Mobility of Oil Particles as a Function of the Depth of the Tube.

located on the walls of the channel. However, as pointed out by Ellis, a correction for this electro-osmotic effect may be made. Since the whole apparatus is closed, solvent that moves due to electro-osmosis must also return, so that the net flux of solvent must be zero. For a rectangular channel this may be stated in the form

$$\frac{1}{x_1}\int_0^{x_1} v_o dx = 0 \tag{8}$$

in which x_1 is the depth of the channel and v_0 is the mobility of the solvent, *i. e.*, the velocity of the solvent at the level x under unit potential gradient.

We also have the equation

$$u_{obs} = u_e - v_o \tag{9}$$

in which u_e is the true electrophoretic mobility and may be assumed to be constant. The mean mobility u_M is evidently given by

$$u_M = \frac{1}{x_1}\int_0^{x_1} u_{obs}\, dx \tag{10}$$

which, with equation (9), gives

$$u_M = u_e - \frac{1}{x_1}\int_0^{x_1} v_o dx \tag{11}$$

Thus since the last term is zero for a closed cell by equation (8), the electrophoretic mobility is equal to the mean mobility u_M. The electrophoretic velocity may thus be obtained by making measurements at a series of levels, and obtaining a mean value from a graphical or other integration of the results with aid of equation 10.

Furthermore v. Smoluchowski [18] has shown that under the conditions of the experiment outlined, the velocity of the solvent at different levels, x, is given by the equation

$$v_o = u_o\left[1 - 6\left\{\frac{x}{x_1} - \left(\frac{x}{x_1}\right)^2\right\}\right] \tag{12}$$

in which u_0 is the electro-osmotic mobility due to the charge on the surface of the cell and x/x_1 is the ratio of the depth to the total depth x_1 of the channel.[19] From equation (12) the velocity v_0 is zero at values of x/x_1 equal to 0.214 and 0.786 so that at these levels $u_{obs} = u_e$. However, these are the correct factors only if the channel is quite broad compared with its depth.

Combining equations (9) and (12) we have

$$u_{obs} = u_e - u_o\left[1 - 6\left(\frac{x}{x_1} - \left(\frac{x}{x_1}\right)^2\right)\right] \tag{14}$$

which may be put into the form

$$u_{obs} = u_e - u_o + b\left(\frac{x}{x_1} - \left(\frac{x}{x_1}\right)^2\right) \tag{15}$$

in which b is a constant. A plot of this equation is represented by the smooth curve in Fig. 9. It will be seen that it represents Ellis' results within the limit of the experimental error. From this equation the electrophoretic mobility, u_e, of the oil droplets was 3.18×10^{-4} whereas the electro-osmotic mobility, u_0, was 4.30×10^{-4}.

[18] M. v. Smoluchowski, Graetz, "Handbuch der Elektrizität und des Magnetismus," v. II, p. 366, Barth, Leipzig, 1921.

[19] Equation (12) is obtained by integrating the equation (Lamb, "Hydrodynamics," p. 550, Cambridge Univ. Press, 1924)

$$\eta \frac{\partial^2 V_o}{\partial x^2} = \frac{dP}{dz} \tag{13}$$

for lamellar flow between parallel walls, the direction x being perpendicular to the walls and z in the direction of the flow, and P the pressure. For a steady state $\partial P/\partial z = \text{const.}$ The effect of varying the width and depth of the channel has been investigated theoretically by Komagata (*Researches Electrotech. Laboratory* (Tokyo), No. 348 (1933)) who has shown that equation (12) is the limiting expression for plates of infinite extent. The error arising from the use of equation (12) for channels in which the width is, say, twenty times the depth is within the experimental error.

Some particularly interesting observations in this field have been made by Abramson.[20] He found that particles when coated with a film of proteins have electrophoretic mobilities which are characteristic not of the particle but of the protein. This was found to be true for glass, collodion, quartz particles, and droplets of mineral oil, using various proteins. Fig. 10 represents some typical results of measure-

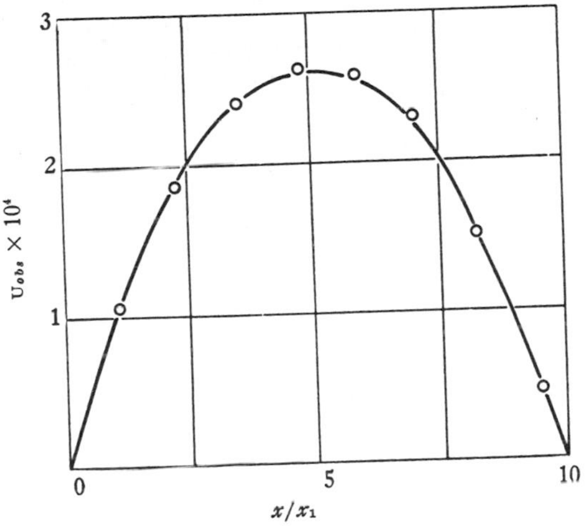

Fig. 10. The Mobility of Gelatin-Coated Quartz Particles.

ments on gelatin-covered quartz particles. The relative mobilities, U_{obs}, are plotted as ordinates and the corresponding relative depths, x/x_1, as abscissae. The smooth curve is a plot of equation (15). However, the curve corresponds to the condition that $U_e - V_0 = 0$ or $U_e/V_0 = 1$. This means, therefore, that the electro-osmotic velocity along the wall of the channel is equal to the electrophoretic mobilities of the particles. To obtain this result it is, however, apparently necessary for enough of the protein to be present to cover completely the suspended particles and the glass walls of the channel.

It is, of course, of importance to determine whether the electrophoretic mobilities determined by the moving boundary and microscopic methods yield the same values. The electrophoretic mobility of serum albumin as a function of the pH has been determined by Tiselius[21]

[20] H. A. Abramson, *J. Am. Chem. Soc.*, **50**, 390 (1928); *J. Gen. Physiol.*, **13**, 169 (1929). See H. Davis, *J. Physiol.*, **57**, XVI (1923).

[21] A. Tiselius, *J. Biol. Chem.*, **31**, 313 (1937).

by the moving boundary method and the mobilities of the same material adsorbed on quartz particles and on collodion particles by the microscopic method have been determined by Abramson and by Moyer. The results of the three sets of measurements are plotted in Fig. 11 which is from a paper by Moyer.[22] It will be seen that, within the rather large

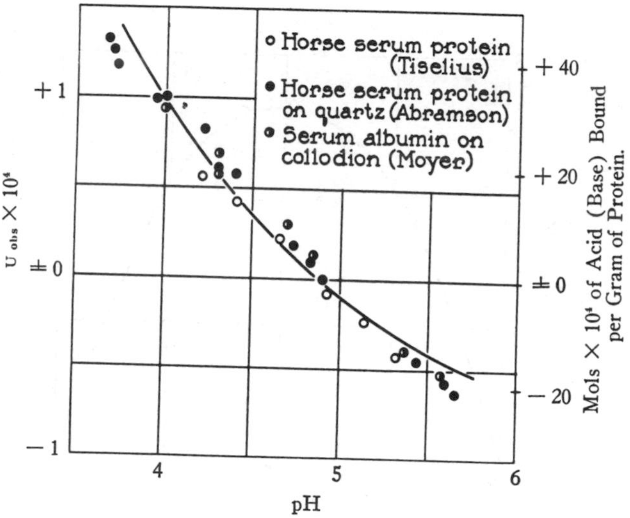

Fig. 11. The Electrophoretic Mobilities of Serum Proteins and Albumin.

experimental error, the three sets of measurements agree. A similar test, using egg albumin, shows less agreement. The smooth curve in Fig. 10 is a plot using the mols of added acid or base per gram of protein necessary to bring the protein solution to the corresponding pH value. This titration curve passes through the point of zero mobility of the protein (the "isoelectric point"). The scale of ordinates has been adjusted to make the titration curve pass through the experimental points as nearly as possible. If the effect of the added acid or base is to produce, by reaction and subsequent ionization, increasing charges on the protein molecules, the relation between the mobility variation and the titration curve, as represented in Fig. 11, would be expected.

If the moving colloidal particle is assumed to be cylindrical in shape, surrounded by a Helmholtz double layer, as shown in Fig. 12, and is situated in a motionless liquid in which there is a unit potential gradient, the forces acting on it will be just those effective on a thread of liquid

[22] L. S. Moyer, *Biochem. J.*, 122, 641 (1938).

in a capillary tube. However, in this case it is the particle which is free to move rather than the liquid. Without repeating the derivation it can be readily seen from these considerations that

$$U_e = \frac{\zeta D}{4\pi\eta} \tag{16}$$

This is the "classical" equation of Helmholtz for electrophoresis. Further comment on this equation will be made at the end of this chapter.

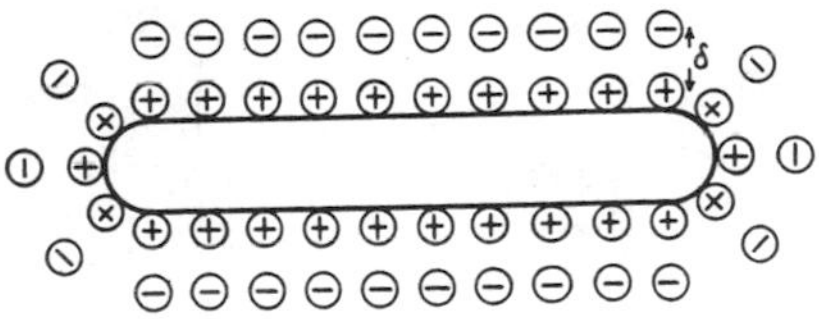

Fig. 12.

The Streaming Potential. Electro-osmosis, it will be recalled, is the motion of fluid which occurs when an electric potential is applied to the ends of a capillary tube, or permeable diaphragm, filled with the fluid. If the fluid is forced by pressure through the diaphragm or capillary an electromotive force is generated which is called the *streaming potential.* It is evident that this phenomenon is the reverse of electro-osmosis. Measurement of streaming potentials is, as we shall see, another method for studying the electrokinetic properties of surfaces.

A typical apparatus for studying streaming potentials due to Kruyt [23] and modified by Freundlich and Rona [24] is shown in Fig. 13. Due to the difference in hydrostatic head in the vessels B and B' liquid flows through the capillary tube, K. Electrical connection of the liquid with the reversible electrodes E and E' is made through the tubes T and T' which are filled with agar jelly containing potassium chloride. Measurements of the resulting potential between the two electrodes are made with the aid of an electrometer.

The "classical" expression due to Helmholtz, relating the streaming potential to the zeta potential, ζ, and measurable quantities may be simply obtained as follows. Fig. 2 represents a capillary tube of radius r with a fixed surface charge of density, σ, and at a distance δ an opposite movable charge of the same density. If the fluid in the tube is forced by

[23] H. R. Kruyt, *Kolloid-Z.*, **22**, 81 (1918).
[24] H. Freundlich and P. Rona, *Sitzngsb. Preuss. Akad. Wiss.*, **20**, 397 (1920).

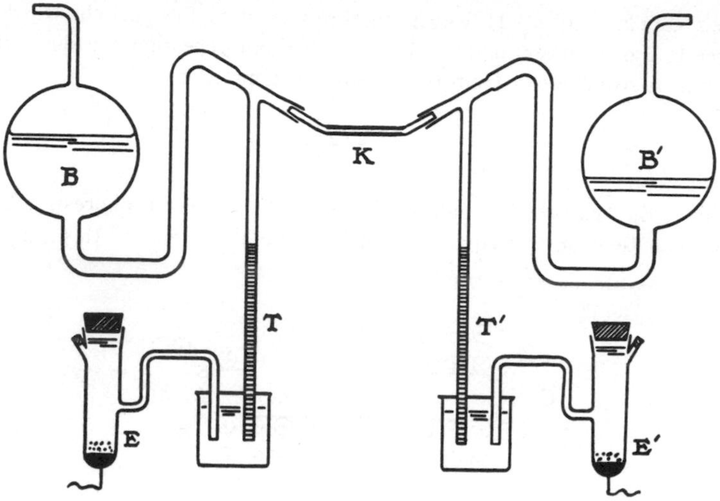

Fig. 13. An Apparatus for Studying Streaming Potentials.

pressure to move with the velocity, *u*, the electric current, *I*, will be

$$I = 2\pi r\sigma u \tag{17}$$

Now if the velocity of the fluid is assumed to change linearly from zero at the surface to *u* at the distance δ we have the relation

$$\frac{PQ}{L} = 2\pi r\left(\frac{u}{\delta}\right)\eta \tag{18}$$

in which P is the difference in pressure at the two ends of the capillary, Q is its cross-section, and L its length. With equation (17) this yields

$$I = \frac{PQ\sigma\delta}{L\eta} \tag{19}$$

Assuming once more that the Helmholtz double layer has the properties of a condenser for which

$$\zeta = \frac{4\pi\delta\sigma}{D} \tag{4}$$

equation (19) becomes

$$I = \frac{PQD\zeta}{4\pi L\eta} \tag{20}$$

The relation of the current I to the potential E (the streaming potential) at the ends of the tube when it contains a fluid of specific conductance L is given by

$$E = \frac{IL}{Q\text{L}}$$

with which equation (20) is

$$E = \frac{PD\zeta}{4\pi \text{L} \eta} \tag{21}$$

Helmholtz and v. Smoluchowski obtained the same equation as (21) using more rigorous hydrodynamic theory than that involved in equation (18).

If the theory outlined above and represented by equation (21) is valid, the streaming potential E should, for one thing, be proportional to the pressure P, *i. e.*, the ratio E/P for any capillary or diaphragm should be a constant, other experimental conditions remaining the same. Roughly at least this appears to be true. The data contained in Table I

TABLE I. THE EFFECT OF PRESSURE ON THE STREAMING POTENTIAL IN GLASS CAPILLARY TUBES

	Capillary A		Capillary B	
Pressure, P, (centimeters Hg)	Streaming potential, E (millivolts)	E/P	Streaming potential, E (millivolts)	E/P
40	486	12.2	340	8.5
50	605	12.1	412	8.2
60	725	12.1	473	7.9
70	834	11.9	534	7.6
80	943	11.8	592	7.4

from the measurements of Kruyt and van der Willigen[25] are more or less typical of results obtained in this field. It will be seen that although the values of E/P are approximately constant there is a decided drift with increasing pressure in both sets of values.

From measured values of E/P and other measurable quantities the zeta potential, ζ, of the capillary surface may be computed with the aid of equation (21). Kruyt and van der Willigen have found average values of E/P resulting from forcing solutions of potassium chloride of different concentrations through capillary tubes. The results for the same capillary B as in Table I are given in Table II. It is an interesting fact that though the hypothetical zeta potential, ζ, computed in this manner, changes with the salt concentration it remains of the same order of magnitude. The computed zeta potential, however, varies greatly even for surfaces of apparently identical material. Different workers obtain quite different values for the same types of interfaces.

[25] H. R. Kruyt and P. C. van der Willigen, *Kolloid-Z.*, **45**, 307 (1928).

TABLE II. THE EFFECT OF THE CONCENTRATION OF POTASSIUM CHLORIDE ON THE COMPUTED ZETA POTENTIAL, ζ.

Concentration of KCl (mols)	E/P average	Specific conductance, L, at 18°	Zeta potential (millivolts)
0.01	272	0.24	73
.025	149	0.43	72
.050	81	0.76	70
.10	42.6	1.40	67
.25	16.7	3.28	62
.50	7.9	6.47	58
1.00	3.6	12.9	52

Since the dimensions of the capillary do not enter into equation (21) the computed zeta potential should be independent of the diameter and length of the tube through which the liquid is forced. Data bearing on this question are given in Table III from the work of White, Urban, and van Atta.[26]

These data indicate that the value of the streaming potential, per unit of pressure, is independent of the size of the tube within the limits given, as long as the capillary bore is not too small. With fine capillaries the streaming potential, in this case at least, was less reproducible, and decreased to nearly zero. The reason for the lower values of the streaming potential for the smaller capillaries is, as yet, unknown.

TABLE III. THE EFFECT OF THE DIMENSIONS OF THE CAPILLARY ON THE STREAMING POTENTIAL

Dimensions		
Length (cm.)	Diameter (mm.)	E/P (mv./cm. Hg)
1.1	0.0964	31.4
4.66	.0405	32.0
1.97	.039	31.0
0.25	.0058	1.6
0.94	.0055	5.8
0.76	.0053	0.0

Briggs[27] substituted a diaphragm of cellulose pulp for the capillary, and found the ratio E/P to be constant for a given diaphragm. When the amount of pulp packed into a given volume was varied the value of E/P also changed. However, this apparent deviation from the theory was found to be due to the fact that the conductance of the liquid in the pores, at low electrolyte concentrations, differed from that in the bulk of the liquid. A correction was made by determining the "cell constant" of the diaphragm with a sufficiently concentrated potassium chloride solution of known specific conductance. If the apparent conductance of the solution in the pores was utilized in equation (21) the

[26] H. L. White, F. Urban, and E. A. van Atta, *J. Phys. Chem.*, **36**, 3152 (1932).
[27] D. R. Briggs, *J. Phys. Chem.*, **32**, 641 (1928).

computed zeta potentials were found to be reasonably independent of the density of packing of the diaphragm.

An interesting comparison is possible of the ζ potentials computed from electrophoretic measurements by Abramson [28] and from streaming potential measurements by Briggs.[29] The electrophoretic determinations were made on quartz particles coated with egg albumin, and the streaming potentials were obtained from a diaphragm of quartz powder saturated with the same albumin. Fig. 14, which is from Briggs'

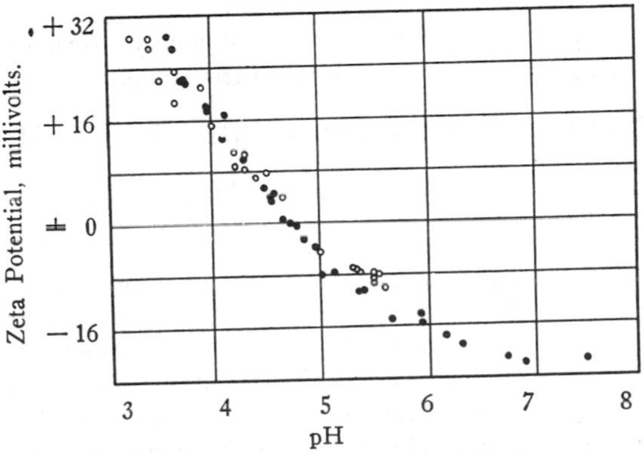

Fig. 14. A comparison of ζ-Potentials from Electrophoretic and from Streaming Potential Measurements.

article, shows that the zeta potentials obtained in the two different ways show the same variation with pH within the rather large experimental error.

Some General Comments on Electrokinetic Phenomena. It is quite impossible in a short chapter to do justice to the vast literature on electrokinetic phenomena, particularly as much of the discussion is highly controversial. Another difficulty in dealing with the subject is that experimental precision and reproducibility have been achieved in very few researches. Theoretical deductions from inadequate data are, and will remain, unsatisfactory. Rapid progress in the direction of increased precision is, however, being made in electrophoretic studies, and important advances are to be expected in the near future.

The origin of the surface charges, and of the double layers is probably different for different materials. Thus colloidal ferric hydroxide

[28] H. A. Abramson, *J. Am. Chem. Soc.*, **50**, 390 (1928).
[29] D. R. Briggs, *ibid.*, **50**, 2358 (1928).

particles, which may be represented by $[Fe(OH)_3]_x$, possibly ionize as follows:

$$[Fe(OH)_3]_x = [Fe(OH)_3]_{x-n}[Fe(OH)_2^+]_n + n(OH^-)$$

Surface charges may also arise from preferential adsorption of ions of one charge. If, as is frequently assumed, protein suspensions consist, at their isoelectric points, of dipolar or multipolar ions, which may be represented in the simplest case by

$$^+NH_3 \cdot R \cdot COO^-$$

(in which the group R may be of any complexity) then combination with acids, such as hydrochloric, results in the reaction

$$^+NH_3 \cdot R \cdot COO^- + H^+ + Cl^- = {}^+NH_3 \cdot R \cdot COOH + Cl^-$$

leaving the protein with a net positive charge. The corresponding reaction with a base is

$$^+NH_3 \cdot R \cdot COO^- + Na^+ + OH^- = NH_2 \cdot R \cdot COO^- + Na^+ + H_2O$$

Thus the surface charges of proteins, and of other ampholytes, may be explained by the addition and removal of hydrogen ions, *i. e.*, protons. Charges on surfaces are frequently said to arise from "preferential adsorbtion" of ions of one charge. This is, however, inadequate as explanation as long as the reason for the adsorbtion remains in doubt.

An important modification of Helmholtz's concept of the electrical double layer was made in 1910 by Gouy.[30] According to Gouy's ideas the charges on a solid in contact with a liquid are more or less fixed, but the balancing charge in the liquid consists of a diffuse layer, disturbed by thermal agitation, of net charge equal to that on the solid. The diffuse layer is the same in principle as the "ion atmosphere" of Debye and Hückel, who, in fact, applied Gouy's theory for surfaces to solutions of electrolytes. The meaning of the Helmholtz double layer, and of the ζ potential in terms of the interionic attraction theory may be shown as follows.[31] Let *a-b-c-d* of Fig. 15 represent a unit surface with a surface charge of density σ fixed to the solid surface of (theoretically) infinite extent. The surface charge, σ, will be balanced by the net charge in the ion atmosphere contained in the volume of liquid extending perpendicularly from the area *abcd* in the direction x in which ions may move freely. As in the corresponding case for solutions of electrolytes discussed in Chapter 7, the ion atmosphere will give rise to a volume charge density ρ, and a potential ψ, for which the Boltzmann equation yields, if we assume

[30] A. Gouy, *J. de phys.*, (4), **9**, 457 (1910).
[31] See H. Müller, *Cold Spring Harbor Symposia on Quantitative Biology*, **1**, 1 (1933).

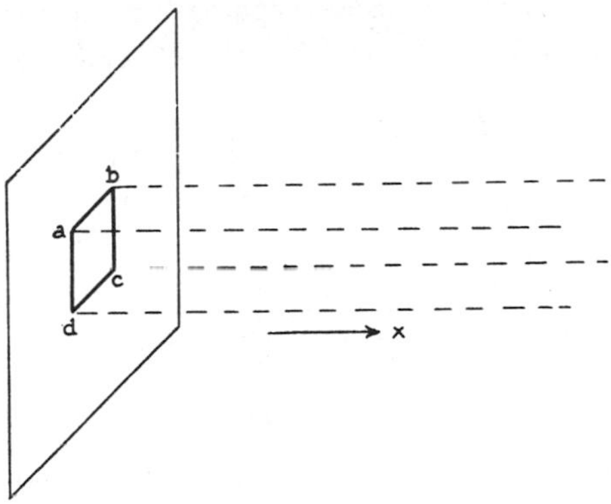

Fig. 15.

univalent ions only in the liquid and neglect higher terms in the expansion,

$$\rho = -\frac{2n\epsilon^2\psi}{kT} \tag{22}$$

(Equation (5) of Chapter 7). The general Poisson equation

$$\frac{\partial^2\psi}{\partial x^2} + \frac{\partial^2\psi}{\partial y^2} + \frac{\partial^2\psi}{\partial z^2} = -\frac{4\pi}{D}\rho$$

reduces for this case to

$$\frac{d^2\psi}{dx^2} = -\frac{4\pi}{D}\rho \tag{23}$$

Eliminating ρ from equations (22) and (23) and recalling the definition of κ given in equation (8) of Chapter 7 we have

$$\frac{d^2\psi}{dx^2} = \frac{8\pi n\epsilon^2}{DkT}\psi = \kappa^2\psi \tag{24}$$

This equation may be readily integrated to

$$\psi = Ae^{-\kappa x} + A'e^{+\kappa x} \tag{25}$$

in which A' must be zero since ψ must approach zero for large values of x. From the considerations given above

$$\sigma = -\int_a^\infty \rho\, dx \tag{26}$$

in which a is the distance of closest approach of the ions to the surface. With equations (23), (24) and (25), equation (26) yields

$$\sigma = \int_a^\infty \frac{D\kappa^2\psi}{4\pi}\,dx = \int_a^\infty \frac{D\kappa^2 A e^{-\kappa x}}{4\pi}\,dx$$

which on integration becomes

$$\sigma = -\left[\frac{D\kappa A e^{-\kappa x}}{4\pi}\right]_a^\infty = \frac{D\kappa A e^{-\kappa a}}{4\pi} \tag{26a}$$

Using the value of A from (26a) in equation (25) we have

$$\psi = \frac{4\pi\rho}{D\kappa} e^{\kappa(a-x)}$$

Expansion of the exponential term yields for the potential ψ

$$\psi = \frac{4\pi\sigma}{D\kappa}\left[1 + \kappa(a - x) + \frac{\kappa^2(a - x)^2}{2} \cdots\right] \tag{27}$$

It will be recalled from Chapter 7 that κ has the dimensions of a reciprocal length and may thus be set equal to $1/\delta$, δ being the thickness of the Helmholtz double layer. Thus it may be seen that the expression utilized by Helmholtz for the potential of a parallel plate condenser

$$\zeta = \frac{4\pi\sigma\delta}{D} \tag{28}$$

is the first term in the expression for the potential of a diffuse double layer as conceived by Gouy if the potential ψ existing in the ion atmosphere is substituted for the zeta potential, ζ. However, if the charge density, σ, of the surface is large it is probable that the approximations made in obtaining equation (22) are not justified, and that the terms corresponding to the extended theory of Debye and Hückel should be considered.

The influence of the size and shape of particles on their electrophoretic mobilities has been the subject of a number of experimental and theoretical researches. From Helmholtz and v. Smoluchowski the electrophoretic mobility u_e is, it will be recalled,

$$u_e = \frac{\zeta D}{4\pi\eta} \tag{16}$$

an expression which ignores the dimensions of the particle but assumes that it is cylindrical in shape. For a given solvent the mobility will according to these conceptions depend only upon the zeta potential, which in turn depends upon the nature and condition of the surface of the particles. Debye and Hückel[32] have obtained, for spherical particles, a limiting expression for the electrophoretic mobility that differs from the foregoing by the substitution of 6π for 4π. Briefly the argument is as follows. For a spherical particle of radius a the relation between the force, F, acting on it and the velocity, v, is given by Stokes' law which has the form

$$\mathrm{v} = \frac{F}{6\pi\eta a} \tag{29}$$

If, however, the particle carries an electrical charge, q, and is in a gradient of potential dE/dx, then

$$\mathrm{v} = \frac{q}{6\pi\eta a}\frac{dE}{dx} \tag{30}$$

For unit potential gradient, the velocity v is the mobility U_e, so that[33]

$$\mathrm{U}_e = \frac{q}{6\pi\eta a} \tag{31}$$

The relation between the charge, q, and the potential ψ of a spherical particle is, if distantly removed from other charges, $\psi_0 = q/Da$, so that, if ψ is identified with ζ equation (31) becomes

$$\mathrm{U}_e = \frac{\zeta_0 D}{6\pi\eta} \tag{32}$$

which differs in the factor, 6, from Helmholtz's relation, equation (16). For the case of a spherical particle in a solution of an electrolyte, *i.e.*, surrounded by an ion atmosphere, the potential ψ is, by equation (15) of Chapter 7

$$\zeta = \psi = \frac{q}{Da(1 + \kappa a)} = \frac{\zeta_0}{(1 + \kappa a)}$$

thus

$$\mathrm{U}_e = \frac{\zeta D}{6\pi\eta} = \frac{\zeta_0 D}{6\pi\eta(1 + \kappa a)} \tag{33}$$

[32] P. Debye and E. Hückel, *Physik. Z.*, **25**, 49 (1924).
E. Hückel, *ibid.*, **25**, 204 (1924).

[33] This may be compared with Onsager's equation for the electrophoretic effect on ionic conductance (equation (11) of Chapter 18),

$$V = -\frac{\epsilon E\kappa}{6\pi\eta}$$

in which V is the velocity of retardation of a particle. Since κ is a reciprocal length the electrophoretic effect on the motion of an ion may be regarded as the motion of a sphere of solution of radius $1/\kappa$, carrying the enclosed ion with it.

According to this picture, equation (33) will tend to approach equation (16) (but not as a limit) for large particles and higher ionic strengths. Henry[34] and Sumner and Henry[35] have made an extended theoretical investigation of this matter and conclude that (a) Helmholtz and Smoluchowski's relation, equation (16) is valid provided the radius of curvature is large compared with the effective thickness of the electrical double layer, *i. e.*, with $1/\kappa$, and that (b) the electrophoretic mobility should vary with size, tending toward the value given by equation (32), for colloidal particles whose size is not large compared with $1/\kappa$. Experimentally, it has been found, in general, that if the surfaces are the same the particles will move with the same velocity, independently of size and shape.[36] It will be recalled that protein-covered particles of various materials all move at the same velocity, and that in this case the electrophoretic and electro-osmotic mobilities are the same. On the other hand, Mooney[37] has found that small oil droplets do not move at the same rate as larger ones. There was, however, doubt in his experiments that the surfaces of the droplets were in equilibrium with the medium in which they were suspended. Protein-covered droplets were all found to move at the same rate.

[*Comment to 1961 Edition*] In no other field covered by this book has progress been so rapid and publication been so voluminous as has been the case for electrophoresis and related topics. A useful summary of the recent research is given in *Electrophoresis: Theory, Methods, and Applications* edited by Milan Bier and published by Academic Press in 1959.

[34] D. C. Henry, *Proc. Roy. Soc., (A)*, **133**, 106 (1931).
[35] C. G. Sumner and D. C. Henry, *ibid., (A)*, **133**, 130 (1931).
[36] H. A. Abramson, *J. Phys. Chem.*, **35**, 289 (1931); *J. Gen. Physiol.*, **16**, 1 (1932).
[37] M. Mooney, *J. Phys. Chem.*, **35**, 331 (1931).

Chapter 24
Irreversible Phenomena
Passivity and Overvoltage

In the major portion of the discussion of electrochemical reactions in this book it has been possible to assume that the reactions occur under substantially equilibrium conditions. However, the actual conditions under which such reactions take place are usually far from being those of equilibrium. It is not the unimportance of phenomena occurring when equilibrium cannot be assumed that has relegated the discussion of them to a short chapter of the present treatise. Rather it is the fact that little real information is available concerning the kinetics of electrochemical reactions. There is much of a speculative nature, some of which will certainly be productive in the future, in the subject matter that has appeared in the literature of chemistry and physics. The proper medium of publication for such speculation appears to the author to be the current scientific journals and monographs, rather than in a more general treatise such as this. The following discussions of passivity and overvoltage do not aim to be more than an introduction to the vast literature dealing with the subjects.

The Passivity of Metals. It has been known for a long time that a number of metals can exist in two states, in one of which they exhibit greater activity, at least with respect to certain types of chemical reactions, than the other. In the first case the metals are said to be in the "active" and in the second to be in the "passive" condition, the two terms being due to Schönbein [1] who made some of the early studies of the phenomena. The most familiar case of passivity is that of iron. A piece of iron when placed in dilute nitric acid will normally dissolve with the evolution of hydrogen, and is in its active state. If it is transferred to concentrated nitric acid and returned once more to the dilute acid little or no reaction will take place. It is now in its passive condition. The active state may be regained by touching the surface with a piece of the active metal, and by other means. In the passive state the metal behaves as if it is more "noble," *i. e.*, has a position lower in the electromotive series, than in the active condition.

[1] C. F. Schönbein, *Pogg. Ann.*, **37**, 390, 590; **38**, 444, 492 (1836) and later papers.

In spite of the fact that investigations of passivity have been in progress for over a hundred years no fully convincing explanation of the phenomenon has been found. A quite formidable number of papers have appeared dealing with the subject.[2] Precise measurements have not yet been found possible in this field, with the result that qualitative descriptions of phenomena are in general all that are available. Even these are not always consistent in different descriptions. In this chapter an attempt will be made to outline the important and typical experimental results and to present an account of some of the theories that have been advanced to explain the observations.

The facts concerning passivity, common to all metals that show the phenomenon, may be outlined in a few words. Passivity is produced by oxidizing conditions. Such conditions may arise from reagents such as nitric or chromic acids, from atmospheric oxygen, or from causing the metal to be an anode in a galvanic cell. The rate of solution of the metal in acids is very much slower in the passive than in the active state, though, in general, solution of the metal does not cease entirely.[3] Anodic corrosion of a metal is much more rapid in the active than in the passive state. In the passive condition a metal may fail to displace from its salts another metal normally lower in the electromotive series. Thus active iron will be rapidly covered with a coating of copper when placed in a copper sulfate solution, whereas with passive iron no reaction will take place. Certain ion constituents in solution, notably hydrogen and the halides, tend to reduce the tendency of metals to go into the passive state. In certain cases the metal is attacked anodically in the passive state. Thus chromium, for instance, enters solution normally at an anode as the divalent ion Cr^{++} whereas in the passive condition and at high current densities the electrode reactions tend toward the formation of compounds, such as chromic acid, in which the element is hexavalent.[4]

Metals in the passive state may be brought back to the active state in various ways, differing somewhat from metal to metal. Thus passive iron may be re-activated by simply touching it with a piece of the active metal, or another active metal such as zinc. The effect may also be produced by making the metal a cathode. Chromium and iron tend to lose their passivity with time, and more quickly at higher temperatures. The active state can frequently be restored by abrasion of the surface of the passive metal, and by reducing agents such as sugar.

[2] The following reviews of the literature will be of service to readers wishing to follow the subject of passivity further: H. G. Byers, *J. Am. Chem. Soc.*, **30**, 1718 (1908); C. W. Bennett and W. S. Byrnham, *Trans. Am. Electrochem. Soc.*, **29**, 217 (1916); H. Gerding and A. Karsen, *Z. Elektrochem.*, **31**, 135 (1925).

[3] If hydrogen is actively evolved from the surface of the metal during this process the phenomenon may be complicated by differences of hydrogen overvoltage in the two conditions of the metal. This question will be discussed later in this chapter.

[4] W. Hittorf, *Z. Elektrochem.*, **4**, 482 (1898); **6**, 6 (1899); **7**, 168 (1900).

Some of the most interesting experiments on passivity have been made by Lillie[5] who has found many analogies between the passage along a wire of the boundary between the active and passive iron and the transmission of impulses along a nerve. As already stated, iron in dilute nitric acid reacts with effervescence. However, after placing the metal in strong nitric acid and returning it to dilute acid, little or no attack takes place. If the end of a passive iron wire is touched with an active sample or a piece of zinc, activation passes along the wire at the rate of several hundred centimeters per second. The return to the active state is made visible by the formation of gas and of loose oxidized material. However, above a certain critical concentration of nitric acid, varying from sample to sample of the wire, the return of the metal to the active state is temporary, and is followed by regeneration of the passive state. If reactivation is again attempted before adequate time has passed, the passage of activity along the wire will go a short distance and stop. It is evident that an active spot on a piece of passive metal has the power of activating the adjacent surface. This effect will be limited if the passivating agent has not had full opportunity to react.

The spread of activity over a passive surface may be followed even more clearly than is described above by immersing the sample in a solution of copper sulfate. The passive iron does not react with the salt. However, by touching a point on the surface with active iron, or by abrasion, a coating of copper starts from the point and rapidly spreads.

The Theories of Passivity. The most obvious explanation of the facts concerning passivity is that it is due to a protecting film. This was the suggestion of Faraday[6] to whom Schönbein submitted his experimental observations. Faraday wrote, "My strong impression is that the surface of the iron is oxidized, or that the surface particles are in such relation to the oxygen of the electrolyte as to be equivalent to an oxidation." The suggestion was strongly opposed by Schönbein, and also by many of the chief workers in the field during the century that followed. Most of the other theories were, however, not very helpful as they were difficult or impossible to test experimentally. The following short and necessarily incomplete description of some of these theories is due to Byers.[7] Mousson (1837) considered that the passive state is due to a protective layer of nitrous acid. Schönbein (1837) thought that the solutions in which iron is passive exert a "negative catalytic effect" on the chemical activity of the iron. On the other hand

[5] R. S. Lillie, *Science,* **48,** 51 (1918); *J. Gen. Physiol.,* **3,** 107, 129 (1920).

[6] M. Faraday, *Phil. Mag.,* **9,** 53 (1836); **10,** 175 (1837); Experimental Researches, **2,** 231 (1859).

[7] H. G. Byers, *loc. cit.*

Hittorf (1900) ascribed passivity to an induced change in the nature of the metallic surface (Zwangzustand). According to Finkelstein (1902) the metal in the passive state is transformed into a "nobler" modification, whereas LeBlanc (1903) ascribed the state to a change in the velocity of ionization of the metal.

Some of the objections to the oxide film theory were as follows: (a) the passivity of metals must be assigned to unknown oxides, as the known oxides do not appear to have the necessary properties; (b) the oxide must, for instance, be a good conductor of electricity; (c) the assumption of an oxide film does not explain the negative potential of the passive metal; (d) the anodic solution of the passive metal does not entirely cease, but changes in nature, as has been explained above for the case of chromium. Added to these objections is the most important one that no film is visible, and that early attempts to discover its presence were not successful. From our present point of view none of these objections entirely rules out the protective film theory. Thus a monomolecular film may well have quite different properties from the same compound in mass. Certain oxides, such as manganese dioxide, are known to have metallic conductance. The negative potential may be due to the presence of oxidized material on the surface of the metal. When the current is flowing the effect of a film covering most of the metal may have the effect of increasing the current density, under which condition the nature of the chemical reaction at an electrode may change.

The most convincing evidence in favor of the protective surface film theory has been offered by Evans and associates.[8] Although the film, in the cases of iron, nickel and cobalt at least, must be exceedingly thin and transparent, these workers have been able, by anodic attack and by chemical action on passive metals, to separate films that retain the shape of the metal surface. A piece of iron, for instance, was made passive by potassium chromate, or potassium nitrite, and an edge sheared away. This edge was then immersed in a chloride solution and subjected to anodic attack. This procedure yielded a thin film which soon broke away in small pieces but which, under a microscope, was seen to consist of two parallel membranes. Other procedures, such as dissolving the underlying metal with iodine, also yielded transparent membranes. The reason that these films are not visible on the surface of the metal is that they are transparent and that they are too thin to yield interference colors. Somewhat paradoxically, metal surfaces holding films thick enough to be visible, or to show interference of light,

[8] U. R. Evans, *Nature*, **128**, 1062 (1931). U. R. Evans and J. Stockdale, *J. Chem. Soc.*, **1929**, 2651. L. C. Bannister and U. R. Evans, *ibid.*, **1930**, 1361. S. C. Britton and U. R. Evans, *ibid.*, **1930**, 1773.

are only slightly passive, if at all. The explanation given by Evans of this observation is that the oxides or other compounds forming the film have greater volumes than the metal from which they are formed so that the film is under stress, and if too thick, tends to crack and separate from the underlying surface.

There are, however, some results of a research by Russell, Evans and Rowell,[9] that appear to be significant in this connection, and cannot be readily explained by the protective film theory of corrosion. These workers mixed amalgams of two metals and subjected the mixture to various oxidizing agents, such as potassium permanganate and potassium chromate. It was found that in each case one of the metals was nearly quantitatively removed from the amalgam before the other was affected. The order of removal, which was independent of the nature of the oxidizing reagent, was found to be

Zn, Cd, Tl, Sn, Pb, Cu, Cr, Fe, Co, Hg, Ni

whereas the order of the standard potentials is

Zn, Cr,* Fe,* Cd, Tl, Co, Ni, Sn, Pb, Cu, Hg

(The order is that given in Table II of Chapter 14. The order of those metals indicated by an asterisk is based on less accurate data.) It will be observed that the order of the two series is the same with the exceptions of iron, chromium and nickel, which are shifted in each case to the position of a more noble metal. It would thus appear that the metals that exhibit the usual type of passivity to the most marked extent are also made passive to oxidizing agents by amalgamation. This is an interesting fact that should receive further investigation. These authors suggest as possible explanations (a) the formation of complexes between the mercury and the metal, and (b) shifts in the orbits of the electrons in the outer electronic shells of the atoms. Their highly speculative ideas may eventually explain the two types of passivity that have been considered above.

Overvoltage. The *overvoltage,* or overpotential, of an electrochemical reaction may be defined as the difference between the potential of an electrode (a) at which the reaction is actively taking place and another electrode (b) which is at the equilibrium potential for the same reaction. As will be seen later, some authors employ the term in a more restricted sense. The most important and most investigated case of overvoltage is that connected with the evolution of hydrogen. Hydrogen overvoltage is the difference of emf between a reversible hydrogen

[9] A. S. Russell, D. C. Evans and S. W. Rowell, *J. Chem. Soc.*, **1926,** 1872.

electrode and an electrode at which the electrochemical reaction

$$2H^+ + 2e^- = H_2 \tag{1}$$

is proceeding with the production of finite amounts of molecular hydrogen. If the reaction represented by equation (1) takes place reversibly there is, of course, no overvoltage. Actually more or less overvoltage is always found if appreciable current is flowing. The large values of overvoltage so far reported are limited to electrochemical reactions involving the evolution of the gases: hydrogen, oxygen and chlorine. Although the effect has been known for a long time the first use of the term, as its German equivalent "Überspannung," appears to have been by Caspari [10] who carried out an early investigation of the phenomenon. Other early workers were Nernst and Dolezalek,[11] Thiel and Breuning,[12] and Haber and Russ.[13]

Whereas there is no concensus of opinions as to the cause of overvoltage, its existence helps to explain quite a number of electrochemical phenomena. Thus if the (usually comparatively small) effect of ion activities is ignored, any metal above hydrogen in the electromotive series given in Table II of Chapter 14 should react, in an acid solution, with the evolution of hydrogen gas. Actually, as is well known, the metals, from zinc downward in the series, if moderately pure, react comparatively slowly in acid solutions. For instance, if a piece of pure zinc is placed in dilute sulfuric acid the reaction

$$Zn + H_2SO_4 = ZnSO_4 + H_2 \tag{2}$$

takes place very slowly. The reaction can be greatly hastened by the addition of a small amount of a copper or platinum salt. The results of such an addition are the precipitation of copper or platinum at points on the surface of the zinc, and an immediate and great increase in the velocity of evolution of the hydrogen gas. The usual explanation is that the particles of the foreign element serve as cathodes in small, short-circuited galvanic cells. This is a necessary part of the interpretation of the phenomenon, but the explanation is incomplete. Thus if a mercury salt is substituted for the copper or platinum salts, the effect is that the rate of the reaction is decreased rather than hastened. To accelerate the reaction represented by equation (2) in the manner just described the metal must have a lower hydrogen overvoltage than zinc itself. As we shall see this is true of copper and platinum. Mercury,

[10] W. A. Caspari, *Z. phys. Chem.*, **30**, 89 (1899).
[11] W. Nernst and F. Dolezalek, *Z. Elektrochem.*, **6**, 549 (1900).
[12] A. Thiel and E. Breuning, *Z. anorg. Chem.*, **83**, 329 (1914).
[13] F. Haber and R. Russ, *Z. phys. Chem.*, **47**, 257 (1904).

on the other hand, has a very high overvoltage. For this reason metals are frequently amalgamated to protect them from direct attack by acids. Zinc is amalgamated when in use in several types of primary cell in order to reduce "local action."

It is also true that the primary cells in ordinary use would not be of practical utility if the hydrogen overvoltage of zinc were not relatively high. For instance, the Le Clanché or "dry" cell may be represented by

$$Zn; ZnCl_2, NH_4Cl, MnO_2; (C)$$

However, the potential of a galvanic cell of the form,

$$Zn; ZnCl_2, NH_4Cl; H_2 \text{ (1 atm.) (Pt)}$$

is about 0.5 volt.[14] Thus if the hydrogen overvoltage of zinc were not at least 0.5 volt, hydrogen would evolve on the surface of the metal. The zinc would thus be involved in an uncontrolled "side reaction." Due to the hydrogen overvoltage the zinc is chemically attacked mainly when current is being drawn from the cell.

Again ignoring the effects of ion activities, no metal higher than hydrogen in the electromotive series could be deposited from an aqueous solution were it not for hydrogen overvoltage. Thus hydrogen overvoltage makes possible, for instance, zinc and nickel plating, and the electrorefining of iron. Furthermore, many other electrochemical oxidations and reductions are made possible by the overvoltage of hydrogen or oxygen.

The Measurement of Overvoltage. Two methods, which differ markedly in principle, and in the results obtained, have been used in measuring overvoltage. These are the *direct* and the *commutator* methods. In the direct method the exciting current is passing while the measurements are being made, while in the commutator method the exciting current and the measuring circuit are alternately connected to the electrodes.

The apparatus used in the direct method for measuring hydrogen overvoltage, as used by Thiel and Breuning and most of the workers in this field, is shown diagrammatically in Fig. 1a. The electrode whose overvoltage is desired, represented by E_x, is observed with a low-power microscope if the connection of overvoltage with gas evolution is being studied. The electrode E_1 may be of zinc, in which case the combination E_1 and E_x is a primary cell, the discharge of current through which may be regulated by the adjustable resistance R. The electrode E_1 may also be an inert electrode such as platinum, in which case the

[14] D. A. MacInnes, *Trans. Am. Electrochem. Soc.*, **29**, 315 (1916).

resistance R must be replaced by adjustable source of potential. The overvoltage is determined by obtaining the potential between the electrode E_x and the reversible hydrogen electrode E_h by means of the potentiometer P. Hydrogen is bubbled through the vessels B and C, as shown. A criticism that has been made of the method just described is that the potential of the electrode E_x may be affected by potential difference due to ohmic resistances in the neighborhood of the electrode, or by so-called "transfer resistances" on the surface of the electrode.

To overcome such apparent difficulties the "commutator" method for measuring overvoltages was devised. For that purpose the electrical connections shown in Fig. 1a may be replaced by those shown in

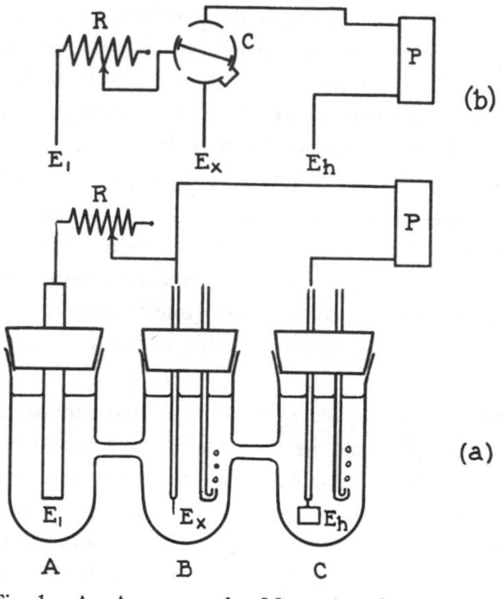

Fig. 1. An Apparatus for Measuring Overvoltage.

Fig. 1b. The rotating commutator alternately connects the electrode E_x with (a) the electrode E_1 and then (b) with the potentiometer P and the reference electrode E_h. Thus the exciting current is not acting when the potentiometer measurements are made. Many overvoltage determinations have been made with an apparatus of this kind by Newbery [15] who uniformly found lower values than have been obtained with the direct method. Although the results obtained with the commutator

[15] E. Newbery, *J. Chem. Soc.*, **105**, 2419 (1914); **109**, 1051, 1107, 1359 (1916); **111**, 470 (1917).

method have been found to be reasonably reproducible, and independent of the speed of the commutator above a certain minimum, the method has been the subject of a number of criticisms.

It has, for instance, been shown by Knobel,[16] Ferguson and Chen,[17] and others, that the potential is not constant during the time that the commutator is connected with the potentiometer, but, in general, varies rapidly, depending upon the experimental conditions. Knobel demonstrated this by measuring the potential only while the commutator passed through a very small angle, and the position of this small angle could be shifted to different positions in the path of rotation. At known rates of revolution of the commutator the difference of potentials of the electrode E_x and E_h could be measured as a function of the time. Some typical results obtained by Knobel are shown in Fig. 2, in which the

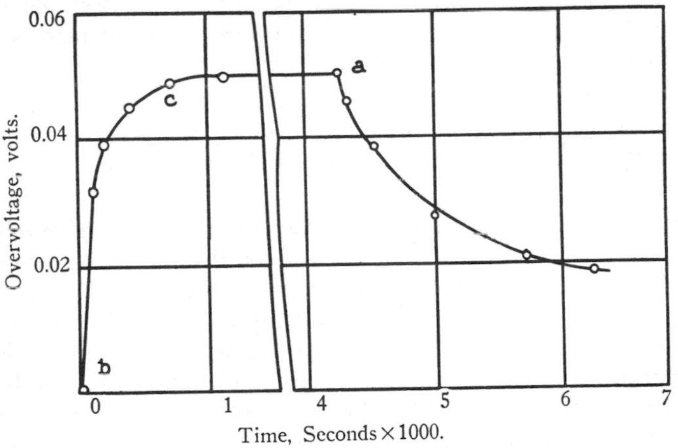

Fig. 2. The Hydrogen Overvoltage on a Platinum Electrode as a Function of Time.

ordinates represent the overvoltage of a platinum electrode and the abscissae the corresponding time. It will be seen at the point *a*, where the exciting current is interrupted, that the potential does not drop instantly to a constant value as it would be expected to do if the difference between the measurements due to the direct and commutator methods were due to an ohmic "transfer resistance." The fall of potential is as a matter of fact comparatively slow. The relatively slow rise of the potential when the exciting current is connected to the electrode

[16] M. Knobel, *J. Am. Chem. Soc.*, **46**, 2613 (1924).
[17] A. L. Ferguson and G. M. Chen, *J. Phys. Chem.*, **36**, 1156, 1166, 2437 (1932).

is also indicated by the curve between the points *b* and *c* in the figure. The constant readings obtained by Newbery and others with the commutator method may be, largely at least, due to the inertia of the moving parts of their measuring instruments.

Evidence that the commutator method yields results that are too low has been obtained by Tartar and Keyes.[18] These workers determined the hydrogen overvoltage of a platinum electrode simultaneously by the direct and commutator methods. The electrode was in contact with a solution containing an acid and a zinc salt. As the current density was increased, the overvoltage, as measured by both ways, also increased. When, however, the overvoltage obtained by the direct method reached about 0.7 volt, which was sufficient for the deposition of zinc from the solution, that metal appeared on the electrode. On the other hand, the overvoltage found with the commutator method had risen only to about 0.3 volt. This indicates clearly that the potential at the electrode, effective for carrying out electrochemical reactions, is higher than is indicated by the commutator method.

The conclusion from these and similar experiments is, therefore, that the direct method for measuring overvoltage gives values which are at least more nearly correct than those found with the commutator method.

Values of the Overvoltages of Metals. Although the overvoltages of the commonly occurring metals have been determined by a number of workers there is little agreement in the values obtained. This is due to the fact that overvoltages depend greatly on the surface conditions of the metals, and are much affected by the purity of the metal itself, the presence or absence of minute amounts of substances in solution, and other variables. In addition there has been no agreement as to the definition of overvoltage. Early workers mostly defined the overvoltage of an element as the minimum potential, measured by the "direct" method described above, at which visible evolution of hydrogen gas occurs. The data of Caspari [10] given in Table I were obtained in this manner. The order of the overvoltage values for different metals as determined by other workers is much the same as that given in the table, though, as has been mentioned, the actual values differ widely.

TABLE I. HYDROGEN OVERVOLTAGE OF VARIOUS METALS

Metals	Volt	Metals	Volt
Platinized platinum	0.005	Copper	0.23
Iron (in NaOH)	0.08	Tin	0.53
Smooth platinum	0.09	Lead	0.64
Silver	0.15	Zinc	0.70
Nickel	0.21	Mercury	0.78

[18] H. V. Tartar and H. E. Keyes, *J. Am. Chem. Soc.*, 44, 557 (1922).

The variation of the overvoltage with the current density has been studied by a number of workers including Tafel,[19] Lewis and Jackson,[20] Harkins and Adams,[21] and by Bowden.[22] The main conclusion from these studies is that the hydrogen overvoltage, which may be represented by E_x, varies with the current density i according to the equation

$$E_x = a + \beta \log i \tag{3}$$

in which a and β are constants. Bowden finds about the same value of β (between 8 and 9) for silver, nickel and mercury at 18°, and that β is roughly, at least, inversely proportional to the absolute temperature. However, measurements on a given piece of metal may yield quite different values of β at different times. Equation (3) also holds, from Bowden's observations, for the overvoltage of oxygen. It must be realized, however, that equation (3) is essentially empirical, and, particularly, cannot be used for extrapolation since for very small values of the current i the computed overvoltage approaches $-\infty$. Actually, of course, the overvoltage must tend toward a value of zero as the current is decreased.

Theories of Overvoltage. A large number of theories of overvoltage have been proposed. As in other fields in which experimental precision and reproducibility have not been yet obtained, any one of a number of theories is equally useful for explaining the observed facts. It also seems probable, as will be outlined below, that an adequate explanation is difficult to find because different mechanisms may be involved for different metals and at different current densities.

The theory that hydrogen overvoltage is due to more or less stable hydrides on the metal surfaces has been supported principally by Newbery.[23] Although hydrides, and solid solutions of hydrogen, are undoubtedly formed on the surface of some metals during hydrogen evolution, they appear, to this writer at least, to be the *result* rather than a cause of overvoltage.

The fact, already mentioned, that large values of overvoltage are encountered only with electrochemical reactions, such as the formation of oxygen, hydrogen, and chlorine, involving the evolution of a gas, has, again in the opinion of the writer, been far too much overlooked in attempts to explain the phenomenon. Since large numbers of gas bubbles must be formed in the process it has appeared to several authors that the work of overcoming the surface tension of the gas-

[19] J. Tafel, *Z. phys. Chem.*, **50**, 641 (1904).
[20] G. N. Lewis and R. F. Jackson, *ibid.*, **56**, 193 (1906).
[21] W. D. Harkins and H. S. Adams, *J. Phys. Chem.*, **29**, 205 (1925).
[22] F. P. Bowden, *Proc. Roy. Soc.*, **126A**, 107 (1929).
[23] E. Newberry, *loc. cit.*

electrolyte interface must have an influence on the overvoltage. How large an influence this has in the case of most metals it is, at present, not possible to say. However, some experiments by MacInnes and Adler[24] on small platinized platinum electrodes bear directly on this point. The evidence may be represented, schematically, by Fig. 3 in

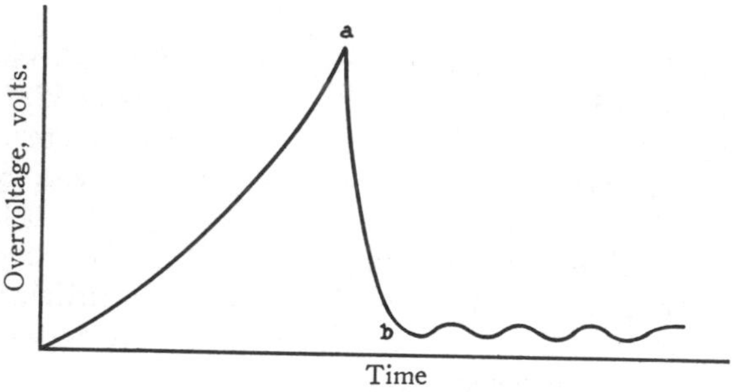

Fig. 3.

which the ordinates represent overvoltage and the abscissae, time. The apparatus used was essentially that shown in Fig. 1a. The electrode, E_x, in the experiment to be described was a small platinized platinum wire. At equilibrium the overvoltage is, of course, zero. However, if a gradually increasing exciting current is allowed to pass, the measured overvoltage will gradually rise, without the formation of bubbles, until it has reached the point *a* in the figure, of about 16 millivolts (or much higher if the electrolyte is stirred), at which point the evolution of bubbles is observed over most of the surface of the electrode. If the exciting current is now slowly reduced the overvoltage will decrease with decreasing evolution of gas, and finally at an overvoltage of about 1.5 millivolts (*b*, in the figure) the potential commences to fluctuate about 0.5 millivolt. It is an interesting fact that one bubble evolves during each of these fluctuations, appearing always from the same point on the electrode. These observations hold only for a small electrode and for platinized platinum. They appear to be important, however, since they seem to indicate that small values of hydrogen overvoltage can be explained by simple concentration polarization. Thus when the reaction

$$H^+ + e^- = H_2$$

takes place the hydrogen dissolves in the electrolyte. If the hydrogen gas phase is not present the solution surrounding the electrode becomes supersaturated, and a fairly high supersaturation is necessary before

[24] D. A. MacInnes and L. Adler, *J. Am. Chem. Soc.*, **41**, 194 (1919).

bubbles appear. Once the bubbles are formed on the electrode, continued evolution of the gas is possible at much lower overvoltages.[25] Since small bubbles have a greater solubility than the gas in its normal state this overvoltage is not zero.

However, this assumption of concentration polarization by dissolved hydrogen is certainly not adequate to explain higher values of overvoltage which are observed at high current densities and for metals with less active surfaces than that of platinized platinum. In such cases it seems probable, as has been suggested by many authors, that the reaction

$$2H^+ + 2e^- = H_2$$

takes place in two steps, *i. e.*,

$$H^+ + e^- = H \tag{4}$$

and

$$2H = H_2 \tag{5}$$

Most writers consider that the formation of molecular hydrogen from atomic hydrogen is slow compared with the rate of discharge of the hydrogen ion, and that the overvoltage is due to an accumulation of atomic hydrogen on an electrode. Among such writers are Lewis and Jackson [26] and Butler.[27] Butler [27] and Erdey-Grusz and Volmer [28] have given modifications of equation (3) to meet the condition that the overvoltage E_x should approach zero for very small values of the current i. In an interesting and important paper Hammett [29] has developed equations based on the assumption that overvoltage depends upon the rates of the two reactions given in equations (4) and (5) at high overvoltages the first of these being the most effective.

Gurney [30] and Fowler [31] have shown that Bowden's experimental results may be accounted for in terms of quantum mechanics. It appears to the author that these theoretical investigations, which are in active progress, will yield much of value in the future.

[25] An analogy is the superheating of a liquid. A higher temperature than the boiling point is frequently necessary to start the evolution of vapor. However, if, by the use of porous material or capillary tubes, the vapor phase is kept in contact with the liquid the evolution of vapor proceeds with little or no superheating.

[26] G. N. Lewis and R. F. Jackson, *Z. physik. Chem.*, **56,** 193 (1906).

[27] J. A. V. Butler, *Trans. Farad. Soc.*, **19,** 734 (1924); **28,** 379 (1932).

[28] T. Erdey-Grusz and M. Volmer, *Z. physik. Chem.*, **150A,** 203 (1930).

[29] L. P. Hammett, *Trans. Farad. Soc.*, **29,** 770 (1933).

[30] R. W. Gurney, *Proc. Roy. Soc.*, **134A,** 137 (1931); **136A,** 378 (1932).

[31] R. H. Fowler, *Trans. Farad. Soc.*, **28,** 368 (1932).

Appendix

The Derivation of Planck's Equation for Liquid Junction Potentials

If, in a tube of unit cross-section containing a "constrained diffusion" boundary (Fig. 7 of Chapter 13), there is a gradient of salt concentration in the direction x, a potential difference dE will be set up, and the ion, i, for instance, will be subject to two influences, (a) the gradient of chemical potential $\frac{\partial \mu_i}{\partial x}$, and (b) the potential gradient $\frac{\partial E}{\partial x}$. If the mobility of a *univalent* ion is U_i, the ion constituent will move with a velocity v equal to

$$v_i = -\mathrm{U}_i\left[\frac{\partial \mu_i}{\partial x} \pm \mathbf{F}\frac{\partial E}{\partial x}\right] \tag{1}$$

The flux, *i. e.*, the velocity times the concentration of the constituent *i*, at the point x in Δs seconds will be

$$\mathrm{C}_i v_i \Delta s \tag{2}$$

and at $x + \Delta x$

$$\left[\mathrm{C}_i v_i + \frac{\partial}{\partial x}\mathrm{C}_i v_i \cdot \Delta x\right]\Delta s \tag{2a}$$

The accumulation of this constituent in the region between x and $x + \Delta x$ in Δs seconds is the difference between these two expressions, *i. e.*,

$$-\frac{\partial}{\partial x}(\mathrm{C}_i v_i)\Delta x \cdot \Delta s \tag{3}$$

This quantity, divided by the volume (Δx in this case since the tube is of unit cross-section) in which the change occurs is the increment of concentration so that

$$\Delta C_i = -\frac{\partial}{\partial x}(C_i v_i)\Delta s \tag{4}$$

or in the limit

$$\frac{\partial C_i}{\partial s} = -\frac{\partial}{\partial x}(C_i v_i) \tag{5}$$

which is the "Equation of Continuity." Together with equation (1) this gives

$$\frac{\partial C_i}{\partial s} = \frac{\partial}{\partial x}\left[C_i U_i\left(\frac{\partial \mu_i}{\partial x} \pm \mathbf{F}\frac{\partial E}{\partial x}\right)\right] \tag{6}$$

Assuming now (a) that the ions are normal solutes, (b) that the mobilities are independent of the concentration and (c), as already mentioned, that the ions are all univalent, this gives

$$\frac{\partial C_i^+}{\partial s} = u_i^+\left[RT\frac{\partial^2 C_i^+}{\partial x^2} + \mathbf{F}\frac{\partial}{\partial x}\left(C_i^+\frac{\partial E}{\partial x}\right)\right] \tag{7}$$

$$\frac{\partial C_i^-}{\partial s} = u_i^-\left[RT\frac{\partial^2 C_i^-}{\partial x^2} - \mathbf{F}\frac{\partial}{\partial x}\left(C_i^-\frac{\partial E}{\partial x}\right)\right] \tag{8}$$

or positive and negative ions respectively. Since, because of the requirements of electrical neutrality $\Sigma C_i^+ = \Sigma C_i^- = C$, the total (9a) concentration, and

$$\Sigma\frac{\partial C_i^+}{\partial s} - \Sigma\frac{\partial C_i^-}{\partial s} = 0 \tag{9b}$$

equations (7) and (8) become

$$\Sigma u_i^+ RT\frac{\partial^2 C_i^+}{\partial x^2} + \Sigma u_i^+ \mathbf{F}\frac{\partial}{\partial x}\left(C_i^+\frac{\partial E}{\partial x}\right) - \Sigma u_i^- RT\frac{\partial C_i^-}{\partial x^2} + \Sigma u_i^- \mathbf{F}\frac{\partial}{\partial x}\left(C_i^-\frac{\partial E}{\partial x}\right) = 0 \tag{10}$$

which, on integration gives,[1]

$$\Sigma u_i^+ RT\frac{\partial C_i^+}{\partial x} + \Sigma u_i^+ \mathbf{F}\, C_i^+\frac{\partial E}{\partial x} - \Sigma u_i^- RT\frac{\partial C_i^-}{\partial x} + \Sigma u_i^- \mathbf{F}\, C_i^-\frac{\partial E}{\partial x} = 0 \tag{11}$$

For simplification let

$$\Sigma u_i^+ C_i^+ = \mathbf{U} \quad\text{and}\quad \Sigma u_i^- C_i^- = \mathbf{V} \tag{12}$$

then equation (11) may be rearranged to

$$\frac{\partial E}{\partial x} = -\frac{RT}{\mathbf{F}}\,\frac{\dfrac{\partial(\mathbf{U} - \mathbf{V})}{\partial x}}{\mathbf{U} + \mathbf{V}} \tag{13}$$

[1] The integration constant is zero because, for any interval, Δx, throughout which there are no concentration changes, all the partial derivatives vanish simultaneously.

Since the transference numbers, t_i^+ and t_i^- are

$$t_i^+ = \frac{u_i^+ C_i^+}{\mathbf{U} + \mathbf{V}}, \quad t_i^- = \frac{u_i^- C_i^-}{\mathbf{U} + \mathbf{V}} \tag{14}$$

equation (13) may be readily seen to be the equivalent of equation (3), Chapter 13, with the restrictions *a, b* and *c* outlined above. It is of interest that equation (13) is obtained kinetically, whereas equation (3), Chapter 13, involves thermodynamic reasoning.

When the diffusion through the boundary layer, formed as described, has reached a steady state there will be no further change with time so that equations (7) and (8) become

$$0 = RT\frac{\partial^2 C_i^+}{\partial x^2} + \mathbf{F}\frac{\partial}{\partial x}\left(C_i^+\frac{\partial E}{\partial x}\right) \tag{15a}$$

$$0 = RT\frac{\partial^2 C_i^+}{\partial x^2} - \mathbf{F}\frac{\partial}{\partial x}\left(C_i^+\frac{\partial E}{\partial x}\right) \tag{15b}$$

These are valid for $x = 0$ to $x = \delta$, the thickness of the boundary. Outside the boundary the differential coefficients will be zero.

Integrating, equations (15a) and (15b) become

$$A_i = RT\frac{\partial C_i^+}{\partial x} + \mathbf{F}C_i^+\frac{\partial E}{\partial x} \tag{16a}$$

$$B_i = RT\frac{\partial C_i^-}{\partial x} - \mathbf{F}C_i^-\frac{\partial E}{\partial x} \tag{16b}$$

Setting $$\Sigma_n A_i = A \quad \text{and} \quad \Sigma_n B_i = B \tag{17}$$

and recalling equation (9a), equations (16a) and (16b) yield

$$A = RT\frac{\partial C}{\partial x} + \mathbf{F}C\frac{\partial E}{\partial x}; \quad B = RT\frac{\partial C}{\partial x} - \mathbf{F}C\frac{\partial E}{\partial x} \tag{18}$$

Adding equations (18)

$$2RT\frac{\partial C}{\partial x} = A + B \tag{19}$$

and integrating gives

$$2RT\,C = (A + B)x + const. \tag{20}$$

i. e., the total concentration varies, through the boundary, linearly with x. If $C = C_0$ at $x = 0$ and $C = C_\delta$ at $x = \delta$ then

$$2RT(C_\delta - C_0) = (A + B)\delta \tag{21}$$

and

$$C = \frac{C_\delta - C_o}{\delta} x + C_o \tag{22}$$

Using equation (18) and the result just obtained leads to

$$2\mathbf{F}C \frac{\partial E}{\partial x} = A - B \tag{23}$$

$$\frac{\partial E}{\partial x} = \frac{(A - B)\delta}{2\mathbf{F}[(C_\delta - C_o)x + C_o\delta]} \tag{24}$$

Integrating from $x = 0$ to $x = \delta$ gives

$$E_L = \frac{(A - B)\delta}{2\mathbf{F}(C_\delta - C_o)} \ln \frac{C_\delta}{C_o} \tag{25}$$

in which E_L is the liquid junction potential. If we define a quantity ξ by

$$\xi = \left(\frac{C_\delta}{C_o}\right)^{\frac{(A - B)\delta}{2(C_\delta - C_o)RT}} \tag{26}$$

then

$$E_L = \frac{RT}{\mathbf{F}} \ln \xi \tag{27}$$

From equations (12) and (16)

$$u_1^+ A_1 + u_2^+ A_2 + \cdots\cdots = RT \frac{\partial \mathbf{U}}{\partial x} + \mathbf{FU} \frac{\partial E}{\partial x} \tag{28a}$$

$$u_1^- B_1 + u_2^- B_2 + \cdots\cdots = RT \frac{\partial \mathbf{V}}{\partial x} - \mathbf{FV} \frac{\partial E}{\partial x} \tag{28b}$$

By elimination of $\partial E/\partial x$ with the aid of equation (13) it can be readily shown that

$$u_1^+ A_1 + u_2^+ A_2 + \cdots\cdots = u_1^- B_1 + u_2^- A_2 + \cdots\cdots = K \tag{29}$$

in which K is a constant

With equations (24) and (29), equations (28a) and (28b) become

$$\frac{\partial \mathbf{U}}{\partial x} + \frac{\mathbf{U}(A - B)\delta}{2RT[(C_\delta - C_o)x + C_o\delta]} = \frac{K}{RT} \tag{30}$$

$$\frac{\partial \mathbf{V}}{\partial x} - \frac{\mathbf{V}(A - B)\delta}{2RT[(C_\delta - C_o)x + C_o\delta]} = \frac{K}{RT} \tag{31}$$

These equations are of the form

$$\frac{dU}{dx} + Uf_1(x) = f_2(x)$$

of which the solution is

$$U = e^{-\int f_1 dx}\left(\alpha + \int e^{\int f_1 dx} f_2 dx\right)$$

Accordingly, the solutions of equations (30) and (31) are

$$\mathbf{U} = \frac{2K[(C_\delta - C_o)x + C_o\delta]}{2(C_\delta - C_o)RT + (A - B)\delta} + [(C_\delta - C_o)x + C_o\delta]^{-\frac{(A-B)}{2(C_\delta - C_o)RT}} \times const. \tag{32}$$

$$\mathbf{V} = \frac{2K[(C_\delta - C_o)x + C_o\delta]}{2(C_\delta - C_o)RT + (A - B)\delta} + (C_\delta - C_o)x + C_o\delta^{\frac{(A-B)}{2(C - C_o)RT}} \times const. \tag{33}$$

Now when $x = 0$, $\mathbf{U} = \mathbf{U}_1$ and when $x = \delta$, $\mathbf{U} = \mathbf{U}_2$ so that

$$\xi\mathbf{U}_2 - \mathbf{U}_1 = \frac{2K\delta(\xi C\delta - C_o)}{2(C_\delta - C_o)RT + (A - B)\delta} \tag{34}$$

$$\mathbf{V}_2 - \xi\mathbf{V}_1 = \frac{2K\delta(\xi C\delta - C_o)}{2(C_\delta - C_o)RT - (A - B)\delta} \tag{35}$$

Through division to eliminate the constant K

$$\frac{\xi\mathbf{U}_2 - \mathbf{U}_1}{\mathbf{V}_2 - \xi\mathbf{V}_1} = \frac{2(C_\delta - C_o)RT - (A - B)\delta}{2(C_\delta - C_o)RT + (A - B)\delta} \cdot \frac{\xi C_\delta - C_o}{C_\delta - \xi C_o} \tag{36}$$

Since from equation (26)

$$\ln \xi = \frac{(A - B)\delta}{2(C_\delta - C_o)RT} \ln \frac{C_\delta}{C_o}$$

$$\frac{\xi\mathbf{U}_2 - \mathbf{U}_1}{\mathbf{V}_2 - \xi\mathbf{V}_1} = \frac{\ln \dfrac{C_\delta}{C_o} - \ln \xi}{\ln \dfrac{C_\delta}{C_o} + \ln \xi} \cdot \frac{\xi C_\delta - C_o}{C_\delta - \xi C_o} \tag{37}$$

This establishes the value of ξ from which the liquid junction potential may be computed with the aid of equation (27) in terms of the other variables contained in the equation.

AUTHOR INDEX

SUBJECT INDEX

The more important references are in bold face type.

A CATALOGUE OF SELECTED DOVER SCIENCE BOOKS

FUNDAMENTAL FORMULAS OF PHYSICS, edited by Donald H. Menzel. Most useful reference and study work, ranges from simplest to most highly sophisticated operations. Individual chapters, with full texts explaining formulae, prepared by leading authorities cover basic mathematical formulas, statistics, nomograms, physical constants, classical mechanics, special theory of relativity, general theory of relativity, hydrodynamics and aerodynamics, boundary value problems in mathematical physics, heat and thermodynamics, statistical mechanics, kinetic theory of gases, viscosity, thermal conduction, electromagnetism, electronics, acoustics, geometrical optics, physical optics, electron optics, molecular spectra, atomic spectra, quantum mechanics, nuclear theory, cosmic rays and high energy phenomena, particle accelerators, solid state, magnetism, etc. Special chapters also cover physical chemistry, astrophysics, celestian mechanics, meteorology, and biophysics. Indispensable part of library of every scientist. Total of xli + 787pp.
60595-7, 60596-5 Two volumes, Paperbound $6.00

INTRODUCTION TO EXPERIMENTAL PHYSICS, William B. Fretter. Detailed coverage of techniques and equipment: measurements, vacuum tubes, pulse circuits, rectifiers, oscillators, magnet design, particle counters, nuclear emulsions, cloud chambers, accelerators, spectroscopy, magnetic resonance, x-ray diffraction, low temperature, etc. One of few books to cover laboratory hazards, design of exploratory experiments, measurements. 298 figures. xii + 349pp.
(EBE) 61890-0 Paperbound $3.00

CONCEPTS AND METHODS OF THEORETICAL PHYSICS, Robert Bruce Lindsay. Introduction to methods of theoretical physics, emphasizing development of physical concepts and analysis of methods. Part I proceeds from single particle to collections of particles to statistical method. Part II covers application of field concept to material and non-material media. Numerous exercises and examples. 76 illustrations. x + 515pp. 62354-8 Paperbound $4.00

AN ELEMENTARY TREATISE ON THEORETICAL MECHANICS, Sir James Jeans. Great scientific expositor in remarkably clear presentation of basic classical material: rest, motion, forces acting on particle, statics, motion of particle under variable force, motion of rigid bodies, coordinates, etc. Emphasizes explanation of fundamental physical principles rather than mathematics or applications. Hundreds of problems worked in text. 156 figures. x + 364pp. 61839-0 Paperbound $2.75

THEORETICAL MECHANICS: AN INTRODUCTION TO MATHEMATICAL PHYSICS, Joseph S. Ames and Francis D. Murnaghan. Mathematically rigorous introduction to vector and tensor methods, dynamics, harmonic vibrations, gyroscopic theory, principle of least constraint, Lorentz-Einstein transformation. 159 problems; many fully-worked examples. 39 figures. ix + 462pp. 60461-6 Paperbound $3.50

THE PRINCIPLE OF RELATIVITY, Albert Einstein, Hendrick A. Lorentz, Hermann Minkowski and Hermann Weyl. Eleven original papers on the special and general theory of relativity, all unabridged. Seven papers by Einstein, two by Lorentz, one each by Minkowski and Weyl. "A thrill to read again the original papers by these giants," *School Science and Mathematics*. Translated by W. Perret and G. B. Jeffery. Notes by A. Sommerfeld. 7 diagrams. viii + 216pp.
60081-5 Paperbound $2.25

EINSTEIN'S THEORY OF RELATIVITY, Max Born. Relativity theory analyzed, explained for intelligent layman or student with some physical, mathematical background. Includes Lorentz, Minkowski, and others. Excellent verbal account for teachers. Generally considered the finest non-technical account. vii + 376pp.
60769-0 Paperbound $2.75

PHYSICAL PRINCIPLES OF THE QUANTUM THEORY, Werner Heisenberg. Nobel Laureate discusses quantum theory, uncertainty principle, wave mechanics, work of Dirac, Schroedinger, Compton, Wilson, Einstein, etc. Middle, non-mathematical level for physicist, chemist not specializing in quantum; mathematical appendix for specialists. Translated by C. Eckart and F. Hoyt. 19 figures. viii + 184pp.
60113-7 Paperbound $2.00

PRINCIPLES OF QUANTUM MECHANICS, William V. Houston. For student with working knowledge of elementary mathematical physics; uses Schroedinger's wave mechanics. Evidence for quantum theory, postulates of quantum mechanics, applications in spectroscopy, collision problems, electrons, similar topics. 21 figures. 288pp.
60524-8 Paperbound $3.00

ATOMIC SPECTRA AND ATOMIC STRUCTURE, Gerhard Herzberg. One of the best introductions to atomic spectra and their relationship to structure; especially suited to specialists in other fields who require a comprehensive basic knowledge. Treatment is physical rather than mathematical. 2nd edition. Translated by J. W. T. Spinks. 80 illustrations. xiv + 257pp. 60115-3 Paperbound $2.00

ATOMIC PHYSICS: AN ATOMIC DESCRIPTION OF PHYSICAL PHENOMENA, Gaylord P. Harnwell and William E. Stephens. One of the best introductions to modern quantum ideas. Emphasis on the extension of classical physics into the realms of atomic phenomena and the evolution of quantum concepts. 156 problems. 173 figures and tables. xi + 401pp. 61584-7 Paperbound $3.00

ATOMS, MOLECULES AND QUANTA, Arthur E. Ruark and Harold C. Urey. 1964 edition of work that has been a favorite of students and teachers for 30 years. Origins and major experimental data of quantum theory, development of concepts of atomic and molecular structure prior to new mechanics, laws and basic ideas of quantum mechanics, wave mechanics, matrix mechanics, general theory of quantum dynamics. Very thorough, lucid presentation for advanced students. 230 figures. Total of xxiii + 810pp.
61106-X, 61107-8 Two volumes, Paperbound $6.00

INVESTIGATIONS ON THE THEORY OF THE BROWNIAN MOVEMENT, Albert Einstein. Five papers (1905-1908) investigating the dynamics of Brownian motion and evolving an elementary theory of interest to mathematicians, chemists and physical scientists. Notes by R. Fürth, the editor, discuss the history of study of Brownian movement, elucidate the text and analyze the significance of the papers. Translated by A. D. Cowper. 3 figures. iv + 122pp.
60304-0 Paperbound $1.50

THE PRINCIPLES OF PSYCHOLOGY, William James. The famous long course, complete and unabridged. Stream of thought, time perception, memory, experimental methods—these are only some of the concerns of a work that was years ahead of its time and still valid, interesting, useful. 94 figures. Total of xviii + 1391pp.
20381-6, 20382-4 Two volumes, Paperbound $8.00

THE STRANGE STORY OF THE QUANTUM, Banesh Hoffmann. Non-mathematical but thorough explanation of work of Planck, Einstein, Bohr, Pauli, de Broglie, Schrödinger, Heisenberg, Dirac, Feynman, etc. No technical background needed. "Of books attempting such an account, this is the best," Henry Margenau, Yale. 40-page "Postscript 1959." xii + 285pp. 20518-5 Paperbound $2.00

THE RISE OF THE NEW PHYSICS, A. d'Abro. Most thorough explanation in print of central core of mathematical physics, both classical and modern; from Newton to Dirac and Heisenberg. Both history and exposition; philosophy of science, causality, explanations of higher mathematics, analytical mechanics, electromagnetism, thermodynamics, phase rule, special and general relativity, matrices. No higher mathematics needed to follow exposition, though treatment is elementary to intermediate in level. Recommended to serious student who wishes verbal understanding. 97 illustrations. xvii + 982pp. 20003-5, 20004-3 Two volumes, Paperbound $6.00

GREAT IDEAS OF OPERATIONS RESEARCH, Jagjit Singh. Easily followed non-technical explanation of mathematical tools, aims, results: statistics, linear programming, game theory, queueing theory, Monte Carlo simulation, etc. Uses only elementary mathematics. Many case studies, several analyzed in detail. Clarity, breadth make this excellent for specialist in another field who wishes background. 41 figures. x + 228pp. 21886-4 Paperbound $2.50

GREAT IDEAS OF MODERN MATHEMATICS: THEIR NATURE AND USE, Jagjit Singh. Internationally famous expositor, winner of Unesco's Kalinga Award for science popularization explains verbally such topics as differential equations, matrices, groups, sets, transformations, mathematical logic and other important modern mathematics, as well as use in physics, astrophysics, and similar fields. Superb exposition for layman, scientist in other areas. viii + 312pp.
20587-8 Paperbound $2.50

GREAT IDEAS IN INFORMATION THEORY, LANGUAGE AND CYBERNETICS, Jagjit Singh. The analog and digital computers, how they work, how they are like and unlike the human brain, the men who developed them, their future applications, computer terminology. An essential book for today, even for readers with little math. Some mathematical demonstrations included for more advanced readers. 118 figures. Tables. ix + 338pp. 21694-2 Paperbound $2.50

CHANCE, LUCK AND STATISTICS, Horace C. Levinson. Non-mathematical presentation of fundamentals of probability theory and science of statistics and their applications. Games of chance, betting odds, misuse of statistics, normal and skew distributions, birth rates, stock speculation, insurance. Enlarged edition. Formerly "The Science of Chance." xiii + 357pp. 21007-3 Paperbound $2.50

ELEMENTARY MATHEMATICS FROM AN ADVANCED STANDPOINT: VOLUME II—GEOMETRY, Feliex Klein. Using analytical formulas, Klein clarifies the precise formulation of geometric facts in chapters on manifolds, geometric and higher point transformations, foundations. "Nothing comparable," *Mathematics Teacher.* Translated by E. R. Hedrick and C. A. Noble. 141 figures. ix + 214pp.
(USO) 60151-X Paperbound $2.25

ENGINEERING MATHEMATICS, Kenneth S. Miller. Most useful mathematical techniques for graduate students in engineering, physics, covering linear differential equations, series, random functions, integrals, Fourier series, Laplace transform, network theory, etc. "Sound and teachable," Science. 89 figures. xii + 417pp. 6 x 8½.
61121-3 Paperbound $3.00

INTRODUCTION TO ASTROPHYSICS: THE STARS, Jean Dufay. Best guide to observational astrophysics in English. Bridges the gap between elementary popularizations and advanced technical monographs. Covers stellar photometry, stellar spectra and classification, Hertzsprung-Russell diagrams, Yerkes 2-dimensional classification, temperatures, diameters, masses and densities, evolution of the stars. Translated by Owen Gingerich. 51 figures, 11 tables. xii + 164pp.
60771-2 Paperbound $2.50

INTRODUCTION TO BESSEL FUNCTIONS, Frank Bowman. Full, clear introduction to properties and applications of Bessel functions. Covers Bessel functions of zero order, of any order; definite integrals; asymptotic expansions; Bessel's solution to Kepler's problem; circular membranes; etc. Math above calculus and fundamentals of differential equations developed within text. 636 problems. 28 figures. x + 135pp.
60462-4 Paperbound $1.75

DIFFERENTIAL AND INTEGRAL CALCULUS, Philip Franklin. A full and basic introduction, textbook for a two- or three-semester course, or self-study. Covers parametric functions, force components in polar coordinates, Duhamel's theorem, methods and applications of integration, infinite series, Taylor's series, vectors and surfaces in space, etc. Exercises follow each chapter with full solutions at back of the book. Index. xi + 679pp.
62520-6 Paperbound $4.00

THE EXACT SCIENCES IN ANTIQUITY, O. Neugebauer. Modern overview chiefly of mathematics and astronomy as developed by the Egyptians and Babylonians. Reveals startling advancement of Babylonian mathematics (tables for numerical computations, quadratic equations with two unknowns, implications that Pythagorean theorem was known 1000 years before Pythagoras), and sophisticated astronomy based on competent mathematics. Also covers transmission of this knowledge to Hellenistic world. 14 plates, 52 figures. xvii + 240pp.
22332-9 Paperbound $2.50

THE THIRTEEN BOOKS OF EUCLID'S ELEMENTS, translated with introduction and commentary by Sir Thomas Heath. Unabridged republication of definitive edition based on the text of Heiberg. Translator's notes discuss textual and linguistic matters, mathematical analysis, 2500 years of critical commentary on the Elements. Do not confuse with abridged school editions. Total of xvii + 1414pp.
60088-2, 60089-0, 60090-4 Three volumes, Paperbound $9.50

INTRODUCTION TO THE DIFFERENTIAL EQUATIONS OF PHYSICS, Ludwig Hopf. No math background beyond elementary calculus is needed to follow this classroom or self-study introduction to ordinary and partial differential equations. Approach is through classical physics. Translated by Walter Nef. 48 figures. v + 154pp. 60120-X Paperbound $1.75

DIFFERENTIAL EQUATIONS FOR ENGINEERS, Philip Franklin. For engineers, physicists, applied mathematicians. Theory and application: solution of ordinary differential equations and partial derivatives, analytic functions. Fourier series, Abel's theorem, Cauchy Riemann differential equations, etc. Over 400 problems deal with electricity, vibratory systems, heat, radio; solutions. Formerly *Differential Equations for Electrical Engineers*. 41 illustrations. vii + 299pp. 60601-5 Paperbound $2.50

THEORY OF FUNCTIONS, PART II. Single- and multiple-valued functions; full presentation of the most characteristic and important types. Proofs fully worked out. Translated by Frederick Bagemihl. x + 150pp. 60157-9 Paperbound $1.50

PROBLEM BOOK IN THE THEORY OF FUNCTIONS, I. More than 300 elementary problems for independent use or for use with "Theory of Functions, I." 85pp. of detailed solutions. Translated by Lipman Bers. viii + 126pp. 60158-7 Paperbound $1.50

PROBLEM BOOK IN THE THEORY OF FUNCTIONS, II. More than 230 problems in the advanced theory. Designed to be used with "Theory of Functions, II" or with any comparable text. Full solutions. Translated by Frederick Bagemihl. 138pp. 60159-5 Paperbound $1.75

INTRODUCTION TO THE THEORY OF EQUATIONS, Florian Cajori. Classic introduction by leading historian of science covers the fundamental theories as reached by Gauss, Abel, Galois and Kronecker. Basics of equation study are followed by symmetric functions of roots, elimination, homographic and Tschirnhausen transformations, resolvents of Lagrange, cyclic equations, Abelian equations, the work of Galois, the algebraic solution of general equations, and much more. Numerous exercises include answers. ix + 239pp. 62184-7 Paperbound $2.75

LAPLACE TRANSFORMS AND THEIR APPLICATIONS TO DIFFERENTIAL EQUATIONS, N. W. McLachlan. Introduction to modern operational calculus, applying it to ordinary and partial differential equations. Laplace transform, theorems of operational calculus, solution of equations with constant coefficients, evaluation of integrals, derivation of transforms, of various functions, etc. For physics, engineering students. Formerly *Modern Operational Calculus*. xiv + 218pp. 60192-7 Paperbound $2.50

PARTIAL DIFFERENTIAL EQUATIONS OF MATHEMATICAL PHYSICS, Arthur G. Webster. Introduction to basic method and theory of partial differential equations, with full treatment of their applications to virtually every field. Full, clear chapters on Fourier series, integral and elliptic equations, spherical, cylindrical and ellipsoidal harmonics, Cauchy's method, boundary problems, method of Riemann-Volterra, many other basic topics. Edited by Samuel J. Plimpton. 97 figures. vii + 446pp. 60263-X Paperbound $3.00

GUIDE TO THE LITERATURE OF MATHEMATICS AND PHYSICS, INCLUDING RELATED WORKS ON ENGINEERING SCIENCE, Nathan Grier Parke III. This up-to-date guide puts a library catalog at your fingertips. Over 5000 entries in many languages under 120 subject headings, including many recently available Russian works. Citations are as full as possible, and cross-references and suggestions for further investigation are provided. Extensive listing of bibliographical aids. 2nd revised edition. Complete indices. xviii + 436pp.
60447-0 Paperbound $3.00

INTRODUCTION TO ELLIPTIC FUNCTIONS WITH APPLICATIONS, Frank Bowman. Concise, practical introduction, from familiar trigonometric function to Jacobian elliptic functions to applications in electricity and hydrodynamics. Legendre's standard forms for elliptic integrals, conformal representation, etc., fully covered. Requires knowledge of basic principles of differentiation and integration only. 157 problems and examples, 56 figures. 115pp. 60922-7 Paperbound $1.50

THEORY OF FUNCTIONS OF A COMPLEX VARIABLE, A. R. Forsyth. Standard, classic presentation of theory of functions, stressing multiple-valued functions and related topics: theory of multiform and uniform periodic functions, Weierstrass's results with additiontheorem functions. Riemann functions and surfaces, algebraic functions, Schwarz's proof of the existence-theorem, theory of conformal mapping, etc. 125 figures, 1 plate. Total of xxviii + 855pp. 6⅛ x 9¼.
61378-X, 61379-8 Two volumes, Paperbound $6.00

THEORY OF THE INTEGRAL, Stanislaw Saks. Excellent introduction, covering all standard topics: set theory, theory of measure, functions with general properties, and theory of integration emphasizing the Lebesgue integral. Only a minimal background in elementary analysis needed. Translated by L. C. Young. 2nd revised edition. xv + 343pp. 61151-5 Paperbound $3.00

THE THEORY OF FUNCTIONS, *Konrad Knopp. Characterized as "an excellent introduction . . . remarkably readable, concise, clear, rigorous" by the* Journal of the American Statistical Association *college text.*

A COURSE IN MATHEMATICAL ANALYSIS, Edouard Goursat. *The entire "Cours d'analyse" for students with one year of calculus, offering an exceptionally wide range of subject matter on analysis and applied mathematics. Available for the first time in English. Definitive treatment.*

VOLUME I: Applications to geometry, expansion in series, definite integrals, derivatives and differentials. Translated by Earle R. Hedrick. 52 figures. viii + 548pp.
60554-X Paperbound $5.00

VOLUME II, PART I: Functions of a complex variable, conformal representations, doubly periodic functions, natural boundaries, etc. Translated by Earle R. Hedrick and Otto Dunkel. 38 figures. x + 259pp. 60555-8 Paperbound $3.00

VOLUME II, PART II: Differential equations, Cauchy-Lipschitz method, non-linear differential equations, simultaneous equations, etc. Translated by Earle R. Hedrick and Otto Dunkel. 1 figure. viii + 300pp. 60556-6 Paperbound $3.00

VOLUME III, PART I: Variation of solutions, partial differential equations of the second order. Poincaré's theorem, periodic solutions, asymptotic series, wave propagation, Dirichlet's problem in space, Newtonian potential, etc. Translated by Howard G. Bergmann. 15 figures. x + 329pp. 61176-0 Paperbound $3.50

VOLUME III, PART II: Integral equations and calculus of variations: Fredholm's equation, Hilbert-Schmidt theorem, symmetric kernels, Euler's equation, transversals, extreme fields, Weierstrass's theory, etc. Translated by Howard G. Bergmann. Note on Conformal Representation by Paul Montel. 13 figures. xi + 389pp. 61177-9 Paperbound $3.00

ELEMENTARY STATISTICS: WITH APPLICATIONS IN MEDICINE AND THE BIOLOGICAL SCIENCES, Frederick E. Croxton. Presentation of all fundamental techniques and methods of elementary statistics assuming average knowledge of mathematics only. Useful to readers in all fields, but many examples drawn from characteristic data in medicine and biological sciences. vii + 376pp. 60506-X Paperbound $2.50

ELEMENTS OF THE THEORY OF FUNCTIONS. A general background text that explores complex numbers, linear functions, sets and sequences, conformal mapping. Detailed proofs. Translated by Frederick Bagemihl. 140pp. 60154-4 Paperbound $1.50

THEORY OF FUNCTIONS, PART I. Provides full demonstrations, rigorously set forth, of the general foundations of the theory: integral theorems, series, the expansion of analytic functions. Translated by Federick Bagemihl. vii + 146pp. 60156-0 Paperbound $1.50

INTRODUCTION TO THE THEORY OF FOURIER'S SERIES AND INTEGRALS, Horatio S. Carslaw. A basic introduction to the theory of infinite series and integrals, with special reference to Fourier's series and integrals. Based on the classic Riemann integral and dealing with only ordinary functions, this is an important class text. 84 examples. xiii + 368pp. 60048-3 Paperbound $3.00

AN INTRODUCTION TO FOURIER METHODS AND THE LAPLACE TRANSFORMATION, Philip Franklin. Introductory study of theory and applications of Fourier series and Laplace transforms, for engineers, physicists, applied mathematicians, physical science teachers and students. Only a previous knowledge of elementary calculus is assumed. Methods are related to physical problems in heat flow, vibrations, eletcrical transmission, electromagnetic radiation, etc. 828 problems with answers. Formerly *Fourier Methods*. x + 289pp. 60452-7 Paperbound $2.75

INFINITE SEQUENCES AND SERIES, Konrad Knopp. Careful presentation of fundamentals of the theory by one of the finest modern expositors of higher mathematics. Covers functions of real and complex variables, arbitrary and null sequences, convergence and divergence. Cauchy's limit theorem, tests for infinite series, power series, numerical and closed evaluation of series. Translated by Frederick Bagemihl. v + 186pp. 60153-6 Paperbound $2.00

AMERICAN FOOD AND GAME FISHES, David S. Jordan and Barton W. Evermann. Definitive source of information, detailed and accurate enough to enable the sportsman and nature lover to identify conclusively some 1,000 species and sub-species of North American fish, sought for food or sport. Coverage of range, physiology, habits, life history, food value. Best methods of capture, interest to the angler, advice on bait, fly-fishing, etc. 338 drawings and photographs. l + 574pp. 6⅝ x 9⅜.
22383-1 Paperbound $4.50

THE FROG BOOK, Mary C. Dickerson. Complete with extensive finding keys, over 300 photographs, and an introduction to the general biology of frogs and toads, this is the classic non-technical study of Northeastern and Central species. 58 species; 290 photographs and 16 color plates. xvii + 253pp.
21973-9 Paperbound $4.00

THE MOTH BOOK: A GUIDE TO THE MOTHS OF NORTH AMERICA, William J. Holland. Classical study, eagerly sought after and used for the past 60 years. Clear identification manual to more than 2,000 different moths, largest manual in existence. General information about moths, capturing, mounting, classifying, etc., followed by species by species descriptions. 263 illustrations plus 48 color plates show almost every species, full size. 1968 edition, preface, nomenclature changes by A. E. Brower. xxiv + 479pp. of text. 6½ x 9¼.
21948-8 Paperbound $5.00

THE SEA-BEACH AT EBB-TIDE, Augusta Foote Arnold. Interested amateur can identify hundreds of marine plants and animals on coasts of North America; marine algae; seaweeds; squids; hermit crabs; horse shoe crabs; shrimps; corals; sea anemones; etc. Species descriptions cover: structure; food; reproductive cycle; size; shape; color; habitat; etc. Over 600 drawings. 85 plates. xii + 490pp.
21949-6 Paperbound $3.50

COMMON BIRD SONGS, Donald J. Borror. 33⅓ 12-inch record presents songs of 60 important birds of the eastern United States. A thorough, serious record which provides several examples for each bird, showing different types of song, individual variations, etc. Inestimable identification aid for birdwatcher. 32-page booklet gives text about birds and songs, with illustration for each bird.
21829-5 Record, book, album. Monaural. $2.75

FADS AND FALLACIES IN THE NAME OF SCIENCE, Martin Gardner. Fair, witty appraisal of cranks and quacks of science: Atlantis, Lemuria, hollow earth, flat earth, Velikovsky, orgone energy, Dianetics, flying saucers, Bridey Murphy, food fads, medical fads, perpetual motion, etc. Formerly "In the Name of Science." x + 363pp.
20394-8 Paperbound $2.00

HOAXES, Curtis D. MacDougall. Exhaustive, unbelievably rich account of great hoaxes: Locke's moon hoax, Shakespearean forgeries, sea serpents, Loch Ness monster, Cardiff giant, John Wilkes Booth's mummy, Disumbrationist school of art, dozens more; also journalism, psychology of hoaxing. 54 illustrations. xi + 338pp.
20465-0 Paperbound $2.75

A CATALOGUE OF SELECTED DOVER BOOKS IN ALL FIELDS OF INTEREST

VISUAL ILLUSIONS: THEIR CAUSES, CHARACTERISTICS, AND APPLICATIONS, Matthew Luckiesh. Thorough description and discussion of optical illusion, geometric and perspective, particularly; size and shape distortions, illusions of color, of motion; natural illusions; use of illusion in art and magic, industry, etc. Most useful today with op art, also for classical art. Scores of effects illustrated. Introduction by William H. Ittleson. 100 illustrations. xxi + 252pp.
21530-X Paperbound $2.00

A HANDBOOK OF ANATOMY FOR ART STUDENTS, Arthur Thomson. Thorough, virtually exhaustive coverage of skeletal structure, musculature, etc. Full text, supplemented by anatomical diagrams and drawings and by photographs of undraped figures. Unique in its comparison of male and female forms, pointing out differences of contour, texture, form. 211 figures, 40 drawings, 86 photographs. xx + 459pp. 5⅜ x 8⅜.
21163-0 Paperbound $3.50

150 MASTERPIECES OF DRAWING, Selected by Anthony Toney. Full page reproductions of drawings from the early 16th to the end of the 18th century, all beautifully reproduced: Rembrandt, Michelangelo, Dürer, Fragonard, Urs, Graf, Wouwerman, many others. First-rate browsing book, model book for artists. xviii + 150pp. 8⅜ x 11¼.
21032-4 Paperbound $2.50

THE LATER WORK OF AUBREY BEARDSLEY, Aubrey Beardsley. Exotic, erotic, ironic masterpieces in full maturity: Comedy Ballet, Venus and Tannhauser, Pierrot, Lysistrata, Rape of the Lock, Savoy material, Ali Baba, Volpone, etc. This material revolutionized the art world, and is still powerful, fresh, brilliant. With *The Early Work,* all Beardsley's finest work. 174 plates, 2 in color. xiv + 176pp. 8⅛ x 11.
21817-1 Paperbound $3.00

DRAWINGS OF REMBRANDT, Rembrandt van Rijn. Complete reproduction of fabulously rare edition by Lippmann and Hofstede de Groot, completely reedited, updated, improved by Prof. Seymour Slive, Fogg Museum. Portraits, Biblical sketches, landscapes, Oriental types, nudes, episodes from classical mythology—All Rembrandt's fertile genius. Also selection of drawings by his pupils and followers. "Stunning volumes," *Saturday Review.* 550 illustrations. lxxviii + 552pp. 9⅛ x 12¼.
21485-0, 21486-9 Two volumes, Paperbound $10.00

THE DISASTERS OF WAR, Francisco Goya. One of the masterpieces of Western civilization—83 etchings that record Goya's shattering, bitter reaction to the Napoleonic war that swept through Spain after the insurrection of 1808 and to war in general. Reprint of the first edition, with three additional plates from Boston's Museum of Fine Arts. All plates facsimile size. Introduction by Philip Hofer, Fogg Museum. v + 97pp. 9⅜ x 8¼.
21872-4 Paperbound $2.00

GRAPHIC WORKS OF ODILON REDON. Largest collection of Redon's graphic works ever assembled: 172 lithographs, 28 etchings and engravings, 9 drawings. These include some of his most famous works. All the plates from *Odilon Redon: oeuvre graphique complet,* plus additional plates. New introduction and caption translations by Alfred Werner. 209 illustrations. xxvii + 209pp. 9⅛ x 12¼.
21966-8 Paperbound $4.00

DESIGN BY ACCIDENT; A BOOK OF "ACCIDENTAL EFFECTS" FOR ARTISTS AND DESIGNERS, James F. O'Brien. Create your own unique, striking, imaginative effects by "controlled accident" interaction of materials: paints and lacquers, oil and water based paints, splatter, crackling materials, shatter, similar items. Everything you do will be different; first book on this limitless art, so useful to both fine artist and commercial artist. Full instructions. 192 plates showing "accidents," 8 in color. viii + 215pp. 8⅜ x 11¼. 21942-9 Paperbound $3.50

THE BOOK OF SIGNS, Rudolf Koch. Famed German type designer draws 493 beautiful symbols: religious, mystical, alchemical, imperial, property marks, runes, etc. Remarkable fusion of traditional and modern. Good for suggestions of timelessness, smartness, modernity. Text. vi + 104pp. 6⅛ x 9¼.
20162-7 Paperbound $1.25

HISTORY OF INDIAN AND INDONESIAN ART, Ananda K. Coomaraswamy. An unabridged republication of one of the finest books by a great scholar in Eastern art. Rich in descriptive material, history, social backgrounds; Sunga reliefs, Rajput paintings, Gupta temples, Burmese frescoes, textiles, jewelry, sculpture, etc. 400 photos. viii + 423pp. 6⅜ x 9¾. 21436-2 Paperbound $4.00

PRIMITIVE ART, Franz Boas. America's foremost anthropologist surveys textiles, ceramics, woodcarving, basketry, metalwork, etc.; patterns, technology, creation of symbols, style origins. All areas of world, but very full on Northwest Coast Indians. More than 350 illustrations of baskets, boxes, totem poles, weapons, etc. 378 pp.
20025-6 Paperbound $3.00

THE GENTLEMAN AND CABINET MAKER'S DIRECTOR, Thomas Chippendale. Full reprint (third edition, 1762) of most influential furniture book of all time, by master cabinetmaker. 200 plates, illustrating chairs, sofas, mirrors, tables, cabinets, plus 24 photographs of surviving pieces. Biographical introduction by N. Bienenstock. vi + 249pp. 9⅞ x 12¾. 21601-2 Paperbound $4.00

AMERICAN ANTIQUE FURNITURE, Edgar G. Miller, Jr. The basic coverage of all American furniture before 1840. Individual chapters cover type of furniture—clocks, tables, sideboards, etc.—chronologically, with inexhaustible wealth of data. More than 2100 photographs, all identified, commented on. Essential to all early American collectors. Introduction by H. E. Keyes. vi + 1106pp. 7⅞ x 10¾.
21599-7, 21600-4 Two volumes, Paperbound $11.00

PENNSYLVANIA DUTCH AMERICAN FOLK ART, Henry J. Kauffman. 279 photos, 28 drawings of tulipware, Fraktur script, painted tinware, toys, flowered furniture, quilts, samplers, hex signs, house interiors, etc. Full descriptive text. Excellent for tourist, rewarding for designer, collector. Map. 146pp. 7⅞ x 10¾.
21205-X Paperbound $2.50

EARLY NEW ENGLAND GRAVESTONE RUBBINGS, Edmund V. Gillon, Jr. 43 photographs, 226 carefully reproduced rubbings show heavily symbolic, sometimes macabre early gravestones, up to early 19th century. Remarkable early American primitive art, occasionally strikingly beautiful; always powerful. Text. xxvi + 207pp. 8⅜ x 11¼. 21380-3 Paperbound $3.50

ALPHABETS AND ORNAMENTS, Ernst Lehner. Well-known pictorial source for decorative alphabets, script examples, cartouches, frames, decorative title pages, calligraphic initials, borders, similar material. 14th to 19th century, mostly European. Useful in almost any graphic arts designing, varied styles. 750 illustrations. 256pp. 7 x 10. 21905-4 Paperbound $4.00

PAINTING: A CREATIVE APPROACH, Norman Colquhoun. For the beginner simple guide provides an instructive approach to painting: major stumbling blocks for beginner; overcoming them, technical points; paints and pigments; oil painting; watercolor and other media and color. New section on "plastic" paints. Glossary. Formerly *Paint Your Own Pictures*. 221pp. 22000-1 Paperbound $1.75

THE ENJOYMENT AND USE OF COLOR, Walter Sargent. Explanation of the relations between colors themselves and between colors in nature and art, including hundreds of little-known facts about color values, intensities, effects of high and low illumination, complementary colors. Many practical hints for painters, references to great masters. 7 color plates, 29 illustrations. x + 274pp. 20944-X Paperbound $2.75

THE NOTEBOOKS OF LEONARDO DA VINCI, compiled and edited by Jean Paul Richter. 1566 extracts from original manuscripts reveal the full range of Leonardo's versatile genius: all his writings on painting, sculpture, architecture, anatomy, astronomy, geography, topography, physiology, mining, music, etc., in both Italian and English, with 186 plates of manuscript pages and more than 500 additional drawings. Includes studies for the Last Supper, the lost Sforza monument, and other works. Total of xlvii + 866pp. 7⅞ x 10¾. 22572-0, 22573-9 Two volumes, Paperbound $10.00

MONTGOMERY WARD CATALOGUE OF 1895. Tea gowns, yards of flannel and pillow-case lace, stereoscopes, books of gospel hymns, the New Improved Singer Sewing Machine, side saddles, milk skimmers, straight-edged razors, high-button shoes, spittoons, and on and on . . . listing some 25,000 items, practically all illustrated. Essential to the shoppers of the 1890's, it is our truest record of the spirit of the period. Unaltered reprint of Issue No. 57, Spring and Summer 1895. Introduction by Boris Emmet. Innumerable illustrations. xiii + 624pp. 8½ x 11⅝. 22377-9 Paperbound $6.95

THE CRYSTAL PALACE EXHIBITION ILLUSTRATED CATALOGUE (LONDON, 1851). One of the wonders of the modern world—the Crystal Palace Exhibition in which all the nations of the civilized world exhibited their achievements in the arts and sciences—presented in an equally important illustrated catalogue. More than 1700 items pictured with accompanying text—ceramics, textiles, cast-iron work, carpets, pianos, sleds, razors, wall-papers, billiard tables, beehives, silverware and hundreds of other artifacts—represent the focal point of Victorian culture in the Western World. Probably the largest collection of Victorian decorative art ever assembled—indispensable for antiquarians and designers. Unabridged republication of the Art-Journal Catalogue of the Great Exhibition of 1851, with all terminal essays. New introduction by John Gloag, F.S.A. xxxiv + 426pp. 9 x 12. 22503-8 Paperbound $4.50

A HISTORY OF COSTUME, Carl Köhler. Definitive history, based on surviving pieces of clothing primarily, and paintings, statues, etc. secondarily. Highly readable text, supplemented by 594 illustrations of costumes of the ancient Mediterranean peoples, Greece and Rome, the Teutonic prehistoric period; costumes of the Middle Ages, Renaissance, Baroque, 18th and 19th centuries. Clear, measured patterns are provided for many clothing articles. Approach is practical throughout. Enlarged by Emma von Sichart. 464pp. 21030-8 Paperbound $3.50

ORIENTAL RUGS, ANTIQUE AND MODERN, Walter A. Hawley. A complete and authoritative treatise on the Oriental rug—where they are made, by whom and how, designs and symbols, characteristics in detail of the six major groups, how to distinguish them and how to buy them. Detailed technical data is provided on periods, weaves, warps, wefts, textures, sides, ends and knots, although no technical background is required for an understanding. 11 color plates, 80 halftones, 4 maps. vi + 320pp. 6⅛ x 9⅛. 22366-3 Paperbound $5.00

TEN BOOKS ON ARCHITECTURE, Vitruvius. By any standards the most important book on architecture ever written. Early Roman discussion of aesthetics of building, construction methods, orders, sites, and every other aspect of architecture has inspired, instructed architecture for about 2,000 years. Stands behind Palladio, Michelangelo, Bramante, Wren, countless others. Definitive Morris H. Morgan translation. 68 illustrations. xii + 331pp. 20645-9 Paperbound $3.50

THE FOUR BOOKS OF ARCHITECTURE, Andrea Palladio. Translated into every major Western European language in the two centuries following its publication in 1570, this has been one of the most influential books in the history of architecture. Complete reprint of the 1738 Isaac Ware edition. New introduction by Adolf Placzek, Columbia Univ. 216 plates. xxii + 110pp. of text. 9½ x 12¾. 21308-0 Clothbound $10.00

STICKS AND STONES: A STUDY OF AMERICAN ARCHITECTURE AND CIVILIZATION, Lewis Mumford.One of the great classics of American cultural history. American architecture from the medieval-inspired earliest forms to the early 20th century; evolution of structure and style, and reciprocal influences on environment. 21 photographic illustrations. 238pp. 20202-X Paperbound $2.00

THE AMERICAN BUILDER'S COMPANION, Asher Benjamin. The most widely used early 19th century architectural style and source book, for colonial up into Greek Revival periods. Extensive development of geometry of carpentering, construction of sashes, frames, doors, stairs; plans and elevations of domestic and other buildings. Hundreds of thousands of houses were built according to this book, now invaluable to historians, architects, restorers, etc. 1827 edition. 59 plates. 114pp. 7⅞ x 10¾. 22236-5 Paperbound $3.50

DUTCH HOUSES IN THE HUDSON VALLEY BEFORE 1776, Helen Wilkinson Reynolds. The standard survey of the Dutch colonial house and outbuildings, with constructional features, decoration, and local history associated with individual homesteads. Introduction by Franklin D. Roosevelt. Map. 150 illustrations. 469pp. 6⅝ x 9¼. 21469-9 Paperbound $4.00

THE ARCHITECTURE OF COUNTRY HOUSES, Andrew J. Downing. Together with Vaux's *Villas and Cottages* this is the basic book for Hudson River Gothic architecture of the middle Victorian period. Full, sound discussions of general aspects of housing, architecture, style, decoration, furnishing, together with scores of detailed house plans, illustrations of specific buildings, accompanied by full text. Perhaps the most influential single American architectural book. 1850 edition. Introduction by J. Stewart Johnson. 321 figures, 34 architectural designs. xvi + 560pp.
22003-6 Paperbound $4.00

LOST EXAMPLES OF COLONIAL ARCHITECTURE, John Mead Howells. Full-page photographs of buildings that have disappeared or been so altered as to be denatured, including many designed by major early American architects. 245 plates. xvii + 248pp. 7⅞ x 10¾. 21143-6 Paperbound $3.50

DOMESTIC ARCHITECTURE OF THE AMERICAN COLONIES AND OF THE EARLY REPUBLIC, Fiske Kimball. Foremost architect and restorer of Williamsburg and Monticello covers nearly 200 homes between 1620-1825. Architectural details, construction, style features, special fixtures, floor plans, etc. Generally considered finest work in its area. 219 illustrations of houses, doorways, windows, capital mantels. xx + 314pp. 7⅞ x 10¾. 21743-4 Paperbound $4.00

EARLY AMERICAN ROOMS: 1650-1858, edited by Russell Hawes Kettell. Tour of 12 rooms, each representative of a different era in American history and each furnished, decorated, designed and occupied in the style of the era. 72 plans and elevations, 8-page color section, etc., show fabrics, wall papers, arrangements, etc. Full descriptive text. xvii + 200pp. of text. 8⅜ x 11¼.
21633-0 Paperbound $5.00

THE FITZWILLIAM VIRGINAL BOOK, edited by J. Fuller Maitland and W. B. Squire. Full modern printing of famous early 17th-century ms. volume of 300 works by Morley, Byrd, Bull, Gibbons, etc. For piano or other modern keyboard instrument; easy to read format. xxxvi + 938pp. 8⅜ x 11.
21068-5, 21069-3 Two volumes, Paperbound $10.00

KEYBOARD MUSIC, Johann Sebastian Bach. Bach Gesellschaft edition. A rich selection of Bach's masterpieces for the harpsichord: the six English Suites, six French Suites, the six Partitas (Clavierübung part I), the Goldberg Variations (Clavierübung part IV), the fifteen Two-Part Inventions and the fifteen Three-Part Sinfonias. Clearly reproduced on large sheets with ample margins; eminently playable. vi + 312pp. 8⅛ x 11. 22360-4 Paperbound $5.00

THE MUSIC OF BACH: AN INTRODUCTION, Charles Sanford Terry. A fine, nontechnical introduction to Bach's music, both instrumental and vocal. Covers organ music, chamber music, passion music, other types. Analyzes themes, developments, innovations. x + 114pp. 21075-8 Paperbound $1.25

BEETHOVEN AND HIS NINE SYMPHONIES, Sir George Grove. Noted British musicologist provides best history, analysis, commentary on symphonies. Very thorough, rigorously accurate; necessary to both advanced student and amateur music lover. 436 musical passages. vii + 407 pp. 20334-4 Paperbound $2.75

JOHANN SEBASTIAN BACH, Philipp Spitta. One of the great classics of musicology, this definitive analysis of Bach's music (and life) has never been surpassed. Lucid, nontechnical analyses of hundreds of pieces (30 pages devoted to St. Matthew Passion, 26 to B Minor Mass). Also includes major analysis of 18th-century music. 450 musical examples. 40-page musical supplement. Total of xx + 1799pp.
(EUK) 22278-0, 22279-9 Two volumes, Clothbound $15.00

MOZART AND HIS PIANO CONCERTOS, Cuthbert Girdlestone. The only full-length study of an important area of Mozart's creativity. Provides detailed analyses of all 23 concertos, traces inspirational sources. 417 musical examples. Second edition. 509pp.
(USO) 21271-8 Paperbound $3.50

THE PERFECT WAGNERITE: A COMMENTARY ON THE NIBLUNG'S RING, George Bernard Shaw. Brilliant and still relevant criticism in remarkable essays on Wagner's Ring cycle, Shaw's ideas on political and social ideology behind the plots, role of Leitmotifs, vocal requisites, etc. Prefaces. xxi + 136pp.
21707-8 Paperbound $1.50

DON GIOVANNI, W. A. Mozart. Complete libretto, modern English translation; biographies of composer and librettist; accounts of early performances and critical reaction. Lavishly illustrated. All the material you need to understand and appreciate this great work. Dover Opera Guide and Libretto Series; translated and introduced by Ellen Bleiler. 92 illustrations. 209pp.
21134-7 Paperbound $1.50

HIGH FIDELITY SYSTEMS: A LAYMAN'S GUIDE, Roy F. Allison. All the basic information you need for setting up your own audio system: high fidelity and stereo record players, tape records, F.M. Connections, adjusting tone arm, cartridge, checking needle alignment, positioning speakers, phasing speakers, adjusting hums, trouble-shooting, maintenance, and similar topics. Enlarged 1965 edition. More than 50 charts, diagrams, photos. iv + 91pp. 21514-8 Paperbound $1.25

REPRODUCTION OF SOUND, Edgar Villchur. Thorough coverage for laymen of high fidelity systems, reproducing systems in general, needles, amplifiers, preamps, loudspeakers, feedback, explaining physical background. "A rare talent for making technicalities vividly comprehensible," R. Darrell, *High Fidelity*. 69 figures. iv + 92pp.
21515-6 Paperbound $1.00

HEAR ME TALKIN' TO YA: THE STORY OF JAZZ AS TOLD BY THE MEN WHO MADE IT, Nat Shapiro and Nat Hentoff. Louis Armstrong, Fats Waller, Jo Jones, Clarence Williams, Billy Holiday, Duke Ellington, Jelly Roll Morton and dozens of other jazz greats tell how it was in Chicago's South Side, New Orleans, depression Harlem and the modern West Coast as jazz was born and grew. xvi + 429pp.
21726-4 Paperbound $2.50

FABLES OF AESOP, translated by Sir Roger L'Estrange. A reproduction of the very rare 1931 Paris edition; a selection of the most interesting fables, together with 50 imaginative drawings by Alexander Calder. v + 128pp. 6½x9¼.
21780-9 Paperbound $1.25

JIM WHITEWOLF: THE LIFE OF A KIOWA APACHE INDIAN, Charles S. Brant, editor. Spans transition between native life and acculturation period, 1880 on. Kiowa culture, personal life pattern, religion and the supernatural, the Ghost Dance, breakdown in the White Man's world, similar material. 1 map. xii + 144pp.
22015-X Paperbound $1.75

THE NATIVE TRIBES OF CENTRAL AUSTRALIA, Baldwin Spencer and F. J. Gillen. Basic book in anthropology, devoted to full coverage of the Arunta and Warramunga tribes; the source for knowledge about kinship systems, material and social culture, religion, etc. Still unsurpassed. 121 photographs, 89 drawings. xviii + 669pp.
21775-2 Paperbound $5.00

MALAY MAGIC, Walter W. Skeat. Classic (1900); still the definitive work on the folklore and popular religion of the Malay peninsula. Describes marriage rites, birth spirits and ceremonies, medicine, dances, games, war and weapons, etc. Extensive quotes from original sources, many magic charms translated into English. 35 illustrations. Preface by Charles Otto Blagden. xxiv + 685pp.
21760-4 Paperbound $4.00

HEAVENS ON EARTH: UTOPIAN COMMUNITIES IN AMERICA, 1680-1880, Mark Holloway. The finest nontechnical account of American utopias, from the early Woman in the Wilderness, Ephrata, Rappites to the enormous mid 19th-century efflorescence; Shakers, New Harmony, Equity Stores, Fourier's Phalanxes, Oneida, Amana, Fruitlands, etc. "Entertaining and very instructive." *Times Literary Supplement.* 15 illustrations. 246pp.
21593-8 Paperbound $2.00

LONDON LABOUR AND THE LONDON POOR, Henry Mayhew. Earliest (c. 1850) sociological study in English, describing myriad subcultures of London poor. Particularly remarkable for the thousands of pages of direct testimony taken from the lips of London prostitutes, thieves, beggars, street sellers, chimney-sweepers, street-musicians, "mudlarks," "pure-finders," rag-gatherers, "running-patterers," dock laborers, cab-men, and hundreds of others, quoted directly in this massive work. An extraordinarily vital picture of London emerges. 110 illustrations. Total of lxxvi + 1951pp. 6⅝ x 10.
21934-8, 21935-6, 21936-4, 21937-2 Four volumes, Paperbound $14.00

HISTORY OF THE LATER ROMAN EMPIRE, J. B. Bury. Eloquent, detailed reconstruction of Western and Byzantine Roman Empire by a major historian, from the death of Theodosius I (395 A.D.) to the death of Justinian (565). Extensive quotations from contemporary sources; full coverage of important Roman and foreign figures of the time. xxxiv + 965pp. 21829-5 Record, book, album. Monaural. $3.50

AN INTELLECTUAL AND CULTURAL HISTORY OF THE WESTERN WORLD, Harry Elmer Barnes. Monumental study, tracing the development of the accomplishments that make up human culture. Every aspect of man's achievement surveyed from its origins in the Paleolithic to the present day (1964); social structures, ideas, economic systems, art, literature, technology, mathematics, the sciences, medicine, religion, jurisprudence, etc. Evaluations of the contributions of scores of great men. 1964 edition, revised and edited by scholars in the many fields represented. Total of xxix + 1381pp. 21275-0, 21276-9, 21277-7 Three volumes, Paperbound $7.75

INCIDENTS OF TRAVEL IN YUCATAN, John L. Stephens. Classic (1843) exploration of jungles of Yucatan, looking for evidences of Maya civilization. Stephens found many ruins; comments on travel adventures, Mexican and Indian culture. 127 striking illustrations by F. Catherwood. Total of 669 pp.
20926-1, 20927-X Two volumes, Paperbound $5.00

INCIDENTS OF TRAVEL IN CENTRAL AMERICA, CHIAPAS, AND YUCATAN, John L. Stephens. An exciting travel journal and an important classic of archeology. Narrative relates his almost single-handed discovery of the Mayan culture, and exploration of the ruined cities of Copan, Palenque, Utatlan and others; the monuments they dug from the earth, the temples buried in the jungle, the customs of poverty-stricken Indians living a stone's throw from the ruined palaces. 115 drawings by F. Catherwood. Portrait of Stephens. xii + 812pp.
22404-X, 22405-8 Two volumes, Paperbound $6.00

A NEW VOYAGE ROUND THE WORLD, William Dampier. Late 17-century naturalist joined the pirates of the Spanish Main to gather information; remarkably vivid account of buccaneers, pirates; detailed, accurate account of botany, zoology, ethnography of lands visited. Probably the most important early English voyage, enormous implications for British exploration, trade, colonial policy. Also most interesting reading. Argonaut edition, introduction by Sir Albert Gray. New introduction by Percy Adams. 6 plates, 7 illustrations. xlvii + 376pp. 6½ x 9¼.
21900-3 Paperbound $3.00

INTERNATIONAL AIRLINE PHRASE BOOK IN SIX LANGUAGES, Joseph W. Bátor. Important phrases and sentences in English paralleled with French, German, Portuguese, Italian, Spanish equivalents, covering all possible airport-travel situations; created for airline personnel as well as tourist by Language Chief, Pan American Airlines. xiv + 204pp.
22017-6 Paperbound $2.00

STAGE COACH AND TAVERN DAYS, Alice Morse Earle. Detailed, lively account of the early days of taverns; their uses and importance in the social, political and military life; furnishings and decorations; locations; food and drink; tavern signs, etc. Second half covers every aspect of early travel; the roads, coaches, drivers, etc. Nostalgic, charming, packed with fascinating material. 157 illustrations, mostly photographs. xiv + 449pp.
22518-6 Paperbound $4.00

NORSE DISCOVERIES AND EXPLORATIONS IN NORTH AMERICA, Hjalmar R. Holand. The perplexing Kensington Stone, found in Minnesota at the end of the 19th century. Is it a record of a Scandinavian expedition to North America in the 14th century? Or is it one of the most successful hoaxes in history. A scientific detective investigation. Formerly *Westward from Vinland.* 31 photographs, 17 figures. x + 354pp.
22014-1 Paperbound $2.75

A BOOK OF OLD MAPS, compiled and edited by Emerson D. Fite and Archibald Freeman. 74 old maps offer an unusual survey of the discovery, settlement and growth of America down to the close of the Revolutionary war: maps showing Norse settlements in Greenland, the explorations of Columbus, Verrazano, Cabot, Champlain, Joliet, Drake, Hudson, etc., campaigns of Revolutionary war battles, and much more. Each map is accompanied by a brief historical essay. xvi + 299pp. 11 x 13¾.
22084-2 Paperbound $6.00

AGAINST THE GRAIN (A REBOURS), Joris K. Huysmans. Filled with weird images, evidences of a bizarre imagination, exotic experiments with hallucinatory drugs, rich tastes and smells and the diversions of its sybarite hero Duc Jean des Esseintes, this classic novel pushed 19th-century literary decadence to its limits. Full unabridged edition. Do not confuse this with abridged editions generally sold. Introduction by Havelock Ellis. xlix + 206pp. 22190-3 Paperbound $2.00

VARIORUM SHAKESPEARE: HAMLET. Edited by Horace H. Furness; a landmark of American scholarship. Exhaustive footnotes and appendices treat all doubtful words and phrases, as well as suggested critical emendations throughout the play's history. First volume contains editor's own text, collated with all Quartos and Folios. Second volume contains full first Quarto, translations of Shakespeare's sources (Belleforest, and Saxo Grammaticus), Der Bestrafte Brudermord, and many essays on critical and historical points of interest by major authorities of past and present. Includes details of staging and costuming over the years. By far the best edition available for serious students of Shakespeare. Total of xx + 905pp. 21004-9, 21005-7, 2 volumes, Paperbound $7.00

A LIFE OF WILLIAM SHAKESPEARE, Sir Sidney Lee. This is the standard life of Shakespeare, summarizing everything known about Shakespeare and his plays. Incredibly rich in material, broad in coverage, clear and judicious, it has served thousands as the best introduction to Shakespeare. 1931 edition. 9 plates. xxix + 792pp. (USO) 21967-4 Paperbound $3.75

MASTERS OF THE DRAMA, John Gassner. Most comprehensive history of the drama in print, covering every tradition from Greeks to modern Europe and America, including India, Far East, etc. Covers more than 800 dramatists, 2000 plays, with biographical material, plot summaries, theatre history, criticism, etc. "Best of its kind in English," *New Republic.* 77 illustrations. xxii + 890pp. 20100-7 Clothbound $8.50

THE EVOLUTION OF THE ENGLISH LANGUAGE, George McKnight. The growth of English, from the 14th century to the present. Unusual, non-technical account presents basic information in very interesting form: sound shifts, change in grammar and syntax, vocabulary growth, similar topics. Abundantly illustrated with quotations. Formerly *Modern English in the Making.* xii + 590pp. 21932-1 Paperbound $3.50

AN ETYMOLOGICAL DICTIONARY OF MODERN ENGLISH, Ernest Weekley. Fullest, richest work of its sort, by foremost British lexicographer. Detailed word histories, including many colloquial and archaic words; extensive quotations. Do not confuse this with the Concise Etymological Dictionary, which is much abridged. Total of xxvii + 830pp. 6½ x 9¼. 21873-2, 21874-0 Two volumes, Paperbound $6.00

FLATLAND: A ROMANCE OF MANY DIMENSIONS, E. A. Abbott. Classic of science-fiction explores ramifications of life in a two-dimensional world, and what happens when a three-dimensional being intrudes. Amusing reading, but also useful as introduction to thought about hyperspace. Introduction by Banesh Hoffmann. 16 illustrations. xx + 103pp. 20001-9 Paperbound $1.00

LAST AND FIRST MEN AND STAR MAKER, TWO SCIENCE FICTION NOVELS, Olaf Stapledon. Greatest future histories in science fiction. In the first, human intelligence is the "hero," through strange paths of evolution, interplanetary invasions, incredible technologies, near extinctions and reemergences. Star Maker describes the quest of a band of star rovers for intelligence itself, through time and space: weird inhuman civilizations, crustacean minds, symbiotic worlds, etc. Complete, unabridged. v + 438pp. 21962-3 Paperbound $2.50

THREE PROPHETIC NOVELS, H. G. WELLS. Stages of a consistently planned future for mankind. *When the Sleeper Wakes,* and *A Story of the Days to Come,* anticipate *Brave New World* and *1984,* in the 21st Century; *The Time Machine,* only complete version in print, shows farther future and the end of mankind. All show Wells's greatest gifts as storyteller and novelist. Edited by E. F. Bleiler. x + 335pp. (USO) 20605-X Paperbound $2.50

THE DEVIL'S DICTIONARY, Ambrose Bierce. America's own Oscar Wilde—Ambrose Bierce—offers his barbed iconoclastic wisdom in over 1,000 definitions hailed by H. L. Mencken as "some of the most gorgeous witticisms in the English language." 145pp. 20487-1 Paperbound $1.25

MAX AND MORITZ, Wilhelm Busch. Great children's classic, father of comic strip, of two bad boys, Max and Moritz. Also Ker and Plunk (Plisch und Plumm), Cat and Mouse, Deceitful Henry, Ice-Peter, The Boy and the Pipe, and five other pieces. Original German, with English translation. Edited by H. Arthur Klein; translations by various hands and H. Arthur Klein. vi + 216pp. 20181-3 Paperbound $2.00

PIGS IS PIGS AND OTHER FAVORITES, Ellis Parker Butler. The title story is one of the best humor short stories, as Mike Flannery obfuscates biology and English. Also included, That Pup of Murchison's, The Great American Pie Company, and Perkins of Portland. 14 illustrations. v + 109pp. 21532-6 Paperbound $1.25

THE PETERKIN PAPERS, Lucretia P. Hale. It takes genius to be as stupidly mad as the Peterkins, as they decide to become wise, celebrate the "Fourth," keep a cow, and otherwise strain the resources of the Lady from Philadelphia. Basic book of American humor. 153 illustrations. 219pp. 20794-3 Paperbound $1.50

PERRAULT'S FAIRY TALES, translated by A. E. Johnson and S. R. Littlewood, with 34 full-page illustrations by Gustave Doré. All the original Perrault stories—Cinderella, Sleeping Beauty, Bluebeard, Little Red Riding Hood, Puss in Boots, Tom Thumb, etc.—with their witty verse morals and the magnificent illustrations of Doré. One of the five or six great books of European fairy tales. viii + 117pp. 8⅛ x 11. 22311-6 Paperbound $2.00

OLD HUNGARIAN FAIRY TALES, Baroness Orczy. Favorites translated and adapted by author of the *Scarlet Pimpernel.* Eight fairy tales include "The Suitors of Princess Fire-Fly," "The Twin Hunchbacks," "Mr. Cuttlefish's Love Story," and "The Enchanted Cat." This little volume of magic and adventure will captivate children as it has for generations. 90 drawings by Montagu Barstow. 96pp. (USO) 22293-4 Paperbound $1.95

THE RED FAIRY BOOK, Andrew Lang. Lang's color fairy books have long been children's favorites. This volume includes Rapunzel, Jack and the Bean-stalk and 35 other stories, familiar and unfamiliar. 4 plates, 93 illustrations x + 367pp.
21673-X Paperbound $2.50

THE BLUE FAIRY BOOK, Andrew Lang. Lang's tales come from all countries and all times. Here are 37 tales from Grimm, the Arabian Nights, Greek Mythology, and other fascinating sources. 8 plates, 130 illustrations. xi + 390pp.
21437-0 Paperbound $2.50

HOUSEHOLD STORIES BY THE BROTHERS GRIMM. Classic English-language edition of the well-known tales — Rumpelstiltskin, Snow White, Hansel and Gretel, The Twelve Brothers, Faithful John, Rapunzel, Tom Thumb (52 stories in all). Translated into simple, straightforward English by Lucy Crane. Ornamented with headpieces, vignettes, elaborate decorative initials and a dozen full-page illustrations by Walter Crane. x + 269pp. 21080-4 Paperbound $2.50

THE MERRY ADVENTURES OF ROBIN HOOD, Howard Pyle. The finest modern versions of the traditional ballads and tales about the great English outlaw. Howard Pyle's complete prose version, with every word, every illustration of the first edition. Do not confuse this facsimile of the original (1883) with modern editions that change text or illustrations. 23 plates plus many page decorations. xxii + 296pp.
22043-5 Paperbound $2.50

THE STORY OF KING ARTHUR AND HIS KNIGHTS, Howard Pyle. The finest children's version of the life of King Arthur; brilliantly retold by Pyle, with 48 of his most imaginative illustrations. xviii + 313pp. 6⅛ x 9¼.
21445-1 Paperbound $2.50

THE WONDERFUL WIZARD OF OZ, L. Frank Baum. America's finest children's book in facsimile of first edition with all Denslow illustrations in full color. The edition a child should have. Introduction by Martin Gardner. 23 color plates, scores of drawings. iv + 267pp. 20691-2 Paperbound $2.50

THE MARVELOUS LAND OF OZ, L. Frank Baum. The second Oz book, every bit as imaginative as the Wizard. The hero is a boy named Tip, but the Scarecrow and the Tin Woodman are back, as is the Oz magic. 16 color plates, 120 drawings by John R. Neill. 287pp. 20692-0 Paperbound $2.50

THE MAGICAL MONARCH OF MO, L. Frank Baum. Remarkable adventures in a land even stranger than Oz. The best of Baum's books not in the Oz series. 15 color plates and dozens of drawings by Frank Verbeck. xviii + 237pp.
21892-9 Paperbound $2.25

THE BAD CHILD'S BOOK OF BEASTS, MORE BEASTS FOR WORSE CHILDREN, A MORAL ALPHABET, Hilaire Belloc. Three complete humor classics in one volume. Be kind to the frog, and do not call him names . . . and 28 other whimsical animals. Familiar favorites and some not so well known. Illustrated by Basil Blackwell. 156pp. (USO) 20749-8 Paperbound $1.50

POEMS OF ANNE BRADSTREET, edited with an introduction by Robert Hutchinson. A new selection of poems by America's first poet and perhaps the first significant woman poet in the English language. 48 poems display her development in works of considerable variety—love poems, domestic poems, religious meditations, formal elegies, "quaternions," etc. Notes, bibliography. viii + 222pp.
22160-1 Paperbound $2.00

THREE GOTHIC NOVELS: THE CASTLE OF OTRANTO BY HORACE WALPOLE; VATHEK BY WILLIAM BECKFORD; THE VAMPYRE BY JOHN POLIDORI, WITH FRAGMENT OF A NOVEL BY LORD BYRON, edited by E. F. Bleiler. The first Gothic novel, by Walpole; the finest Oriental tale in English, by Beckford; powerful Romantic supernatural story in versions by Polidori and Byron. All extremely important in history of literature; all still exciting, packed with supernatural thrills, ghosts, haunted castles, magic, etc. xl + 291pp.
21232-7 Paperbound $2.50

THE BEST TALES OF HOFFMANN, E. T. A. Hoffmann. 10 of Hoffmann's most important stories, in modern re-editings of standard translations: Nutcracker and the King of Mice, Signor Formica, Automata, The Sandman, Rath Krespel, The Golden Flowerpot, Master Martin the Cooper, The Mines of Falun, The King's Betrothed, A New Year's Eve Adventure. 7 illustrations by Hoffmann. Edited by E. F. Bleiler. xxxix + 419pp. 21793-0 Paperbound $3.00

GHOST AND HORROR STORIES OF AMBROSE BIERCE, Ambrose Bierce. 23 strikingly modern stories of the horrors latent in the human mind: The Eyes of the Panther, The Damned Thing, An Occurrence at Owl Creek Bridge, An Inhabitant of Carcosa, etc., plus the dream-essay, Visions of the Night. Edited by E. F. Bleiler. xxii + 199pp.
20767-6 Paperbound $1.50

BEST GHOST STORIES OF J. S. LEFANU, J. Sheridan LeFanu. Finest stories by Victorian master often considered greatest supernatural writer of all. Carmilla, Green Tea, The Haunted Baronet, The Familiar, and 12 others. Most never before available in the U. S. A. Edited by E. F. Bleiler. 8 illustrations from Victorian publications. xvii + 467pp. 20415-4 Paperbound $3.00

MATHEMATICAL FOUNDATIONS OF INFORMATION THEORY, A. I. Khinchin. Comprehensive introduction to work of Shannon, McMillan, Feinstein and Khinchin, placing these investigations on a rigorous mathematical basis. Covers entropy concept in probability theory, uniqueness theorem, Shannon's inequality, ergodic sources, the E property, martingale concept, noise, Feinstein's fundamental lemma, Shanon's first and second theorems. Translated by R. A. Silverman and M. D. Friedman. iii + 120pp. 60434-9 Paperbound $1.75

SEVEN SCIENCE FICTION NOVELS, H. G. Wells. The standard collection of the great novels. Complete, unabridged. *First Men in the Moon, Island of Dr. Moreau, War of the Worlds, Food of the Gods, Invisible Man, Timo Machine, In the Days of the Comet.* Not only science fiction fans, but every educated person owes it to himself to read these novels. 1015pp. 20264-X Clothbound $5.00